Mechanism, Experiment, Disease

Mechanism, Experiment, Disease

Marcello Malpighi and Seventeenth-Century Anatomy

DOMENICO BERTOLONI MELI

The Johns Hopkins University Press
Baltimore

This book has been brought to publication with the assistance of endowment funds created and supported by generous friends of the Johns Hopkins University Press.

Printed in the United States of America on acid-free paper
2 4 6 8 9 7 5 3 1

The Johns Hopkins University Press
2715 North Charles Street
Baltimore, Maryland 21218-4363
www.press.jhu.edu

Library of Congress Cataloging-in-Publication Data

Bertoloni Meli, Domenico.
Mechanism, experiment, disease : Marcello Malpighi and seventeenth-century anatomy / Domenico Bertoloni Meli.
p. ; cm.
Includes bibliographical references and index.
ISBN-13: 978-0-8018-9903-4 (hardcover : alk. paper)
ISBN-10: 0-8018-9903-6 (hardcover : alk. paper)
ISBN-13: 978-0-8018-9904-1 (pbk. : alk. paper)
ISBN-10: 0-8018-9904-4 (pbk. : alk. paper)
1. Malpighi, Marcello, 1628–1694. 2. Anatomists—Italy—Biography. 3. Anatomy—History—17th century. 4. Microscopy—History—17th century. I. Title.
[DNLM: 1. Malpighi, Marcello, 1628–1694. 2. Anatomy, Comparative—history—Italy. 3. Embryology—history—Italy. 4. History, 17th Century—Italy. 5. Physiology, Comparative—history—Italy. QS 11 GI8 B546m 2011]
QM16.M33B47 2011
611.0092′2—dc22 2010023458

A catalog record for this book is available from the British Library.

Figure I.1 is reproduced by courtesy of the Fitzwilliam Museum, Cambridge, UK.
Figure 6.6 is reproduced by courtesy of the University Library, Bologna, Italy.
All other illustrations are courtesy of the Lilly Library, Indiana University, Bloomington.

Per Ada e Vasili

CONTENTS

ACKNOWLEDGMENTS

My interest in the history of seventeenth-century anatomy and medicine began several years ago, when research on Giovanni Alfonso Borelli's wide-ranging activities led me to investigate the anatomists and physicians active in his circle. I am grateful to the Wellcome Trust for a fellowship at the Cambridge Unit for the History of Medicine that enabled me to start work on my project. My interests in these topics persisted through the years until the current form of this book crystallized in my mind during a year at the Institute for Advanced Study, Princeton. I am grateful to the IAS for the opportunity to work in such a wonderful and stimulating environment and to all the colleagues and friends who offered suggestions and criticisms during my stay, especially Mechthild Fend, Jonathan Israel, Sarah McPhee, Jeremy Popkin, and above all Heinrich von Staden. I benefited from several discussions in the history of scientific observation group at the Max-Planck-Institut in Berlin; I am grateful to all participants for their comments on my work on the color of blood and to Raine Daston for her encouragement and support. I am also extremely grateful to the John Simon Guggenheim Foundation for a fellowship that provided me with the time necessary to bring the project to conclusion.

A number of friends and colleagues over the years offered advice and support in many ways. I wish to thank Anne Mylott, who read a preliminary draft of the entire manuscript and provided many criticisms and suggestions on smaller and larger issues. I am grateful for advice on a number of topics from Marta Cavazza, Antonio Clericuzio, Maria Conforti, Hal Cook, Silvia De Renzi, Paula Findlen, Michael Friedman, Dan Garber, Kimberly Hart, Gideon Manning, Larry Principe, Sophie Roux, and Nancy Siraisi. I wish to thank colleagues and current and former students at Indiana University with whom I discussed matters relating to my project informally, in graduate seminars, and in reading groups, especially Colin Allen, Tawrin Baker, Ann Carmichael, Martha Crouch, Karin Ekholm, Sandy Gliboff, Kevin Grau, Nicole Howard, Mark Kaplan, Joel Klein, Bill Newman, Carl Pearson, Evan Ragland, Jutta Schickore, and Rebecca Wilkin.

I wish to express my gratitude for assistance from Maria Conforti at the Biblioteca di Storia della Medicina at La Sapienza in Rome and the staffs of the Houghton Library of Harvard University; the National Library of Medicine in Bethesda, Maryland; the New York Academy of Medicine, especially Arlene Shaner; the University Library and the Biblioteca dell'Archiginnasio at Bologna; and the Interlibrary Loan Department at the Wells Library, at Indiana University, Bloomington. A special thanks to the Lilly Library at Indiana University, Bloomington, especially Breon Mitchell and Joel Silver for their expert assistance and Zach Downey for his help with photographic reproductions. Last but not least, I wish to thank Bob Brugger and the Johns Hopkins University Press for their support and commitment to this project.

I claim sole responsibility for all remaining errors and omissions.

This book builds on and substantially expands a number of essays published in the past dozen years: "The New Anatomy of Marcello Malpighi," 17–60, and "The Posthumous Dispute between Borelli and Malpighi," 245–73, in Domenico Bertoloni Meli, ed., *Marcello Malpighi, Anatomist and Physician* (Florence: Olschki, 1997). "The Archive and Consulti of Marcello Malpighi," in Michael Hunter, ed., *Archives of the Scientific Revolution* (Woodbridge: Boydell Press, 1998), 109–20. "Francesco Redi e Marcello Malpighi: ricerca anatomica e pratica medica," in Walter Bernardi and Luigi Guerrini, eds., *Francesco Redi. Un protagonista della scienza moderna* (Florence: Olschki, 1999), 73–86. "Blood, Monsters, and Necessity in Malpighi's *De polypo cordis*," *MH* 45 (2001), 511–22. "Mechanistic Pathology and Therapy in the Medical *Assayer* of Marcello Malpighi," *MH* 51 (2007), 165–80. "The Collaboration between Anatomists and Mathematicians in the Mid-Seventeenth Century with a Study of Images as Experiments and Galileo's Role in Steno's *Myology*," *Early Science and Medicine* 13 (2008), 665–709. "The Color of Blood: Between Sensory Experience and Epistemic Significance," forthcoming in Lorraine Daston and Elizabeth Lunbeck, eds., *Histories of Scientific Observation*. "The Representation of Insects in the Seventeenth Century: A Comparative Approach," *Annals of Science* 67 (2010), 405–29.

Mechanism, Experiment, Disease

Introduction

Anatomy, Medicine, and the New Philosophy

1. Anatomical Research in the Second Half of the Seventeenth Century

This volume examines how anatomical knowledge in the second half of the seventeenth century was gained and transmitted, as well as how these processes interacted with the experimental and mechanical philosophies, natural history, and medical practice. In line with contemporary usage, by "anatomy" I mean the study of structures as well as their actions and purpose, thus including what we today call physiology. I am especially interested in the growing range of techniques and tools of investigation adopted in the study of animals and plants, in visual representation, and in philosophical perspectives about the mechanistic understanding of the body. Moreover, I pay close attention to the mutual relationships among anatomical research, pathology, and therapy. Marcello Malpighi was a key figure in all these developments and therefore occupies a central position in this work.[1]

The last three quarters of the seventeenth century witnessed the revival of classical techniques of investigation, such as vivisection, the striking refinement of others, such as chymical assaying and vascular injections, and the emergence of entirely new ones, such as microscopy. Moreover, the study of the lesions produced by disease in dead bodies shed light on pathology and on the normal operations of the body as well. These developments led to profound and at the same time problematic changes in the understanding of the body and its diseases. The shifting genres of publication are indicative in this regard: the earlier anatomical literature to the beginning of the seventeenth century often consisted of treatises on the entire human body, such as sixteenth-century classics by Andreas Vesalius and Realdo Colombo and the main works at the turn of the century, *Historia anatomica* by André du Laurens, physician to Henry IV, and *Theatrum anatomicum*, by the Basel professor Caspar Bauhin. By contrast, a vast portion of the anatomical literature from the second half of the seventeenth century appeared in short essays dealing with specific organs, vessels, and bodily fluids, such as pancreatic juice or saliva, the main exceptions being successive editions of several textbooks and compendia, as well as Giovanni Alfonso Borelli's *De motu animalium*. These preliminary observations lead to a time frame for my work.[2]

The years 1627 and 1628 witnessed the publication of Gasparo Aselli's *De lactibus sive lacteis venis* and William Harvey's *De motu cordis et sanguinis*, both relying on vivisection: curiously, Aselli claimed the discovery of a new anatomical part—the milky veins carrying chyle from the intestine to the liver—whose existence and role had already been anticipated and conceptualized since antiquity; by contrast, Harvey discovered no new body part but provided a new understanding of the motion of the heart and blood in a circle, views without precedents in antiquity and at best dubious ones during the Renaissance. The 1639 Leiden edition of Harvey's *De motu* announced in the address to the reader the publication of Aselli's *De lactibus*, which appeared in 1640 as a companion to Harvey's work, with which it is often bound. Starting from that time, Aselli's and Harvey's works were often joined as the outstanding anatomical contributions of their time, together with two important letters of 1640 by the Leiden professor Johannes Walaeus confirming their findings. Contemporary anatomical textbooks and treatises included Aselli's and Harvey's findings. Later in the century, the huge *Bibliotheca anatomica* by the Geneva physicians and medical historians Daniel Le Clerc and Jacques Manget, first published in 1685 and then in expanded form in 1699, presented in over two thousand double-column folio pages the most recent findings in human anatomy, often with valuable critical notes, focusing chiefly on works published after 1650. Le Clerc and Manget both reflected and contributed to shaping the anatomical horizon of their time; Aselli's and Harvey's works were included, but they were the exception in a collection focused on the second half of the century. While taking into account Aselli's and Harvey's works, it seems thus justified to consider midcentury as a starting point for my project.[3]

Toward the end of the century, the contradictions emerging from some of the most interventionist techniques of investigation led to doubts and controversies. In the study of many body parts such as the spleen, liver, and cerebral cortex, the refined injections by the Amsterdam anatomist Frederik Ruysch led to a vascular anatomy, as opposed to Malpighi's glandular anatomy resulting from microscopy. Thus, from the perspective of the questions addressed in this book, it seems unhelpful to focus on one technique of investigation alone, be this microscopy or injections, since it was the problematic interference among techniques that contributed to disenchantment with interventionist methods and a shift toward more traditional ones. Moreover, both Ruysch and Malpighi were quite secretive about their methods, which remained private rather than being publicly shared and critically assessed. After Ruysch's and Malpighi's main works at the very end of the seventeenth century, major figures such as Herman Boerhaave at Leiden and Giovanni Battista Morgagni at Padua retreated to more secure methods and expressed a renewed interest in pathology and disease. Thus, the turn of the century, coinciding with the second (1699) edition of the *Bibliotheca anatomica*, marks a turning point in my account of anatomical thinking, a point at which the tide turned.

This brief characterization offers a useful contemporary assessment of the state of the field and provides a preliminary temporal framework for the present study: the very existence of the *Bibliotheca anatomica* and its diffusion show that late seventeenth-century anatomists perceived their time as one of profound transformations and felt the need for an easily accessible collection of the large body of recent contributions. The *Bibliotheca*, with its scope, valuable introductions, and notes, is an exceedingly useful tool on which I shall frequently rely. Its contents, however, raise interesting questions: despite its editors' program to exclude works on animal anatomy and pathology, findings from those areas often crop up between its covers because they were crucially relevant to human anatomy. Demarcation was so problematic that the *Bibliotheca* included a large number of works on animals, my favorite example being a nearly forty-folio-page extract from Johann Conrad Peyer's *Merycologia, sive de ruminantibus et ruminatione* (Basel, 1685), on the stomach of ruminants. But other boundaries were also problematic: the *Bibliotheca* included extracts from works on therapy and disease, such as *De purgantium medicamentorum facultatibus* (Leiden and Amsterdam, 1672) by the Dutch physician Johann Pechlin and *Cicutae aquaticae historia* (Schaffausen, 1679) by the Swiss physician Johann Jakob Wepfer, on hemlock poisoning, both announcing new glands in the digestive tract; and—this time in its entirety—Malpighi's *De polypo cordis* (Bologna, 1666), which used heart polyps to investigate the properties and composition of blood in healthy patients. It is appropriate to recall here that one of the first endorsements of Harvey's circulation came in a work on kidney stones, *De calculo renum* (Leiden, 1638), by the Dordrecht physician Johann van Beverwijck, a work approved by Harvey himself.[4]

These examples are powerful reminders of the porosity and problematic nature of the boundaries surrounding human anatomy: it is one of the aims of my work to probe and trespass those boundaries across medical as well as nonmedical disciplines. Issues such as experimentation and the mechanical understanding of the body are crucial to a number of areas and to a new intellectual history enriched by a number of novel approaches, from the emphasis on practice to the history of the book. Thus, *Mechanism, Experiment, Disease* calls for a recomposition of the intellectual landscape of the seventeenth century, triangulating among the history of medicine, science, and philosophy. Although anatomy occupies center stage in my investigations, my work addresses historians not only of anatomy but also of science, philosophy, and medicine.

2. Malpighi's Role on the Anatomical Stage

This book is neither a history of anatomy in the second half of the seventeenth century nor an intellectual biography of Malpighi. After the nearly 2,500 folio pages of Howard Adelmann's monumental—and at times sprawling—*Marcello Malpighi and the Evolution of Embryology*, including a paraphrase of almost the entire correspondence and translations of many primary sources, a biography may seem redundant. Adelmann's work, however, is closer to a primary source itself than to an analytic study. Whereas the format of a history of anatomy or medicine could be rather rigid, a work studying Malpighi in relation to the medico-anatomical world of his time appears as a coherent and fruitful enterprise for a number of reasons.

Malpighi was a key figure in the anatomical and medical worlds. Readers of the second edition of the *Bibliotheca anatomica* would find at the outset a special section of over one hundred folio pages including Malpighi's recently published and extensive *Vita* and his assessment of the significance of anatomy to pathology and therapy, written in response to Giovanni Girolamo Sbaraglia's attack. Although the *Bibliotheca* obviously excluded major works by Malpighi, such as his study of the silkworm, *De bombyce*, or of the anatomy of plants, *Anatome plantarum* (which was only excerpted), it did contain Malpighi's numerous other works; the editorial preface to the treatise on the liver, *De hepate*, for example, is a panegyric to Malpighi's skills with the microscope and to the novelties of his findings, which had left readers dumbfounded. One of the editors of the *Bibliotheca anatomica*, Manget, had even planned to dedicate one of the three volumes of the *Bibliotheca medico-practica* to Malpighi.[5] In addition, an examination of Carlo Frati's *Bibliografia malpighiana* offers a striking picture of the circulation of his works. All his publications went through edition after edition, from the *Epistolae de pulmonibus* (Bologna, 1661) to the *Opera posthuma* (London, 1697). The *Epistolae*, for example, gained European currency thanks to the Danish anatomist and physician Thomas Bartholin, who reissued them as part of *De pulmonum substantia et motu diatriba* (Copenhagen, 1663). Later editions—no less than fifteen—appeared as far away as Amsterdam, Padua, Frankfurt, and Jena. A French translation appeared in Paris for the benefit of surgeons in 1683 and again in 1687.[6]

Striking as this brief survey may be, it provides only a preliminary justification for choosing Malpighi as a common thread in my narrative. Although in many areas other scholars made major contributions of comparable standing, it is hard to find researchers who contributed to so many fields in such a significant way by contemporary standards. Malpighi's study of the lungs is a classic in the history of anatomy, providing the first microscopic investigation of the inner structure of an organ. Malpighi provided visual evidence for the circulation of the blood and for the notion that blood flows always inside blood vessels, pointing to the anastomoses or junctions between arteries and veins. His contributions to other areas, such as exploring the microstructure of several organs, the anatomy of the silkworm, the process of generation, and the anatomy of plants, were also remarkable, but what is especially noteworthy is their sheer range across human, animal, and plant anatomy. Moreover, Malpighi stands out for the range of investigation techniques he employed, including, but not limited to, microscopy, vivisection, injections, staining, and chymical analysis; his works display a remarkable spectrum of tools and methods typical of his time. Lastly, his publications, from *Epistolae de pulmonibus* to the *Opera posthuma*, span over three decades. By contrast, some of the best anatomists of his time had a much shorter productive period: Jan Swammerdam and Nicolaus Steno, for example, went through religious crises, while others, like Reinier de Graaf, died young.

Therefore, Malpighi looks like an ideal candidate for my study of mechanistic anatomy in the second half of the seventeenth century. Rather than focusing exclusively on his contributions, I wish to use his work as a lens or a probe to investigate and reflect on the anatomical and medical world of his time. My work is structured as a series of thematic investigations and what could be informally called "intellectual triangulations"—characterizing a field without necessarily providing an exhaustive coverage—including Malpighi's publications and relevant ones by his contemporaries. For example, his work on the lungs was followed by a series of investigations on respiration at Oxford and London, notably by Richard Lower and Robert Hooke; his work on the sense organs appeared together with publications by his friend Carlo Fracassati and the Pisa anatomist Lorenzo Bellini; the investigation of insects was carried out also by the Medici archiater Francesco Redi and by the Dutch anatomist Jan Swammerdam; his work on plant anatomy appeared at the same time as that of the physician Nehemiah Grew; large portions of his posthumous *Vita* are devoted to a refutation of *De motu animalium* by Giovanni Alfonso Borelli, his philosophical mentor and professor of mathematics at Pisa; his medical consultations followed a fate similar to Redi's, and the two can be profitably compared. These and a number of other contemporary investigations provide us with the opportunity to work on a rich and flexible canvas offering a broad picture of seventeenth-century themes.

In defining the contours of my research, I rely extensively on contemporary resources. The case of respiration around 1670 provides useful examples: in 1673 Malpighi showed to the Neapolitan lawyer Francesco d'Andrea three recent relevant books, Swammerdam's *De respiratione* (Amsterdam, 1667), Malachi Thruston's *De respirationis usu primario* (London, 1670), and John Mayow's section on respiration in *Tractatus duo* (Oxford, 1669)—the other being on rickets. The *Bibliotheca* set these works together with a common introduction. Further, a Leiden publisher reprinted Thruston's and Mayow's works in 1671 in the same-sized paper as Swammerdam's: a note on the back of Thruston's and Mayow's title pages states that readers could bind all three together in a *Sammelband*. At the time books were usually sold unbound, the binding being provided by the buyer: in this case the publisher suggested an intellectual association, providing us with valuable information on how anatomical books were printed, sold, and used. We shall encounter other examples of *Sammelbände*, showing how contemporary readers and publishers associated and bound books, especially in the Netherlands. I shall rely extensively on the organization of the *Bibliotheca anatomica* and contemporary publishing and reading practices.[7]

My research is not limited to anatomy, but touches on other issues such as techniques of investigation, visual representation, and the relations between anatomical research and disease. Some of these themes emerge only in a rather blurred fashion from Adelmann's magnum opus.[8] But, perhaps more significantly, by situating Malpighi's works alongside the works of his contemporaries, we can gain a deeper sense both of anatomy as part of the medical and intellectual worlds in the second half of the seventeenth century and of the significance of Malpighi's contribution to those worlds. I hope that this format will stimulate further studies and reflections on this extraordinarily rich period.

3. Medical Locations: The Sites of Malpighi's Work

In order to provide a physically and geographically situated perspective on Malpighi's work, I wish to explore the sites and contexts of his research and activities in the medical arena and the way he moved across them at Bologna, Pisa, Messina, and Rome. For the most part, the sites of Malpighi's professional activity were far from exceptional, but rather can be seen as representative of the field; therefore, thinking about the spaces and situations within which he operated helps us to visualize the settings within which many of his colleagues moved.

For the greatest part of his life Malpighi lived in Bologna; despite the conflictual atmosphere at the university, he had profound Bolognese roots and regularly sought to return to his land after stints at other universities. After a brief period in 1656 when he lectured on logic at his alma mater, Malpighi moved to Pisa, where between 1656 and 1659 he held the chair of theoretical medicine and was a colleague and friend of Borelli's. Back in Bologna he held chairs first of the theory of medicine and then from 1660 of the practice of medicine. Between 1662 and 1666 he was primary professor of medicine at Messina, a post Borelli secured for him thanks to his local contacts. Lastly, at the end of his life, between 1691 and 1694, Malpighi left Bologna for Rome, where he was archiater to the pope.[9] The sites where we find him and his contemporaries at work require us to think in terms not simply of mechanistic anatomy, but more broadly of mechanistic medicine, thus including also pathology and therapy.

The single most significant center for anatomical research and teaching at the time was the anatomist's private house, or more precisely a space therein equipped with the necessary instrumentation, referred to at times in the literature as "house laboratory." Malpighi's anatomical apprenticeship took place with the *Coro anatomico*, an informal gathering of students who met in the house of Professor Bartolomeo Massari, where they performed dissections and vivisections and discussed anatomical novelties, such as Harvey's circulation of the blood. Moving from Bologna to Pisa, Malpighi found himself in a similar setting, this time in Borelli's house, where he developed his mechanistic approach. Lastly, Malpighi's own house was equipped with an area for anatomical and microscopic research that was destroyed in a fire early in 1684: he lamented the loss of his papers and microscopes in several letters of that period. Nor were such spaces limited to anatomical research: Franciscus Sylvius at Leiden had several chymical laboratories in his house, and Malpighi also relied on chymical apparatus, to which he referred on several occasions; whether this was in his own house or elsewhere, such as in the apothecary's shop of his friend Angelo Michele Cantoni, is unclear.[10] In addition, Malpighi relished country retreats, away from his main residence, as several of his letters testify. After having rented a few places over the years, in 1682 he acquired a villa in Ronchi di Corticella, near Bologna, trying to escape from medical practice, the harassment of his enemies, and other professional annoyances. There he devoted himself to reading and research, relying on the services of a friendly butcher who provided him with animal parts for his investigations.[11]

While the house laboratory and the country villa served as venues for research and private teaching to a select number of students, the university lecture hall was the venue for official university teaching. Whereas at his house Malpighi had free choice as to the students or colleagues he wished to invite and subjects he wished to investigate, within university walls, as professor of the practice of medicine, Malpighi had to lecture on topics indicated by university statutes to students he could not select. His ability to form a school including Antonio Vallisneri (1661–1730), Ippolito Francesco Albertini (1662–1738), Antonio Maria Valsalva (1666–1723), and Giacomo Sandri (1657–1718), among others, built on his university and private teaching. In a letter of 1663, Borelli urged Malpighi to follow the method Gassendi outlined in the preface to *Exercitationes paradoxicae adversus Aristoteleos* (1649), where he argued that when he was professor of philosophy, at the end of each lesson he would raise some doubts about the accepted peripatetic doctrine. Those doubts did not trouble weak minds but aroused the interest of the best students. Possibly Malpighi followed a similar approach, formally complying with university statutes, yet raising doubts and introducing novelties. His lecture notes, still awaiting careful study, show that he discussed in his lectures results from contemporary research. Be that as it may, Malpighi's association with the university, as well as that of the overwhelming majority of anatomists and physicians discussed in this work, calls into question the claim that universities were not sites of intellectual innovation at the time.[12]

A site related to university life but at the same time open to a much wider audience than professors and students was the anatomical theatre. The annual public dissections performed usually at carnival were the occasions for major confrontations among professors belonging to rival schools. Indeed, the very architecture of the Bologna theatre, with a *cathedra* whence the professor put forward his theses far removed from the dissecting table, suggests that the purpose of the public anatomy centered on the display of the human body as well as on the disputation. The Bologna theatre, however, was often not a forum for novelties, but rather for attack against novelties. Although Malpighi never performed a public dissection because he was not the professor of anatomy, he attended such performances and was stimulated to develop his research partly in response to the challenges put forward against him at such functions. He published *De structura glandularum* (London, 1689), for example, in response to attacks on his opinions put forward at public dissections. In all likelihood, Malpighi made use of another university site, the botanic garden, a venue for teaching simples to medical students and presumably a relevant site for the author of *Anatome plantarum* (London, 1675–79). It is probably there that he saw specimens of the *mimosa* family, the sensitive plant that curls when touched, which attracted the attention of several seventeenth-century philosophers.[13]

In the seventeenth century anatomy was also a courtly entertainment performed in order to amuse and instruct. Malpighi claimed that the Medici, for example, were very interested in anatomy and that the Accademia del Cimento they promoted originated from anatomical dissections. Unlike Francesco Redi, however, who used to entertain the Medici with his dissections and other performances, Malpighi was no courtier and probably lacked Redi's social skills in this area. Far more important for him was the postmortem dissecting table for autopsies. The walls within which that table was located included churches, hospitals, and private houses. Much like William Harvey, Malpighi believed morbid anatomy to be enormously valuable, and from 1666 he started compiling a booklet with notable reports of autopsies he had witnessed or performed. Those reports are often mentioned in his publications and present a vivid image of his investigations.[14]

Travel was an important aspect of the physician's intellectual formation; a large number of physicians from northern Europe visited Italy and especially Padua in the seventeenth century, although their numbers declined from the time of William Harvey around 1600 to that of Thomas Bartholin in midcentury. Italian physicians were less enterprising in their travels, despite Gerolamo Cardano's celebrated trip to Scotland in the mid-sixteenth century. Malpighi and most of his colleagues

Figure I.1. Ferdinand de Saint Urban: reverse of Malpighi's medal

never left Italy, the most notable exception being the Sicilian naturalist Paolo Boccone, whose travels took him as far as England. Malpighi traveled in the Bologna region to assist his teachers in their medical work, to cure patients himself, and to view celebrated paintings. In addition, his stays at Naples and Rome on his way to and from Messina were significant episodes that cemented his contacts with the Naples Investiganti Academy, for example, and enabled him to become personally acquainted with Steno; at Naples Malpighi tried to attend the medical lectures by several professors—especially Carlo Pignataro, the conservative protomedico of the Kingdom of Naples who was opposed to the Investiganti Academy—but was barred from entering because he was not enrolled in the university. In 1667 he also made a trip to Padua with his friend Silvestro Bonfiglioli in order to listen incognito to Antonio Molinetti lecture on anatomy; this time he had more luck in that he was allowed to attend but found the lectures inadequate.[15]

Malpighi's friendship with Bonfiglioli leads us to another locale of significant intellectual import, the combined museums of Ulisse Aldrovandi and Ferdinando Cospi, of which Bonfiglioli was made director in 1675. The museum housed a collection of curiosities ranging from Etruscan urns and ancient coins to animal skeletons and coral. Like his teacher Andrea Mariani, Malpighi had antiquarian interests and collected ancient coins. The medal he had Ferdinand de Saint Urban cast in 1691 on the occasion of his call to Rome as pontifical archiater and election to the Bologna College of Physicians shows his portrait and, on the other side (fig. I.1), a female figure,

Figure I.2. Ferdinando Legati, *Museo cospiano*: the Cospi Museum at Bologna

probably representing curiosity and the new philosophy, observing with a lens plant specimens while reclining over a stack of books on a parallelepiped on which a worm and insect are crawling. The words inscribed on the parallelepiped, "STAT SOLIDO," may allude to the reliability of Malpighi's investigations; the image is framed by the inscription "TUTISSIMO LUMINE EXHIBITO," or "having shown the light most safely," possibly suggesting the care he had devoted to his observations. The overall arrangement strongly resembles an Etruscan urn, such as those in the Bologna Museum, visible under the cases in figure I.2. In the dedication of *Anatome plantarum* to the Royal Society, Malpighi talked of the Etruscans as the ancestors of his people, comparing himself to a haruspex who is dissecting plants in the hope of uncovering nature's secrets rather than slaughtering animals in order to divine the future. In this light the resemblance of the image on the medal to Etruscan funerary urns seems more than accidental: since the figures on the lids of those urns often symbolize the deceased's trade, Malpighi's medal ties his antiquarian interests to his activity as a naturalist and microscopist. Malpighi donated remarkable anatomical specimens to the museum, such as a horn grown from the neck of a cow, a petrified bone showing blood vessels, a petrified aorta, and an ivory-like testicle of a ram. Whereas in the Renaissance such specimens were seen as jokes of nature, he conceived them in a mechanistic fashion as products of the same laws of nature operating in the healthy animal and used them as tools of investigation. Thus, those specimens were key items in his research.[16]

The existence of a further site relevant to Malpighi's activities—of considerable interest to our discussion—can be inferred from circumstantial evidence. In several medical consultations Malpighi recommended an apothecary's shop, the "Spezieria del Sole" in the Galliera quarter of Bologna. The owner was Cantoni, who had been best man at Malpighi's wedding in 1667. When Pope Innocent XII chose Malpighi as his archiater in 1691, it was deemed appropriate to approach the anatomist through his best friends, namely, Cantoni and Bonfiglioli. Thus, the deep friendship with Cantoni spans several decades and suggests a familiarity with his shop that is otherwise difficult to document, given that the two lived in the same town and their correspondence is virtually nonexistent. Malpighi's consultations mention not only the shop but also specific preparations to be found there, thus suggesting a familiarity between the two medical men extending to professional issues. Apothecary's shops played an important role as venues for experiments and debates in the seventeenth century; one has only to think of the apothecaries John Clarke, Harvey's apothecary and the host of the Oxford meetings, and John Crosse, both at Oxford, who hosted Robert Boyle and whose shop was the venue for several experimental investigations. Thus, Malpighi's familiarity with the Spezieria del Sole and friendship with Cantoni appear to be important aspects of his Bologna life.[17]

Lastly, the bedside was a key site at the intersection between anatomical research and medical practice, where Malpighi met established therapeutic traditions and vested interests. This site was especially prominent in the last years of Malpighi's life, when he was archiater to Pope Innocent XII, but he started practicing with his teachers from a young age. Medical practice implied both visiting the patient and acting "at a distance" through the *consulti*, which offer us a glimpse of Malpighi's pathological and therapeutic thinking. Malpighi's medical consultations in response to requests of advice on difficult cases often offered a deep pathological framework for his therapies. Conversely, his anatomical publications contain several therapeutic insights; it is through those publications, those devoted to medical practice, and the *consulti* that we gain a sense of Malpighi's thinking at the bedside, even if they lack information on the physical examination of the patient. The sites of Malpighi's activities point to the importance of medical practice to his daily life and, as we will see, anatomical research.

4. Mechanism and Mechanics

A scholarly tradition intersecting the history of seventeenth-century science and philosophy has long been established; classic works such as Edwin Arthur Burtt, *The Metaphysical Foundations of Modern Science* (1924), Gerd Buchdahl, *Metaphysics and the Philosophy of Science* (1969), and more recently Gary Hatfield, "Metaphysics and the New Science" (1990), and Daniel Garber, *Descartes' Metaphysical Physics* (1992) join terms like "metaphysics" with "science" and "physics" in the title. Until recently, however, the status of medicine and anatomy was problematic in this regard, as if they had few intersections with the history of philosophy and of the physical sciences. An authoritative example is *The Cambridge History of Seventeenth-Century Philosophy* (1998), edited by Michael Ayers and Daniel Garber, to a large extent reflecting the state of the field at the time when it was produced: Harvey is mentioned once, most other anatomists, from Steno to Lower, never. In the dust jacket blurb we read: "the 'philosopher' was as likely to be peering through a microscope or preaching on divine justice as discussing skepticism, consciousness, or the concepts of good and evil."[18] Yet Hooke's *Micrographia*—although mentioned in the bibliography—is never mentioned in the text, and the other leading microscopists of the century, such as Malpighi, Swammerdam, Redi, Grew, and Antoni van Leeuwenhoek, either are not mentioned at all or, when they are, make cameo appearances. Whereas the intellectual links between Thomas Sydenham and John Locke are referred to, the leading anatomists of the time are ignored. Fortunately, this situation is rapidly changing, with a growing number of studies addressing medical and philosophical issues together, such as the notion of mechanism, for example.[19]

The notion of mechanism, together with its cognates, such as "*artificium mechani-*

cum," or mechanical contrivance, occupies center stage in seventeenth-century anatomy and especially in Malpighi's work. This notion is significant to contemporary thinking in biology too, despite lack of agreement as to its exact meaning. Although the concept itself is much older, the term "mechanism"—"*mechanismus*" in Latin—is a seventeenth-century neologism especially prominent in English. An especially early reference occurs in *The Immortality of the Soul* of 1659 by the Cambridge divine Henry More; it was in the same work that More employed the expression "the mechanical philosophy," by which he meant to a large extent Descartes' philosophy, and indeed Descartes had used the very same expression to characterize his own views. In *The Passions of the Soul* Descartes had claimed that blinking in response to a friend moving a hand in front of our eye does not require the mediation of a soul but occurs because of the mechanical organization of the body; we blink even if we know that the friend intends no harm. More challenged Descartes' view (emphasis in the original): "the Soule is to be entitled to the action, and not the meer *Mechanisme* of the Body." Here More means by "Mechanisme" an action or a structure tied to a material process. In contrasting the notion of "Mechanisme" with the action of the soul, More captured a key dichotomy of the time. Soon thereafter, in 1664, the physician and experimental philosopher Henry Power argued that the microscope will reveal "the curious Mechanism and organical Contrivance" of little animals, while in the 1665 *Micrographia* Hooke referred to "the *Mechanism* of the muscle." In both occurrences the term "mechanism" was associated with structures, including microscopic ones.[20]

Here my aim is to discuss how the notion of mechanism was used in the seventeenth century and to examine how it helps us understand important features of anatomical research and medicine. Besides the tension between structure and process, there are other meanings at the center of the concerns of philosophers and anatomists. The first is machine-like, with a special emphasis on the relation between the component parts and the whole. In his *Discours* on the anatomy of the brain, Steno argued that it is generally impossible to understand how a machine is assembled by observing its motions because those same motions could be performed in different ways; rather, it is necessary to take it apart and examine all its minute components—*ressorts* is his term: "Now since the brain is a machine, we should not hope to find its artifice [*artifice*] by other ways than those one uses to find the artifice of other machines. There is therefore nothing left to do besides what would be done to any other machine, I mean to dismantle piece by piece all its components [*ressors*, sic] and consider what they can do separately and together."[21] Here Steno compares the brain to a machine and applies the way to understand machines to it, seeking to grasp the relations between individual components and the whole: the juxtaposition of the qualifications "separately" and "together" is especially noteworthy. By "mechanical" he and other anatomists understood "machine-like" rather than based on the laws of mechanics; this interpretation goes hand in hand with a view of seventeenth-century mechanics according to which objects take center stage and embody more abstract relations. As in mechanics, in anatomy too understanding a complex structure meant decomposing it and recognizing in it elements associated with simpler, known objects that could be understood and handled separately.

The notion of machine at the time was quite a complex and heterogeneous one: typical examples include clocks, mills (including water mills), organs, fountains, and other pneumatic devices of the Alexandrian tradition. Hybrid chymico-mechanical devices, such as a gun or containers for the fermentation of fluids, were also considered pertinent to mechanistic anatomy. Besides these more complex or compound machines, we should consider the simple machines of the classical tradition such as levers and wedges, as well as filters and sieves, and new seventeenth-century devices such as pendulums and springs, which were mentioned in connection with motion, such as the contraction of the heart and arteries or respiration. These simpler devices were especially significant in connection with the notion of mechanism, because they were involved in a process of localization that was central to mechanistic anatomy: traditionally the main organs of the body were thought of as consisting of parenchyma, a term used since antiquity by Erasistratus and Galen, meaning extravasated blood—thus suggesting a largely undifferentiated structure. Since organs worked mainly because of and in connection with the faculties of the soul, there was not much point—nor were suitable tools available—in searching for minute structures and microstructures in order to understand their operations. Mechanistic anatomists, by contrast, believed that precisely their structures and microstructures were crucial to understanding how organs operated. The process of localization was central to their efforts: by localization I mean the identification of the internal organization of an organ enabling the investigator to determine the precise site where the key operations occurred and the nature of the mechanical processes involved. The idea that the body could be understood only by examining its structure in its minutest—at times microscopic—details was a characteristic and contested one in the seventeenth century.[22]

An especially instructive example comes from Nehemiah Grew's *Anatomy of Plants* (1682); discussing the discharge of seeds in coded arsmart, Grew identified the mechanism at play in a curled membrane working like a bow; or, as he put it (emphases in the original), "From this *Mechanism,* the manner of that violent and surprising *Ejaculation* of the *Seeds,* is intelligible" (see fig. 9.23). Here Grew employed the term mechanism and associated it with not only localization but also the notion of intelligibility, as if mechanization went hand in hand with understanding. Of course, at times such mechanisms were found or thought to be microscopic: in the former case magnification made them accessible to sight; in the latter the issue was one of the belief in their existence in principle. In *Principes de la philosophie,* Descartes stated that he saw no fundamental difference between the machines made by artisans and bodies made by nature, except that the tubes or springs in the latter case would be much smaller than those in the former and would not be perceptible by our senses.[23]

Other discussions of mechanism placed emphasis on different aspects. The notion of mechanism can be contrasted with the idea that the soul—or nature—and its faculties were responsible for bodily operations such as generation, growth, nutrition, and selective attraction, for example. According to this view, growth occurred because of a faculty or *dynamis* of plants and animals that is intrinsically different from how a machine operates. According to Aristotelian and Galenic doctrines, the animal was the instrument of its soul and the operations of its organs were governed by the faculties, either of the soul, as for Aristotle, or of nature and the soul, as for Galen. In *On the Natural Faculties* Galen had denied that the soul was involved in the lowest level of activities in living organisms, that of vegetation; therefore, in his opinion the faculties associated with it, primarily growth, nutrition, and generation, were more correctly linked to nature rather than the soul. But while rejecting that plants have souls, Galen still accepted a division between nonliving and living organisms—including both plants and animals; Galen's natural faculties differed from mechanical operations and pertained specifically only to animals and plants. During the Renaissance, the doctrine of the faculties of the soul was a major area of study, as shown by the many commentaries on Aristotle's *De anima.* Aristotle and his followers, however, did not argue that each and every bodily operation must necessarily occur because of the faculties: both in principle and in practice there was no problem in accounting for the operations of specific organs in mechanical terms. Such types of explanations had become more widespread in the third century BCE at Alexandria, in conjunction with the flourishing of mechanical studies there. By contrast, mechanistic anatomists were especially concerned with providing an account of the actions of a given organ exclusively in terms of its structure and of physical laws common to the whole of nature, excluding other types of explanation. Mechanistic anatomists sought to account for growth in terms of processes such as the superimposition of layers or unfolding. According to those defending the role of the soul, the kidneys had a faculty selectively to attract urine from blood, whereas the mechanists believed the kidneys to work like sieves, simply filtering urine. The notion of mechanism was presented by contrasting soul-like and machinelike explanations and by an attempt at localizing the site of secretion. As we will see in chapter 8, generation was especially problematic for mechanistic anatomists.[24]

At times the notion of mechanism is contrasted with teleology or some form of design, which are rather complex notions in their own right. It is helpful here to distinguish two different notions of teleology. According to one of them, the world and all creatures are created by God and therefore display His wisdom; this view was widespread in the seventeenth century, for all anatomists we shall discuss were Christians and believed in a provident creator. This view was not necessarily in contrast with mechanistic explanations, in that God could and would have created organisms working in a machinelike fashion. It is helpful once again to choose Steno as my source; introducing his discovery of the lachrymal glands and duct, Steno compared God to a mechanician and tears to the grease used to lubricate the moving parts of a machine; tears were filtered from arterial blood by glands without the need for faculties. The danger of this approach was that by attributing too many operations to mechanisms, organisms could be explained independently of God and one could therefore think of disposing of Him altogether, at least potentially. The other notion of teleology is immanent to individual living organisms and seeks to explain their operations in terms of a goal or end guiding their operations or acting to their benefit rather than through machinelike operations; this notion does not rely on a rational agent or creator. This was a view Malpighi, for example, tentatively found objectionable: arguing in *De polypo cordis* that nature operates in similar ways following the same laws in health and disease, he stated (my emphasis), "Moreover, we can consider whether all these things happen by the *sole necessity of matter and motion*, without a guiding mover for the animal's benefit."[25]

The notion of mechanism in anatomy and medicine was central to the works of many scholars, such as Descartes, Borelli, and Steno. Among practicing anatomists, however, Malpighi stands out for his sustained and systematic efforts over more than three decades to identify and explain mechanisms responsible for all bodily operations except for rational thinking in humans. Although the mechanistic program itself predates him, Malpighi made it a centerpiece of his agenda not just at the philosophical level but also in the detailed explanation of operations such as sense perception, secretion, growth, nutrition, and generation, as well as in understanding disease.

5. Experiment and Collaboration

Like the notion of mechanism, the practice of experimentation ties the medico-anatomical disciplines to other areas. The extensive collaboration among anatomists, physico-mathematicians, and experimental philosophers challenges the confinement of experimentation to any one domain at the expense of others. Historians of art and science have long appreciated the significance of the collaboration between medical men and artists in the sixteenth century, leading to striking new representations of plants and the human body in such works as *Herbarum vivae eicones* (1530), a work conceived by the publisher Johann Schott with illustrations by Albrecht Dürer's student Hans Weiditz and text by the Bern town physician Otto Brunfels; *De historia stirpium* (1542), conceived and written by the Tübingen professor of medicine Leonhard Fuchs with illustrations drawn by Albrecht Meyer, transferred to woodblocks by Heinrich Füllmaurer, and cut into wood by Veit Rudolf Speckle; and *Humani corporis fabrica* (1543) by the Padua professor of anatomy and surgery Andreas Vesalius, soon to become imperial physician, with illustrations by an unknown artist in Titian's circle. It is now time to put more firmly on the map a new intellectual alliance in the seventeenth century.[26]

At the beginning of the century, Galileo, while professor of mathematics at Padua, was a friend, colleague, and patient of the anatomy professor Hieronymus Fabricius and of the professor of theoretical medicine Santorio Santorio. Galileo intended to write on the motion of animals, vision and color, and sound and hearing, topics on which both Fabricius and his assistant Giulio Casserio published important works. Although Galileo did not complete those projects, his writings, such as *Two New Sciences*, include several references to anatomical issues. In 1603 Santorio mentioned an instrument based on the pendulum for timing the heartbeat, suggesting a contact with Galileo, who in those years was investigating the properties of the pendulum. In the same vein, in 1666 the physician Giovanni Battista Capucci called Malpighi "un 2° Galileo," probably comparing Malpighi's expertise with the microscope and Galileo's with the telescope, as well as the broad philosophical impact of their investigations.[27]

Many of Descartes' works, starting from *Discours de la méthode* (1637), deal with anatomical matters joining mechanics, philosophy, and anatomy. Descartes was a leading proponent of mechanistic anatomy; his works, culminating with the posthumous *De homine* (1662) and its original French *L'homme* (1664), with extensive annotations by the physician Louis de la Forge, set the scene for a sustained investigation by several anatomists. Descartes corresponded and debated with several physicians and anatomists, but it appears that he carried out a large portion of his research on his own, which led to a lack of anatomical accuracy. Often, however, Descartes did not pretend to do anatomy proper; rather, he introduced his anatomical reflection through the rhetorical device of a statue resembling a human body. Among his immediate correspondents and followers were the physicians and medical professors Vopiscus Fortunatus Plemp, Cornelis van Hogelande, and Henricus Regius.[28]

In the mid-seventeenth century we witness several examples of collaborations among philosophers, mathematicians, and anatomists. In Paris, for example, the anatomist Pecquet debated matters and collaborated with other scholars, including Pierre Gassendi and the mathematicians Adrien Auzout and Gilles Personne de Roberval. In 1651 Pecquet published his groundbreaking *Experimenta nova anatomica*, in which he announced the discovery of the thoracic duct and argued that chyle entered the bloodstream just before the heart. Pecquet reported several experiments on the vacuum relying on the Torricellian tube in a section of his work titled "physico-mathematica." It was as part of those experiments that Pecquet introduced the notion of spring or elasticity of the air or *elater*. The circle around Borelli at Pisa involved at different stages the Lorraine anatomist Claudius Auberius, Malpighi, Bellini, and Fracassati. Borelli served as philosophical mentor to the group and relied on his physico-mathematical knowledge and on the experiments performed at the Accademia del Cimento to interpret the anatomists' findings. Theirs was not the only case of collaboration at the Tuscan court. The Danish anatomist Nicolaus Steno was active in Florence in the late 1660s and collaborated with the mathematician Vincenzo Viviani in his work on muscles. Steno stated that anatomy had to be studied mathematically.[29]

In England we find notable examples of cross-disciplinary collaboration. John Wallis's autobiography mentions the discussions of novelties among scholars with different disciplinary affiliations prior to the establishment of the Royal Society in 1660 and lists areas such as Harvey's circulation of the blood, the lymphatic system, and astronomical matters. Notable collaborations among scholars in the mathematical and medical disciplines involved Thomas Willis, Christopher Wren, Richard Lower, and Robert Hooke. Newton collaborated with ophthalmologist Thomas Briggs and was followed by a number of physicians who wished to apply the Newtonian method to medicine. Robert Boyle was active in many medical areas. As Hal Cook has forcefully argued, physicians represented a large portion of the intellectual world in the seventeenth century and figure prominently among the members of the Royal Society: they read widely and experimented on matters pertaining to natural and the mechanical philosophy. The extensive collaborations among scholars from different disciplines call into question any strong demarcation between experimentation in the medical and physical sciences and draw attention to the vitality and cross-fertilization among these fields.[30]

At the same time, anatomical and medical experimentation poses specific problems—besides ethical issues—requiring careful handling; they can be captured by a series of dichotomies, such as the generalization from an individual to a general case within the same species; differences between male and female bodies; transfer of knowledge across species, including humans; reliance on minimalist versus interventionist techniques of investigation, affecting the object of study; the merits of dead versus living organisms; and—as we shall see in the following section—the study of healthy versus diseased states. As Nancy Siraisi has recently argued for Vesalius, the peculiar features of individual bodies posed a challenge to the identification of what is "normal" and "canonical" and to generalization. Not only did female and male bodies differ from each other, but the very same body could differ depending on age, season, and geographical location. The issue involved a range of variations within the normal, as well as monsters, which attracted in their own right a large literature with collections of rare and remarkable cases, such as hermaphrodites.[31]

Since antiquity, the issue of dissection and vivisection has attracted controversy and debate. On the one hand, the need to see actions in addition to structures and the rapid decay of the body—especially some parts—spoke in favor of vivisection. On the other hand, vivisection was more challenging to perform and in some circumstances may have affected the object of study, since nature is "put into such disorder by the violence offer'd" that its course could be altered, as Hooke put it. He contrasted vivisection with microscopy in the case of the water gnat, whose pellucid membranes allowed one to observe its inside without cutting, or, as he put it, "quietly peep in at the windows." Moreover, in line with his study of the water gnat, Hooke's microscopy focused primarily on outer surfaces and was relatively non-interventionist, this being a notable difference with respect to other microscopists like Malpighi, who was considerably more interventionist.[32]

Since vivisection could not be performed on human subjects, animals were used in some cases. Alas, animals can differ quite significantly from humans and among themselves, as the classic case of the *rete mirabile* shows: whereas Galen had attributed a major role to this structure at the base of the brain, as the locus where animal spirits are generated from arterial blood, first Berengario da Carpi and then Vesalius challenged its existence in humans—and rightly so—although in 1664 Thomas Willis still believed that it could be found in some humans, but only those "of a slender wit." Other cases also show significant variations, which could be used to the anatomist's advantage when a body part in an animal is larger or easier to study than in humans. Such procedures, however, required careful handling.[33]

Anatomical investigations were elaborate enterprises in the seventeenth century, when traditional dissections and vivisections were accompanied by new methods of injection and microscopy. Traditional dissection often required remarkable skills going well beyond mere observation, as testified by the many ways of dissecting the brain, for example; vivisection was often accompanied by ligatures and other more or less interventionist techniques; injections reached an unprecedented level of refinement and involved elaborate preparation methods, special syringes and other devices, and new mixtures of fluids ranging from ink and mercury to alcohol and colored wax; lastly, microscopy required fixation and staining methods, as well as lighting techniques that had to be patiently and painstakingly learned and tested over generations. Possibly the study of insects highlights most effectively the technical skills attained by seventeenth-century anatomists. These techniques were problematic and contested: Malpighi's finding of glands in the cerebral cortex, for example, was challenged as an artifact of the microscope and his preparation techniques, but also Ruysch's injections were questioned in that pressure was believed to compress soft parts around the vessels, making organs appear exclusively vascular.[34]

6. Disease and Anatomy

The rise of a chymical and mechanistic understanding of the body and new anatomical findings profoundly interacted with changing notions and new explanations of disease. The tension between individual cases and generalizations was problematic for pathology and therapy, since a patient's individual constitution meant that disease was routinely conceptualized on a case-by-case basis, involving the patient's temperament and the balance of the humors. In the sixteenth century chymical healers had already challenged traditional views by introducing major changes conceptually as well as in the pharmacopœia and therapeutics; the new pharmacopœia, for example, included preparations based on antimony and mercury. Some of those innovations became progressively part of traditional medicine, as testified by the many editions of *antidotarii* issued in many European cities. However, early chymical healers and physicians accepted soul-like agents and *archei*, vital principles resembling traditional faculties that were to be banned by mechanistic physicians. Thus, whereas for the Flemish physician Johannes Baptista van Helmont, for example, disease and therapy hinged on the *archeus*, mechanistic physicians conceived disease in terms of a malfunctioning machine and therapy more like cleaning and mending it. In this case too localization, often in the glands, played a key role. The issue of whether mechanistic anatomy led to changes in pathology and therapeutics is quite complex and was the object of debate already in the seventeenth century. Unlike chymical medicine, mechanistic approaches did not lead to significant changes in the available pharmacopœia. However, it would be misleading to conclude from this that no change occurred, since pathological thinking and therapy changed significantly in dealing with many diseases—from apoplexy to aneurysms. In this work I strictly adhere to a seventeenth-century perspective of disease in general and of individual diseases: whenever I talk of the plague or jaundice I mean what Malpighi and his contemporaries understood by those terms, even when their views differ widely from our own.[35]

The emerging glandular and hydraulic understanding of the body led to a notion of disease located at the juncture between solid and fluid parts, especially glands and blood. From the 1660s, by glands anatomists understood organs of filtration. In a healthy state, specific fluids are filtered at different locations in the body by appropriate glands depending on their inner conformation: bile in the liver, pancreatic juice in the pancreas, semen in the testicles, nervous fluid in the cerebral cortex, urine in the kidneys. When a bad diet, for example, alters over time a balanced composition of blood, glands become partly obstructed or damaged; something similar could happen as a result of specific unsuitable agents, as in the French pox. In such cases blood is improperly filtered in those glands and the fluid they secrete is affected. At times such a fluid is merely excreted, as in the case of urine, for example; at other times, however, the fluid could be an important part of the animal œconomy, as in the case of all the fluids pertinent to digestion, for example, or nervous fluid. In these cases improper filtration had a double effect, in that blood was not properly cleansed and could become too acidic, while at the same time fluids playing a role in the body were defective and therefore could not work properly. In both cases therapy was long and difficult, consisting usually of remedies balancing the composition of blood in the hope of reversing or correcting the damage and obstructions in the glands. In addition, the new anatomy affected the understanding of specific diseases because new findings made older explanations obsolete, especially when the plumbing of the body was revised. Pecquet's finding of the thoracic duct, for example, deprived the liver of its faculty of making blood and required a reconceptualization of its pathology. Steno's finding of glands and ducts in the mouth led him to challenge the traditional doctrine according to which catarrh descends from phlegm in the brain, since no passageway from the brain to the mouth had been found, whereas his new ducts supported his novel view. Aneurysms were explained in mechanistic—specifically hydraulic—fashion in terms of pressure.[36]

Disease and anatomy were tied in other ways. The study of disease was relevant to anatomy, in that disease could lead to the hardening and enlargement of body parts, enabling their investigation, and to the production of excess fluid that could be chymically assayed. Malpighi relied extensively on the hardening of body parts, such as those he donated to the Bologna Museum, which became permanent witnesses to his views, and those he mined in the literature. Others, however, such as Ruysch, questioned this method and argued that bony or stony concretions in no way enlighten us about the constitution of an organ. At times disease can enlarge a body part, such as a gland, but in this case too caution is required to prevent attributing features of diseased states to healthy ones. The same caution is required in the case of the accumulation of fluids due to disease, which may render its collection easier but can also alter its state and make its study problematic. Commenting on Thomas Willis's research on the brain, medical historian Kenneth Dewhurst discussed two cases in which Willis relied on postmortems and clinical speculations: in the first, he noticed during a postmortem that despite an occlusion in the left carotid artery, a man still lived and blood reached all the parts of his brain. Together with Lower, they confirmed their suspicion of an anastomosis among the four arteries supplying the brain by means of vivisection experiments involving colored injections on a dog. Thus, in this case a "clinico-pathological" observation led to an anatomical discovery, later known as the circle of Willis. The other case led to rather different results. Unable to find evidence of a fluid in the nerves through traditional methods of cutting and ligating, Willis relied on clinical observations: he noticed that the urine of patients affected by nervous complaints turns clear and watery when they recover, a symptom he attributed to the discharge of a watery fluid from the nerves. Since, as Dewhurst put it, Willis "was primarily a busy doctor whose anatomical research was merely a spare-time occupation," it is not surprising to find anatomy and disease closely tied, at times confusingly so, according to Dewhurst. Willis's case was representative of the seventeenth century and speaks loudly against any sharp demarcation between anatomy and medicine more broadly, however confusing their intermingling may appear to us. In Malpighi's case too we find discussions of disease and therapy in all his anatomical treatises, starting from *Epistolae de pulmonibus*. It is as problematic to study the history of anatomy ignoring pathology and clinical practice as it is to study disease and therapy ignoring anatomical knowledge and theoretical conceptualizations: these areas were mutually related and cannot be disentangled by the historian.[37]

7. Structure and Organization

The book consists of four parts, with three chapters each, organized around key moments, texts, and themes in the history of anatomy and related areas. I adopt historically sensitive strategies for selecting, grouping, and reading those texts. Each chapter analyzes works by Malpighi in the context of contemporary research, following a combined chronological and thematic order from his *Epistolae* on the lungs to his posthumous medical consultations. In addition, for each of the four parts I have found it helpful to select one theme that figures especially prominently in its chapters.

Part I takes the lead from the friendship between Borelli and Malpighi to discuss the rise of mechanistic and microscopic anatomy in the mid-1660s, reading the works Malpighi composed after his return from Pisa to Bologna in 1659 and at Messina until 1666 in a wider setting. After an introduction on Malpighi's apprenticeship in the context of broader European developments, chapter 1 discusses his *Epistolae de pulmonibus* and his exchanges with Borelli in relation to the work on respiration carried out in England, chiefly by Malachi Thruston, Richard Lower, and Robert Hooke. Chapter 2 moves to a more medical theme, focusing on the literature of fevers, Borelli's 1649 treatise of malignant fevers, the 1661 Pisa epidemic, and the 1665 controversy between Malpighi and Lipari at Messina. Although Malpighi's *Risposta* to Lipari first appeared posthumously in 1697, over thirty years after it was written, it is helpful to discuss it in conjunction with other works from the time when it was composed. In 1664 and 1665 a series of works on the brain and the sense organs were published and discussed. Chapter 3 discusses works by Willis, Steno, Malpighi, and Fracassati on the brain, as well as related works by Malpighi, Fracassati, Bellini, and Rossetti on the tongue and the skin as organs of taste and touch and their nervous connections to the brain. In addition, Malpighi's *De cerebro* deals with vision and challenges Descartes' views. All these works are central to the efforts to provide a mechanistic understanding of sensory perception. The theme running through this material is the relation between corpuscularism and sensations, especially the atomistic doctrine of sensory perception and the role of color, which occurs in discussions on color change of blood in respiration, in the study of disease, and in experiments on color indicators.

Part II focuses on the problem of secretion and the anatomy of glands as key loci of seventeenth-century conceptualizations of the body and of Malpighi's investigations, setting his *De viscerum structura* (Bologna, 1666–68) and the later *De structura glandularum* in the context of contemporary research and debates. Chapter 4 deals with the revival of the glands in works by Francis Glisson, Thomas Wharton, Steno, and Franciscus Sylvius. I study their contributions in relation to Malpighi's work on the glandular structure of the viscera. Chapter 6 extends the study of anatomical and experimental investigations by a number of anatomists and physicians, such as Pechlin, Peyer, and Brunner, to Anton Nuck in the 1680s. Whereas chapters 4 and 6 deal with glands, chapter 5 takes the lead from two intriguing works by Malpighi on fat and blood to reflect on the mechanical organization of the body, comparing his reflections on fat to Descartes'. There are reasons for considering these works together, because in both of them Malpighi used the suggestive notion of *necessitas materiae* in his reflections on the body's organization. Malpighi studied the composition of blood and used disease as a way to achieve his aim at the time of the first blood transfusions, which provide a novel medical context for this work. The theme running through these chapters is the growing role of unusual techniques of anatomical investigation, such as those involving disease and pathological states that hardened and enlarged body parts, making them permanent witnesses to a given structure and allowing them to be more closely inspected.

The links among anatomy, natural history, and the study of lower animals are the main focus of part III, which covers the period of Malpighi's association with the Royal Society from 1668. Major contributions in these years include Malpighi's stunning microscopic investigation of the silkworm and his studies on generation and the anatomy of plants. Some of these investigations may appear to be unconnected, yet they share common concerns and agendas of considerable philosophical import. By exploring the problem of generation and the anatomy of plants, Malpighi sought to provide a viable mechanistic understanding of processes such as generation and growth, areas to which Hooke also offered original and surprising contributions.[38] Chapter 7 examines the literature on insects around 1665–69 in the works by Hooke, Redi, Malpighi, and Swammerdam. Chapter 8 deals with the study of generation, setting Malpighi's study of chicken eggs against the background of Harvey's 1651 *De gen-*

eratione animalium and other investigations by Steno, de Graaf, and Swammerdam. Chapter 9 is devoted to Malpighi's *Anatome plantarum* and Grew's 1682 *The Anatomy of Plants*, the first microscopic studies in this area. Moreover, I examine shifting conceptions of sexual reproduction in nature in works by Swammerdam on snails and by Camerarius on plants. The theme emerging with special prominence from these chapters is the role of illustrations, which posed major problems in the case of insects and was especially prominent in the study of generation and plants.

Lastly, part IV deals with the connections among anatomical research, pathology, and therapy; I examine the nature and contents of Malpighi's posthumous publications, starting from the *Opera posthuma*, published by the Royal Society in 1697. The *Opera posthuma* differs considerably in style from his previous publications; it includes the extensive *Vita*, as well as the *Risposta* to Malpighi's colleague at Bologna Giovanni Girolamo Sbaraglia. Chapter 10 deals with the *Vita*, in which Malpighi listed all the abuses and injustices he believed he had suffered from his student years to the last day of his life and defended his entire production. Since a large portion of this material is devoted to Borelli, I reconsider their relationship with special attention to Malpighi's response to *De motu animalium*. In addition, I study the controversy with Paolo Mini, as well as Ruysch's investigations based on elaborate injections, which he started to develop at the end of the century and which challenged Malpighi's results. More broadly, this chapter provides the opportunity to reconsider a number of themes at the center of Malpighi's work and seventeenth-century debates, from respiration to the generation of plants. The last two chapters share a concern with the connections among the new anatomy, pathology, and therapy, thus addressing this important topic that is often ignored by historians of anatomy. Chapter 11 is devoted to the controversy with Sbaraglia, including Morgagni's 1705 defense of Malpighi. Having been a student of Malpighi's students, Morgagni saw himself as his intellectual grandson and as a key defender of the role of anatomy to medical practice. Whereas the appearance of the *Opera posthuma* was carefully and craftily engineered by Malpighi to secure his reputation and that of his followers, the medical consultations that appeared between 1713 and 1747 were published against his wishes. Yet they are an integral and important part of his work: they shed light not only on his methodology and pathological thinking but also on his medical practice, establishing a bridge between anatomy and therapy, and offer a picture of the medical world in which Malpighi and his contemporaries operated. Chapter 12 examines the structure of Malpighi's consultations in relation to Redi's, whose peculiar jovial style provides an ideal counterpoint to Malpighi's deadly serious and scholarly demeanor, and Morgagni's learned ones. This strategy is helpful in analyzing a difficult genre like the *consulti* relying on contemporary texts while avoiding the pitfalls of retrospective diagnoses. The theme running through these chapters is the role of styles of writing as vehicles for intellectual and philosophical messages; this topic has broad implications, yet a careful study of Malpighi's *Vita*, *Risposta* to Sbaraglia, and consultations will prove especially fruitful from this perspective.

Finally, a brief epilogue sets some of the key results attained in a sharper light.

PART ONE

THE RISE OF MECHANISTIC AND MICROSCOPIC ANATOMY

Malpighi's Formation and Association with Borelli

The rise of mechanistic and microscopic anatomy went hand in hand with several examples of collaborations among anatomists and physico-mathematicians. Such collaborations were a common and important phenomenon characterizing most of the seventeenth century, much like the collaborations between artists, botanists, and anatomists in the sixteenth century. Traditionally Descartes is considered a key figure in mechanistic anatomy for his anatomical interest and publications, starting in 1637 with the *Discours de la méthode*, and for his debates and association with several anatomists. Similar associations and collaborations, however, extended well beyond Descartes and his immediate disciples.

The decade-long collaboration between Borelli and Malpighi was one of the most visible and productive of the whole century. Although Malpighi learned the art of dissecting and vivisecting at Bologna, it was largely at Pisa that he transformed his philosophical outlook under the tutelage of the mathematics professor Borelli and learned microscopy. After exploring Malpighi's intellectual and technical formation at Bologna and Pisa in relation to similar settings elsewhere in Europe, part I studies the key period of that collaboration, focusing on Malpighi's first publications on the lungs, his defense of neoteric medicine against the most traditional Galenism, and the 1665 works on the brain and the organs of sensation—notably the tongue and skin. I set his research against both traditional doctrines and some of the most significant contemporary investigations and philosophical reflections, paying special attention to seventeenth-century perceptions, as they are documented in contemporary debates and in the pages of the *Bibliotheca anatomica*. I focus in particular on the research carried out by other anatomists in Borelli's circle and several English anatomists and physicians. Chapter 3 is especially devoted to the interplay between atomism and the organs of sensation; the links among corpuscularism, sensory perception, and the role of color in particular involve philosophical, experimental, chymical, and medical issues and emerge as a common thread tying all the chapters in part I.

The young anatomist Jean Pecquet was a key figure in the anatomical world of the time for his finding of the thoracic duct, mechanistic standpoint, and collaboration with physico-mathematicians. Pecquet's discovery of the thoracic duct denied the liver's traditional role and, much like Harvey's discovery of the circulation of blood, shook the foundations of traditional anatomy and medicine. Unlike Harvey, however, Pecquet was sympathetic to the new mechanistic anatomy, and, unlike Descartes, he was a careful and skillful dissector. The figure below reproduces the title page of a dedication copy of the 1654 edition of Pecquet's *Experimenta nova anatomica* with

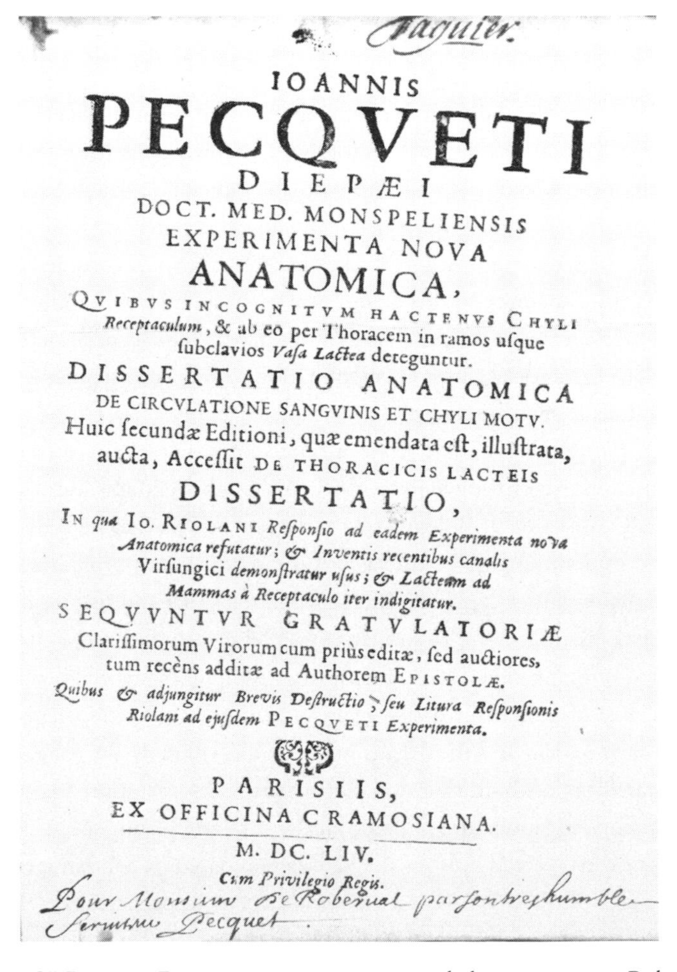

IOANNIS
PECQVETI
DIEPÆI
DOCT. MED. MONSPELIENSIS
EXPERIMENTA NOVA
ANATOMICA,
QVIBVS INCOGNITVM HACTENVS CHYLI *Receptaculum*, & ab eo per Thoracem in ramos uſque ſubclavios *Vaſa Lactea* deteguntur.
DISSERTATIO ANATOMICA
DE CIRCVLATIONE SANGVINIS ET CHYLI MOTV.
Huic ſecundæ Editioni, quæ emendata eſt, illuſtrata, aucta, Acceſſit DE THORACICIS LACTEIS
DISSERTATIO,
IN *qua* IO. RIOLANI *Reſponſio ad eadem Experimenta nova Anatomica refutatur; & Inventis recentibus canalis* Virſungici *demonſtratur uſus; & Lacteum ad Mammas à Receptaculo iter indigitatur.*
SEQVVNTVR GRATVLATORIÆ
Clariſſimorum Virorum cum priùs editæ, ſed auctiores, tum recèns additæ ad Authorem EPISTOLÆ.
Quibus & adjungitur Brevis Deſtructio, ſeu Litura Reſponſionis Riolani ad ejuſdem PECQVETI *Experimenta.*
PARISIIS,
EX OFFICINA CRAMOSIANA.
M. DC. LIV.
Cum Privilegio Regis.

Figure PI. Pecquet, *Experimenta nova anatomica*: dedication copy to Roberval

Pecquet's inscription to the Paris mathematician Gilles Personne de Roberval. Pecquet witnessed and reported Roberval's celebrated experiment with the air bladder of a carp, arguing that air was elastic, and explained the motion of chyle according to mechanical principles without having recourse to attraction. Thus, this dedication copy—"Pour Monsieur De Roberval par son tres humble Serviteur Pecquet"—embodies one of the most significant and representative seventeenth-century collaborations.

The New Anatomy, the Lungs, and Respiration

1.1 Changing Anatomical Horizons

The middle decades of the seventeenth century witnessed major changes in the methods of anatomical investigation and profound transformations in the understanding of the body. The works by Aselli and Harvey introduced subtle shifts in the way vivisection was carried out; these shifts became more established and were developed in the ensuing investigations of the 1640s and 1650s on the circulation of the blood, the thoracic duct, and the lymphatics. Although microscopy and injections had been applied to anatomy in the past, from the 1660s they took a more prominent role on the anatomical stage in a number of areas, starting from Malpighi's first publication, the 1661 *Epistolae* on the lungs. We shall move from the relations among anatomical apprenticeship, techniques of investigation, results attained, and philosophical views in the first half of this chapter to the structure of the lungs and the problem of respiration in the second half.

With regard to the lungs, major questions arise on the differences between Borelli and Malpighi at Pisa and Bologna on the one hand and Hooke and Lower at Oxford and London on the other. Traditionally, Malpighi's *Epistolae* have been studied as a striking example of his application of microscopy to anatomy and, to a lesser extent, of his reliance on mercury injections. Besides dissection, however, Malpighi also used vivisection, a technique he had learned in Bologna, and chymical analysis. Relying on his atomistic standpoint and experiments at the Accademia del Cimento, Borelli denied any role to color as providing information on a substance. As a result of these views, Borelli and Malpighi ignored color change in blood in the presence of air and in respiration. Whereas Malpighi visualized microscopic structures and motions as a way to understand the purpose of the lungs, Lower and Hooke challenged Malpighi's views on the purpose of respiration and sought to grasp the mode of operation of the lungs and pinpoint the site of color change in blood through vivisection, eschewing structural investigations. Their work provides an instructive anatomical and philosophical contrast to Malpighi's and Borelli's approach: not only were techniques such as vivisection, microscopy, injections, and chymical analysis used differently, but the philosophical underpinnings of the investigations differed too.

The following two sections take the lead from Malpighi's training at Bologna and Pisa to investigate the issue of anatomical apprenticeship among his contemporaries. Section 1.2, on Malpighi's training at Bologna, compares the *Coro anatomico* with the *Collegium privatum* of Amsterdam and then moves on to document the revival of vivisection in the middle decades of the seventeenth century in the works by Aselli, Harvey, and their followers, such as Walaeus and Pecquet. Section 1.3 uses Malpighi's philosophical apprenticeship at Pisa and his friendship with Borelli to investigate the rise of microscopic anatomy and the role of the new philosophy. Section 1.4 looks at how this background played out in Malpighi's first publication, the 1661 *Epistolae* on the lungs. I study his techniques of investigations and Borelli's role in advising him and interpreting his anatomical findings, neglecting the role of color. Lastly, section 1.5 takes a broader look at the problem of respiration with special emphasis on the English scene. Thomas Bartholin reissued Malpighi's *Epistolae* in 1663, giving them a broader European circulation. Further, the *Bibliotheca anatomica* included them in a large section on the thorax including also Harvey's *De motu cordis et sanguinis*, Malachi Thruston's *De respirationis usu primario*, Lower's *De corde*, and several other treatises and excerpts on the heart and lungs. Bartholin's edition and the *Bibliotheca anatomica* provide us with valuable information on how texts circulated and were read. In this large body of literature I focus on the work by Thruston, which is especially relevant here because he tried to replicate Malpighi's microscopic observations, and Lower, who refuted Malpighi.

1.2 Malpighi's Bologna Apprenticeship: Anatomical Venues and Vivisection

In 1653 Malpighi gained his degree in philosophy and medicine from Bologna University. The double degree was standard at the time and testifies to the profound link between the two disciplines. Malpighi mentions among his medical and philosophical sources the professors of medicine Caspar Hofmann at Altdorf and Andrea Cesalpino at Pisa and the philosophers Francesco Buonamici at Pisa and Cesare Cremonini at Padua, not exactly the most up-to-date list in the mid-seventeenth century. In the 1650s Malpighi was learning to be a physician. He was instructed in the "metodo di medicare" by Bartolomeo Massari and Andrea Mariani, who relied on Hippocrates, *De ratione victus*, and the sixteenth-century author Leonardo Giacchini. Malpighi must have faced the problem of reconciling ancient therapies with novel findings at Bologna, since this was one of the standard issues associated with the reception of the circulation of the blood. It appears that contrasts over different therapeutic practices preceded the emergence of mechanistic anatomy and Malpighi's meeting with Borelli. At Bologna, for example, such contrasts can be documented in the generation of Malpighi's teachers Massari and Mariani.[1]

Malpighi's activities in his early years and especially his participation in the *Coro anatomico* of Massari and Mariani shaped his interest and skills in anatomy. In his *Vita* Malpighi recounts that around 1650 his teacher Massari instituted an informal gathering at his home of nine subjects, who in turn devoted themselves to the study of anatomy and to performing autopsies of executed criminals, as well as vivisections and dissections of animals. Informal groups meeting in private houses in order to perform dissections were not uncommon at the time, as shown by the example of the Oxford group studied by Robert Frank. Here I wish to highlight the similarities between the *Coro anatomico* and a later group of anatomists and physicians active in the mid-1660s and early 1670s, called *Collegium privatum* of Amsterdam, devoted to anatomy and led by the Amsterdam professor of medicine and town physician Gerardus Blasius and by Matthew Slade, physician to the municipal St. Peter's Hospital. The *Collegium* counted among its members the young Jan Swammerdam, then still a student. In both informal groups the number nine seems to have played a role—possibly for its association with the number of the muses—with nine auditors selected by Massari and nine members at Amsterdam; both groups included university professors of medicine and physicians interested in anatomy, as well as younger students. Some of the activities of the Amsterdam *Collegium* are known through two small pamphlets issued in 1667 and 1673 owing to the initiative of the publisher, Caspar Commelin. The name of the group suggests that it was an informal society that in all probability met in private houses. The 1673 pamphlet mentions a meeting in the house of a member of the *Collegium*, Hermann van Friessem, in order to study sturgeons. Much like the *Coro anatomico*, their investigations were inspired by the recent literature. The first volume refers frequently to Steno, including his work on the salivary duct; moreover, several observations concern the kidney, the topic of Lorenzo Bellini's first publication, which had been reprinted in Amsterdam in 1665 with additions by Blasius; they examined various animals, including frogs and several types of fishes. The 1673 pamphlet discusses novel techniques such as mercury injections and deals with the pancreas of fishes and its juice, with special emphasis on whether it was acid, a claim they denied; Reinier de Graaf, who had pioneered similar investigations, was quoted and criticized. Malpighi is mentioned only in the second part of 1673, but it is plausible that he inspired the investigation of frogs and mercury injections discussed in the first part. One could only wish that a Bolognese publisher had been as solicitous as Commelin in Amsterdam; lacking a Bolognese Commelin, we have to rely on plausible conjectures.[2]

Malpighi stated that Massari instituted the *Coro, curiositate motus,* in order to assess different opinions on recent findings, including the circulation of the blood and other novelties, *aliaque nova experimenta anatomica.*[3] Whereas the reference to Harvey is unequivocal, the rest of the quotation is ambiguous, but curiously Malpighi's words echo precisely the title of another major anatomical work of the time, Pecquet's 1651 *Experimenta nova anatomica*: indeed, the works by Aselli, Pecquet, and their followers were the obvious other major anatomical novelty. Aselli's findings in the posthumous *De lactibus sive lacteis venis* were often quoted in the literature, but it is exceedingly interesting from our perspective to learn of attempts to replicate them at Bologna in the 1640s by the Bologna anatomist Giovanni Antonio Godi. Although he failed to confirm them, the report of his attempt by the professor of medicine and member of the Bologna college of physicians Bartolomeo Bonaccorsi in the 1647 *Della natura de polsi* is a rare testimony on anatomical research in Bologna at the time.[4] Thus, the available data on Malpighi's education highlight his knowledge on the circulation of the blood and possibly the milky veins and thoracic duct. The two other young members of the *Coro anatomico* who we know of, namely, Carlo Fracassati and Giovanni Battista Capponi, found themselves in the neoteric camp with Malpighi.[5]

The works by both Aselli and Harvey on the milky veins and the motion of the heart and blood rely on vivisection, yet they used it for strikingly different purposes. Aselli set out to investigate traditional subjects for vivisection, such as the role of the recurrent nerves—controlling the larynx—and the motion of the diaphragm in respiration; quite accidentally, however, he ended up finding a structure whose existence had been envisaged by Galenic anatomy but had not been properly observed and described, namely, the milky veins allegedly carrying chyle from the intestine to the liver. The vivisection of a dog a few hours after it had eaten enabled him to detect vessels so ephemeral as to have escaped the attention of previous anatomists; thus, his work involved not only a new structural finding but also the realization of the importance of the timing of the animal's last meal. By contrast, Harvey's work presented no new anatomical structures: his focus was on motions, and vivisection enabled him to study the systole and diastole of the heart in warm- and cold-blooded animals and the direction of blood flow by means of ligatures. Whereas Aselli's illustrations show the new structures he had uncovered, Harvey's illustrations show a process rather than structures. The significance of timing the animal's last meal, the observation of the dying heart, and the use of ligatures to investigate directionality go back to Galen; with their novel findings, however, Aselli and Harvey showed the vitality and significance of these techniques, which were to become standard tools of investigation in the following decades.[6]

The reception of Harvey's views is inextricably bound with a number of themes, including the emergence of a mechanical understanding of the body. The way Har-

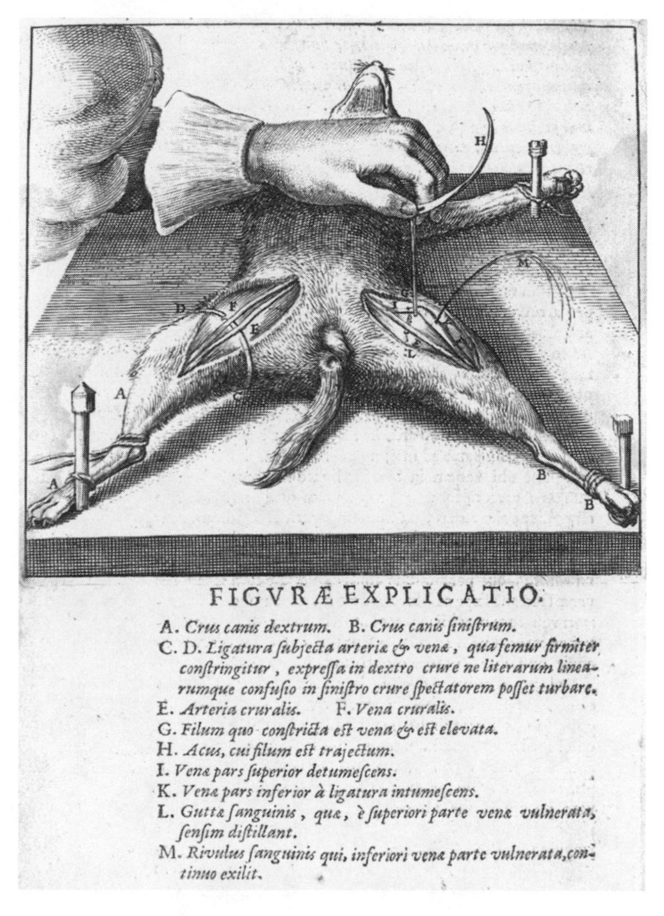

Figure 1.1. Walaeus, *Epistola*: vivisection experiment on the circulation

vey was read and the nature of the debates and controversies on the circulation of the blood are of great intrinsic interest and at the same time shed light on themes to which Malpighi would have likely been exposed.[7] Although we lack precise information on the Bologna practices, we do have some data on the contemporary Padua scene. *De motu cordis et sanguinis* went through several editions following its first publication, including one at Padua in 1643 with two epistles in support of Harvey by Johannes Walaeus; they had first appeared two years earlier in Caspar and Thomas Bartholin's *Institutiones anatomicae*, a widely circulating summa of anatomical knowledge that went through many editions. Walaeus's letters describe two important vivisection experiments: in one of them, he ligated the crural vein of a dog and showed that blood spurts out forcefully from a distal puncture, on the side away from the heart, whereas hardly any blood spurts out from a proximal puncture, on the side of the heart (fig. 1.1); this experiment was inspired by a passage in chapter 11 of Har-

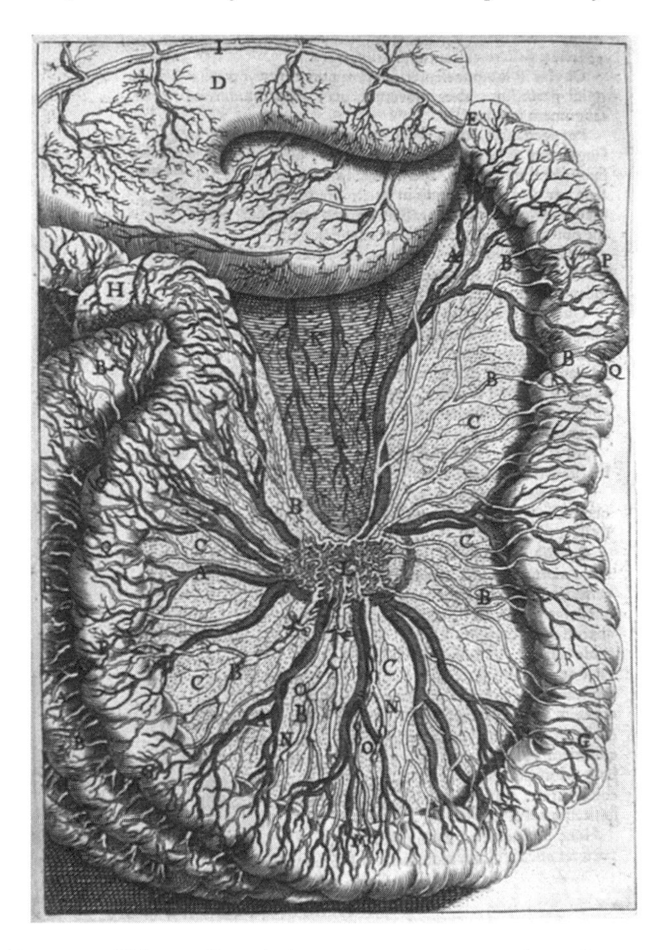

Figure 1.2. Walaeus, *Epistola*: ligature experiment on the milky veins

vey's *De motu cordis*. The other experiment consists of applying a ligature to Aselli's milky veins to show that the fluid they contain comes from the intestine (fig. 1.2). In 1644 the Padua mathematician Andrea Argoli published an astronomical work, *Pandosion sphaericum*, in which he referred to experiments on the circulation performed by Johann Georg Wirsung (called "Verden"), who had recently discovered the pancreatic duct. Wirsung's experiments, as well as Walaeus's and Argoli's accounts, rely heavily on Harvey's quantification argument that the body cannot provide from ingested food the huge amount of blood pumped by the heart. Moreover, two letters of 1655 from Heinrich van Moinichem at Padua to Thomas Bartholin report that the anatomists Antonio Molinetti and Pietro Marchetti performed experiments on blood circulation; Molinetti defended it while vivisecting a dog during the annual anatomy dissection in January, whereas later in the year Marchetti gave two lectures in a hospital, one on circulation and the other on the lymphatics. He defended the circulation by injecting with a siphon warm water into the artery of a female cadaver and showing the water emerging from the vein, as he had mentioned in his *Anatomia* of 1652. In 1647 Bartolomeo Bonaccorsi discussed Argoli's account in *Della natura de polsi*. Bonaccorsi referred to Wirsung's experiment and discussed the quantification argument, although he did not accept it.[8]

In 1651 the young Pecquet, then still a student in Paris, published *Experimenta nova anatomica*, a work that sent shock waves across the medico-anatomical community with his discovery of the thoracic duct that carried chyle directly to the subclavian vein, bypassing the liver; in this way the largest internal organ in the body was deprived of its traditional role, held since antiquity, of making blood. Like Aselli's finding, Pecquet's was also stimulated by an extemporaneous observation while vivisecting a dog to study the motion of the heart, but then it required an extensive investigation lasting three years and involving the dissection and vivisection of many animals. Figure 1.3 shows on the right a dissected dog with the thoracic duct, closely resembling in posture and shading figure 1.1; on the left the duct is shown enlarged. Notice the receptacle of the chyle at the bottom, the ladderlike structure for carrying chyle, and the insertion in the subclavian vein near the heart. The novelty of the structure made the choice of a bipartite figure especially adroit: that on the right shows the location and relative size of the thoracic duct inside the animal; that on the left, by contrast, shows the thoracic duct in much greater detail, unencumbered by the surrounding structures. Within ten years of the Paris *editio princes*, Pecquet's work was printed at least seven more times, including a partial Genoa edition in 1654 by the Lorraine physician Jean Alcide Munier with a dedication by the publisher to the mathematician Giovanni Battista Baliani. Pecquet was an early defender of mechanistic anatomy who joined anatomical skill with interest in contemporary debates in the sciences: his work included a section titled *Experimenta physico-mathematica de vacuo* in which he discussed a number of experiments on the Torricellian tube and argued that air is elastic. The issue was quite pertinent because Walaeus had argued in favor of attraction for the motion of blood and chyle. Pecquet's additional section was rather unusual in an anatomical text, and the Genoa edition, for example, omitted it. Figure 1.4 from Pecquet's *Experimenta* shows Roberval's inflated air bladder of a carp inside the Torricellian tube. In an experiment performed in Paris and witnessed by Pecquet, Roberval showed that an air bladder from a carp emptied of air expanded in the space at the top of the Torricellian tube. Pecquet saw this experiment as confirming his views about the elater of the air, which was able to expand once it had been compressed, like a sponge or wool: elasticity was the subject of extensive experimentation in those years and was becoming a key notion in mechanics. The experiment was intended to show that no attraction was necessary to explain the motion of the

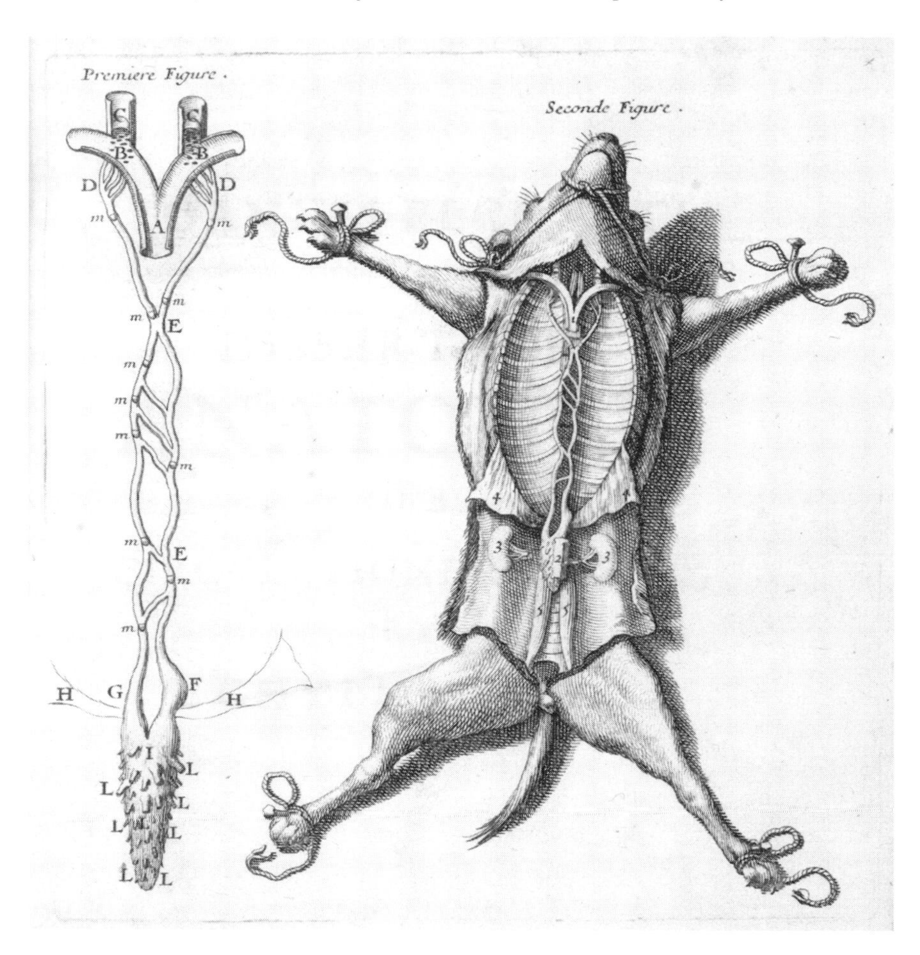

Figure 1.3. Pecquet, *Experimenta nova anatomica*: thoracic duct in a dog

chyle. It is unclear whether Pecquet's work was known at Bologna, and if so in which edition. In a letter of January 1654 to Famiano Michelini—Borelli's friend and predecessor on the mathematics chair at Pisa—Baliani announced the Genoa reprint, and in 1656 Vincenzo Viviani ordered a copy from Paris. Pecquet's work circulated in Florence, where it inspired several experiments of the Accademia del Cimento.[9]

In 1652 the Danish anatomist Thomas Bartholin published *De lacteis thoracicis*, in which he extended Pecquet's findings to the human body but challenged his claim that the liver was entirely bypassed and that the receptacle of the chyle always existed in man. Bartholin argued that the thinnest part of chyle was carried to the heart, whereas the thickest part went to the liver. In the following year, however, in *Vasa lymphatica*, he dramatically changed his mind and announced both that all the chyle does indeed reach the subclavian vein, as claimed by Pecquet, and the discovery of a

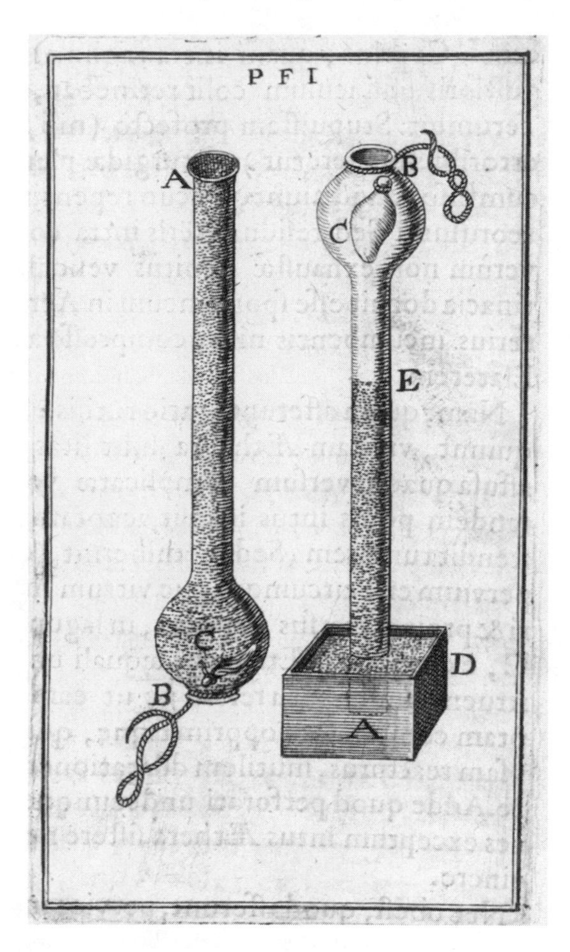

Figure 1.4. Pecquet, *Experimenta*: Roberval's experiment with the carp bladder

new set of vessels carrying lymph, a very pure transparent fluid. There seemed to be a correlation between the visibility of the thoracic duct and the time of the animal's last meal before being dissected, four hours being an optimum time to highlight them. By contrast, the lymphatic vessels were less variable but were also so delicate that Bartholin compared them to the fibers of a spider web. He discovered them in an animal that had eaten six hours before, when the chyliferous vessels had subsided. The Swedish anatomist Olaus Rudbeck made similar findings to Bartholin's.[10]

These considerations provide valuable background on the state of the field before Malpighi's move to Pisa in 1656 and his anatomical and philosophical formation. In addition, one may consider the role of chymistry and chymical remedies. Unfortunately, at this stage both this issue and that of the potential connections between chymistry and some form of the corpuscular philosophy at Bologna in those years remain matters for speculation. Bologna, however, appears to have been more tolerant than Paris in this regard, since Pierre de la Poterie or Pietro Poteri, expelled from Paris because of his chymical remedies, lived and practiced at Bologna for decades, dying probably in the 1640s. Some of his chymical remedies are mentioned by Malpighi.[11]

1.3 Malpighi's Pisa Apprenticeship: Microscopy and the New Philosophy

After some years practicing with his mentors, Malpighi obtained a chair of theoretical medicine at Pisa. The account provided by Malpighi himself in his *Vita*, and generally accepted thereafter, is that he arrived at Pisa instructed in the peripatetic philosophy and left "imbued with the precepts of the free philosophy." In an earlier draft Malpighi had used the qualification "Democritean," as to emphasize atomism. The key figure in this conversion was Borelli, a former student of Benedetto Castelli who had been called to teach mathematics at Pisa in the same year as Malpighi. Borelli was a leading proponent of atomism and mechanistic anatomy. However, it appears that Malpighi was probably fertile soil for many of Borelli's philosophical tenets and may have already been familiar with some of them before 1656, since mechanistic anatomy was rapidly gaining momentum at the time, as shown by the reception of Harvey's circulation of the blood and Pecquet's influential work on the thoracic duct, for example. At Pisa Malpighi was instructed by Borelli in philosophical matters and performed many dissections of animals provided by the Medici in Borelli's house and, occasionally, in hospitals. The two most famous anatomical findings of those years concerned the structure of testicles by the Lorraine anatomist Claudius Auberius and the spiral structure of the heart muscle by Malpighi. Both observations exerted a considerable impact on Malpighi, but whereas muscle anatomy remained at the margin of his interests, the structure of the testicle was closer to the center because of its association with the problem of secretion. Auberius was professor of anatomy at Pisa, colleague of Malpighi, and one of Borelli's early collaborators; in 1658 he published a single sheet titled *Testis examinatus*, in which he compared the testicle of a wild boar to a human testicle. Both the size of the wild animal, provided by the Grand Duke, and the timing of the dissection during the mating season allowed Auberius to uncover a finer structure than had previously been noticed. He discovered that the inner structure of the testicle was not glandular, as Nathaniel Highmore had believed, but consisted of vessels, or is "omnino vasata." *Testis examinatus* did not refer to Wharton's 1656 *Adenographia*, the comprehensive treatise on glands that was to become a standard reference work on the subject and that ascribed testicles among them. Figure 1.5, from Auberius's sheet, shows a human testicle alongside a boar's one, as it were an enlargement of the former, as a "microscope of nature," to use Brun-

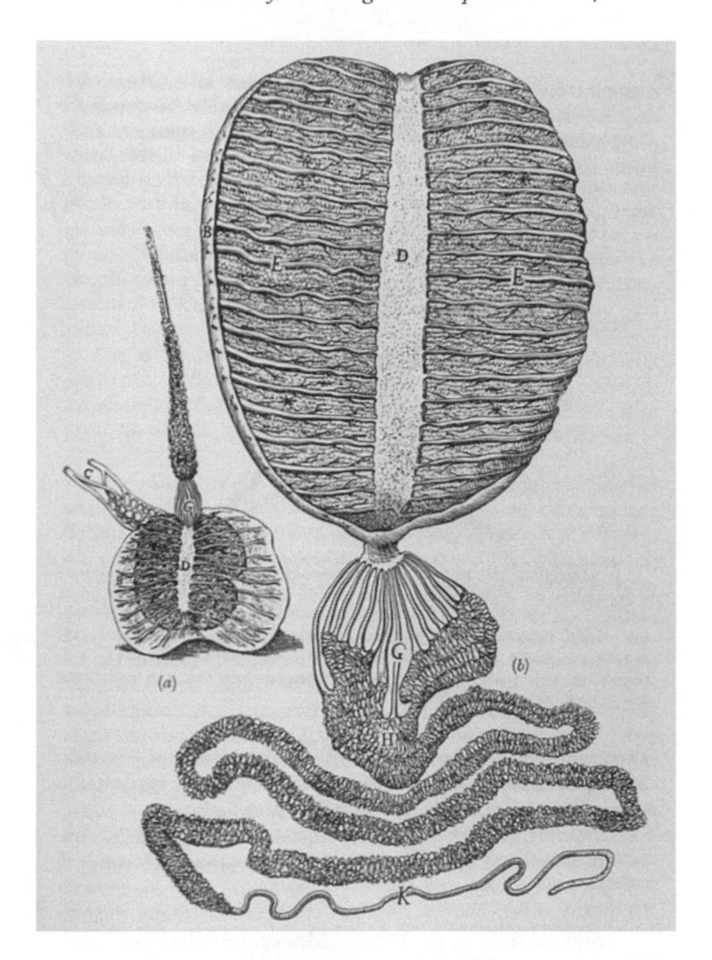

Figure 1.5. Auberius, *Testis examinatus*: testicles of a man and wild boar

ner's and Peyer's later expression. It is worth reflecting, however, on the lack of the use of the microscope, suggesting that microscopy was not an established practice in Borelli's circle at this stage.[12]

References to magnification tools can be found in the works of Galileo and several Lincei, Harvey, Gassendi, Marco Aurelio Severino—who was probably also in contact with the Lincei—and Nathaniel Highmore. Luigi Belloni has identified Giovanni Battista Odierna and Pierre Borel as "pioneers" of microscopy. In *L'occhio della mosca*, Odierna provided a summary account of how to make a microscope and then, crucially, discussed some observation techniques. He relied on a compound microscope or *occhialino* with two semispherical or convex lenses, thicker than a lentil, stating that he mounted his device on a tripod to keep it at the right distance from the object. However, he argued that a good single lens as large as a chickpea would be suitable

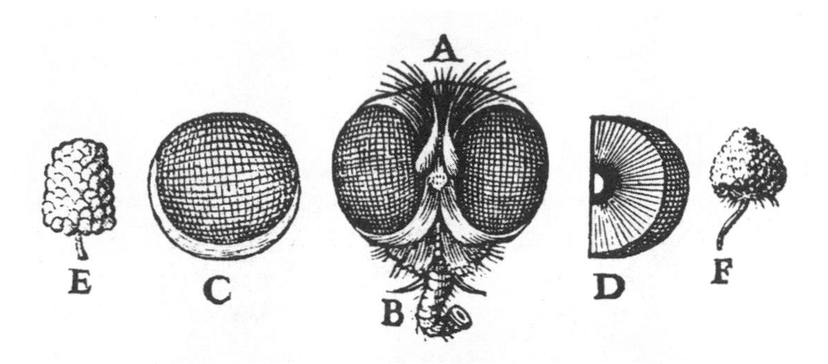

Figure 1.6. Odierna, *L'occhio della mosca*: eye of the fly

too. The eyes of crabs and other animals, especially the fly, had to be boiled and then dried in the sun in order for the cornea to be removed, with the help of a very thin and sharp knife, and eventually sectioned. In his description of the eye Odierna relied on analogies with the vegetable world, such as a white mulberry, a pomegranate, or a strawberry. His woodcut (fig. 1.6) shows at far left (marked E) a white mulberry and at far right (marked F) a strawberry, with three images of the eye in the middle. Odierna's work (Palermo, 1644) was possibly known to Borelli, then living at Messina, and is a likely source for some of Malpighi's techniques, which are quite similar. Moreover, in 1656 the French physician Pierre Borel published *Observationum microscopicarum centuria* in an appendix to *De vero telescopio inventore*, a book likely to attract the attention of the Medici court. Two observations of the *Centuria* refer to the use of the microscope in anatomy, in order to study both fluids, such as chyle and urine, and several organs, such as the heart, kidneys, testicles, liver, and lungs.[13]

Probably Malpighi started using the microscope for anatomical purposes around 1656–59, although little is known about his sources and training in this area. He would have found available microscopes at the Tuscan court, since we have reports of Grand Duke Ferdinand II having made an observation on vinegar in 1655. Later examples, such as the microscopic observations of oak galls, are more common; Redi reports that the Medici had an extensive collection. In a letter of 1661 Borelli stated that he had a microscope by the leading maker Eustachio Divini, intimating that it was his own. A few years later he advised Malpighi that Iacopo Ruffo at Messina had several lenses and that by mounting two in a tube at the appropriate distance, which could be easily found by trial and error, one could fabricate microscopes of different sizes; thus, it seems plausible that Borelli was acquainted with microscopy from his time at Messina before 1656. All these observations point to Pisa as the site where Malpighi started to use the microscope.[14]

In 1659, the same year that he left Pisa for Bologna, Malpighi sent Borelli two dialogues. Unfortunately, they are now lost, but from his *Vita* we know that they took place among a Galenist physician, a *chirurgus mechanicus* or barber-surgeon, and a neutral observer. The first dialogue dealt with the nature of blood, the second with purgation, nutrition, and the existence of the faculties. We also know that most of their contents were used by Malpighi in his 1665 *Risposta* to Michele Lipari, which I shall discuss in section 2.5. This summary account is of interest in several respects. It shows, for example, that at Pisa Malpighi faced the problem of reconciling the medical knowledge acquired by his Bologna teachers with a novel picture of the human body. At the very least, even if remedies and cures were left unchanged, he would have had to provide novel justifications for them in terms of the mechanical philosophy. Moreover, Malpighi's choice of a *chirurgus mechanicus* as his mouthpiece was remarkable for intellectual and social reasons. Clearly inspired by the opening of Galileo's *Discorsi*, in which both Sagredo and Salviati praise the knowledge of the technicians in the Venice arsenal, Malpighi went as far as to present the barber-surgeon as refuting a learned physician. This move would have appeared daring to his contemporaries, and even Borelli, by no means a likely one to be shocked by these matters, found it objectionable, since, he argued, a "cerusico idiota" could hardly be expected to force into submission a learned Galenist physician. Rather, Borelli suggested that Malpighi should either have the barber-surgeon state that he had been instructed by a "uomo grande" or introduce a fourth personage, i.e., a physician and philosopher follower of the "free philosophy"—much like himself, one may add. Borelli's reference to a "cerusico idiota" shows that Malpighi's *chirurgus mechanicus* was not university trained; hence, Malpighi was stressing the role of observation, practice, and dissecting skills over bookish learning. This episode suggests that Malpighi had become more Galilean than his Galilean mentor and points to a potential tension between the skillful anatomist and the philosopher unable to dissect.[15]

We can now try to assess the impact of different philosophical traditions on Malpighi, besides his Aristotelian training. Luigi Belloni has advocated a Galilean root for Malpighi's program with regard to corpuscularism, the study of sense perception, and microscopy, pointing to several passages from Galileo's writings as direct sources of inspiration for Malpighi. In addition to areas such as taste or touch and the structure of the tongue and skin, explicit reference to Galileo can be found in *De omento et pinguedine*, where Malpighi referred to a passage from Galileo's *Discorsi* on the floating of fishes.[16] By contrast, Theodore Brown has stressed the Cartesian roots of Malpighi's mechanistic program, despite important differences in specific instances. Descartes used a version of the circulation as a cornerstone of his new understanding of the body already in the *Discours de la méthode*. He conceived the motion of the heart differently from Harvey, however: Harvey believed the active phase to be systole or contraction, whereas Descartes thought it to be diastole or expansion and argued that the heart was very hot and blood escaped from it like milk boiling over out of a pan. Borelli was fiercely opposed to the notion of the heat of the heart and sought other mechanical explanations for the heart's contractions; while at Pisa, he inserted a thermometer in the viscera of a live stag to show that the temperature was not higher in the heart than elsewhere.[17] However, many Cartesian themes became widespread in the mid-seventeenth century and were transmitted to Malpighi through several intermediate sources. Robert Frank identified several passages from Descartes and Gassendi lending themselves as philosophical sources for the mechanistic program of the Oxford physiologists. Although Frank did not establish those connections with regard to Malpighi, it is not difficult to do so if one considers the great reputation Gassendi enjoyed at the Tuscan court and especially with Borelli and his extensive references to the circulation, milky veins, and thoracic duct. For example, in a letter to Malpighi of 1661 Borelli called him a "galantuomo" and set him as a model for his frankness in retracting views he had previously held in light of new evidence. Gassendi witnessed several dissections, including one by Pecquet in 1654 in which he saw the thoracic duct, and admitted having changed his mind with regard to chyle and the circulation of the blood. Malpighi probably shared Gassendi's atomistic program, mitigated skepticism, strong empirical inclinations, and attempts to retrieve and reinterpret ancient sources, all aspects lacking in Descartes. Malpighi's views echo several passages in Gassendi. In *Syntagma philosophicum*, for example, one reads:[18]

> For just as a man born in the forest and ignorant of all human arts, if shown a clock enclosed in the setting of a ring, would only wonder at the delicacy and elegance of its structure, and the long duration, regularity, and spontaneity of its motions and would never guess how the little machine could be made so perfect, so too do our powers fail completely when we are confronted by these achievements of Nature, at the elaboration of which we were present neither as spectators nor as participants, and like untutored woodsmen we can only be struck with wonder but cannot divine or conceive by what artifice they have been accomplished; for, indeed, each one of them is a little machine within which are enclosed in a way impossible to comprehend almost innumerable [other] little machines, each with its own little motions.

The problem of the philosophical sources in Malpighi and his contemporaries is a complex one. In the following chapters we shall encounter several references to Galilean, Cartesian, and Gassendist themes.

1.4 Malpighi's *Epistolae* on the Lungs

In 1661 Malpighi published two *Epistolae* on the lungs addressed to Borelli, containing anatomical microscopic observations unprecedented in their significance. The *Epistolae* are quite brief and report observations with the help of the microscope and other techniques of investigation, including vivisection, drying, insufflation, cooking, and injection with water and mercury. He relied on a range of optical tools and techniques: he used both single-lens and two-lens microscopes, placing the lungs on a glass plate lit from below, increasing magnification gradually by using more and more powerful instruments—a point Malpighi emphasized at a later stage in response to the criticisms from the Dutch anatomist Theodor Kerckring—and employing different illumination techniques. Malpighi compared the process of seeing through the microscope with reading, as when we identify the individual letters on a page. He showed his skill with techniques of preparation and observation already from his first publication. Even Borelli, who was receiving detailed instructions and had excellent microscopes—probably better than Malpighi's—had difficulties in replicating his observations.[19]

Malpighi showed that the substance of the lungs was not fleshy, as was commonly believed, but consisted of communicating *alveoli* or small cavities separated by thin membranes, as in a sponge. The pulmonary membranes were covered by a network tentatively identified by Malpighi as composed of nerves. He reported that according to the ancients the purpose of the lungs was to cool down the excessive heat of the heart, but such ideas were then viewed with increasing skepticism. As a result of his structural findings, Malpighi put forward the view that the lungs serve to mix the components of blood with the help of air pressure and fermentation, since chyle and lymph could not mix adequately in the right ventricle of the heart. He sought to confirm his view with a number of observations on human activities: for example, he included a detailed corpuscular analysis of the cause of fluidity in terms of small particles of fire or water working like ball bearings and of the chymical analysis of blood, including distillation in an alembic; he mentioned women who beat fresh blood with their hands or with a stick in order to prevent the separation of its components, women who mix flour with water, as well as the network of blood vessels in the incubated egg and the gills of fishes. At the end of the first *Epistola* he mentioned a well-known vivisection experiment of insufflation, whereby if one blows air through a pipe into the lungs of the animal after they had collapsed, the heart starts beating again, even after it had almost ceased to move; Malpighi believed that air pressure was responsible for this process. This experiment also had a pathological significance for him since he argued that the obstruction of pulmonary vessels is followed by arrhythmia and then death.[20]

In the second *Epistola* Malpighi corrected an embarrassing mistake and added new important observations. Borelli had urged him to proceed on the example of others, such as Gassendi and Thomas Bartholin, whom he praised for having honestly changed their views. By observing with Fracassati's assistance the lungs of frogs and tortoises, whose microstructure is easier to detect than in higher animals, Malpighi reinterpreted the network previously thought to consist of nerves as vascular, observed during vivisection the contrary motion of blood in arteries and veins, and under higher magnification detected their anastomoses or junctions, showing that blood flows always inside vessels. The observation of the motion of blood in contrary directions in arteries and veins is a noteworthy example of the microscope used to study actions rather than structures. By relying on analogy and on the widely held assumption of nature's uniformity and simplicity, Malpighi generalized his results on the lungs of some animals to the anastomosis of arteries and veins in the other parts of the body and all animals: he relied extensively on this crucial assumption in the rest of his work.[21]

Malpighi also interpreted the purpose of lungs on the basis of a suggestion put forward by Borelli, attributing it to his philosophical mentor in a form that today would probably involve multiple authorship. This example is representative of a typical form of collaboration in which the anatomist described the structure and Borelli explained its operation and purpose. Borelli put forward a *capriccio*—greatly admired by the Medici—based on his observation of grafting jasmine and vine twiglets on a lemon tree. This example was particularly appealing because of the contrast between the sweetness of jasmine and grapes and the sourness of lemon. He argued that plant lymph is refined in different ways as to produce such different species as lemons, jasmine, and vines only on the basis of the sizes and shapes of the vessels. In the same way as lymph can be arranged so as to produce leaves, flowers, and fruits, so blood can be arranged in the lungs to produce flesh, ducts, and spirits. Borelli's idea was that the internal arrangement of the vine and jasmine vessels transforms the particles of the acid juice of the lemon into a sweet one. Analogously, the lungs would rearrange the particles of blood mixed with chyle and make them ready to form all the body parts. Borelli's suggestion relied on a corpuscular view of matter whereby its properties depend on the arrangement of its constituent parts. According to his new interpretation, fermentation was no longer required and was quietly dropped from the second *Epistola*. Malpighi added that the structure of the lungs resembles that of the testicles, as if the animals' nutrition were a form of regeneration. Two recent anatomical findings were at the basis of his novel interpretation: Pecquet's discovery of the thoracic duct and Auberius's discovery of the structure of the testicle. Inspired by them, Borelli offered a purely mechanical interpretation of respiration whereby air pressure would help the mixing of the blood through the motion of the lungs, but air would not mix

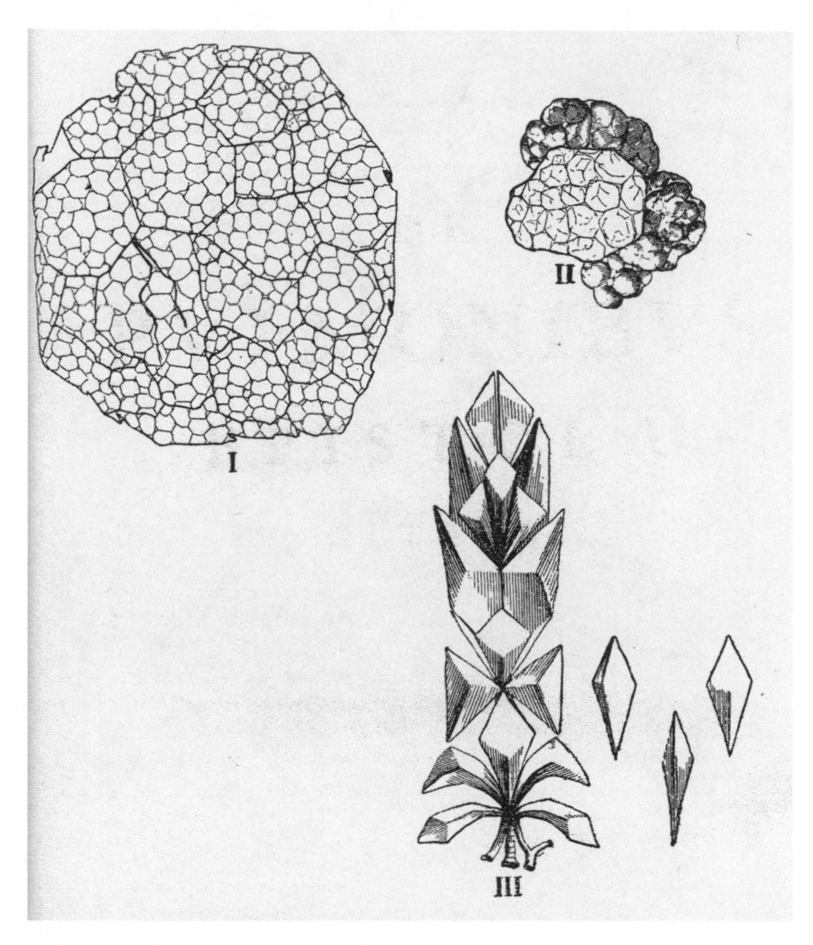

Figure 1.7. Malpighi, *Epistolae*: pulmonary lobules

with blood, which always flowed inside vessels. Following Borelli's views, however, it is difficult to see why one could not breathe the same air time and again. Later both Borelli and Malpighi changed their mind on this point and attributed a different role to air in respiration.[22]

Malpighi's *Epistolae* included rather crude copper engravings—the first printed images of the microstructure of an organ. They show a network covering external vesicles (I) and a few internal vesicles (II) and different modes of insertion of the pulmonary lobules (III; fig. 1.7) and the lungs of frogs (I) with an elementary alveolus with arteries (C) and veins (D) (II; fig. 1.8). In an important letter Borelli had advised Malpighi to follow Descartes' example by including illustrations of his findings, even if he had to alter dimensions and positions for clarity's sake: this passage offers valuable insights into Borelli's art of anatomical representation, according to which it is

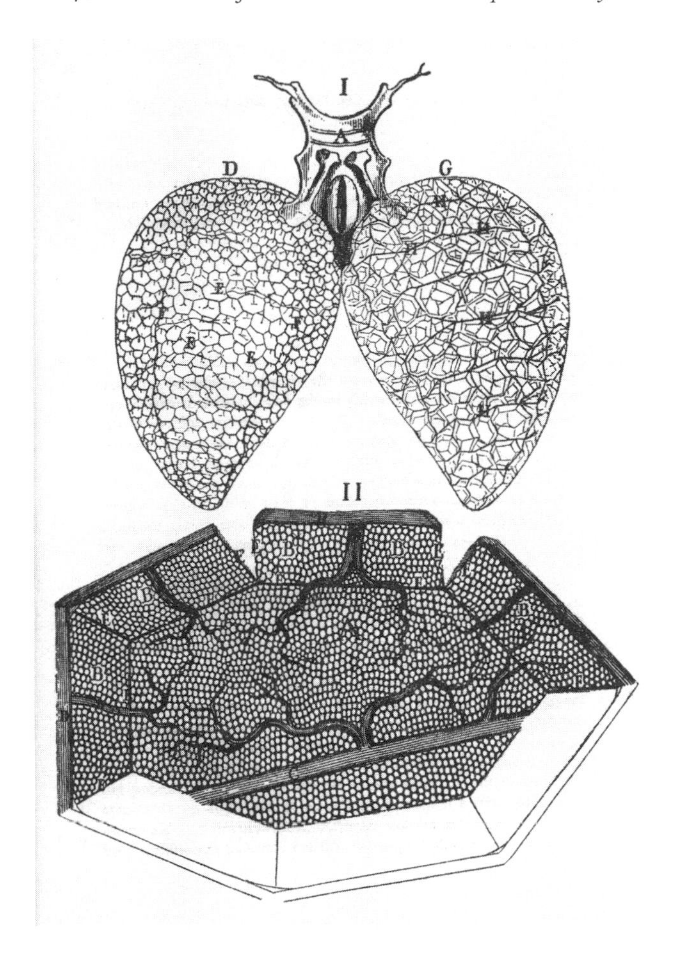

Figure 1.8. Malpighi, *Epistolae*: lungs with elementary alveolus enlarged

necessary to sacrifice a strict realism in order to gain a better understanding of the parts. This letter emphasizes the novelty of Malpighi's approach and the problems he was facing: he was presenting not only new findings but also new methods of inquiry based on microscopy that had to be justified or at least made plausible visually. Borelli wrote:[23]

> Nor should the things be considered too difficult to be drawn and cut because they are very small, because Vostra Signoria can enlarge them by claiming that, for the sake of greater clarity, it is necessary to alter the dimensions of those small cavities, or membranes, and of their positions. Descartes did something similar in his philosophy, or in his work on meteors, and with his beautiful and artificial way of explaining and presenting himself, he has enchanted many good people.

One surprising aspect missing from Malpighi's letters and the correspondence with Borelli concerns the change of color of blood. I believe that this lack of interest in color change stemmed from two sources, one philosophical and the other experimental. In *The Assayer*, Galileo had put forward a distinction between qualities such as shape and size on the one hand and tastes, odors, and colors on the other; these qualities were later known as primary and secondary, respectively. The former would still exist without a perceiving subject, whereas the latter reside only in the perceiving subject and would disappear if the living creature were removed. This distinction had an atomistic ancestry and privileged shape and size over other qualities, in that they were indicative of a body's fundamental properties. Galileo sought to provide an account of the shapes of the corpuscles responsible for specific tastes and odors—sharp or smooth—but in the case of colors he did not explain how they originate, arguing that he was insecure over the matter and an explanation would be too complex. In those years the Accademia del Cimento was engaged in the study of color change in solutions as parts of chymical tests on various mineral waters. Some of those experiments, involving dramatic shifts from red to green and back to red, for example, were reported in the *Saggi* of the Accademia. In a revealing letter of 1660 to Malpighi, Borelli commented on the change of color in the blood serum of a friend of Malpighi's afflicted by pain in the joints. Borelli stated that the change of color of some callous particles from white to that of a putrid humor was not a matter of great interest, "knowing that the colors of things can be very easily changed." In the following two chapters we shall see other instances in which Borelli downplayed the philosophical significance of color and color change. Whereas Malpighi, coming from the medical tradition, saw color as a valuable tool that had been used since the Hippocratic corpus to assess disease, Borelli's philosophical stance was different: the tradition he followed did not see color or taste change as revealing of substantive transformations. In this tradition the far from dramatic change of color in blood from dark red to bright red and back seemed unworthy of investigation; Borelli and Malpighi may have attributed it to the rearrangement of blood components in the lungs, but they did not test where and under what circumstances it occurred. Their perspective is especially striking since Malpighi knew well that air changes the color of blood. Borelli was by no means alone in distrusting colors, however: Steno reached similar conclusions by noticing the ease with which colors can be changed in solutions.[24]

Harvey's reception depended on a wide range of themes involving the role of experiment, vivisection, the meaning of anatomy and natural philosophy, quantification, and the relations between anatomy and medicine. Probably no single observation or experiment would have been universally seen as compelling evidence in favor of his views. From some perspectives, however, Malpighi's stunning visual evidence would have appeared very powerful. This is certainly the case with respect to a follower of Gassendi, who had been influenced by the "observation" of pores in the septum of the heart. Gassendi had been interested in the issue of circulation for several decades and discussed it in a critical fashion in the anonymous *Discours sceptique*, published in 1648 at Leiden, and later in *Syntagma*. The first of his many objections to Harvey concerned precisely the issue of the junctions between arteries and veins: if these could be seen, the matter regarding circulation would be resolved not simply in terms of likelihood, but with virtual certainty, argued Gassendi. Although the existence of anastomoses between veins and arteries was not strictly necessary for the circulation of the blood, Gassendi saw it as sufficient evidence in its favor. Thus, Malpighi's observations of the junctions between arteries and veins in the lungs of frogs and of blood moving in opposite directions in veins and arteries provided powerful evidence in favor of the circulation. Moreover, Borelli exhorted Malpighi to give due emphasis in his *Epistola* to that "most beautiful observation" of the motion of blood, which would refute the few surviving Galenists opposing the circulation of the blood. Strictly speaking, the Galenists believed blood to move outward in both arteries and veins. Following Realdo Colombo's work on pulmonary transit, however, blood was understood to move in the lungs as observed by Malpighi even before Harvey's finding of the circulation. Malpighi, however, sought to generalize his findings to the rest of the body by analogy, and by observing the same phenomenon in the bladder of frogs.[25] Thus, Malpighi's findings were perceived by his contemporaries as relevant to the reception of the circulation.

1.5 The Purpose of Respiration: Thruston, Lower, and Hooke

Borelli's role did not end with providing an explanation of the operations of the lungs. As soon as the *Epistolae* were published, he had Carlo Dati send copies to France, Flanders, and England. Dati was a member of the Accademia della Crusca and Accademia del Cimento who entertained correspondences with scholars across the Alps. One of the copies was sent to the renowned anatomist Thomas Bartholin at Copenhagen; Bartholin may have also met Borelli when he had visited Messina in 1644, where Borelli was professor of mathematics.[26] Bartholin was struck by Malpighi's work and reissued it as an appendix to *De pulmonum substantia et motu diatribe* (Copenhagen, 1663), a short treatise in which he commented and expanded on Malpighi's work, praising Dati, Borelli, and Malpighi. There are some curious analogies in the works by Bartholin and Malpighi: in their careers both published in close succession two tracts, the second correcting portions of the first, and—as Bartholin himself noticed—both had their works reprinted shortly afterward, Bartholin's on the lymphatics by Hemsterhuis and Munier, Malpighi's by Bartholin himself. Bartholin stated that he wished to be of help to Malpighi as others had been to him. Bartholin tried to situate Malpighi's findings within the anatomical literature. He referred to Descartes' recently published *De homine* (Leiden, 1662), Gassendi, and Cornelis van Hoghelande, who appeared to have adumbrated Malpighi's discovery with his claim that the parenchyma of the lungs consists of extremely thin pellicles. Bartholin also corroborated Malpighi's finding by pathological means, relying on the peculiar structure of the lungs of a four-year-old boy he had previously dissected.[27]

The reissue of his first publications by one of the leading anatomists of the time was a notable success for thirty-three-year-old Malpighi. From Copenhagen, Bartholin sent a copy of his *Diatribe* to his student Nicolaus Steno at Leiden, who procured a microscope to replicate Malpighi's observations and discussed related anatomical matters with his friend Jan Swammerdam. Bartholin's correspondence in the years 1662–63 contains several references to Malpighi's work, which he advertised across central Europe. By August 1663 a copy of the *Diatribe* was available to James Allestry, one of the printers for the Royal Society in London; others in England had presumably already received some of the original copies sent by Dati. Thus, Malpighi's *Epistolae* reached England while Hooke was working at his *Micrographia* and was presenting his curious observations at meetings of the Royal Society.[28]

Late in 1664 Malachi Thruston, fellow at Gonville and Caius at Cambridge University, discussed a medical thesis on respiration in which, according to his friend and colleague Walter Needham, he explained and confirmed Malpighi's views about the structure of the lungs to much applause. Thus, Malpighi's work was being discussed in England at the time in the context of an academic disputation. Thruston replicated some of Malpighi's microscopic observations but, unlike Malpighi, believed that the subtlest portion of air does enter the lungs and mixes with the blood. Thruston's work was published in 1670 together with a set of criticisms by George Ent and Thruston's replies. Ent challenged Malpighi's views, arguing that the vesicles in the lungs of frogs have no communication with the bronchi or among themselves. His observation highlights the difficulties in replicating Malpighi's microscopic observations. On receiving a copy of the book from Oldenburg, Malpighi was stimulated to repeat and expand his previous investigations, which were published in a brief note in the *Philosophical Transactions*, in which he expressed astonishment at Ent's claims.[29]

In the 1660s, at the same time that Thruston defended his Cambridge thesis, a remarkable series of experiments performed at Oxford and at the Royal Society in London profoundly changed the understanding of respiration: some of them were already known and mentioned by Thruston. The protagonists of these experiments were the anatomist and physician Richard Lower and Robert Hooke, professor of geometry at Gresham College and curator of experiments at the Royal Society. As we now know, John Locke was also involved and collaborated with Lower. There is a striking similarity with the Italian scene, in that both cases involved the collaboration between scholars from different fields. But whereas Borelli offered analogies in the belief that nature operates and is best understood that way, Hooke's contribution was experimental. Moreover, whereas microscopic investigations could be carried out by one person at a time, vivisection experiments could be performed publicly, in front of the entire Royal Society. In October and November 1664 the Royal Society debated whether air enters the body through the lungs. The fact that during the vivisection of a dog it was possible to revive the heartbeat by blowing air into the *receptaculum chyli*, whence it reached the heart through the thoracic duct, suggested a role for air in heart pulsation. On November 7 Hooke, Jonathan Goddard—a former student of Francis Glisson at Cambridge—and Oldenburg inserted a pair of bellows into the trachea of a dog and inflated its lungs. Hooke opened the thorax and cut the diaphragm, observing the heart beating regularly for over one hour as long as air was in the lungs. Hooke could not determine whether air entered the lungs, but he could establish that the motion of the heart was related to the inflation of the lungs, even though the two were not synchronous.[30]

The English virtuosi soon elaborated on this experiment and went beyond this initial result, relying on Lower's superior skill with vivisection. Following his teacher Thomas Willis, Lower attached great importance to blood, its fermentation, and color change. Emphasis on experimentation was a hallmark of both the Royal Society and the Accademia del Cimento, but in this case the English virtuosi asked questions about issues the Italians had deemed of no significance, such as the color of blood. As we will see in sections 2.5 and 3.4, it was Fracassati's report of an experiment similar to Malpighi's turning caked blood upside down that had alerted English readers to the significance of air in the changing color of blood. For Malpighi and Fracassati, however, the change of color of blood as a result of mixing with air was irrelevant to understanding respiration. Interestingly, Boyle performed several experiments on color indicators similar to those performed at the Accademia del Cimento, although he reached different conclusions. According to him and Hooke, for example, color was related to the texture of bodies and therefore provided some information on the nature of a body. Initially in the *Vindicatio* (London, 1665) Lower had argued that blood changes color in the heart as a result of a ferment in the left ventricle. Colombo had performed a similar vivisection in the previous century, but his aim was to determine whether the pulmonary vein contains blood or smoky waste, an easier task than determining the precise color and nature of blood. Lower believed that blood in the lungs was venous, probably because in his early trials the animal's lungs had collapsed and were empty of air. But additional experiments refuted his initial view. On 10 October 1667 Hooke and Lower performed an experiment at the Royal Society analogous to that of 1664, but this time they relied on two pairs of bellows instead of one, producing a continuous airflow. An incision in the pleura allowed air to exit the lungs, which thus remained inflated but motionless. By cutting a portion of the lungs, they could observe the blood moving through the lungs whether they were inflated or not. In this way the animal was kept alive, thus showing that their motion was not indispensable as long as fresh air was supplied. Traditionally vivisection was used in order to investigate motions, including those associated with respiration. In this case, however, Lower and Hooke used vivisection in order to keep the lungs still: only in this way could they investigate whether their motion was a necessary component of respiration, in mixing the blood, or rather fresh air was all that was needed; thus, Borelli's and Malpighi's view of respiration was refuted. In *Tractatus de corde* Lower reported his vivisection experiments with the two pairs of bellows and thanked Mr. Robert Hooke for his assistance in a passage echoing Malpighi's acknowledgement to Borelli: "I acknowledge my indebtedness to the very famous Master *Robert Hooke* for this experiment—by which the lungs are kept continuously dilated for a long time without meanwhile endangering the animal's life." It would be reductive, however, to see Hooke merely as the operator of the second pair of bellows; rather, I would suggest that Hooke contributed to those anatomical experiments a mode of mechanical thinking and operating for which he was one of the recognized masters.[31]

Lastly, Hooke and Lower performed yet another experiment in two parts on a dog. First, in the initial vivisection, they closed the trachea and showed that the blood coming from the cervical artery, after the blood had gone through the left ventricle of the heart, was venous. Thus, the change of color of the blood did not occur in the heart. Then the animal died and they kept the lungs inflated by means of the two pairs of bellows with the insufflation experiment we have seen above, obtaining bright red arterial blood from the pulmonary vein. Thus, it was not the motion of the lungs, or a ferment in the heart, or the animal's heat that was responsible for the change of color of blood, but only air.[32] A similar experiment performed by William Harvey on a throttled man in order to show that the interventricular septum in the heart was not porous was known to George Ent and may have inspired Lower and Hooke. Their experiment strikes me as especially significant in showing that the change of color of blood was not due to the soul or one of its faculties, because the animal was dead. Performing a vivisection in order to keep an organ still was not the only paradox in this case. Respiration had been traditionally associated with life and motion, and this is why it had been investigated through vivisection since antiquity. By blowing air through the lungs of a dead animal, Lower and Hooke enacted respiration, implicitly showing that one of the key operations associated with life involved purely chymical and mechanical processes. Here the notion of mechanism as a process acquires a new connotation.[33]

Experiments and debates on respiration occupied the English virtuosi for several years and involved instruments and techniques ranging from the air pump to chymical analysis. Yet I believe the collaboration between Lower and Hooke to be an especially instructive term of comparison with the Italian scene. Their experiments showed that the motion of the lungs was not necessary to keep the animal alive, thus disproving the purely mechanical view of respiration, and that the heart played no role in the changing color of blood, because blood changed color in the lungs as a result of the presence of air. Lower attached great significance to the change of color of blood, so much so that the title of his work, *Tractatus de corde item de motu & colore sanguinis et chili in eum transitu* (1669), highlights precisely this point. In his work Lower implicitly challenged the interpretation of the lungs' purpose offered by Malpighi and Borelli, arguing that blood cannot be better fragmented in the lungs than in muscles. Rather, he believed that through the lungs blood was being mixed with air; fresh air was therefore essential to the animal not because of a mechanical operation, but rather because of a chymical one, as Lower hinted at when he argued that "wherever therefore a fire can burn sufficiently well, there we can equally well breathe." Many of the subsequent investigations by the English virtuosi were devoted to the chymistry of air and respiration.[34]

Focusing on the relationships among apprenticeship, techniques of investigation, results attained, and philosophical views has enabled us to investigate Malpighi's formation and first publications, highlighting a number of similarities and differences with other cases. Informal gatherings of senior and younger scholars, often alongside university life or scientific academies, were major venues for apprenticeship and collaboration in anatomy, as exemplified by the *Coro anatomico* in Bologna and the *Collegium privatum* in Amsterdam, or the Accademia del Cimento and the Royal Society. The collaboration of anatomists with scholars with different disciplinary affiliations, including experimental philosophy and the physical-mathematical disciplines, was a key mode of investigation in the seventeenth century.

The examples we have seen show many striking parallels between the Italian and English scene: in both cases professors of mathematics and leading proponents of a mechanistic worldview, such as Borelli and Hooke, collaborated with anatomists, such as Malpighi and Lower, in venues and contexts related to the new scientific societies, the Accademia del Cimento in Tuscany and the Royal Society in London. However, there were significant differences: whereas Malpighi relied primarily on microscopy to grasp the structure of the lungs and the purpose of respiration, Lower and Hooke paid no special attention to structure and relied instead on the variation of key features of respiration—such as the motion of the lungs and the role of air—in vivisection experiments. Thus, they radically subverted Malpighi's and Borelli's interpretation of the lungs' operation and purpose. Relying on experiments carried out at the Accademia del Cimento, Borelli dismissed the philosophical and chymical significance of color change, whereas Willis and Lower believed in its importance, especially for blood.

In the early 1660s advanced microscopy was still in its infancy: few significant observations predate Malpighi's *Epistolae*. Although Steno at Leiden and Thruston at Cambridge could replicate some of Malpighi's results, very few anatomists tried to do so and could compete with him in this area. Injection techniques were not yet very refined and involved mainly mercury and insufflation. Although Hooke's reliance on two pairs of bellows was strikingly ingenious, the main difficulty in the experiment probably came from the vivisection side. Most anatomists and their collaborators and mentors discussed in this chapter shared a mechanistic stance that shaped significant aspects of their investigations: Lower's and Hooke's enaction of respiration in a dead animal showed that one of the key operations—or we could say mechanisms—associated with life involved purely chymical and mechanical processes.

Epidemic Fevers and the Challenge to Galenism

2.1 Galenic Traditions and New Medical Thinking

Changing conceptualizations on the causes and cures of epidemic fevers and controversies on the status of Galenism bore problematic relations to new anatomical findings. The relevant texts we are going to investigate discuss the decline of traditional qualities—hot, cold, wet, and dry—and the distinction between primary and secondary qualities with regard to color and taste in understanding disease and therapy. More generally, they address the relations between philosophy and medicine, tradition and innovation, and theory and practice.

The fortunes of Galenism in the early modern period underwent major changes. While the sixteenth century saw the revival of Galen's doctrines and methods of dissection and more accurate editions of some of his major texts, the seventeenth century witnessed a decline in the interest of his works. Nonetheless, Galen was a complex and prolific scholar whose works remained a centerpiece of academic education: a leading anatomist such as Giovanni Battista Morgagni still lectured on them in the eighteenth century. Already in the sixteenth century we can identify two main strands of Galenism, one more philological, and the other more practical and hands-on. The Paris professors Guinther von Andernach and Jacobus Sylvius were enthusiastic followers of Galen and edited *On Anatomical Procedures* (Paris, 1531) and *On the Usefulness of the Parts of the Body* (Paris, 1538), respectively. Vesalius, by contrast, was more critical and in the preface to *De humani corporis fabrica* stated that Galen had erred more than two hundred times. Vesalius's rejection of the interventricular pores (allowing venous blood to seep through from the right to the left ventricles of the heart) and of the *rete mirabile* (the site where arterial blood was allegedly refined into animal spirits flowing through the nerves) was potentially devastating to Galenic anatomy. The *Fabrica*, however, was largely true to its title in its focus on structures, and Vesalius did not draw those devastating consequences.[1] While some key features of Galen's understanding of the body were challenged in the seventeenth century, many physicians retained a belief in the importance of anatomy, natural philosophy, and the search for causes of disease—all Galenic tenets.

The rise of Paracelsianism and chymical medicine more broadly challenged traditional views and specifically Galenic medicine at the end of the sixteenth and into the seventeenth century. The understanding of bodily operations in health and disease and the subsequent editions of *antidotarii*—or lists of officially sanctioned remedies—testify to the changing medical orthodoxy. In England too we witness shifting allegiances at the London College of Physicians away from Galenism, toward chymical and mechanical doctrines. In addition, the discovery of the circulation of the blood, the thoracic duct, and the lymphatic system dealt serious blows to traditional anatomy, pathology, and therapy. The angry reaction by the leading Paris Galenist Jean Riolan the Younger to Harvey's, Pecquet's, and Thomas Bartholin's anatomical findings and their implications for medical therapy is emblematic of the state of disarray of traditional medicine at the time.[2]

My focuses are not London or Paris but Messina and Pisa, especially the works Borelli and Malpighi produced there from 1649 to 1665; they are of considerable interest in themselves and for the light they shed on the local scene and also on the broader circulation of texts, practices, scholars, and ideas. Pisa was firmly on the intellectual map, with Medici patronage and scholars of the caliber of Borelli, Malpighi, Bellini, and Fracassati, but Messina too was a notable intellectual and political center, as we will see: in 1644 Thomas Bartholin visited the city, met the professor of medicine and prior of the medical college Pietro Castelli, and was offered a chair at the university, for example, and in 1670 Fracassati followed Malpighi in the first chair of medicine.[3]

In 1649 Borelli, who held the chair of mathematics at Messina, published a treatise on the malignant fevers of the previous years, *Delle cagioni de le febbri maligne*, the focus of section 2.2. Borelli's work is a valuable document on the rise of chymical thinking in medicine and the decline of astrology, an aspect that has often been neglected as part of the decline of traditional Galenic medicine. Moreover, *Delle cagioni* offers an early picture of the medical thinking of one of the leading mechanists of the seventeenth century. It is instructive to compare *Delle cagioni* with Borelli's correspondence with Malpighi on the Pisa epidemic of 1661; indeed, Borelli also compared the two cases and drew lessons from them. Although *Delle cagioni* is a formal treatise addressed to a learned but diverse academy and the correspondence between Borelli and Malpighi was a private exchange between two allied scholars, the two cases offer several motives for reflection, as section 2.3 shows. The 1665 controversy between the Messina Galenists and the Neoterics is discussed in section 2.4 as a useful complement to the earlier material: it consists of a series of conclusions and refutations culminating with Lipari's *Galenistarum triumphus* and Malpighi's *Risposta* to it.

Malpighi's extensive work requires a detailed study to be found in section 2.5. When Malpighi's *Risposta* eventually appeared in the 1697 *Opera posthuma*, it was over thirty years old and was seen as out of date by contemporaries: the editors of the *Bibliotheca anatomica* justified its omission by claiming that Lipari's rejection of the circulation of the blood made his views so obsolete that they were no longer of interest. Although Adelmann described Malpighi's *Risposta* to Lipari as "rather dull," I find it an invaluable document on Malpighi's early thinking on a range of themes, such as the certainty of medicine, the relations between new anatomy and therapy, the intersections between philosophical and medical thought, and the circulation of the medical and anatomical literature. Moreover, it provides an instructive term for comparison with Malpighi's other *Risposta*, this time to Sbaraglia, which also appeared in the *Opera posthuma* and which I will discuss in section 11.3.[4]

2.2 Borelli and the Sicilian Epidemics of 1647–48

Before analyzing Borelli's treatise, I wish to provide a brief characterization of the Messina scene in the middle decades of the seventeenth century. Whereas the situations in Tuscany under the Medici and at Bologna as part of the Papal States are relatively better known, that of Messina—at the time the major intellectual center in Sicily and, together with Naples, the major intellectual center in Italy south of Rome—requires a brief account. Spanish rule in the Kingdom of Sicily dated back to the 1282 Vespers, when the Sicilians rebelled against the French and offered the crown to the King of Aragon. Because Sicily was not conquered by the Spaniards, it retained an unusually high degree of autonomy; Messina in particular was an enclave with such extensive privileges that within Spanish domains it could be compared only to Barcelona. The wide-ranging powers of the senate extended to taxation and the appointment of university chairs; senate elections occurred every year. In 1639 the city senate took control of the university, the Fucina Academy was established, and Borelli became professor of mathematics, a position he retained until his move to Pisa in 1656. His *Delle cagioni* stemmed from debates within the academy, was dedicated to it, and its publication was financed by the senate. At Messina there was a link between political and intellectual affairs, with the party most active in defending civic autonomy being also sympathetic to intellectual novelties. Borelli was a key figure in this regard and in 1642–43 was sent by the senate on a trip across Italy in order to attract notable professors for the main chairs at Messina University. Borelli maintained close contacts with his Messina friends even after his move to Pisa: through them he was able to arrange for many prestigious appointments, such as Malpighi in 1662 and Fracassati in 1670. Several letters by Borelli state that his friends' victory in the forthcoming senate elections would secure intellectual affairs to his own and Malpighi's satisfaction;

Borelli's letter to Malpighi of 1 April 1662 shows how Domenico Catalano, professor of medicine at the university, maneuvered behind the scenes to secure Malpighi's call to the first chair of medicine, and the official letter of appointment from the senate to Malpighi mentions Borelli's favorable report.[5]

In the years 1646 and 1647 malignant fevers affected first Palermo and then Messina. The Fucina Academy became the venue for an extensive debate over the causes of the fevers and their treatment: Borelli's work provides a rare and revealing account of his own early views and, indirectly, a picture of the debates within the academy. His work consists of three parts with an appendix; in the first two parts Borelli refuted rival opinions about the cause of the epidemics that, no doubt, reflected those of his opponents, possibly within the academy; in the last part he presented his own views about the real "cagioni." The appendix contains Borelli's explanation of the nature of fevers and an account of digestion. Despite the claim that Borelli's treatise is his "iatromechanical manifesto," there is very little indeed in *Delle cagioni* suggesting a mechanistic worldview, apart from a few references to Santorio Santorio's *De statica medicina* and insensible transpiration; rather, Borelli relied mostly on a chymical understanding of disease and therapy with strong empirical overtones.[6]

Borelli set out to refute the claim that the fevers, which he defined as epidemic and contagious, were due to the corruption of the humid and hot air. The fevers were epidemic in that they affected people of different complexion, age, sex, and modes of life; they were contagious in that a close contact led to infection, although not as severely as with the plague, when everything that is touched by the sick is contaminated. In characterizing the course of the disease, Borelli relied on the examination of urine and other signs: initially the urine of the sick resembles that of the healthy, whereas at a more advanced stage it becomes thin and muddy; later the frequency of the pulse increased and other symptoms included delirium, heart complaints, and a black tongue, while some developed a rash or petechiae before dying ("muoiono con le petecchie"). On this basis Borelli reasoned that the cause of the disease did not depend on the corrupt humors, as shown by the urine that is initially healthy and blood that is well colored and of good consistency.[7] The climax of Borelli's refutation of the rival opinion came with an experiment, which he claimed he had originally performed with a different end: he took a large round glass container and put it head down on a pot of pure boiling water, until the air in the container became saturated with vapor. Then he recommended sealing it with clay and keeping it warm for several days or even weeks; he claimed that the enclosed air would not become smelly, corrupt, or poisonous and could be breathed without problems. Borelli liked no-nonsense, straightforward experiments, this one being typical of his style: we have seen that in order to prove that the heart is not hotter than the rest of the body he compared its temperature to that of other viscera with a thermometer in a vivisection experiment on a deer; further, in order to prove that air impedes a projectile rather than propelling it from behind, as claimed by some Aristotelians, Borelli attached some hairs to a sphere and threw it in the air, showing that the hairs pointed backward rather than forward.[8]

Borelli devoted the second part of his treatise to a refutation of traditional astrological explanations of the epidemic; such explanations went hand in hand with views on the corruption of the air that we have seen above. The use of astrology in medicine dates from antiquity and was prominent in Galen's work as well. As a former student and follower of the unorthodox philosopher Tommaso Campanella, who had called him "ex virtute filius" in *Astrologicorum libri VIII* (Frankfurt, 1630), Borelli was likely versed in astrological computations. Borelli, however, did not follow his erstwhile mentor in this regard and came to reject astrology and its medical applications that were accepted by many at the Fucina Academy and elsewhere. His main line of reasoning in this section was to highlight many ambiguities and logical inconsistencies of astrology: he pointed to temporal discrepancies between celestial and terrestrial phenomena; he argued that from each constellation one could predict both favorable and unfavorable horoscopes; further, he showed that often astrologers argued that a planet is dry or hot, for example, but if it is in conjunction with the sun that is dry and hot, instead of having a stronger effect, it loses its power. As he put it, astrology would lead us to believe that two thieves or two assassins in isolation are evil, but together they would lead to justice and temperance.[9]

Coming to the third part of his treatise, Borelli moved from the *pars destruens* to the *pars construens*, in which he put forward his own and his allies' views on the causes of the fevers and the effective therapies. He attributed the epidemic to poisonous exhalations that, explicitly borrowing from Lucretius, he called "semi di pestilenza." The reference to Lucretius testifies to the renewed interest in his work and classical atomism more broadly in those years: in 1647 at Florence the physician Giovanni Nardi—a correspondent of Harvey—published *De rerum natura* by Lucretius, and Gassendi published at Lyon *De vita et moribus Epicuri*, while in the same year as *Delle cagioni* Gassendi published in Lyon *Animadversiones in decimum librum Diogenis Laertii, qui est de vita, moribus, placitisque Epicuri*, containing a virtually complete edition of Lucretius's poem.[10] Borelli attributed the origin of such seeds to exhalations from the ground, which he compared to those forming comets, a clear sign that at that stage Borelli shared the opinion of Galileo and others that comets originate from the earth rather than from the heavens.[11] In order to account for the apparently random and unpredictable nature of the epidemic, Borelli had recourse to an analogy with lightning: the seeds of the plague coalesce in the sky above Sicily and occasionally descend and infect men and women in a random fashion, much like lightning hits the ground and only occasionally humans. Whereas in the first part Borelli had provided only a general description of the symptoms during the course of the disease, in the third part he identified a specific lesion found in the lungs of victims of the fevers, namely, the lungs appeared putrid and rotten, of a dark color and with dark spots or petechiae; the other organs, such as the brain—even those of patients who were affected by delirium—the liver, the diaphragm, and the heart, showed no sign of disease. Borelli and his colleagues had witnessed postmortems, possibly performed by Pietro Castelli, who had anatomical skills Borelli lacked and who had performed public dissections. Crucially, Borelli could report that the Palermo physician Giuseppe Galeano had written to him having observed in postmortems exactly the same feature in the lungs of fever victims in that city.[12]

Together with the symptoms previously observed, Borelli could outline the course of the disease, which originates from the lungs and only later spreads to the other parts of the body through the circulation of the blood demonstrated by Harvey: this is one of two references to Harvey in *Delle cagioni*, testifying to the knowledge and acceptance of Harvey's views at Messina. Curiously, in a later reference Borelli stated that the heart pushes blood to the termination of the arteries, whence it is "sucked" by the termination of the veins, a term that one would have liked to have been elucidated.[13]

On the basis of these findings, Borelli drew from the chymical literature the idea that sulphur would be an antidote to the fevers, a view confirmed by his colleagues Placido Reina, professor of philosophy, and Pietro Castelli, then professor of the theory of medicine. Castelli in particular was a leading figure in the introduction of chymical remedies, first at Rome and then at Messina, who had equipped the botanic garden he had established at Messina in the late 1630s with a chymical laboratory. Castelli was a prolific author who had contributed to the literature on the Messina fevers with a short treatise, *Praeservatio corporum sanorum*, in which he also attacked astrology, reported penances and prayers, and advocated *exiccantia* in the form of medicinals and diet. Borelli's sources included a rather long list of iatrochymical texts, such as works by Jean Béguin, Oswald Croll, and Andreas Libavius.[14] Allegedly sulphur proved to be a good antidote, in that of one hundred people to whom it was administered no one contracted the disease; of those who were already sick, however, it proved effective only in the cases when the disease was in the early stages. To the question about the temperament of sulphur, whether it is hot or cold, dry or moist, by virtue of which it can work as an antidote, Borelli answered by challenging traditional and Galenic doctrines, arguing that poisons do not act according to their traditional qualities, and conversely these qualities are not effective ways to assess a substance's properties against poisons. The wet saliva in the bite of a dog affected by rabies, Borelli continued, produces striking effects, but if it is hot and dry why doesn't fire produce the same effects? And if it is cold, why doesn't snow produce the same effects? Likewise, he argued that we do not know the reason why mercury cures syphilis or a magnet attracts iron, yet we know such things from experience; as it happens, he had direct experience of the former, since in 1643 when he was allegedly affected by syphilis, after having tried in vain several remedies, he had recourse to the "suffumigio del mercurio" or mercury fumigations, which he claimed was a very common practice at Messina. In conclusion, Borelli argued that experience and the testimony of the authors he had cited above were the only guides in this area. *Delle cagioni* contains a reference derived from the atomist tradition and Galileo's *Assayer*, in which Borelli argued that neither tastes, nor smells, nor colors are reliable or indeed viable ways to tell apart poisons from healthy ingredients; as we know from the previous chapter, he believed that these features could be easily changed, leaving the underlying substance largely unaltered, and vice versa, one could change the substance, leaving perceptible features largely unchanged. In *Delle cagioni* Borelli advocated experience, arguing that there is no other way to ascertain the effect of a substance, in his case sulphur. The issue of color was to be a matter for debate in Messina in subsequent years. In 1654 Pietro Castelli published a short tract, *Responsio chimica. De effervescentia, et mutatione colorum in mixtione liquorum chimicorum*, in which he discussed the nature of color and color change in solutions and in other circumstances, namely, animals that change color such as the chameleon. Castelli discussed color change in tincture of roses, which, as we shall see, was a matter for debate at the Accademia del Cimento as well. Castelli reviewed the literature according to the principles of both the chymists and the philosophers, arguing that the way colors are produced seems like a labyrinth.[15]

In the final remarks of the third part of *Delle cagioni*, Borelli advocated the use of wine to cure fevers, a relatively unorthodox remedy. His rationale came from his erstwhile mentor Campanella, whose *Medicinalia juxta propria principia* (Lyon, 1635) was in Borelli's library; Borelli argued that fever was not a disease but rather the body's reaction to the disease and as such it seemed reasonable to encourage it with wine; similar views had been put forward in the mid-sixteenth century by the anti-Galenist physician Gomez Pereira and later by Johannes Baptista van Helmont. Urged to clarify this matter, Borelli added an extensive appendix on the nature of fevers in which he touched on several topics. On apoplexy, for example, Borelli argued that its cause was the impediment of blood in the veins and the remedies were removing those impediments either with phlebotomy or by increasing the motion of the heart. Several pages at the end of the appendix deal with digestion, a topic for which Borelli relied again on chymical notions when he concluded that it is due to acid juices in the stomach.[16]

In *Delle cagioni* we have seen Borelli attack astrology and the traditional doctrine of the qualities that played a key role in Galenic medicine, propose therapies for fevers, and discuss the nature of and cure for apoplexy. We shall encounter again many of these themes in the dispute between the Neoterics and the Galenists and in later works by Borelli and his circle.

2.3 Borelli, Malpighi, and the Pisa Epidemics of 1661

It is extremely instructive to compare *Delle cagioni* with Borelli's reasoning and strategy in the case of the Pisa epidemic of 1661; we are fortunate to have not only his letters to Malpighi but also occasionally Malpighi's replies, offering a rare and detailed account of their thinking over such a dangerous and problematic case. Borelli's letter to Malpighi of 18 November 1661 conveys the dramatic news of a "stragie" or massacre at Pisa due to acute fevers. Although those like Borelli who had arrived from elsewhere after October 31 "ognissanti" seemed to have been spared, his first concern was to ask for permission from the Grand Duke to leave town and to arrange for postmortems at the local hospital of those who had died of the disease, since, as Malpighi put it, the cadavers were the only basis on which to philosophize in such cases.[17]

As it happens, Borelli's continuous health due to—as he put it—a most severe lifestyle, the desire to find out the outcome of the postmortems, his high visibility due to his contacts at court, and his concern for three students of his who had fallen ill led him to remain in Pisa despite the fact that he had received legal permission to leave in the form of a pass with blank dates. By November 25 Borelli reported that the fevers, which he called epidemic, had started to affect even those who had arrived after "ognissanti" and three university lecturers had already died. Alas, he feared that since the fevers were widespread in England, France, Rome, and Venice, there was no point in escaping. Moreover, postmortems performed by an unknown student from Sicily and another from Bologna, Antonio Laurenti, who sadly was to pass away at the end of December,[18] started giving results. Borelli hastened to state that in four cadavers opened in his presence no pulmonary lesion could be found, except some dryness due to the fever; only the gall bladder and the stomach, and in some cases the intestines, were found overflowing with bile, hence the epidemic differed from that of 1648 or a later one of 1654—about which unfortunately we have few details—in which the lungs were affected and the other viscera were intact. In his reply, Malpighi wondered whether bile was the cause or rather the effect of the disease due to a ferment in the blood, which would be very hard to cure; moreover, he pointed out that by itself the excessive presence of bile in blood does not cause fevers, as proved by jaundice.[19]

Borelli was interested in the course of the disease and therapy as well. The standard course started with a simple tertian, with confusion of the mind, stomach pain, bitter taste in the mouth; according to the hospital camerlingo, all those attacked by the disease were affected by pains in the body. On the seventh day the fevers became continuous, on the eleventh they became malignant, and those who did not have natural evacuations died on the fourteenth day or even earlier. Borelli was relying on a classic Hippocratic taxonomy of fevers and apparently on some form of the doctrine of critical days as a guide to the progression of the disease. As to therapy, he believed phlebotomy to be useless, since all those who had died had been bled; in a comment on the medical implications of Harvey's circulation of blood, he doubted whether leeches had a different effect from phlebotomy from the arm, given that from the circulation it should not really matter whence blood is taken out.[20]

The exchange between Borelli and Malpighi highlights the existence of local therapeutic traditions influenced by previous epidemics. Malpighi, for example, claimed that at Pisa it was an "inviolable method" to use mild purgative at the onset of the disease. In his reply, Borelli argued that in fact the habit of purging was common among Padua-trained physicians; local physicians, on the contrary, scared of the events of 1654, when all those who were purged died, began the therapy with phlebotomy. As it happens, following his Bologna mentors, Malpighi was rather skeptical of the efficacy of purging because it acts on the effects rather than the causes of illnesses.[21]

In his letter of 20 December 1661 on the fever epidemic, Borelli returned to the problem of color and taste. In response to Malpighi's surprise that an excess of bile in the digestive tract did not necessarily lead to spontaneous evacuations, Borelli mentioned several experiments leading him to distrust taste and especially the correlation between acidity and a substance's ability to produce evacuations, as well as other properties. Vinegar is quite sour on the tongue, yet it turns quite sweet when it is distilled several times; however, vinegar is not very corrosive, whereas once distilled it is considerably more corrosive. Therefore, taste is not a reliable indicator of a substance's properties. Tastes could be deceptive in other instances too, as he had just experienced by noticing the similarity in taste between two liquids with very different properties, such as that in which olives are kept—known as "salamoia"—and that found in the stomach of fishes, which he compared to an *aqua fortis*, or between milk and the liquid found in the stomach of hawks: here too the similarity in taste is deceptive in that hawks can eat bones. Hence, there are very corrosive juices that do not lead to evacuations. Addressing Malpighi's claim that excessive bile in the blood may lead to jaundice but does not generate fevers, Borelli proceeded to question whether jaundice is truly caused by an excess of bile and more broadly to speculate on the significance of color; he mentioned an experiment in which he noticed that a piece of paper placed in the urine of a patient with the face and eyes very yellow came out with its color unchanged rather than yellow, a sign that there was no excess of bile in the blood. He then discussed some experiments performed at the Accademia del Cimento and later published in the *Saggi*. A few drops of oil of tartar or spirit of sulphur could turn a large amount of liquid obtained from red roses and spirit of vitriol from red to green and back to red. The red liquids, however, had such different properties that the first had a pleasant taste and was healthy, whereas the final product could have killed a man. Hence, nature could easily change colors and tastes without changing a body's substance and, vice versa, make very different substances look and taste similar. We encounter again here the doctrine he had already put forward in *Delle cagioni* with regard to sulphur and poisons, whereby sensory experience is deceptive. Borelli's philosophical stance contrasted Malpighi's medical perspective, although Borelli sought to justify his views by means of observations and experiment of a medical nature as well. In the conclusion of the letter he urged caution about the Pisa epidemic, since nature can produce similar effects from different causes.[22]

The Pisa epidemic subsided by the end of the year; although Borelli and Malpighi were only partially successful in dealing with it, their exchange highlights how the search for appropriate symptoms and therapies, philosophical views about matter and its qualities, and experiments on color and disease were interwoven in intriguing ways.

2.4 The 1665 Controversy between the Neoterics and the Galenists

Soon after the exchange with Borelli, Malpighi moved to Messina, where he lived from 1662 to 1666. One of the major issues he faced was whether to practice medicine. Borelli urged him to, and, as primary professor of medicine, he was probably expected to by his local supporters. Malpighi, however, felt reluctant to act in such a sensitive area and to challenge established interests not just intellectually but also economically, and even to declare himself as a supporter of novelties. At his arrival he was welcomed by several scholars and aristocrats, including Viscount Iacopo Ruffo, Placido Reina, and the learned pharmacist Lorenzo di Tommaso, but the reception was mixed. Malpighi's belief in the beneficial role of phlebotomy in fevers and apoplexy was especially controversial; as he reports in an earlier fragment of his *Vita*, upon his arrival in Messina the physician and professor of medicine Francesco Avellini asked him whether he would perform phlebotomy in apoplexy, and when he replied affirmatively, Avellini retorted that if he were viceroy he would have physicians like him hanged—not the most encouraging welcome to the new primary professor of medicine.[23]

In 1665 Messina was the arena of a controversy between the Galenists and the Neoterics. The occasion for the exchange was the annual election of the protomedico and his junior assistant; the latter position was contested by two candidates with opposite views on the new medicine, Michele Lipari and Francesco Maria Giangrandi, who held a public disputation on preassigned conclusions. Giangrandi, however, who was trained by Malpighi and his ally Catalano, did not prepare sufficiently and failed to satisfy Malpighi, who refused to preside over the function, while Catalano was apparently indisposed, "caput afflictum habere." Lipari's supporters were the professors of medicine Avellini and Paolo Varvesi; the latter was a member and sometime prior of the college of physicians who also taught Greek. The controversy unfolded around four texts: two sets of conclusions to be discussed by the contestants, of which Lipari's survive only in part, whereas Giangrandi's—drafted mostly by Malpighi, especially for the anatomical and physiological parts, with contributions by Catalano for the medical ones—survive in their entirety. The event was not a success for Giangrandi; thus, we have our third text in the form of a pamphlet by Lipari, significantly titled *Galenistarum triumphus*, which survives in some manuscript copies and one single recently discovered printed version; lastly, we have our fourth text, the *Risposta* to Lipari written by Malpighi under the name of his student Placido Papadopoli.[24]

In his conclusions Lipari defended astrological medicine, a sign that Borelli's attack from sixteen years earlier had not eradicated it; nor was Messina an exception, since in the same years at Bologna Ovidio Montalbani also defended it. Lipari argued, "it is a sign of great rashness and ignorance to employ phlebotomy when the moon is in conjunction, quadrature, or opposition to the sun, Saturn, and Mars, and also during the malign aspects of the malign planets among themselves." Moreover, he argued that the pulse does not depend on the circulation of the blood but on the generation of vital spirits and communication of heat. Further, Lipari argued that apoplexy results from the phlegm's total obstruction of the flow of vital spirits in the brain, whereas a partial obstruction produced epilepsy. While sharing the belief in the association between apoplexy and epilepsy, Malpighi believed in a different etiology.[25]

The conclusions put forward by Malpighi and Catalano are a valuable document on the nature of texts discussed at such functions, the state of neoteric medicine, and Malpighi's views at the time. Out of forty-six conclusions, the first twenty-four are mainly anatomical, the following twenty-two medical. After challenging the division of medicine into theoretical and practical, for example, the third conclusion provides a statement of mechanistic anatomy and pathology:[26] "We explain the natural constitutions, actions, and morbid dispositions of the parts most easily and appropriately as those that consist of certain machines and organs, and are carried out mechanically by the directing soul." Here we find a rather primitive formulation of the relations between body-machine and the soul, whose role is not spelled out; Malpighi will address the same problem again in his posthumous *Vita* and *Risposta* to Sbaraglia. The conclusions then move to specific organs, starting with the heart, which is described ostensibly following Hippocrates as a mere muscle rather than a parenchyma; in the late 1650s the heart had been an object of study by Borelli and Malpighi at Pisa, where they discovered its spiral muscular structure. It is of interest that Malpighi sought to defend modern positions by finding precedent in Hippocrates.[27] Following Malpighi's 1661 *Epistolae*, the lungs are described as consisting of innumerable membranes covered with blood vessels; moreover, the sixth conclusion adds a statement about their motion, which does not occur because of the "fuga vacui," of which the lungs, and even more so inanimate things, have no perception, but rather from the weight of the air pressing them and the force of the entering air. Contrary to Malpighi's later opinions, the liver, kidneys, and testicles are not described as glandular: the liver does not make blood but filters bile in its parenchyma, a term suggesting that the views Malpighi was to put forward in 1666 had not yet matured. While attributing a mechanical role of filtration to the liver, he located this process in its parenchyma rather than glands. The kidneys do not attract urine but filter it with the help of the conformation of the fistulae—or tubules—that imbibe it, an implicit reference to Bellini's finding of 1662, which will be discussed in section 4.5. The testicles, as claimed by Auberius, are not glandular but consist of a mass of tubules, but Malpighi was unsure whether there were many separate ones or only one folded several times. Against Galen, he argued that the purpose of glands was not to support bifurcating blood vessels, but to provide nutrition, excretion, and the reduction of juices.[28] Malpighi proceeded to challenge other Galenic tenets, such as that the purpose of respiration is to cool the heart and that the artery-like vein transports air or sooty vapors. The last three anatomical conclusions concern generation and employ textile analogies with terms like "stamen" or warp and "tela" or cloth that are typical of Malpighi's language.[29]

The medical conclusions, twenty-two in all, start from a description of fevers as an inordinate motion of blood with an excessive fermentation and effervescence. I provide a brief selection largely guided by some themes we have already discussed or we are about to encounter. As Adelmann suggests, these conclusions may well be Catalano's. Apoplexy, which as we have seen was a contested area at Messina, was defined as the lack of motion of blood in the brain; unless phlebotomy was performed immediately, as the Roman physician Aulus Cornelius Celsus recommended, there was no hope of recovery. The humors entering the mouth in cases of catarrhs do not descend from the brain through the pituitary gland—as claimed by ancient doctrine, one may add—since there is no suitable passage, but rather come from salivary vessels. We witness here a significant link between the new anatomy and pathology; in all probability Malpighi and Catalano were referring to Steno's finding in the mouth glands and ducts, which he tentatively considered as the source and passageway, respectively, of saliva and catarrhs, although van Helmont and the Wittenberg physician Conrad Victor Schneider had also challenged traditional doctrines. Steno's findings were already known in Italy early in 1663, when Bellini displayed the parotid and salivary ducts at the Tuscan court, as Borelli reported in a letter to Malpighi—although Louis de la Forge in his 1664 comment on Descartes' *L'homme* still claimed that saliva comes from the brain. The conclusions touched on the role of bile in jaundice as well: although the yellow coloring of the skin in jaundice is likely caused by bile, Catalano and Malpighi argued, jaundice does not necessarily result from an obstruction of the bile duct, but may be due to the corruption of blood serum. The claim found here that the empirical method should not be disapproved of in medicine, since experience is a better and safer guide than reason, differs noticeably from Malpighi's later statements, as does the claim that in employing the rational method, the physicians should not rely on indications alone but should take into account a broader set of factors; "*indicationes*" is a "terminus technicus" meaning the therapeutic strategy based on the knowledge of the causes of the disease. By contrast, as we will see in section 11.3, in his *Risposta* to Sbaraglia Malpighi highlighted the role of *indicationes*, arguing that in many cases therapy could be established a priori. Fully consonant with Malpighi's therapeutic strategy is the belief that whereas in normal circumstances, when the body is full of excrements, purgation is useful, when fever is present it should be avoided, even if Hippocrates allows it in some cases; this doctrine stems from Galen's *De methodo medendi*, the same text in which one finds an extensive discussion of *indicationes*.[30]

Sadly, we do not know what happened at the actual disputation at the end of February or the beginning of March. Lipari and his allies, however, felt sufficiently emboldened to publish *Galenistarum triumphus*, a pamphlet that reads like a scholastic exercise by an eager but mediocre student in which he rejected in seven sections many of the conclusions by the Neoterics about both anatomy and medical practice. The title page shows an engraving with four bearded men on a carriage pulled by two elephants: it was customary at the time for traditional physicians to grow a beard; hence, the four men in the carriage can be identified with the triumphant Galenists. Lipari relied extensively on ancient authority, as when he started by arguing that medicine is not conjectural but is a science. While defending the claim that liver makes blood, Lipari relied on the Dutch amateur anatomist Lodewijk de Bils, who, besides having devised a way to preserve cadavers and body parts and to perform dissections without shedding blood, had recently defended the traditional role of liver in making blood.[31] Lipari also defended the notion that blood consists of the traditional four humors by relying on Hippocrates, Galen, and the fact that there are four seasons in the year, one for each humor, and that each humor abounds in its appropriate season. Moreover, Lipari states that blood let from a healthy man and left standing in a bowl shows a stratification with bile at the top, blood proper, phlegm, and melancholia at the bottom, something he claimed was known even to barbers. A version of this observation can be found already in the *Canon* by the Persian philosopher and physician Avicenna, who argued that blood separates into a red part or "cholera rubea," one called "fex" or melancholia, one resembling egg white that is phlegm, and a watery part whose excess is expelled in urine. As we shall see, Malpighi radically reinterpreted this separation in his *Risposta*. As to the medical conclusions, Lipari challenged the definition of fevers mentioned above and argued instead that it is necessary to define them in terms of preternatural heat, as shown by the etymology of the term "fever" in different languages. In support of his claim that apoplexy is an obstruction to the flow of vital spirits in the brain due to phlegm, Lipari had recourse to "experiments," by which he meant postmortems, showing that the ventricles and substance of the brains of cadavers of patients who had died of apoplexy presented an obstruction from a thick, viscid, and slow humor. He proudly supported his claim by quoting not Hippocrates, Galen, or a Galenist, but the Leiden professor of medicine Walaeus, who had argued that apoplexy was treated with purging as a rule and only in the case of redness in the face and plethora was phlebotomy allowed: "The Novatores are slain with their own sword," gloated Lipari, who quoted one of Thomas Bartholin's letters as well in defense of his own views on pleurisy, thus highlighting disagreement within the neoteric camp.[32] Even such an undistinguished pamphlet as Lipari's highlights the problems at the intersection between anatomical research and medical practice, as well as between novelties and tradition, faced by physicians, especially when the confrontation was not purely a scholarly affair but a public and to some extent political one as well.

In this regard it is worth recalling that in 1674 Messina rebelled against Spanish rule and for four years resisted with the help of French military support several attempts by the Spaniards to regain control. Although it is by no means possible to draw simplistic conclusions about political and intellectual allegiances in the Spanish domains in southern Italy, at Messina a large number of the Neoterics sided with the rebellion and with France, including the painter and naturalist Agostino Scilla, the pharmacist di Tommaso, Borelli, and many members of the ruling class, such as Malpighi's sponsor Alberto Tuccari; Lipari, by contrast, was hanged as a Spanish spy.[33]

2.5 Malpighi's *Risposta* to *Galenistarum triumphus*

The Neoterics had a few options: ignore Lipari's pamphlet, publish a formal rebuttal, or reply in manuscript form. The Messina senator Alberto Tuccari urged Malpighi and Catalano to reply, but upon Catalano's refusal, Malpighi hastily drafted on his own a manuscript response or *Risposta* that, in line with his style to avoid publishing polemic works, remained unpublished in his lifetime, although it did circulate in manuscript form and, according to Max Fisch, inspired a section on the triumph of spagyric or chymical medicine in the work by the Neapolitan iatrochymist Sebastiano Bartoli. Malpighi drafted his *Risposta* in Italian, so that he may more effectively address Lipari's local constituency; at the end he defended the use of the vernacular, on the example of Galileo, Descartes, Boyle, and others. In drafting his response Malpighi sought advice from Borelli, who as usual was more than happy to offer it. Commenting on a draft of Malpighi's reply, Borelli urged him to expand his treatment and to learn how to treat controversies from Galileo's *Assayer*.[34] Indeed, echoes of the *Assayer* can be found in specific passages, as where Malpighi used the metaphor of the book of nature and of the need to understand the characters with which it is written. Here of course he was referring to the celebrated passage in which Galileo had employed the metaphor of nature as a book whose characters are geometric figures like triangles and circles. In the same work Malpighi praised the telescope and the microscope, arguing that through the telescope Galileo had discovered more in astronomy than had been found in the previous millennia, and that the microscope too had revealed many mechanical contrivances in animals—a reference to his own discoveries. In this way Malpighi was implicitly presenting himself as the Galileo of medicine, since he had made the most significant anatomical finding through the microscope. It was probably in response to this passage that the Crotone physician Giovanni Battista Capucci, on receiving a manuscript copy of the text, called Malpighi "un 2° Galileo."[35]

On the other hand, Malpighi did not follow Borelli's advice blindly, and in one major respect he departed from his mentor's suggestion. In a letter of 1 August 1665, Borelli predicted that Lipari would quote only very common books and would rely on ancient doctrines, urging Malpighi to state that sensory experiences and anatomical experiments had refuted those doctrines; thus, Borelli saw a break between ancient and modern medical authors. In some places Malpighi followed this strategy and argued that novel findings had subverted classical medical doctrines, as when he claimed that in order to save Galen at all costs, Jacobus Sylvius had stated that the human body must have changed from Galen's time to his own when he found a discrepancy between text and nature. On other occasions, however, Malpighi adopted an irenic approach in displaying his knowledge of the classical medical literature and exegetical skills; he frequently argued that Lipari had misinterpreted and only partially read the texts, which, if properly studied, would not support his views. Overall, Malpighi often chose a more cautious tone emphasizing continuity rather than rupture, stating that in medicine it is helpful to learn the truth whether from the ancients or the moderns. He cautioned against Lipari's overconfident view of medicine, emphasizing its conjectural nature in its reliance on both experience and reason or a priori. This position differs significantly from the one he was to espouse in his *Risposta* to Sbaraglia, which is the subject of section 11.3.[36]

Apoplexy was the subject of an analysis extending over a dozen pages; bearing in mind Avellini's challenge upon his arrival and the subsequent exchanges, Malpighi must have felt the need to compose almost a treatise within the *Risposta* just to address this issue. He tried to link the disease to lesions in the anatomical structure of the brain and mentioned his own observations of the brains of fishes in support of his views: we witness here a revealing link between anatomical research on animals and pathology. As we will see in section 3.2, in *De cerebro* Malpighi identified in the brains of fishes an aggregation of nervous fibers forming the marrow and terminating in the gray matter with its extensive blood vessels. Malpighi challenged Lipari's claim that the animal spirits are generated in the ventricles of the brain and that apoplexy is due to the accumulation of phlegm in those ventricles; rather, from the structure of the brain he had uncovered and from the examination of cadavers, he identified the gray matter as the site where animal spirits are generated and where apoplectic lesions occur. He further dismissed Lipari's report of postmortems, arguing that humors are often found in the ventricles of patients who have died of fevers, whereas several anatomical reports show that at times the ventricles of those patients who have died of apoplexy are empty. On the basis of his analysis and of a careful reading of Hippocratic, Galenic, Arab, Renaissance, and contemporary texts, Malpighi came to support the use of phlebotomy in apoplexy. He further expressed the hope that remedies that render the blood more fluid may be beneficial, especially if they are administered in the way recommended by Fracassati in *De cerebro*, where he had reported his own and Bonfiglioli's early experiments on venous injections or *chirurgia infusoria*, as discussed in section 3.4.[37]

Although Malpighi's agenda in the *Risposta* was dictated by the previous texts in the dispute, in some cases he was able to rely on his own previous unpublished works. He questioned whether viscera and muscles can be called sanguineous, since their first components are all different and once they are drained of blood they appear white: Malpighi was about to turn to the microstructure of those viscera. As we have seen in section 1.2, while at Pisa Malpighi had started composing two dialogues in Galilean style that remained unpublished at the time and are now lost; a portion of their contents, however, was used in the *Risposta* to Lipari. Malpighi challenged traditional views on blood with a few simple "esperienze." Referring to the experiment on the caked blood reported by Lipari, Malpighi recounted that if some blood is left to stand in a dish, the top is bright while a darker portion forms at the bottom. Malpighi challenged the common belief that this was a melancholy humor—one of the constituents of blood—since if congealed blood is turned upside down, when the darker portion is exposed to air it becomes bright red and the red one that is now at the bottom becomes darker; the entire caked blood turns dark if it is kept under water and becomes bright red when salt is added. According to Malpighi, this showed that the difference in color was due to some accidents that did not change the substance of blood. Malpighi used this experiment to challenge the traditional doctrine of the four humors; he justified his claim with the freedom to philosophize that characterized his century, new observations, and the resurgence of the doctrines of Leucippus and Democritus, arguing that the components of blood exceed forty. We encounter here a medical application of atomism and Galileo's doctrine of primary and secondary qualities, filtered by Borelli and the experiments of the Accademia del Cimento that we have seen above. It is worth highlighting that Malpighi was fully aware of the role of air in the changing color of blood: in another passage dealing with pulmonary disease, he argued that blood spits are bright red because blood is mixed with air, whereas blood in the rest of the body can be quite different in color and texture. Thus, Malpighi did not see a connection between the color of blood in the lungs, whose structure he had recently investigated, and their operations and purpose: in all probability, besides Borelli's stance, his finding that blood in the lungs flows always inside blood vessels did not suggest a role for air in its color change in the body's healthy state; moreover, air was only one among many substances changing the color of blood.[38]

Purgation, a topic of the second Pisa dialogue that was also touched on in the medical conclusions, was also discussed at some length in the *Risposta*; Malpighi was eager to reconcile the therapeutic principles he had acquired from his Bologna teachers with the new philosophy. Following the admiration of his teachers Massari and Mariani for the views of Leonardo Giacchini, Malpighi argued that acute fevers deprived the body of precious humors by opening the extremities of the arteries. Malpighi relied on classical sources, including several works by Galen, such as *Ars parva* and *De praecognitione*—the latter translated by Giacchini (Venice, 1533)—and Hippocrates' *Epidemics*, from which he reported an extensive list of case histories supporting his views. In his *Vita* Malpighi praised Giacchini and added a mechanical explanation of his and Galen's views in terms of the opening of *meati* and the consequent loss of precious humors; he was suspicious about purgation because it acts on the effects rather than the causes of illnesses. Here Malpighi was grafting modern mechanistic thinking onto the traditional body of Galenic doctrine.[39] Moreover, Malpighi drew a rather interesting, if somewhat forced, simile with a clock, the most emblematic and commonly used example of a machine at the time. Malpighi used it in an unusual way by drawing an analogy with pathology and therapy, arguing that when a clock malfunctions the skillful artisan cleans and mends its parts rather than throwing some away; similarly, it would be foolish to dispose of precious fluids in acute diseases rather than mending the body.[40]

Malpighi's reflections on fevers are of considerable interest for the light they shed on the Neoterics' thinking about their nature and therapy; moreover, they resonate with Borelli's attack on the significance of traditional primary qualities in *Delle cagioni*. In *Galenistarum triumphus* Lipari had argued that fevers are cured by humidifying and refrigerating remedies, which must have contrary qualities to them, hence fevers must be hot and dry passions. Malpighi retorted that fevers are traditionally cured by milk, which according to Galen is not quite cold and dry because butter and cheese that are made from it are hot, from which it should follow that fevers are cold passions, clearly an absurd conclusion. Malpighi then proceeded to his key argument. He claimed that truth is so powerful that it cannot be hidden, since it emerges even from a fallacious reasoning like Lipari's. Thus, it is indeed true that milk cures fevers, particularly hectic fevers, but the correct reason is that such fevers consist of an excessive fermentation of the blood. Milk, and especially its buttery and cheesy particles, has the power to bind the blood particles and prevent excessive fermentation. Malpighi's medical practice documented in his *consulti* shows that he prescribed routinely different types of milk, especially jenny-ass milk. I believe this example to be representative of a broader class of cases in which Malpighi reformulated the theoretical justification of standard therapies. The role of fermentation in fevers was gaining a more prominent position in those years thanks to Thomas Willis's 1659 *De fermentatione* and *De febribus*, a work Malpighi referred to. In the case of pleurisy and pulmonary affections, he relied on his own anatomical findings, as he had done in his study of apoplexy, and Steno's refutation of the existence of a passageway from the brain to the lungs through which phlegm would flow; he explained affections of the lungs in terms of their microstructure and the components of blood. In an important passage on pathology he questioned the significance of the classification of diseases, which he called a "medical metaphysics," and the use of terms such as "exquisitum" and "non exquisitum," which he saw as signs of our ignorance and artificial supports of our precarious knowledge. He further denied that excessive phlegm is to be found in the blood of patients affected by pleurisy; rather, the white part found in the lungs and heart of cadavers of patients affected by pleurisy occasionally forming polyps is not phlegm but blood serum or coagulated nervous juice. As we will see in section 5.4, Malpighi returns to this problem in his 1668 *De polypo cordis*.[41]

A number of themes in seventeenth-century medicine intersect the relations between political power and knowledge at Messina: the dichotomies of theory versus practice and tradition versus innovation stand out for their centrality. We have encountered several instances of the interrelations among new anatomy, disease, and therapy in works by Steno and Malpighi: new anatomical findings, such as Steno's glands and ducts in the mouth, challenged traditional views on phlegm's origin from the brain, while the notion of fermentation required a reconceptualization of therapies for fevers. Moreover, Borelli relied on physical experiments to challenge traditional views on corrupt humors as the cause of malignant fevers and, more generally, engaged with their etiology and therapy.

The role of Galenism was complex. On the one hand, the Galenic understanding of the body's operations was rapidly crumbling, and key notions about pathology and therapy, such as those based on the qualities hot, cold, wet, and dry, were rejected by both Borelli in *Delle cagioni* and Malpighi in the *Risposta* to Lipari; moreover, astrological medicine was also challenged. On the other hand, Malpighi defended Galen's approach to medicine based on finding the causes of diseases and knowledge of natural philosophy, praised Galen's *De methodo medendi* as an altogether golden book, and adopted Giacchini's neo-Galenist views on treating fevers. Thus, the Galenic legacy showed obsolescence and vitality at the same time.

As one would expect from debates on malignant fevers and other ailments, therapy took center stage. Purgation in acute fevers and phlebotomy in apoplexy were controversial subjects: in both instances the therapeutic tools were entirely traditional; the issue was when to apply them. Borelli, however, had recourse to new chymical remedies to cure his suspected syphilis, and during the fevers of 1647–48 he and his Messina associates relied extensively on chymical literature and remedies. In line with what we have seen in the previous chapter, we find again Malpighi tentatively relying on color and taste as valid diagnostic tools, only to be rebuffed by Borelli's reflections and experiments: some, which he had presumably performed alone, were medical; others were part of the activities of the Accademia del Cimento and can be characterized as chymical. In cases as diverse as the failure of the senses to distinguish poisons from medicinal substances, the investigations on jaundice and malignant fevers, and the study of the change of color in blood, a connection has emerged among atomism, experimental endeavors, and the medical and anatomical traditions. We shall encounter analogous themes in the following chapter.

CHAPTER THREE

The Anatomy of the Brain and of the Sensory Organs

3.1 Atomism and the Anatomy of the Senses

The brain and the sensory organs pose some of the most philosophically significant problems in anatomy, as testified by the extensive references by seventeenth-century anatomists to ancient and modern philosophers who wrote on these matters, from Plato and Aristotle to Descartes and Gassendi. Although some recent historians have argued that the mechanical philosophy merely replaced the Aristotelian qualities—such as the dormitive virtue of opium—with the size and shape of invisible particles that were equally inaccessible to empirical investigation, evidence shows otherwise: in the mid-1660s anatomists sought to determine the structure and mode of operation of the brain and organs of perception, such as the eyes, tongue, and skin, through challenging and innovative techniques of anatomical and chymical investigation. Sense perception was both a tool and subject of inquiry: color—as an important component of sight—was investigated in relation to the properties of matter, whereas taste, touch, and smell were studied with regard to the size and shape of corpuscles and composition of matter, at the intersection between anatomy and atomism. The link between what came to be called primary and secondary qualities was key to several investigations.[1]

Before delving into this material, I wish to offer a preliminary justification for the sources I have selected based on contemporary perceptions. Works on the brain include Willis's 1664 *Cerebri anatome*, a lecture delivered by Steno in Paris in 1665, Malpighi's *De cerebro*, and Fracassati's treatise with the same title, both of 1665. These works include extensive cross-references and were grouped together in the *Bibliotheca anatomica*, underscoring that the links among them were clear at the time. In addition, Descartes' 1662 *De homine*—published in 1664 in the original French as *L'homme*—contains important sections on the brain and sense perception. This work's ambiguous status in anatomy was highlighted by its exclusion from the *Bibliotheca anatomica*, although Willis's and especially Steno's extensive discussions justify its inclusion here.[2]

Whereas research on the brain had a European dimension, anatomical research on the tongue and the organ of touch was carried out mainly between Messina and Pisa. We have already seen instances of the interplay between medical-anatomical research and a number of experiments at the Accademia del Cimento, such as those on capillary tubes and color indicators. We are now going to encounter further instances of interplay between experiments carried out at the Accademia del Cimento and more generally at the Tuscan court and anatomical research in the works by Malpighi, Fracassati, and Bellini. Malpighi's *De cerebro* is an epistolary treatise addressed to Fracassati that was part of the 1665 *Tetras anatomicarum epistolarum*, a collection including also *De lingua*, on the tongue, addressed by Malpighi to Borelli, and treatises on the tongue and brain—also titled *De lingua* and *De cerebro*—by Fracassati, addressed to Borelli and Malpighi. In addition, also in 1665, Malpighi published separately a treatise on the skin as the organ of touch, *De externo tactus organo*, and Bellini published *Gustus organum* on taste. The plan for the collective volume and Bellini's work was due to Borelli, who encouraged his colleagues to build on Malpighi's findings. While Malpighi's originality was curtailed by the other contributions, Borelli's plan led to an enhanced visibility and a public display of mutual encomia, although the anatomists were by no means in agreement on every detail. The *Bibliotheca anatomica* also grouped together all the publications by Malpighi, Fracassati, and Bellini on sense perception, notably taste and touch.[3]

The following section studies the key texts on the brain, notably Willis's, Steno's, and Malpighi's. Given their profound connections conceptually and in the techniques of investigation, section 3.3 analyzes Malpighi's *De lingua* and *De externo tactus organo*, a work frequently bound with *Tetras anatomicarum epistolarum* and added to its second edition in 1669.[4] Fracassati's *De lingua* and *De cerebro*, discussed in section 3.4, deal in fact with a wider range of problems, and they too are best treated together for the light they shed on the philosophical debates and experimental research carried out at Pisa and at the Tuscan court. Lastly, section 3.5 discusses Bellini's *Gustus organum* and the work of his Pisa colleague Rossetti, *Antignome*, a treatise containing valuable observations on sense perception that precipitated the break between Borelli and Malpighi.

3.2 Brain Research in the 1660s: Willis, Steno, and Malpighi

The correspondence between Richard Lower and Robert Boyle documents from the very beginning of 1662 the inception of the treatise by "the doctor" Thomas Willis, Sedleian Professor of Natural Philosophy at Oxford since 1660. Dissatisfaction with

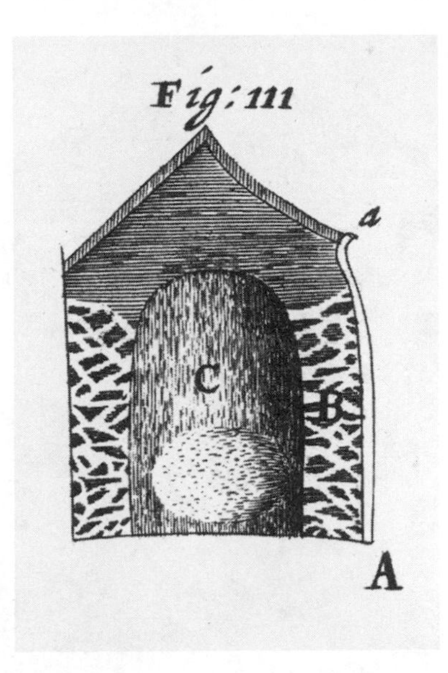

Figure 3.1. Willis, *Cerebri anatome*: the *rete mirabile*

the current state of knowledge of the brain was apparently one of the reasons to embark on the extensive project, one that extended beyond the 1664 *Cerebri anatome* to the 1667 *Pathologiae cerebri, et nervosi generis specimen*; in fact, anatomy and pathology were tied from the start. Willis's impressive treatise consists of twenty chapters on the brain and nine on the nerves, starting from a method for dissecting the brain and including chapters on the brains of birds and fishes and on the *rete mirabile*, which he believed slowed the motion of blood into the brain and could be found in some animals such as calves and also in humans, although only in those "of a slender wit" or "destitute of all force and ardor of the mind." Figure 3.1 shows the *rete mirabile* of a calf in what is, as Willis admitted, not the most handsome figure in his treatise.

The most celebrated finding of *Cerebri anatome* is that the arteries form a loop at the base of the brain, later called the "circle of Willis." Figure 3.2 was probably drawn by the Savilian Professor of Astronomy and architect Christopher Wren and shows, in the middle, the dark loop of arteries at the base of a human brain. In the preface to his work Willis acknowledged Wren for having drawn many pictures and for the discussions on the uses of the brain. Willis showed that colored liquid injected into an artery on one side is soon seen to descend from the artery on the other side; thus, the anastomoses among the arteries regulate blood flow to the brain, preventing it from being deprived of—or engorged with—blood: injections in this case were used not so much to reveal a structure as to investigate its mode of operation and purpose. Wren was adept at this aspect as well. We notice here a similar pattern to the collaboration

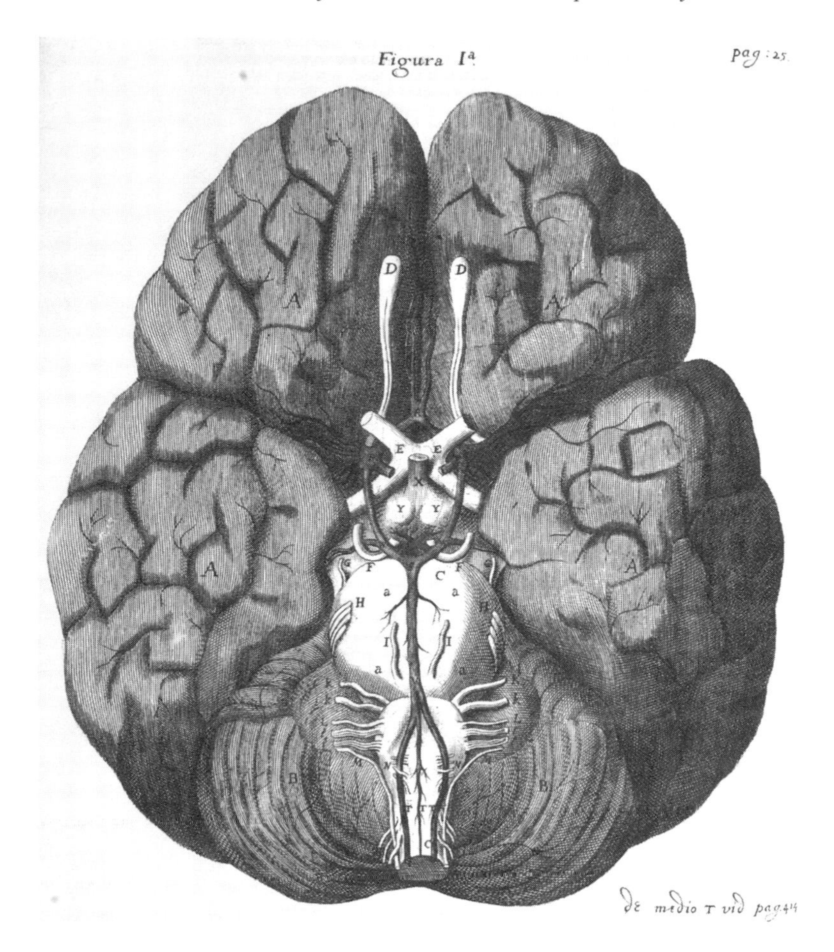

Figure 3.2. Willis, *Cerebri anatome*: anastomosis of arteries at the base of the brain

between Borelli, Malpighi, and Bellini, where the anatomists study the structure and the physico-mathematician helps with its interpretation. Willis was able to confirm the purpose of the arterial loop he had detected in a notable study of a diseased state based on a postmortem, in which he found the right carotid artery bony and almost entirely obstructed, yet the man had lived a normal life because of the abnormal enlargement of the compensating vertebral artery of the same side: here a concealed diseased state shed light on the normal purpose of the arterial anastomoses. The report of injection of a colored liquid highlights one of the characteristic techniques of investigation of *Cerebri anatome*, one that was reported in the very first letter by Lower to Boyle announcing Willis's intention to syringe a liquid "tincured with saffron, or other colours," into the carotid artery in order to detect its ramifications in the brain; exactly the same technique of injection of water colored with saffron—or even milk—had been used by Francis Glisson in *Anatomia hepatis*. Other remarkable injection experiments were reported in subsequent letters by Lower and in the published text.[5]

Willis did not refrain from attributing specific roles to portions of the brain, arousing the skepticism of Borelli, who in a letter to Malpighi wondered how much one could know about such matters. He probably had chapter 10 in mind, in which Willis proposed an explication of the uses of the parts of the brain and tried to localize its operations: imagination would be a wavering motion of the animal spirits from the inside outward, whereas memory would be a contrary motion of the animal spirits from the outside inward. As to fantasy, he stated that "sometimes a certain sensible impression, being carried beyond the callous Body, and striking against the *Cortex* of the Brain it self, raises up other species lying hid there, and so induces Memory with Phantasie." Although by cutting a nerve no fluid exuded and by ligating it no swelling occurred, Willis still believed in its existence; he justified the failure of traditional tests by arguing that the fluid is very subtle and provided additional evidence from pathology about its existence and role in nutrition.[6]

Willis and his collaborators dissected and vivisected a large number of animals of different species, systematically comparing animals and humans. One such comparison had a philosophical import: Willis noticed that the pineal gland is found not only in humans and four-footed animals but also in birds and fishes, animals that in his opinion seemed so destitute of higher functions such as imagination and memory that it looked most unlikely that the pineal gland could be the seat of the soul. Clearly Descartes was his target here, although his name was not mentioned. In a letter to Malpighi, however, Borelli professed to be scandalized at Willis's "Cartesian rashness" that he had detected right at the opening of his treatise, which at that point he had looked at only superficially and "a salti." Possibly Borelli had in mind the passage in the dedication to Gilbert Sheldon, Archbishop of Canterbury, in which Willis expressed his intention to "unlock the secret place of Mans Mind." Thus, by "Cartesian rashness" Borelli seems to have meant Willis's attempt to account for the higher operation of the brain through anatomy rather than the adherence to key aspects of Descartes' system, such as the role of the pineal gland.[7] We are going to witness other challenges to Descartes' lack of mastery of anatomical detail and his philosophical theorizing in the works by Steno and Malpighi.

Nicolaus Steno was one of the leading anatomists of the time, one whose works of the early 1660s on the glands and muscles had left a mark on the anatomical world. Steno had been a pupil of Thomas Bartholin in Copenhagen and Franciscus Sylvius at Leiden. Sylvius in particular had proposed a new method for dissecting the brain by removing the cranial vault and proceeding first with the right and then with the left half, in situ, contributing important findings. Steno himself, while a student at

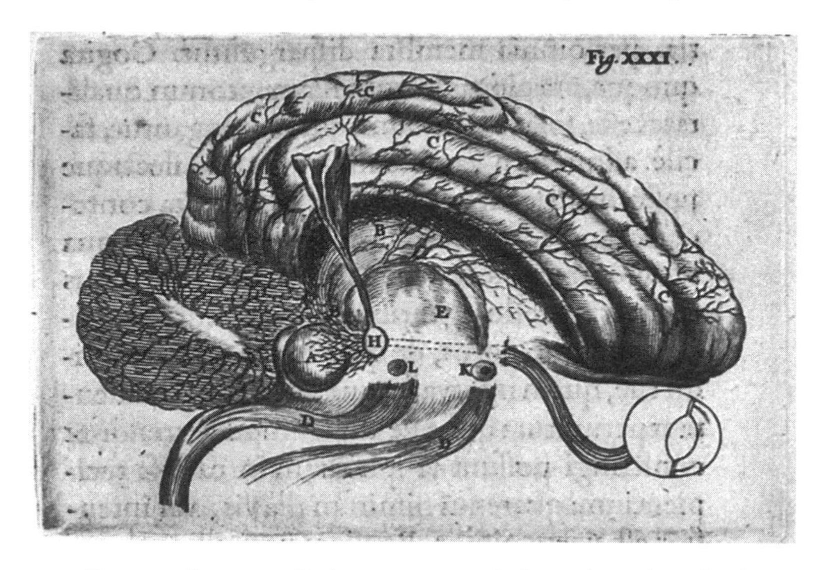

Figure 3.3. Descartes, *De homine*: section of a brain drawn by Schuyl

Leiden, was one of the first readers of Descartes' newly published *De homine*. In the letter to his teacher Thomas Bartholin, Steno expressed his admiration for the illustrations in *De homine* and for Descartes' brain, doubting however whether the objects represented could be found in any brain. Bartholin, by contrast, was more sympathetic to Descartes and—amusingly—thought that the dark color of the pineal gland supported Descartes' views about vision, since external images are clearer on the wall of a dark room. The copper engravings to which Steno referred in *De homine*, however, were not due to Descartes—since the originals had been almost entirely lost—but to the work's editor, the philosopher turned medical professor and curator of the botanic garden at Leiden, Florentius Schuyl: figure 3.3, showing a so-called medio-sagittal section of the brain with the pineal gland at H, seems to be a compromise between a realistic anatomical drawing and an "expository schema" for a theory, as if Schuyl had wished to dress Descartes' mechanistic views with anatomical robes.[8]

In 1665 at Paris Steno delivered a lecture on the brain at the academy of the learned diplomat Melchisédec Thévenot, *Discours sur l'anatomie du cerveau*, later published in 1669. The topic was ripe in the aftermath of Descartes' and Willis's works, and Steno was the right person to address it. In a Socratic manner, he addressed the members of Thévenot's academy by professing his ignorance. Steno highlighted the difficulties in dissecting the brain and the pros and cons of different methods. Steno was reluctant to go beyond admitting the presence of a white substance continuous with the nerves and a gray matter or cortex enveloping it and separating some of its filaments. He expressed dissatisfaction with the state of knowledge about the nature of the com-

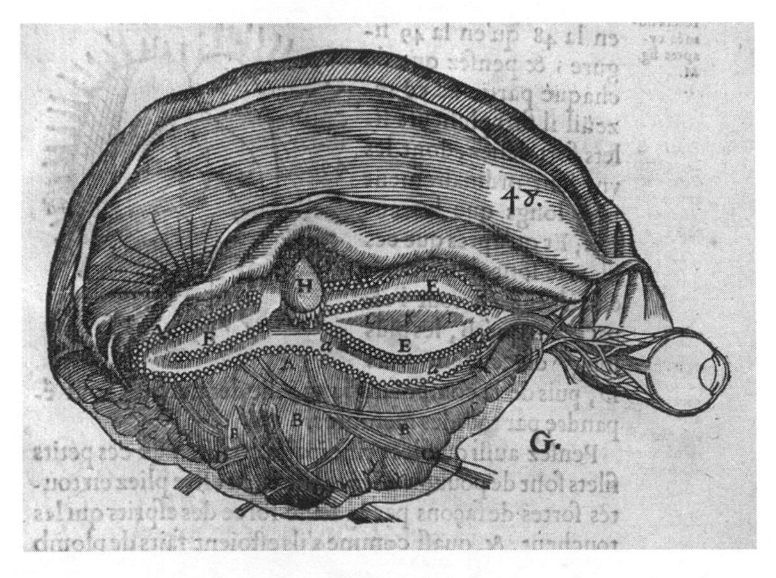

Figure 3.4. Descartes, *L'homme*: section of a brain drawn by van Gutschoven

ponents of the brain, the connections between the nerves and the white substance, and the nature of the ventricles. In a striking analogy, Steno argued that the brain is like a machine and in order to understand how it works it is necessary—unless the artificer were to reveal the artifice—to disassemble it into its minutest components and see what they can do separately and together: the belief that the brain works like a machine led him to advocate a method of inquiry appropriate for machines. As to the figures in the available literature, Steno judged Willis's to be the best, but even those were inaccurate in many respects, as he proceeded to discuss in considerable detail. By that point Schuyl's 1662 Latin translation of Descartes' work had been supplemented by Claude Clerselier's 1664 edition of the French original, with woodcuts by the la Flèche professor Louis de la Forge and the Louvain anatomist Gerard van Gutschoven. Figure 3.4 shows another rendering of a section of the brain by van Gutschoven, with the pineal gland H, where D are threads, which can be taut or relaxed depending on whether one is awake or asleep, for example. Schuyl's engraving seems to have inspired van Gutschoven's woodcut, although the former attempted a more realistic representation, as suggested also by the use of engraving, whereas the latter seemed more concerned with illustrating a theory, as highlighted by the use of woodcut and the omission of the cerebellum at the back of the brain, on the left. Indeed, Clerselier challenged the figures in Schuyl's edition, arguing that Schuyl had provided images giving the erroneous impression that they were anatomically correct, whereas in fact they should have been clearly presented as illustrative schemas. Steno's own rather enigmatic figures were produced in pairs, with a line drawing and a shaded version for

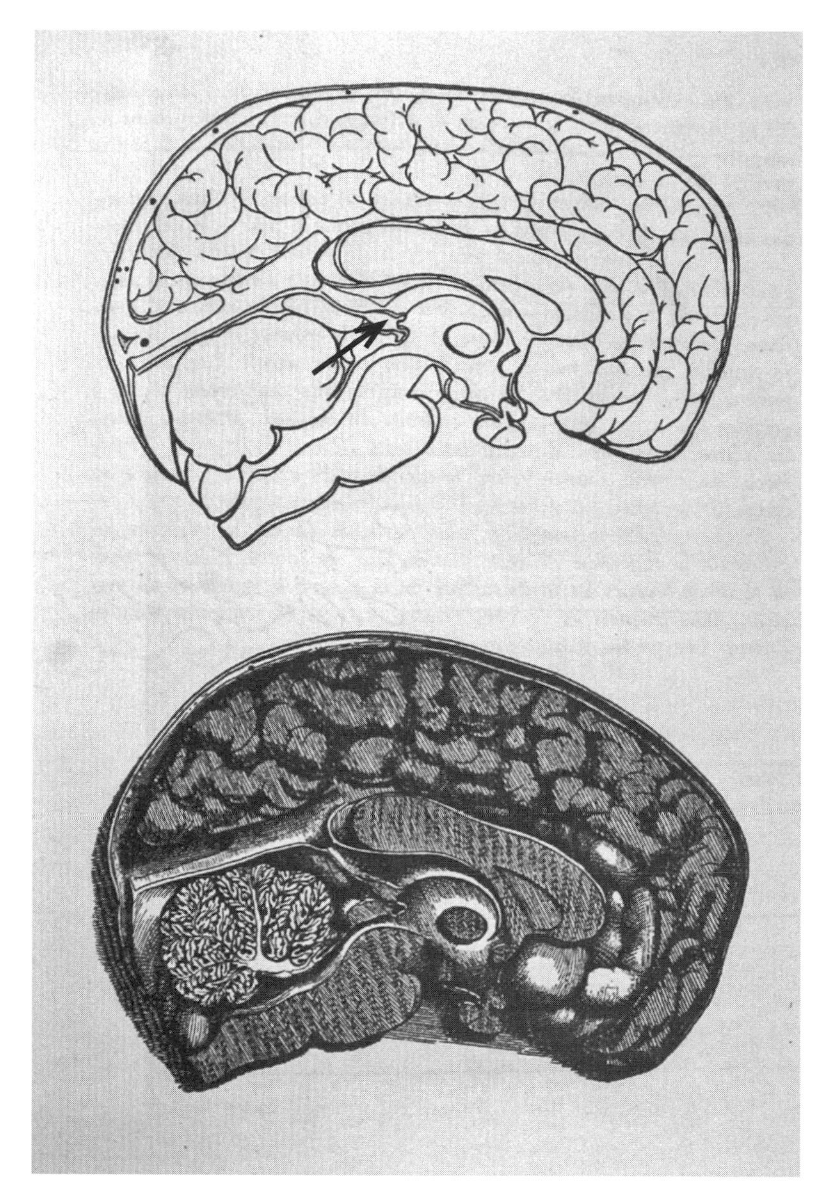

Figure 3.5. Steno, *Discours sur l'anatomie du cerveau*: section of a human brain

each image. They appeared without captions in the 1669 *editio princeps* of his lecture and were altogether omitted from the 1671 Latin edition. Figure 3.5, from Steno's *Discours*, shows a section of the human brain, in which I have indicated the pineal gland. Although figure 3.5 is quite accurate and represents Steno's own views, figure 3.6—also from Steno's *Discours*—appears as an attempt at an anatomical rendering of

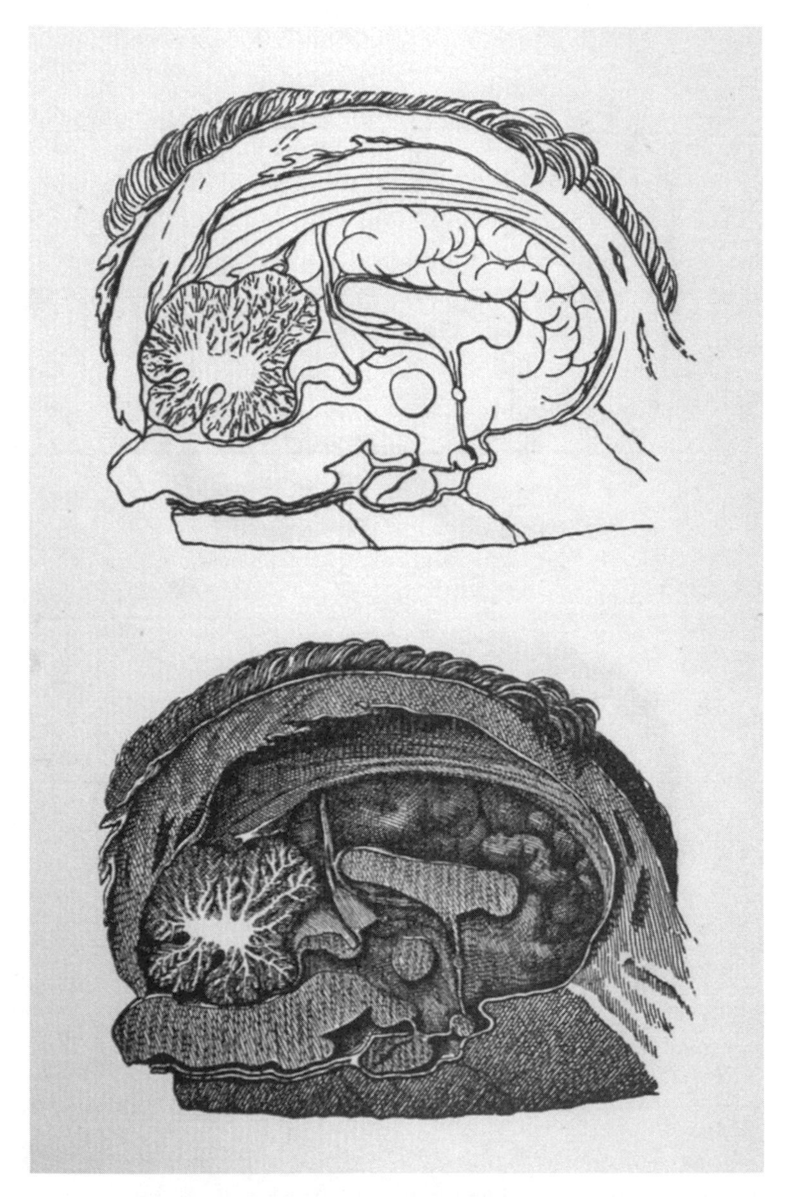

Figure 3.6. Steno, *Discours*: section of a human brain in Cartesian fashion

figure 3.3 from *De homine*, highlighting several inaccuracies: notice in particular the erroneous inclination of the pineal gland. Thus, figures 3.3–3.6 form a set of related images and display a wide range of epistemological roles and forms of representation requiring careful examination: figure 3.5, by Steno, is an attempt at an accurate anatomic drawing; figure 3.4, by van Gutschoven, seeks to provide an illustration of

Descartes' views, with no pretense toward anatomical accuracy; figure 3.3 by Schuyl and figure 3.6 by Steno are hybrids, partly didactic or illustrating a theoretical point and partly anatomical, yet they had opposite purposes—Schuyl gave the impression that his rendering of Descartes was anatomically sound, whereas Steno wished to highlight the contrast between anatomical accuracy and Descartes' fanciful views.[9]

Like some of his contemporaries, Steno focused on the pineal gland, which had a key role in Descartes' system at the interface between the immaterial soul and the body. Steno tried to present himself as a defender of Descartes, arguing that the Frenchman was no anatomist and had put forward only an ingenious proposal with the idea of a statue entirely of his invention resembling the human body. Although not questioning Descartes' mechanistic standpoint, Steno questioned the specific account provided in *L'homme*. Alas, Steno argued, many are taking Descartes' treatise as a faithful description of the human body, and this was something he was not prepared to do. He quoted key passages on the pineal gland from *L'homme*, refuting them one by one: for example, he argued that the gland cannot incline in any direction because it is fully blocked on all sides; the arteries surrounding it, which would have an important role in the separation and motion of the spirits, are in fact veins. He further questioned methods of insufflation and other invasive procedures that altered the internal configuration of the brain. Concluding his treatise, Steno highlighted that both the structure and the operations of the brain remain obscure. Figures are especially difficult to draw because the brain collapses or subsides; therefore, it is necessary to have several brains to draw a single figure, a sign that he did not rely on fixation methods. Generation, diseased state, and injections can be profitably used in the investigation of the brain; Steno advocated dissecting especially animals because some portions of their brains can be clearer than in humans and because he argued that we can treat animals the way we wish, performing vivisections and surgical procedures and administering drugs and various substances to study their effects.[10] Steno's conclusion is a valuable account of the tools and procedures in the arsenal of a seventeenth-century anatomist.

Steno's reflections were unknown to Malpighi when he published *De cerebro*. Malpighi, however, was aware of Descartes' *De homine* and Willis's treatise, which he feared may preempt his own findings. In the end, however, Borelli reassured him that *Cerebri anatome* did not anticipate his work and Descartes was not worth reading. As Malpighi put it in a letter to his friend Silvestro Bonfiglioli, his own work consisted of two main themes: the structure of the brain and of the optic nerve. In both cases the swordfish proved crucial because the fibers of its brain are especially visible and its optic nerve has a very peculiar structure: the swordfish was to the brain what the frog had been to the lungs. Malpighi argued that the brain consists of two parts, a white medullary substance and the cortex. The swordfish enabled him to determine that the

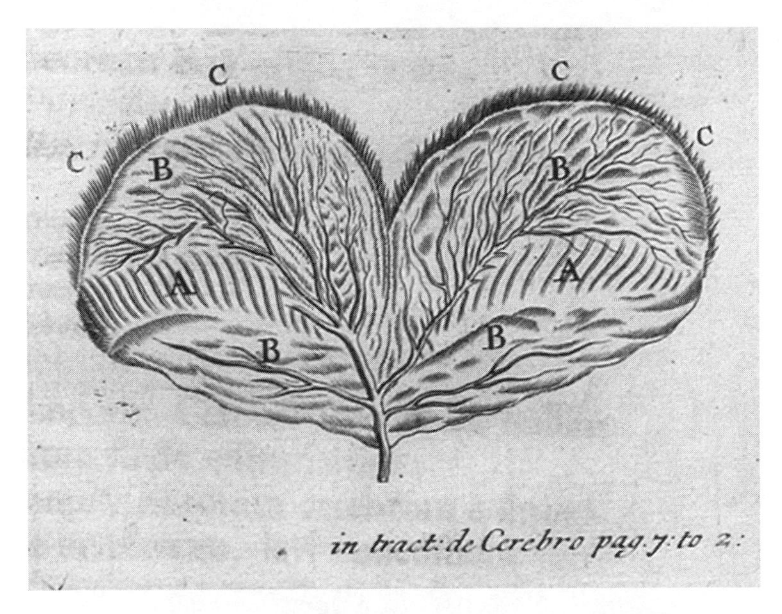

Figure 3.7. Malpighi, *De cerebro*: ventricles of the fish brain

former consisted of a mass of white fibers resembling "small intestines" or even those structures identified by Auberius in the testicles. The trunk of those white fibers is the spinal medulla, terminating in the cerebral cortex. In his treatise Malpighi compared those fibers to an ivory comb or the parallel reeds of an organ: figure 3.7 from *De cerebro* shows the fibers at A. Malpighi sought to determine whether those fibers were canals, and whether some juice is filtered in the brain, notably in the cerebral cortex. We witness here another connection between anatomy and the study of disease: as in the *Risposta* to Lipari he had mentioned the brain of fishes, in *De cerebro* he mentioned several case histories pointing to the separation of serum in the brain; one of those cases involved epilepsy, which he believed was related to apoplexy (see section 2.4). Malpighi further observed that, if not one-half, at least one-third of the blood in the body goes to the brain, and he suggested a purpose for such an abundant supply, namely, the serum of blood percolates through the small fibers of the brain into the nerves. Although the white fibers were the main focus of *De cerebro*, the reference to the possible role of the cerebral cortex envisaged the project that was to be developed in *De cerebri cortice*.[11]

Malpighi may have consulted the chapter on vision in Gassendi's *Syntagma*, which relied on large fishes like tuna to provide a description of the structure of the eye. Malpighi's discovery, his discussion of the problems of vision, and the copper plate accompanying his text attracted the attention of contemporaries: the copper plate from *De cerebro* with a portion of the optic nerve of the swordfish shows an outer

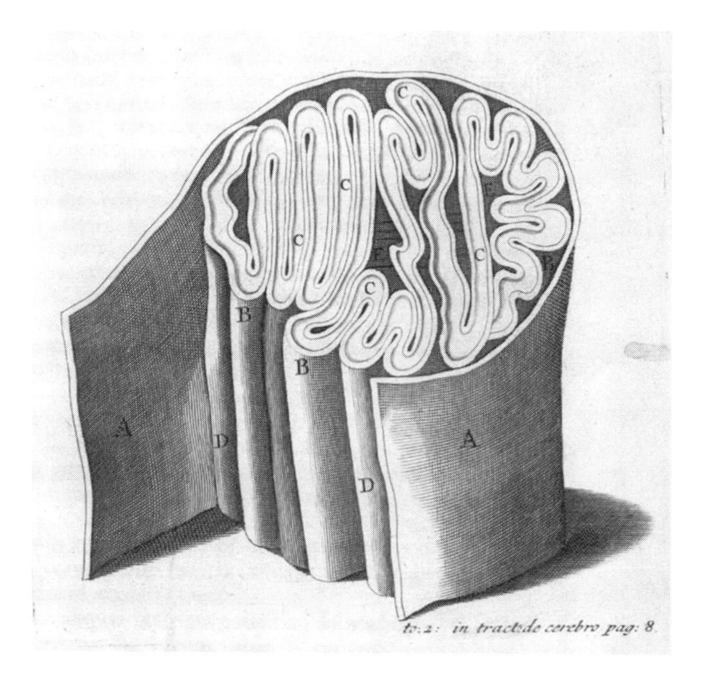

Figure 3.8. Malpighi, *De cerebro*: optic nerve of the swordfish

membrane enclosing the optic nerve, which appears folded like a piece of cloth (fig. 3.8). In his first letter to Malpighi, the secretary of the Royal Society Henry Oldenburg mentioned only this one among Malpighi's many findings, highlighting its impact.[12]

Malpighi's discovery was also discussed at the Medici court, where his originality was challenged by the English anatomists John Finch and Thomas Baynes, who reported a passing claim by Bartolomeo Eustachio. It was probably as a result of that controversy of 1664 that Malpighi started a new project, a historical defense of the primacy of Italian anatomy, or *Anatomia italica.* Malpighi was soon persuaded by Borelli and Fracassati to abandon historical writing and return to the dissecting table and the microscope. This episode is significant in showing the oscillations in Malpighi's projects and styles, as well as how much he relied on external guidance in his work. Indeed, the crucial passage on the optic nerve of the swordfish also owed much to Borelli, who prompted Malpighi to present a detailed and unambiguous refutation of the theory of vision put forward by Descartes' 1637 *La dioptrique.* First, Malpighi had to clear the way from a striking anomaly in his belief in the uniformity and regularity of nature. This anomaly consists in the different structure of the optic nerve of the swordfish and of higher animals, where the optic nerve consists of a thin bundle of fibers. Malpighi argued that the considerable size of the eye and retina of the swordfish, as well as tuna and large fishes in general, requires a great number of nervous fibers and this explains their folded structure. He then outlined Descartes' ideas on nervous structure, nervous transmission, and vision. The optic nerve consists of three parts, according to Descartes: first, an external membrane propagating from the brain to the eye; second, thin threads unfolding through the whole length of the optic nerve from the retina to the brain; lastly, the animal spirits, which propagate like a wind from the brain to the eye. Vision ensues when the soul situated in the pineal gland perceives the images produced in the brain by the motions of the threads, such as shaking and vibrating. Malpighi argued that such ingenious views were fraught with difficulties and proceeded to their refutation based on a survey of anatomical structures. The optic nerve, for example, does not terminate in the position of the brain indicated by Descartes' figure, as in figure 3.9, from *La dioptrique*, where points

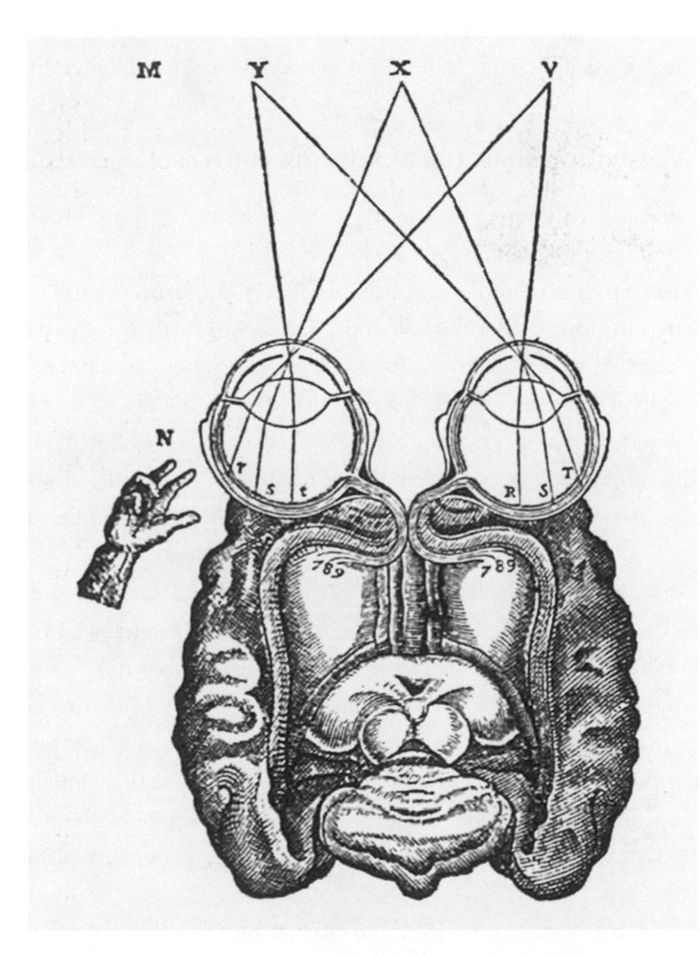

Figure 3.9. Descartes, *La dioptrique*: vision and the brain

VXY would go to points 789; therefore, visual sensations cannot easily reach the pineal gland. Further, inside the optic nerve of the swordfish there are no threads, whether tense or loose, surrounded by animal spirits. Indeed, almost anything can be conceived except distinct threads free to shake and vibrate. Malpighi added that even in more perfect animals if the nervous substance is exposed to the flame, it does not curl in threads, but hardens in a thin membrane: presumably this test would cast further doubts on Descartes' views. Malpighi systematically used visual inspection, taste, fire, and other chymical analyses in both anatomy and the study of morbid affections. Moreover, he argued that just before the retina all portions of the optic nerve or fibers, "si quae sint," or if some exist, were joined together, hence their motions could not be independent, rendering Descartes' mechanism for vision untenable. Much as Steno had done in *Discours*, in *De cerebro* Malpighi challenged Descartes, thus showing how deeply he knew Descartes' works and how closely related were the problems they were addressing.[13] Although once again Descartes' views were refuted by the anatomical evidence, the process of nervous transmission for vision remained a mystery.

3.3 Malpighi's Anatomical Findings on Taste and Touch

Malpighi's publication strategy shows a growing distancing from Borelli. While Malpighi discussed the contents of *De omento* and *De cerebro* with his Pisa mentor, Borelli received a copy of the manuscript *De lingua* already in rather advanced form, although in time to allow Borelli and Fracassati to carry out observations on the tongue inspired by Malpighi. As for *De externo tactus organo*, Malpighi had it published without informing Borelli or even his Messina friends such as Catalano, although it was dedicated to Viscount Ruffo, also a close friend of Borelli's but someone Malpighi must have trusted to keep things to himself; Borelli's letter poorly conceals his astonishment on receiving a publication by Malpighi he had heard nothing about.[14]

Their correspondence reveals important exchanges on the structure and mode of operation of the tongue. In a letter of 19 July 1664 Borelli had proposed an experiment to ascertain whether the top layers of the tongue are porous or not: once removed from the tongue, he intended to wet the upper side with a very bitter or salty juice, tasting with his own tongue the opposite side to ascertain whether any portion was seeping through. As in the case of the lungs, Borelli referred to the Cimento experiments on thin tubes or "sifoncini" to argue that water can rise or move inside them; therefore, all that was required for a fluid to enter the inside portions of the tongue were pores.[15]

It is possible that Malpighi wished to investigate another sense organ in the wake of his research on the optic nerve. To this end, he studied the tongue as a whole, whose nature and mode of operation were obscure; in this case, however, unlike that of sight, his anatomical findings enabled him to draw some positive conclusions on taste, especially the identification and localization of taste buds. The engraving in *De lingua*—possibly due to the Messina painter and naturalist Agostino Scilla, whose help Malpighi sought at the time—shows at the top a bovine tongue with five lines across its width indicating the exact places where it has been cut (fig. 3.10 reproduces Malpighi's original plate: see *figura* I): the subsequent five *figurae* show the structure of muscle fibers at the precise locations of the cuts; this is one of the few instances in which he studied myology. The different orientations of the muscles enable the tongue to move in many different ways; the muscles in *figura* II.1 show a striking analogy to a textile, as pointed out by Malpighi. *Figura* III.1 shows the base in transverse section. While denying any peculiar nature to the tongue, Malpighi was prepared to admit that the presence of fat and some glands made its taste—in a curious twist—especially pleasing.[16]

Having ruled out a glandular structure for the tongue, or one peculiar to that organ, Malpighi proceeded to investigate the mode of operation of taste. Boiling the tongue enabled him to remove two surface layers; the outer one is covered with several tiny protuberances, cartilaginous in bovines and bony in fishes. Such protuberances are of two types: some, at the apex of the tongue, are shaped like wild boar fangs; others, at the base of the tongue, are flattened. The latter present a cavity in the middle, whereas the small size of the former prevents the eye from determining this issue with certainty. Removing with the nails this first outer layer, Malpighi detected an additional reticular layer or membrane with holes of different sizes and shapes, the smallest of which are accessible only through the microscope by tearing it and observing it against the light. This reticular layer is white toward the outside and tends to the black on the other side.[17] Under this reticular layer Malpighi identified a yellowish and whitish papillary body or membrane covering the upper portion of the tongue with papillae or nervous receptors, which in oxen, goats, sheep, and humans can be divided into three different orders (fig. 3.10, *figura* III.2): those of the first order, resembling snail horns or fungi, are connected at the base to a nerve, as Fracassati had confirmed, and open through the two covering layers to the surface of the tongue; those of the second order, more numerous than the preceding ones, resemble horns and correspond to the protuberances of the outer layer. His *figura* III.3 shows a papilla of the first order and other papillae with two covering layers. Notice also at the bottom left in *figura* II.4 the branch *I* of the nerve ending in the nervous papillae, visible in *figura* III.2 as well. Having compressed the papillae, Malpighi rejected the claim that those of either the first or second order emit any fluid—hence they are not glands. After careful observation under the microscope, he detected in the larger ones small appendices resembling hairs. Lastly, he detected a huge number of papillae of

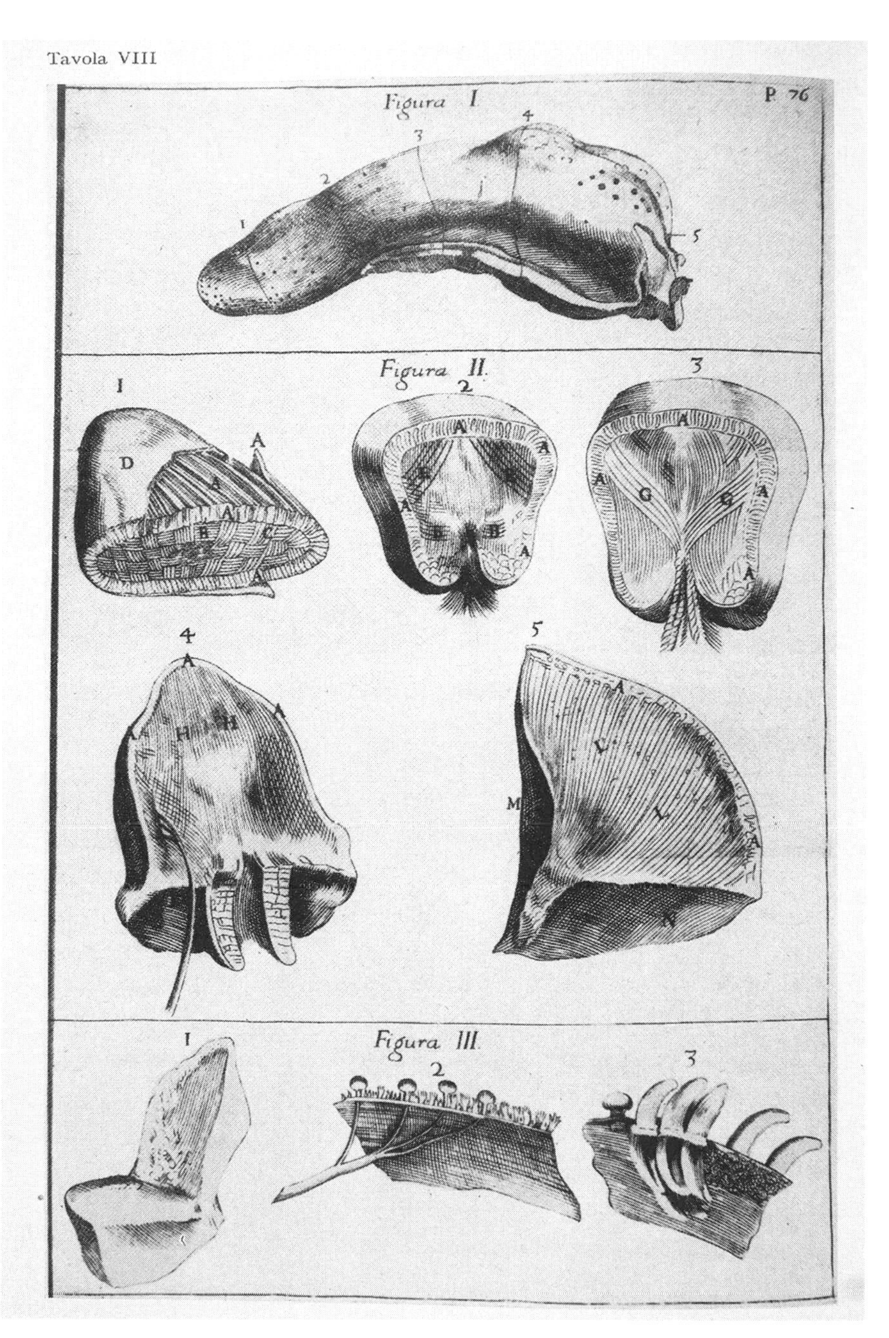

Figure 3.10. Malpighi, *De lingua*: muscles and papillae of the tongue

the third order, reaching the same height as the previous ones but thinner and with a conical figure.[18]

Having discussed the structure of the tongue, Malpighi denied that its flesh could be responsible for sensation, both because of its muscular nature and because of the layers covering it. He further denied on the basis of a careful analysis Thomas Wharton's view, according to which the tonsils are the organ of taste, because of the lack of suitable nervous connections, and the view put forward by André du Laurens, according to which taste pertained to the outer layers, because nature would not have made them pervious: Malpighi relied on a close analysis of structure in order to determine the mode of operation.[19] He then proceeded to identify the sensory receptors in all three orders of the papillae with their nervous connections to both the surface and basis of the tongue. In an important passage, he argued that salts and similar foods, dissolved in saliva and other fluids, hit the nervous receptors in different ways depending on their motions and shapes. The source Malpighi quoted at this point was Plato's *Timaeus*, which indeed discusses taste along similar lines, albeit through passages from the tongue to the heart. Luigi Belloni identified additional analogies with a celebrated passage from Galileo's *Assayer*, a text with which Malpighi was familiar, as mentioned in the previous chapter. More generally, sources in the atomistic tradition going back to Democritus follow a similar pattern in attributing bitter, salty, or acid tastes to the shape of constituent particles or atoms.[20] Malpighi ended by denying the Peripatetics' claim that a medium is required for taste, since the sapid body touches immediately the nervous receptors with no need for intermediaries, as shown by the case of Augusto Corbetta—reported by Gerolamo Cardano—who felt pain sampling pepper rather than its taste. The issue of whether Aristotle himself believed taste to require a medium is quite entangled and requires careful handling. Aristotle, however, stated that taste is a form of touch, and in this respect he was closer to Malpighi.[21]

De externo tactus organo, the companion to *De lingua*, had a checkered publication history: Malpighi states that it appeared at Messina with a false Naples imprint. Three footnotes in the opening pages may suggest what happened. On all three occasions Malpighi explained that references to "necessity" and "fortune" should be interpreted in terms of the Catholic doctrines of free will and divine providence. It seems plausible that an overzealous local censor raised problems and Malpighi—possibly with the help of the dedicatee Iacopo Ruffo—cut some corners and had the book promptly printed at Messina anyway.[22]

The opening of *De externo tactus organo* reports a curious episode: Malpighi states that in the wake of his observations on the tongue, while observing his fingertips with a microscope, he detected some round transparent bodies. Alas, they turned out to be sweat, but this episode highlights the uncertainty and problems of microscopy: in this case the initial shortcomings were swiftly overcome, although the broader issue of the reliability of the microscope remained. Malpighi proceeded to investigate the structure of the skin along lines similar to those he had used for the tongue: he boiled the hoof of a pig and, having removed the hard nail and an outer layer similar to that observed in the tongue, found oblong and almost pyramidal papillae resembling a sword extracted from its sheath. Malpighi observed other areas of the skin in different animals, such as the hoof and lip of oxen, the foot of birds, and the human hand. He removed the outside layer with a hot iron, then the reticular layer, to reveal the papillary body. On the basis of the analogy with what he had observed in the tongue and the location of the papillae of the skin in areas especially devoted to touch, Malpighi concluded that the structures he had uncovered were indeed responsible for this sense. Relying on the Stoic analogy of the soul with a polyp with seven tentacles responsible for the five senses plus semen and voice, Malpighi offered a material equivalent based on the structure of the brain and its nervous ramifications, which reach out to all the parts of the body. Nor was his research limited to the skin: in two passages he referred to his observations of the papillae and abundant innervations in the nostrils of pigs, involving smell.[23]

Malpighi's research had reached a level enabling him to provide a mode of operation of the sensory organs for taste and touch in considerable detail. He challenged Aristotle's enigmatic claim that the organ of touch is hidden inside the body for its failure to specify the exact configuration of the sense organ, which could not be reached by rough or smooth objects. Malpighi also challenged Gassendi's belief that the tension of the skin enables it to rebound when pressed upon and therefore to alter the motion of the spirits in the nerves. Since many portions of the skin are wrinkled and slack, he argued, the spirits are affected more easily and effectively by a direct action on the papillae, which are not appendices to the skin but nervous organs connected to the brain and spinal cord at one end and erupting from the skin at the other. A key term describing the nervous connection between the central nervous organs and the sensory receptors is that of "fiber," which was central to *De cerebro* and thus connects the three works. Although Malpighi reserved a much better treatment for Gassendi here than he had offered Descartes in *De cerebro*, both philosophers had become the subjects of his anatomical strictures. In the end, however, Malpighi mitigated his views by arguing that the nervous terminations in the skin almost change their nature as to protect the nerve from too direct a contact with the outer world. Thus, he compared the nervous papillae to the stick held in the hand by a blind man, in that they seem to extend in the fibers of the stick. The image of the stick was an old one, occurring in Simplicius's commentary on Aristotle's *De anima* and in the *Liber de oculis* attributed to Galen, besides Gassendi's *Syntagma* and Descartes' *L'homme* (fig. 3.11). Ultimately, however, Malpighi left the issue of the admirable *artificium* or

Figure 3.11. Descartes, *De homine*: blind man illustrating the sense of touch

mechanism whereby the animal feels, or of nervous transmission, to the dedicatee Iacopo Ruffo.[24]

3.4 Fracassati's Far-Reaching Investigations

Whereas Malpighi's works on the tongue and skin are related in the techniques of investigations employed and in their focus on the anatomical aspects of the problem of sensation, the works by Fracassati on the tongue and brain share common features not only for their references to philosophical problems of sensation but also for their far-reaching—and, in the case of *De cerebro*, almost disjointed—discussions of a wide range of anatomical, medical, and philosophical problems, such as the nature of color, the constitution of blood, and the growing interest in injections. His works shed light on the research and debates of Borelli's group at Pisa.[25]

In 1663 Fracassati had joined Borelli and Bellini at Pisa University as professor of theoretical medicine and in the informal meetings at the Tuscan court, where the Medici princes promoted research and debates on anatomical matters. Later, Fracassati, who had been trained in anatomy at the *Coro anatomico* with Malpighi and had also taught surgery at Bologna, performed public dissections at the anatomical theatre in Pisa and moved to the anatomy chair. *Tetras anatomicarum epistolarum* opens with Fracassati's dedication to Prince Leopold, and his two treatises are replete with references to experiments performed at the Tuscan court.[26]

De lingua opens with a discussion of preparation techniques and anatomy of the tongue. Much like Malpighi, Fracassati boiled the tongue of a calf and detected a large number of protuberances of different shape at different locations on its surface. He argued that those with a little head or *capitula* were the true papillae, which he identified as the organs of taste. Fracassati also claimed that Bellini had been the first to detect papillae in the human tongue. *De lingua* discusses the relationships between taste and the other senses, especially touch, and provides a wealth of observations on related matters. For example, Fracassati challenged Aristotle's claim that taste is a modification of the moist due to the dry under the action of heat: if it is so, why can we taste a frozen fruit?[27] Despite its title, the main focus of *De lingua* was not the anatomy of the tongue but rather the study of salts, including the geography of salt mines, and the relationship between their shape and taste. Fracassati believed that larger structures are formed by the aggregation of smaller basic units or modules: the same would be true for salts; snow; the liver, since a cooked pig liver shows rhomboid or cubic elements; the hexagonal cells of a beehive; the lungs, which consist of lobules as found by Malpighi; conglomerate glands, which consist of the aggregation of other glands as found by Steno; and blood, in which Boyle detected rhomboid elements. Thus, Fracassati identified a common mode of operation in nature extending over a remarkably wide domain.[28] Explicitly following Democritus, Fracassati argued that taste came from salt and studied the shapes of different salts, hoping to find a correlation with the corresponding tastes. For example, he argued that common salt is cubic, alum is an octahedron, niter is hexagonal, and vitriol is a rhomboid. Further, he reported the microscopic observations first carried out at the Tuscan court by the Cimento academician and medical professor at Pisa Antonio Oliva on a long list of the salts of plants obtained from the pharmacy of the Grand Duke; while they could determine that such salts purged, the correlation derived from the atomistic tradition between shape and taste failed to materialize. Francesco Redi took over Oliva's experiments, which developed in a different form, and published *Esperienze intorno a' sali fattizi* (Florence, 1674). We encounter here a problematic area at the juncture between philosophical program and experimental investigations. Salt mixtures in particular proved problematic but at the same time offered an explanation for the failure to detect the expected correlation: only original or primitive salts would have to be taken into account.[29]

In *De cerebro* Fracassati shifted his interests from the brain to an even wider range of issues. Whereas Malpighi's *De cerebro* amounts to just over six pages in the *Bibliotheca anatomica*, including the two figures, Fracassati's response reaches almost twenty. Here I am going to focus on just a few issues relating to the material covered in the present and the previous chapters and at the same time instantiating the breadth of Fracassati's interests. We have seen that although Malpighi attacked Descartes' account of the transmission of an image through the optic nerve, he refrained from providing his own account of how sensory nerves work. By contrast, Fracassati proposed a way based on the notion that the brain is a "pneumatic instrument" sensitive to differences of pressure in the sensory nerves, thus instantiating a mechanism for sense perception: his idea relies on a well-known experiment among Galilean circles, one performed in 1648 in Rome by Benedetto Castelli's disciple Raffaello Magiotti, who published in the same year a short treatise, *Renitenza certissima dell'acqua alla compressione*. The experiment consists in observing the behavior of small glass globules or *pillulae* partially filled with air and floating in a tube filled with water, closed at the bottom and with a small opening at the top (fig. 3.12). Pressing with the finger the water at the top, the floating globules move down, because air is compressible and the air they contain comes to occupy a smaller volume, but water is not and therefore their specific gravity increases. One may add that at the time there were similar devices, one of which worked on the basis of temperature rather than pressure differences: in this case an increase in temperature would make the small containers rise in the tube. Fracassati, however, refers to globules moving up and down simply as a result of pressure changes. Fracassati compared the nerves to the tube filled with water and the sensory experiences to the pressing hand making the small containers move and thus transmit the sensory perception to the brain at the other end. This account seems rather crude on several levels: its implicit reliance on gravity makes its application to sense perception problematic; moreover, it is hard to envisage how any sensation might be transmitted this way. However, it opens a window onto the views of a prominent member of Borelli's circle and an alternative to Descartes' shaking and trembling threads. Fracassati went on to argue that air is contained in the brain, as shown by the fact that animals in an air pump go into convulsion when the air is evacuated. In a classic essay on mechanistic anatomy in the seventeenth century, Luigi Belloni reproduced Magiotti's *Renitenza* and discussed its significance. Belloni identified passages from Borelli's *De motu animalium* discussing motor nervous transmission from the brain to the muscle, but whereas Borelli's example concerns only the transmission of pressure and lacks the globules moving up and down, Fracassati's

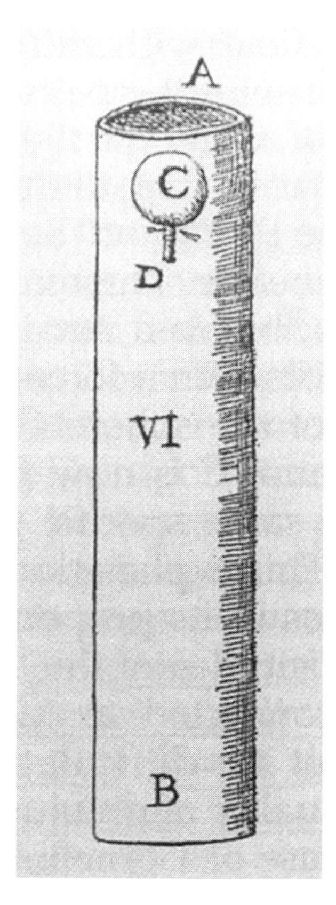

Figure 3.12. Magiotti, *Renitenza certissima dell'acqua alla compressione*

case is a rather faithful rendering of Magiotti's work in anatomy. The notion that the transmission of nervous impulses is electrical was a few decades away.[30]

The second issue I wish to address concerns the change of color of blood. There are two passages in *De cerebro* in which Fracassati addressed the problem. In the first he reported the experiment performed by Malpighi while at Pisa on the color of blood; we have already seen a variant of this experiment in section 2.5. According to the traditional doctrine, blood at the bottom of a container appears dark because it is rich in melancholia. But if it is thrown into a dish and mixed with air, it turns bright red. Fracassati, much like Malpighi, took this simple experiment to disprove the alleged separation of the constituent humors of blood and indeed to question the existence of the four humors. Fracassati and Malpighi attributed the change of color of blood to the mixing with air but, far from considering this as a significant phenomenon, regarded it as showing that color is not a valid indicator of the nature of a substance. Ironically, it was precisely this experiment as reported in the *Philosophical Transactions* that alerted English natural philosophers to the significance of air in the color change of blood. In a later passage from *De cerebro*, Fracassati reported experiments of mixing several substances with blood to make it purple, ending with the italicized words *ne crede colori*, or beware of color: his rejection of its philosophical significance could not be more explicit.[31]

The last issue I am going to introduce concerns the so-called *medicina infusoria* or the new practice of injecting liquids into the veins of animals in order to study their effects on blood and also to find new cures to apoplexy and epilespy. The German physicians Johann Daniel Major and Johann Sigismund Elsholtz published works on the new method, *Prodromus chirurgiae infusoriae* (1664) and *Clysmatica nova* (1665), respectively, about the same time as *Tetras anatomicarum epistolarum*. Their works were eagerly read in England too, where injections had already been performed by Christopher Wren, among others, as a paper in the *Philosophical Transactions* reminded its readers; the review of Fracassati's work in the same journal focused on these aspects. In fact, Fracassati and Silvestro Bonfiglioli had been performing injection experiments for a variety of purposes, such as congealing blood in the veins so that animals could be more easily dissected, but also curing diseases like apoplexy. An initial stimulus to such research had come from the work of the Dutch dilettante Lodewijk de Bils, who had become a celebrity with his new method for preserving cadavers and of dissection without spilling blood. As it happens, despite what Fracassati and his collaborators thought, de Bils bathed bodies in appropriate solutions but does not seem to have relied on injections. The Italian anatomists injected dogs with substances such as *aqua fortis*, spirit of vitriol, oil of sulphur, and oil of tartar and studied the effect on the blood's fluidity. Since postmortems had shown that apoplexy was often caused by a coagulation of blood, it was judged that a timely infusion of a substance rendering blood more fluid could prevent the disease. Fracassati's *De cerebro* discussed the causes and cures for apoplexy and examined several sources from Hippocrates to Prospero Martiano and Johann Jakob Wepfer: in line with Malpighi's *Risposta* to Lipari and *De cerebro*, Fracassati denied a role to phlegm and argued that apoplexy and epilepsy are due to a blockage of blood serum. This research shows a medical link with the study of brain anatomy and at the same time brings into focus an important aspect of the investigations carried out at Pisa under the auspices of the Medici.[32]

3.5 Bellini and Rossetti: Atomistic Anatomy of Taste and Touch

The last works we are going to examine are *Gustus organum* by Borelli's young protégé Bellini, who in 1662, at age nineteen, had already published a celebrated treatise on the kidneys, and Rossetti's *Antignome fisico-matematiche*, containing important passages on sense perception. Whereas Malpighi's *De lingua* occupies about three and a half pages in the *Bibliotheca anatomica*, and Fracassati's nearly twelve, Bellini's work reached thirty-one pages over fourteen chapters. Bellini did not have a lot more to say on strictly anatomical matters—although he did differ from Malpighi on some points—but sought to display his erudition and devoted more space to philosophical matters; the difference in the title reflects Malpighi's greater focus on the tongue as such, whereas Bellini comparatively devoted more space to taste. No doubt, the treatise was written under Borelli's close supervision: in the letter on the tongue addressed to Borelli, Fracassati referred to Bellini's treatise with the words "te consulente." A passage of *Gustus organum* narrates the origins of the work, when Borelli showed Bellini the manuscript treatise he had just received from Malpighi. In fact, Bellini claimed that initially Malpighi had overlooked the papillae, soon thereafter discovered independently by Malpighi, Bellini, and Fracassati, although Malpighi must have felt partly defrauded of this finding by Borelli's actions. *Gustus organum* includes a letter addressed to Malpighi offering support against the Messina Galenists.[33]

Under Borelli's guidance, Bellini built on Malpighi's findings and composed a treatise joining atomistic doctrines, recent anatomical findings, and experiment. Bellini identified two interpretive traditions on taste: one, based on the doctrine of Aristotle, according to which taste is a quality resulting from a mixture of elements; and the other according to which taste is due to the size and shape of particles with no role for qualities. Bellini attributed this opinion to the ancient doctrine going back to Moschus, the ancient Phoenician atomist, who was followed by a long list of philosophers from Pythagoras to Asclepiades of Bithynia, Leucippus, Democritus, and Epicurus, as well as the chymists in his own time, whose experiments were not rejected by Galileo, Gassendi, and Descartes. Bellini accepted Aristotle's view that water is insipid but challenged his claim that taste is a passion produced by the dry in the moist, which thus makes taste from potential to actual—Bellini quoted the relevant passage here. Rather, he believed that taste is due to salt, which is extracted from food through mastication; if salts are removed, ashes have no taste even on a moist tongue. Bellini, like Fracassati, also reported the result of experiments with differently shaped salts from a large number of vegetables obtained thanks to the Grand Duke's munificence, from licorice and black pepper to endive and pumpkin. He still clung to the notion that taste relates directly to shape, since salts with more obtuse angles are smoother to the tongue.[34]

With chapter eight Bellini moved to anatomical matters. He argues that the organ of taste is neither the fleshy part, nor the membrane covering it, nor the glandular portion or the tonsils—as claimed by Wharton—or nerve endings: taste occurs only in the mouth, whereas nerve endings can be found elsewhere, and salt poured onto a wound produces pain rather than taste. Working on a range of animals including wild boar, deer, oxen, and man, Bellini used the same techniques employed by Malpighi and reached similar conclusions in many respects: the tongue is a muscle covered by a membrane that under the microscope appears porous like a sieve or cloth; salty bodies access the fungiform papillae through those pores. Bellini, however, in an interesting comment on the effect of techniques of preparation, argued that there is a single membrane covering the tongue; if it looks double, this is an effect of cooking, which leads to a delamination or separation into multiple layers. In the opening of the last chapter on the identification of the purpose of the fungiform papillae, Bellini mentioned Borelli and the discussions they held: once again, Borelli appears in the role of guide or interpreter of the purpose of the structural findings detected by the anatomist. Relying on the microscope, Bellini studied the papillae and, noticing their nervous connections, came to suspect that they are the instrument of taste. A striking experiment confirmed his view: placing sal ammoniac at the back of the tongue, where there are no papillae, no taste was detected; however, if it is placed at the tip of the tongue, taste did occur. Bellini drew a comparison with other senses too, such as smell, which works through the mammillae in the nose.[35]

In 1667 the Pisa professor of logic Donato Rossetti published *Antignome*, a book in dialogue form whose title can be rendered as "unorthodox opinions" and which indeed contains a number of surprising claims, such as that the earth has a heart. *Antignome* is dedicated to Cardinal Leopoldo de' Medici, although it also contains a dedication letter to Borelli and Bellini, who figures quite prominently in the text. The first dialogue deals largely with the problem of sensation and puts forward some intriguing opinions. After having claimed, following Gassendi, that the senses cannot deceive us, Rossetti went on to argue that there are eleven senses and corresponding sense organs. The first four are unproblematic: they are sight, hearing, smell, and taste; as to taste, he argued that Bellini's *Gustus organum* had left nothing else to be said on the matter—a statement not designed to please Malpighi. Pouring salt onto the wound, Rossetti went on to claim that although Malpighi did discover the dermal papillae—which Rossetti called "papille Malpighie" in his honor—he had rushed into print prematurely in order to gain priority. The next six senses decompose touch into different parts, based on the idea that feeling whether the surface of a body is rough or smooth is quite different from feeling whether it is hot or cold. These senses detect the following: whether a body is rough or smooth, from the ridges on the skin (5); heat and cold, through the porosity of the skin (6); the hardness or softness of a body, through Malpighi's papillae, which work like a nail going through a soft body like wax or being stopped by a hard one (7); size or the breadth, length, and depth of figures (8); the perimeter or contour of figures (9); and the recognition of larger and smaller sizes (10). Lastly, Rossetti introduced a sense dealing with change of place (11), although he argued that the feeling of pain did not require a specific sense because it resulted simply from the excision of a nerve. Although some of the senses he identified may appear unconvincing, his observation that different sensory receptors are at play for feeling the surface, temperature, and hardness of bodies, for example, was not entirely new, since it can be found in Aristotle's *De anima*; one wishes that Rossetti had more securely anchored his intuition to reliable anatomical investigations.[36]

Not surprisingly, Malpighi saw Borelli's hand at work behind Rossetti's rather insulting comments: the senior scholar was taking his revenge at Malpighi's decision to publish *De externo tactus organo* at Messina without consulting with Borelli and his Pisa group, avoiding the ensuing delays and priority issues. Rossetti's *Antignome* led Malpighi permanently to break off his correspondence with Borelli in mid-1668, right when the secretary of the Royal Society Henry Oldenburg had started a correspondence with Malpighi, thus taking over Borelli's role of intellectual guidance.[37]

The discovery of new features in the brain and especially the sense organs helped to provide a more accurate picture of the nervous system and bring the study of sensory perception to a new level of understanding. However, many problems remained. Descartes' views on the pineal gland led to the investigation of its anatomy in relation to its alleged role at the interface between body and soul. Steno's and Willis's critical reflections undermined Descartes' views but did not lead to a consensus on the brain's operations; in addition, Steno's work raised important methodological issues on the techniques of investigation and representation appropriate for the brain.

Atomism emerged as the key philosophical tradition underpinning the research by Italian anatomists, such as Bellini and Rossetti. With atomism came the problem of reconciling the shape and size of atoms with the specific sensations they induce in the sensory receptors. Fracassati's unease with the link between shapes of salts and the taste they induce highlights some of the difficulties faced by seventeenth-century anatomists working in this area. In addition, his statements in *De cerebro* enrich and confirm those passages we have encountered in the previous chapters about the anatomical and medical role of color: in the Pisa tradition sensible experience could rely on size and shape, whereas color and taste were deemed unreliable indicators of the nature of substances. Once again, philosophy, anatomy, and medicine interacted with empirical investigations in creative and problematic ways and in different contexts.

Anatomists were able to pinpoint sense receptors in the tongue, skin, and nostrils and to explain how they were affected by external stimuli. Anatomical research of this period, however, was more successful in undermining current views on the operations of the brain and the mechanisms of sense perception than in establishing viable alternatives. The case of the optic nerve of the swordfish is emblematic: while refuting Descartes' claims about the transmission of the image from the retina to the brain, Malpighi was unable or unwilling to propose an alternative on how vision occurs. Willis's localization of brain functions was considered fanciful and indeed impossible by Borelli, while Fracassati's idea that the brain is a pneumatic machine and views on the operation of sensory nerves based on Magiotti's experiment seem to have been quietly dropped. Thus, mechanistic anatomists found localizing nervous receptors easier to handle than unraveling the operations and mechanisms at play in the operations of the brain and in nervous transmission—a similar situation to the one on the localization and mode of operation of glands, as we shall see in the following chapter.

PART TWO

SECRETION AND THE MECHANICAL ORGANIZATION OF THE BODY

Glands as the Centerpiece of Malpighi's Investigations

At the beginning of the seventeenth century glands occupied a minor area in anatomy. Thomas Wharton's 1656 treatise specifically devoted to them, *Adenographia*, marked a turning point after which glands gained a central position on the anatomical stage and retained it for many decades. Although Wharton's work provided a description and taxonomy of the glands of the entire body and functioned as a catalyst for further investigations, later anatomists provided a different classification and understanding of their operations. Franciscus Sylvius introduced the distinction between conglomerate, or composite, and conglobate, or simple, glands—associated with the lymphatic system. Many anatomists identified a wealth of glands unknown to Wharton, such as the lachrymal glands, the parotids, and other glands in the digestive tract and reproductive organs. Crucially, the revival of glands and the identification of many organs as glandular were major aspects of the new anatomy, enabling a mechanistic explanation of many operations. Whereas Wharton believed that glands selectively attracted a fluid from the nerves, Steno and most later anatomists believed that glands mechanically filtered arterial blood. Malpighi's microscopic investigations led to a reclassification of body parts such as the liver, cerebral cortex, kidneys, and spleen as glandular, thus operating mechanically.

Glands were the main—although not the only—body part attracting the attention of anatomists with regard to the body's organization, however. On the one hand, fat proved a challenge to the mechanistic thinking of Descartes and Malpighi, not so much for its origin, which Malpighi located in the conglobate glands, as for its purpose. On the other hand, blood became a focus of attention both because it was the fluid filtered by glands and because of the recent spate of investigations on blood transfusions involving animals and humans, a procedure with significant anatomical as well as medical implications.

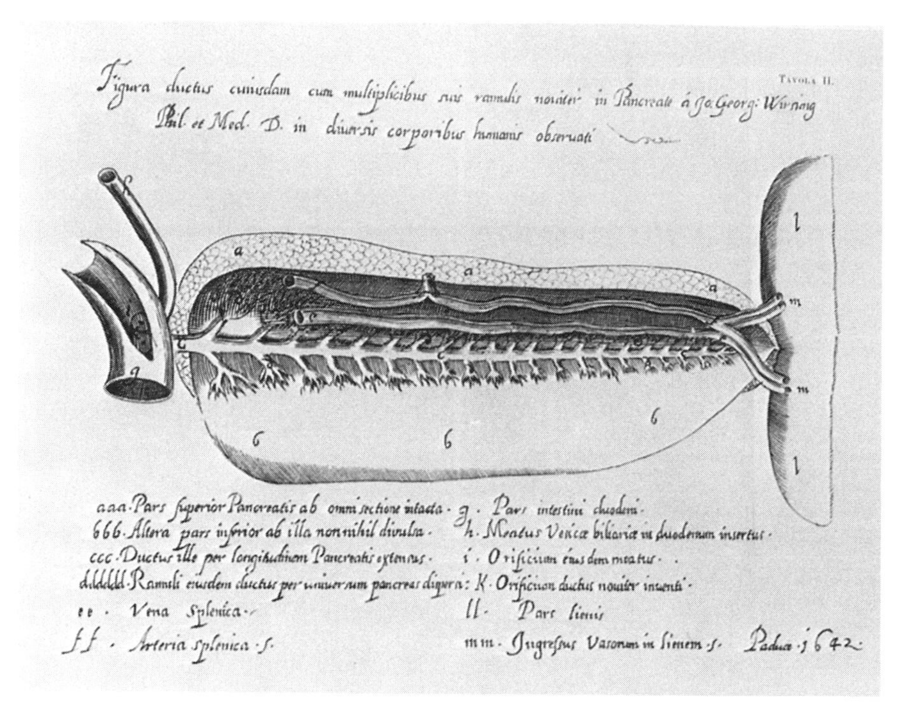

Figure PII. Johann Georg Wirsung: pancreas with its duct

Here we shall witness the growth of a range of techniques of investigation aimed at grasping the structure and mode of operation of glands: besides the refinement and creative deployment of techniques that we have seen in part I, such as microscopy, injections, vivisection, and the microscope of nature, they involved maceration in water and other liquids, staining, chymical analysis, and the reliance on pathology, whereby diseased states hardening and enlarging the affected body parts made them witnesses to the anatomical structure and especially the glandular composition of many organs. I have called these techniques "mining for stones" and "the microscope of disease."

In the 1664 *De musculis et glandulis*, Steno traced the very beginning of the revival of glands to a single sheet, published in 1642 by the Padua prosector Johann Georg Wirsung, announcing the discovery of the pancreatic duct. The sheet, which was circulated among distinguished anatomists of the time, shows a sausage-shaped pancreas between the spleen *ll* on the right and the intestine *g* on the left, where *h* is the bile duct. The pancreatic duct *cc* with its ramifications *dd* runs across the length of the pancreas, below the artery *ff* and the vein *ee*. Although several inadequacies in this rather primitive plate were soon pointed out, following Steno, we can take it as marking a first stage of the revival of glands in the seventeenth century.

The Glandular Structure of the Viscera

4.1 The Revival of Glands

The development of novel techniques of investigation complementing those we have seen above, such as microscopy, was related to the mechanization of anatomy chiefly through the study of glands. Up to the second half of the seventeenth century, glands had not been at the center of attention. Judging from the influential 1600 *Historia anatomica* by du Laurens, for example, their importance was negligible: the chapter dedicated to them occupies about two out of nearly six hundred pages. According to du Laurens, their purpose was to offer support to branching vessels, to absorb phlegm and serum so that they do not rush to more noble parts, and to moisten some body parts so they do not dry. Although occasional references to glands can be found elsewhere in the text, their role remained marginal.[1]

The situation changed in the 1650s and especially in the 1660s: if the 1650s were the decade of the lymphatics, the 1660s were to be the decade of glands. In 1656 the London physician Thomas Wharton published *Adenographia*, the first modern treatise on glands. Wharton's work inaugurated an era of anatomical contributions on glands extending to the end of the century and beyond. Major early contributions include Steno's 1662 *De glandulis oris* and Malpighi's *De viscerum structura*, published between 1666 and 1668. Whereas Wharton relied on the Galenic notion of similar attraction, for Steno and Malpighi glands operated mechanically. Malpighi in particular, seizing on Steno's seminal work, sought to mechanize several organs by identifying microscopic glands in them: in his mind localization was a key step toward mechanization.

In response to recent developments, just before licensing his treatise, the Cambridge professor Francis Glisson added a last chapter to his 1654 *Anatomia hepatis* including his reflections on lymph and glands. Since Thomas Wharton's *Adenographia* developed out of Glisson's research, section 4.2 discusses their works together and then proceeds to Steno's mechanistic challenge to Wharton's interpretation of the glands.

The following four sections take the lead from Malpighi's *De viscerum structura* and examine in turn all the body parts that he studied in that work—the liver, cerebral cortex, kidneys, and spleen—together with relevant works by his contemporaries and their techniques of investigation. Section 4.3 is devoted to the liver: Malpighi responded to recent works on the topic, including Glisson's, whose treatise preceded his own in the *Bibliotheca anatomica*. Section 4.4 is devoted to the cerebral cortex and uses Willis's *De cerebro* as counterpart to Malpighi's own work. Malpighi's attempt to subvert Wharton's taxonomy of body parts is especially evident in his works on the liver and cerebral cortex, where he devoted whole sections to that task. Section 4.5 deals with the kidneys and shows that, despite their common mechanistic allegiance, Bellini and Malpighi detected rather different structures with injections and microscopes. Lastly, section 4.6 is devoted to the spleen and Malpighi's almost desperate use of a strikingly wide range of techniques, such as splenectomy, chymical analysis, and a fistula to collect the juice allegedly secreted by the spleen, on the example of Reinier de Graaf's work on the pancreas.

I found it useful to rely on methods drawn from the history of the book, which enable the historian to anchor intellectual developments to specific reading practices and objects. It is not uncommon to find *Sammelbände* in original binding bringing together between two covers coherent collections of anatomical and medical texts. We have seen examples of this practice in the previous chapter with *Tetras anatomicarum epistolarum* and Malpighi's *De externo tactus organo*. A few notable cases, mostly with a Dutch provenance, are especially telling of reading practices on glands, since they join Wharton's treatise with relevant works by Steno, Malpighi, de Graaf, and the Amsterdam *Collegium privatum*, thus instantiating the claim that glands were a recognized anatomical field and reading topic in those years.[2]

4.2 Changing Perceptions on Glands: Glisson, Wharton, and Steno

Glisson's *Anatomia hepatis* reflects the two stages at which it was composed. The first part, covering the first forty-four chapters, draws mostly from his Gulstonian Lectures on the liver, delivered at the London College of Physicians in 1641; we shall briefly examine this material before moving to the extensive last chapter, which was a response to recent developments on the lymphatics discovered by Bartholin and Rudbeck in 1653. The Gulstonian Lectures were devoted to diseases and the body parts they affect. Several references in the texts make clear that Glisson and Wharton carried out research together with George Ent, who translated Glisson's English manuscript into Latin.[3]

Anatomia hepatis was the major treatise on the liver of its time and provided a

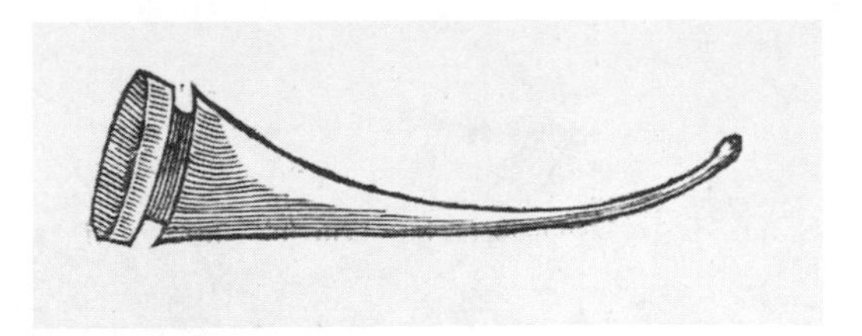

Figure 4.1. Glisson, *Anatomia hepatis*: injection device

detailed description of the liver's parts, including a capsule around the portal vein that is named after him. Glisson's treatise appeared at a crucial juncture, right in the aftermath of Pecquet's and Bartholin's findings. Glisson tentatively endorsed Pecquet's claims that the lacteals do not pertain to the liver, against Bartholin's early opinion that a portion of chyle reached it. As we have seen in section 3.2, Glisson used color injections in order to highlight the vessels of the liver; to this purpose he used a rather primitive device, consisting of a fistula with a bladder attached, as in the instrument to administer enemas (fig. 4.1). Indeed, he used several such tools of different sizes for different vessels, thus borrowing from medical practice a tool for anatomical research. In addition, in order to highlight the vessels, Glisson relied on excarnation or the removal of the parenchyma of the liver, a technique already mentioned by Vesalius. Glisson referred to two methods used by van den Spiegel: boiling the liver, then putting it in water and removing the parts around the vessels with a blunt knife; or having a large quantity of ants do the job for you. Glisson found both methods unsatisfactory and used instead one he had devised himself and discussed around 1642 at the College of Physicians, based on boiling the liver for at least one hour and then removing the unwanted parts with a blunt stick, a procedure similar to the one mentioned by Harvey in his 1648 reply to Jean Riolan the Younger. Harvey had used boiling in order to remove the parenchyma in the liver, spleen, lungs, and kidneys in search for the anastomoses of arteries and veins. The outcome of Glisson's method (fig. 4.2) revealed the arteries, veins, and bile ducts. Glisson wished to establish not only the branching of the vessels but also their connections. His overall conclusion was that the purpose of the liver was to secrete bile from the impure blood entering through the portal vein and, once it is cleansed, drive it through the vena cava. Thus, the liver attracted bile from blood and bile was considered excrement to be discharged.[4]

In *Anatomia hepatis* Glisson talked of the attraction of bile by the gall bladder, bile vessels, and bile itself in terms of *attractio similaris sive magnetica* or even *electrica*. He denied that similar attraction was an occult quality; rather, he argued that it occurred between bodies with an affinity or familiarity, when at least one body has the desire to unite with the other, with an actual endeavor, and, further, that it operates through

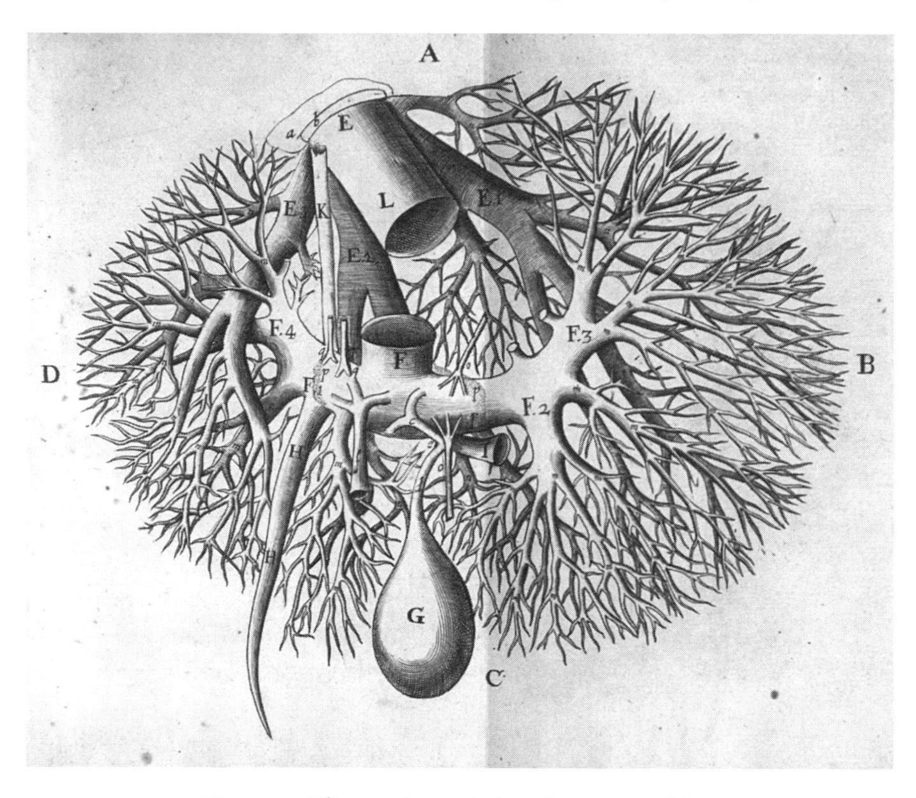

Figure 4.2. Glisson, *Anatomia hepatis*: excarnated liver

effluvia. Glisson drew a distinction between similar attraction on the one hand and organic or mechanical attraction on the other, which he had discussed previously, denying that it applied to bile. He considered four modes of organic or mechanical attraction: in the first mode the attracting body is connected or tied to the attracted one, as in a carriage attached to a horse or when we pull with our hands or teeth; the second mode occurs through suction and requires the attracting body to enlarge its cavity, something that does not happen with bile because the gall bladder is flaccid; in the third mode the cavity of the attracting body insinuates itself imperceptibly into the attracted body, as the intestine by means of its straight fibers with chyle; the last mode is deglutition or peristaltic motion, which is related to suction or the second mode. Glisson offered further evidence that similar attraction occurs in nature, as when animals of the same species tend to congregate. The editors of the *Bibliotheca anatomica* pointed out in the introduction and in notes to the text that the notion of *attractio similaris* had a whiff of old-fashioned philosophy.[5] Glisson provided a comprehensive treatment of the liver; a major focus of his work, however, was on the vessels, whereas Malpighi was to focus on the remaining part, the so-called parenchyma.

It was probably Glisson who provided the topic for the Gulstonian Lectures for 1653—delivered by Wharton on glands—which became the basis for *Adenographia*. In the last chapter (chap. 45) of *Anatomia hepatis* Glisson had envisaged an elaborate system vaguely analogous to the circulation of blood: according to Glisson, blood nourishes the sanguineous parenchymata, whereas a subtle nutritive fluid nourishes the spermatic parenchymata. This subtle fluid came from chyle, was absorbed chiefly by the glands of the mesentery, and was carried by a watery liquid—the lymph—partly in the nerves and partly in the lymphatics. He was relying on the old distinction between two parenchymata, depending on such notions as the amount of blood reaching them, color, size, and nobility; sanguineous parenchymata included muscles and the viscera, such as the heart, lungs, liver, spleen, and kidneys. Spermatic parenchymata included the stomach, intestine, bladder, womb, true skin, glands, and brain. Glisson and Wharton were not entirely in agreement as to the bones, fibrous parts, and membranous parts that were classed as hard or firm, in that Glisson included and Wharton excluded them from parenchymata. Both Glisson and Wharton, however, agreed on similar classification criteria based on color and nobility. Glisson objected to the argument that in vivisections severed nerves do not exude a fluid and ligatures do not swell, suggesting that the motion of the fluid in the nerves occurs through selective attraction, which ceases once the nerve is cut or ligated. The spleen occupies a crucial role in Glisson's system, in that it is the locus of separation of the watery vehicle of the nutritive juice from arterial blood and serves as a receptacle for it; he interpreted its many nerves and fibers, which he thought were also nerves, as vessels for the watery fluid carrying the subtle nutritive juice. Thus, Glisson's system of spermatic parenchymata envisages nerves, the spleen, and the glands responsible for nutrition. These were only one type of glands, the others being the excretory glands, such as the testicles and the maxillary glands, and the reductive glands, such as the parotids and axillary glands. Glisson's account is a powerful reminder of the problems anatomists faced when seeking to conceptualize the newly discovered lymphatics.[6]

In *Adenographia* Wharton fully endorsed Glisson's opinion that there were two types of nutritive juices and their corresponding vessels and three different purposes for glands. It was within this framework that he provided a taxonomy of glands, described how they function, and identified some new anatomical features, claiming[7]

> that there are two kinds of nourishment, one of the blood itself and the sanguineous parenchymata, the other of the spermatic parts; that for the latter purpose, the nourishment is dispensed to the fibrous parts by the ministration of the nerves; and that the nerves first imbibe it from the nutritious glands. Furthermore, I agree with him [Glisson] that there are three primary uses of the glands, namely nutrition, excretion, and restoration; and that these glands of the mesentery and those of the loins are chiefly nutritious.

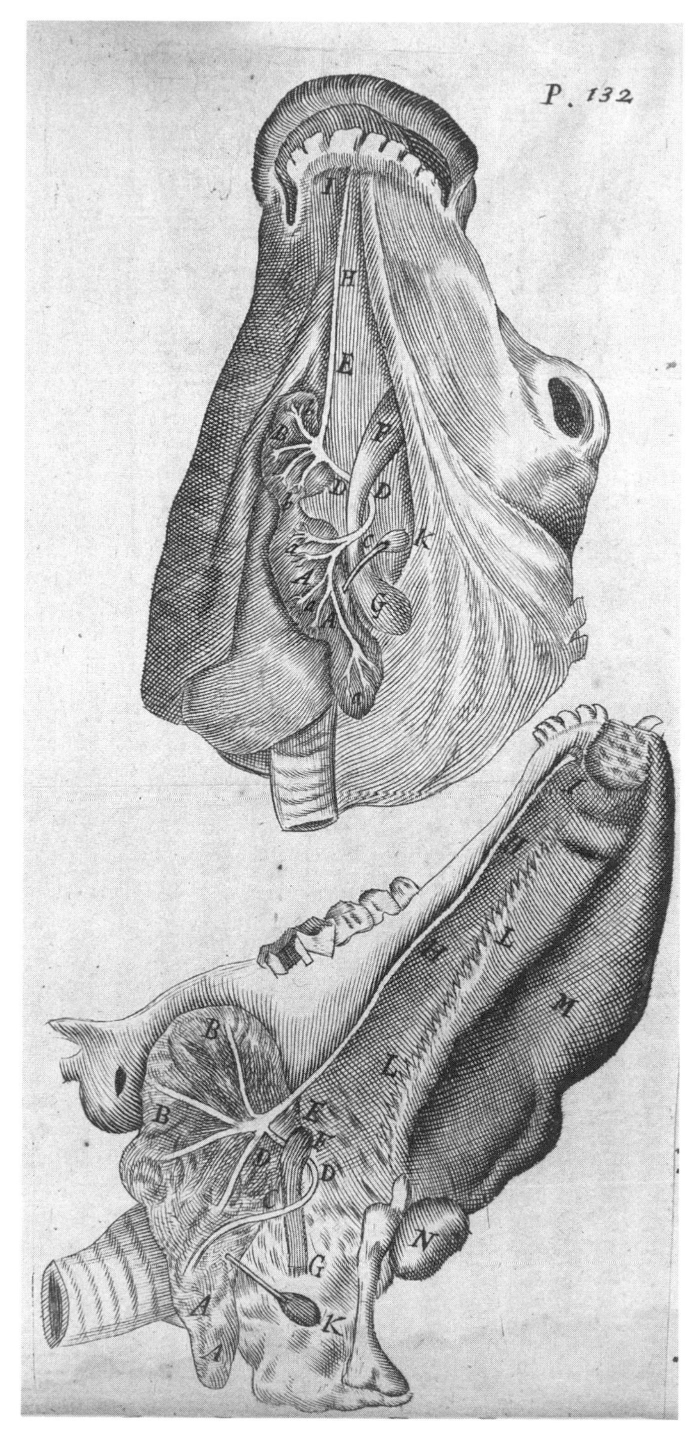

Figure 4.3. Wharton, *Adenographia*: maxillary gland and duct of a horse

According to Wharton, glands were smaller, less noble, and whiter than sanguineous parenchymata. They were subordinate to the nerves and drew fluids from them:[8]

> Glands are spermatic parenchymata with one comprehensive membrane overall, and often enclosed piece by piece in several separate ones: in fact parenchymata of such a kind that, when compared with that of the viscera, they are described as nervous rather than sanguineous, and are subservient to the brain rather than the heart; they are endowed with four kinds of vessels: namely, the arteries and veins, which are the smaller ones; and the nerves and either the lymph ducts or a special excretory duct, which are the most conspicuous ones.

In *Adenographia* Wharton announced the discovery of a previously unknown duct of the maxillary gland—now bearing his name—which he found while performing a dissection of the head of an ox in the presence of Francis Glisson: figure 4.3 shows at A and B the maxillary glands of a horse and at E and H their excretory salival duct.[9]

Wharton's treatment of other glands is also quite revealing. He stated that the purpose of the parotid glands was "to receive the overflows from the harder branch of the nerve of the fifth pair; secondly, to revive the ear and the ear lap with their warmth; thirdly, to fill up the pit in the periphery of the ear, and to make it level." Some, he argued, thought that their purpose was to moisten the surrounding parts, but he rejected this opinion because no appropriate vessel had ever been found. This passage highlights an important feature of Wharton's views, namely, that glands lacked a uniform *usus* and could take on a variety of purposes besides those we have seen above, including filling an empty space. In another important passage later in the book, Wharton argued that phlegm flows from the brain to the tonsils and into the mouth as mucus and is also sucked from the nostrils, thus defending traditional views. Wharton believed that different organs have the property of selective attraction of like with like: the pancreas, for example, excretes a "sweet, or at least insipid, mild, and soothing" fluid, which does not differ much from the pancreas itself, so that the pancreas can selectively attract it from the nerves in view of their similarity and cast it into the intestine. His plate of the pancreas (fig. 4.4) shows a lobular structure A with the ramification of the pancreatic duct, B and C, joining the bile duct D before entering the duodenum F. Wharton's figure is a notable improvement on Wirsung's plate (see page 104).[10]

Wharton's work served as a catalyst for a renewed interest in glands, although those who worked on this topic in the 1660s in no way endorsed his specific opinions or indeed worldview. In 1660 the Dutch anatomist and Leiden professor of medicine Franciscus Sylvius introduced a distinction between conglomerate and conglobate glands. The former seem to be aggregates and have an irregular shape, whereas the latter are smooth, spherical, or bean-shaped. A few years later Sylvius's student

Steno provided a brief account of the seventeenth-century revival of the anatomy of glands. The conglomerate glands owed their revival to the Bavarian anatomist Johann Georg Wirsung, prosector at Padua, who in 1642 found the pancreatic duct during a human dissection witnessed by Thomas Bartholin. Bartholin later discovered the round or conglobate glands associated with the lymphatic system. Steno mentions also Wharton's *Adenographia*, a work that had come under his scrutiny for both its specific claims and general worldview. Steno went on to argue that scholars working on glands seemed to suffer a curse, since Wirsung was murdered, Anton Deusing tried to appropriate Wharton's finding, and Gerardus Blasius tried to appropriate his own findings.[11]

In 1661—and then more fully in 1662 with additional essays—Steno put forward an essay on glands, *De glandulis oris*, in which he announced the discovery of yet another previously unknown duct, that of the parotid, now bearing his name. It was unfortunate for Wharton that only five years after his hasty words Steno had found that vessel: figure 4.5 shows at the top the head of a calf with the conglomerate parotid (marked *a*), the conglobate parotid (*b*, or lymph node), the parotid duct (at *e*), and nerves (at *g*). Moreover, as we have seen in section 2.4, Steno challenged the view that phlegm descends from the brain into the mouth. Steno's approach to anatomy and understanding of glands were quite different from those of his English colleague. Unlike Wharton, Steno was trained in the new chymical and mechanical traditions; despite the few years separating their works, they belonged to different generations. The issue was not simply one of discovering a new duct, but understanding the role of glands in a new way. Steno was skeptical of classifications based on color and nobility and of the notion that glands derive their humors from the nerves. Rather,

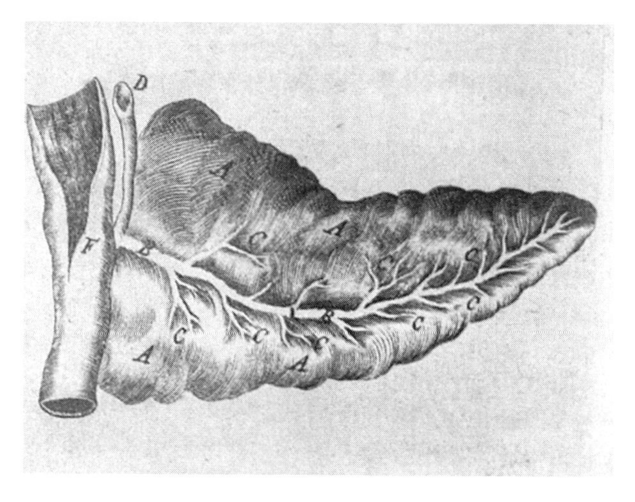

Figure 4.4. Wharton, *Adenographia*: the pancreas with its duct

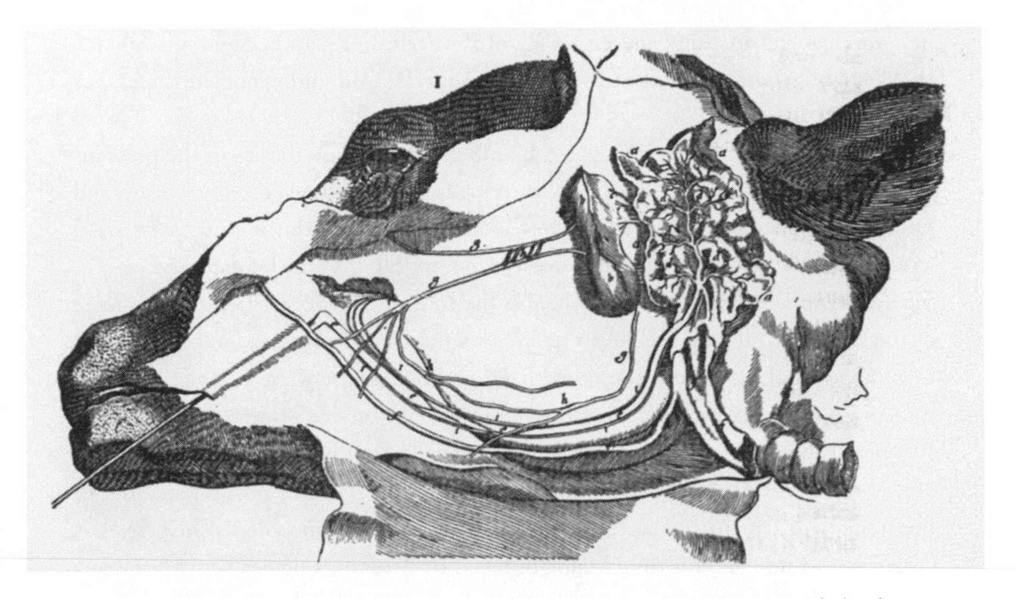

Figure 4.5. Steno, *De glandulis oris*: conglobate and conglomerate parotid glands

he envisaged an ingenious mechanism based on quantitative reasoning rather than visual inspection, whereby saliva is secreted from blood by the salivary glands. Steno argued that the nerves could not supply enough fluid to the glands to produce salivation, yet they must serve a purpose. Salivation was affected by the nervous system, however: when we see delicious foods, for example, salivation increases. Thus, Steno believed that arterial blood reaches the salivary glands, where in normal circumstances its greatest part turns into venous blood; only a small portion of saliva is filtered by the gland and reaches the mouth. The nerves, however, could partially restrict the openings of the veins, which therefore would take in less venous blood, and at the same time more saliva is produced. As we are going to see, understanding the actual mechanism of glandular secretion was a major concern for mechanistic anatomists.[12]

4.3 Malpighi's Treatise on the Liver

In 1666 Malpighi left Messina and returned to Bologna. Although the Messina Senate had reappointed him as primary professor of medicine, he never set foot in Sicily again. On his way back he stopped in Naples, where he met Tommaso Cornelio, Lionardo di Capua, and other members of the Investiganti Academy, and in Rome, where in the house of the surgeon Giovanni Guglielmo Riva he met Steno. It is hard to imagine a more crucial meeting for both anatomists. As a protagonist of the new mechanistic anatomy and especially of the revival of glands, Steno was pivotal to Malpighi's central project at the time, which relied on and extended the Dane's findings.[13]

Fearful as ever of being anticipated by others, in this case too, as in the previous year, Malpighi had a preliminary set of three treatises on the liver, cerebral cortex, and kidneys published in the late summer of 1666 at Bologna with the title *De viscerum structura*. Two years later the definitive version appeared with the date unchanged, but with two additional treatises: one, on the spleen, was closely tied to the theme of the book; the other, on the heart polyp, was sufficiently unrelated to be presented as an appendix.[14] This time too, as with *De externo tactus organo*, Malpighi did not inform Borelli, who was thus prevented from offering advice or orchestrating ancillary publications as in the past. There is another analogy between *De externo tactus organo* and *De viscerum structura*: both contain no illustrations, possibly because of the haste with which they were produced. As we will see, in order to help readers visualize his findings, this time Malpighi relied on several analogies between microscopic glands and different types of fruit, such as a bunch of grapes, pomegranates, dates, cherries, or apples on an apple tree. The introduction offered a justification for his investigations based on the authority of Pliny and Hippocrates and Malpighi's own views about nature. Nature, he argued, adopts in large and notable organs the same mode of operation as in very small parts, whether of plants or animals. Indeed, the largest parts consist of aggregates of *minima*, the smallest components capable of performing the same action; he expressed similar views later in *De liene* as well. In a display of modesty, Malpighi sought to justify his work after the publications on the liver, brain, and kidneys by Glisson, Willis, Fracassati, and Bellini. Although he argued that he wished to strengthen their statements with new observations, in some cases he provided corrections that must have been received, and were probably written, with considerable embarrassment.[15]

In *De hepate* Malpighi offered a major reconceptualization of the structure and role of the liver and a rejection of Wharton's taxonomy. The introduction to his work in *Bibliotheca anatomica* highlights the amazement with which his views had been received. Traditionally the liver had been seen as the site of sanguification, but Harvey's work on the circulation of the blood, Pecquet's discovery of the thoracic duct—showing that chyle traveled to the subclavian vein, bypassing the liver—and its confirmation by Thomas Bartholin undermined current thinking about the largest internal organ in the body. Bartholin went as far as to write a joking obituary for the liver. In the aftermath of the discovery of the pancreatic duct and the rise of interest in glands, Malpighi claimed that the liver was a conglomerate gland. Much like the pancreas and other glands, the liver was also endowed with its excretory duct, known as the bile duct. Moreover, although Glisson had argued that the liver formed a continuous substance, he had to admit that it was not compact but rather friable; in some circumstances Glisson admitted that the texture of the liver did not appear continuous, but he attributed this to disease. Malpighi took the liver's texture as evidence for a different structure. Although this reconceptualization made sense of many features of the liver and provided a much-needed purpose for its existence, it also went against traditional thought and therefore required careful justification.[16]

Because the fine structure of the liver of higher animals is more recondite, Malpighi relied on a combination of the microscope of nature and actual magnification, starting his investigation from lower animals such as snails, lizards, fishes, and mice, moving to cats, oxen, and humans. He argued that the liver was best studied by boiling it and examining it under a microscope, thus highlighting the existence of small glands. As in other circumstances, he did not say much about his microscopes and microscopic techniques. In this regard *De hepate* is quite brief even by his standards, although here Malpighi describes a rich range of other investigation techniques. At one point he wondered how the separation of bile occurred in the glandular structures he had identified, an issue on which he had to admit defeat because the glands were so tiny that they could not be resolved even with his best microscopes; all one could do was to formulate hypotheses and proceed by analogy, using reason where the senses failed. In different animals livers have different shapes, like clover or a pea, but the bunch of grapes remained his favorite simile. He supported it by referring to a specific passage from *De generatione* in which Harvey had studied the generation of the chick. Harvey drew several analogies from the vegetable world and from diseased states to illustrate how the liver appears on the branches of the umbilical vessels: "For just as grapes grow to the cluster, buds to their twigs and the young ear to the corn-stalk, so also does the liver grow to the umbilical vein from which it arises, as fungi from trees and proud flesh in ulcers or diseased fleshy tumors bordering on the branches of arteries from which they are fed, and where, at times, they spread into a vast mass."[17] This passage is likely to have attracted Malpighi's attention also because of Harvey's reference immediately above it to having used a magnifying glass. Malpighi probably saw in that passage a crucial link tying the investigation of an organ's structure to the study of generation and of plants. Thus, the analogy between liver and grapes does not rely simply on superficial appearances, but is anchored to a deep structural analysis prefiguring Malpighi's investigations of generation and the anatomy of plants.

Malpighi's polite challenge to Wharton's distinction between glands and viscera shows that intellectually they belonged to different generations. He rejected Wharton's views about the nobility of body parts and the distinction between sanguineous and spermatic organs based on color, heat, and vitality, arguing that whereas the heart was previously considered as the monarch of the body, now it was seen as an ass at the mill; moreover, many body parts, once they have been cleansed from blood by the anatomist, are white and reveal a similar nature, thus challenging Wharton's classification.[18]

Malpighi had to address another problem, namely, that the lymphatic vessels seemed to originate from the liver and if one ligates them, they fill from its side; thus, if the liver was a gland, the question was to identify whether its true excretory duct was the bile duct or lymphatic vessels. Relying on Steno's figure of the parotid gland (fig. 4.4), he argued that nature joins conglobate glands, associated with the lymphatic vessels, and conglomerate ones; Malpighi claimed that he had observed as much several times in the liver of calves. Since the lymphatic vessels are not connected with the structure of the liver, strictly speaking, the true excretory duct of the liver is the bile duct.[19] Malpighi entered the huge dispute following the publication of the *Epistolica dissertatio* by de Bils, in which he reported a vivisection experiment according to which four or five hours after the mesenteric arteries had been ligated, the arteries below the ligature were found to be empty, whereas the veins were filled with a dark, ashen juice, which could not have come from the arteries and therefore was coming from the intestine. Malpighi objected that, despite the ligature, blood may have reached the veins through micro-arteries in the mesentery.[20]

Next Malpighi addressed the views by the Rotterdam physician Jacques de Back and Franciscus Sylvius, according to whom bile is not separated from blood in the liver as excrement of sanguification, but rather is mixed with blood in the liver. According to this view, bile was filtered by the gall bladder—much like the stomach produces a digestive humor and the kidneys urine—and then went partly to the liver, to enrich blood, and partly to the intestine, where it helped digestion. Recently Sylvius had claimed on chymical grounds that bile could not be separated in the liver. In order to settle the matter, Malpighi performed a vivisection experiment on a young cat of a few months. He ligated the neck of the gall bladder, blocking the entrance to the bile duct to the liver and *ductus choledochus* to the duodenum, and emptied it with an incision. Then he ligated the *ductus choledochus* just before it opened into the duodenum. In this way the only opening of the bile passageways was toward the liver. A few years later the Leiden physician Johann Pechlin claimed to have performed a similar experiment in 1667, but, unlike Malpighi, he added a beautiful engraving showing the arrangement of the bile vessels: in figure 4.6 KL is the gall bladder with its neck ligated, O is a small flask with a narrow neck inserted into the *ductus choledochus* to collect bile—Malpighi only ligated the *ductus*—and NM is the bile flowing from the liver and filling the bile passageways. Once released, the animal lived for a considerable time. When it was opened again, the *ductus choledochus* and bile duct were found to be full of bile, a clear sign that bile came from the liver, not the gall bladder. The same happened after having removed the gall bladder altogether. Malpighi also tried to push the bile up in the duct with his fingers—much like Harvey had done

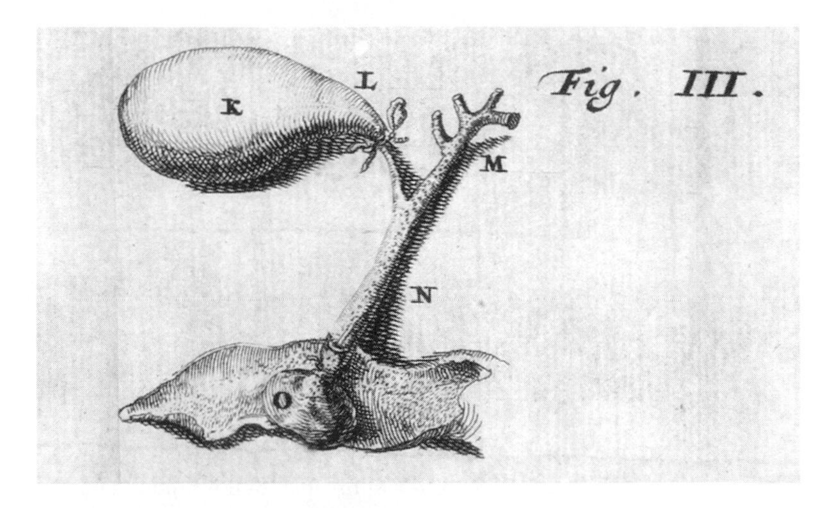

Figure 4.6. Pechlin, *De purgantium medicamentorum facultatibus*: experiment on bile

for venous blood in the arm—or some other device toward the liver, but it returned back with greater force.[21] The cases of de Bils and Sylvius highlight the state of flux anatomy was in at the time, even about basic "plumbing" issues in the body.

In the last section of his treatise Malpighi addressed medical matters, as he had done on other occasions. He devoted special attention to dropsy preceded by jaundice; he saw a link between the two because jaundice results from an obstruction of the flow of bile, whereby digestion is affected and blood corrupted. There follows a damage to the glands of the body, which become filled with a tartar that hardens them. At this point Malpighi reported the observation of the Renaissance physician and naturalist Rembert Dodoens, who, in a dissection of the cadaver of a patient whose jaundice had turned into dropsy, saw the liver filled with very hard stones, as he reported in his *Medicinalium observationum exempla rara* of 1581; Malpighi took this to result from the hardening of the glands. This reference is especially interesting because it comes from Dodoens's annotations on *De nonnullis ac mirandis morborum et sanationum causis* by the Renaissance Florence physician Antonio Benivieni, an early advocate of postmortem pathology. Earlier Malpighi had mentioned the reports by the Jena professor of medicine Gottfried Möbius of a human liver filled with square stones—which he interpreted as lobules hardened by tartar from their glands—and by Glisson on the London physician Assuerus Regemorter, who had found in a human cadaver a liver filled with glands shaped like peas; unlike Regemorter, Malpighi thought that these were not generated but merely rendered manifest and enlarged by disease. These examples show that Malpighi mined remarkable cases by Renaissance and more modern authors for reports of unusual cases, especially petrified or bony parts, which would thus become permanent witnesses to his anatomical views: we shall encounter this technique of mining the literature for reports of stones in other works by Malpighi and in those by his critics. His interest in observation reports seems to have been shared by Capucci, who in a 1668 letter expressed his preference for books containing specific observations rather than doctrine, even by the *novatores*, and mentioned Dodoens's work as one he lacked and wished to purchase.[22]

As a result of the damage to the glands, Malpighi continued, blood serum does not follow its natural routes and opens new ones. Malpighi was able to confirm these views with the postmortem of a man who had died of dropsy.[23] From his collection of postmortem reports, it is possible to identify this case with that of Father Bolognetti, who had died at Bologna on 6 August 1666, since the accounts are virtually identical. It is significant that the report of Bolognetti's death is the first chronologically in Malpighi's collection. From his report we know that he found the abdomen filled with a large amount of fluid; under the action of fire, part of this fluid coagulated like beaten egg and tasted like it, thus proving in his eyes and according to standard views at the time that it originated from blood serum; the residual portion remained fluid. This reference shows that Malpighi relied on taste, in line with the medical tradition, and did not share Borelli's views of a radical decoupling between the composition of a substance and its taste. The liver looked like a parotid gland and overall it resembled a bunch of grapes, with the glands hanging from the blood vessels (notice here the similarity with Harvey's account cited above). Although the glands were not petrified, they appeared dry. In this remarkable report Malpighi must have seen a vindication of his views: his new mechanistic understanding of the structure and purpose of the liver led to a different conception of its pathology, and in turn a pathological state confirmed the anatomical findings. In yet another report from approximately the same time on the dissection of the daughter of the apothecary Pietro Francesco Castelli, Malpighi stated, "I saw the glands in the liver." In *De hepate* he defended his identification of the liver as a gland also because its diseases are typical of glands. The link between anatomy and pathology can be extended to therapeutics by means of a consultation for a case of dropsy first published in 1713, in which Malpighi provided a causal account of the disease analogous to what we have seen here, with the exception that jaundice is not present in the 1666 cases.[24]

These observations show a triangulation strategy involving anatomy, pathology, and therapy reflecting the connections among different areas of Malpighi's medical thinking and reminding us of the dangers of separating anatomy from medical practice. We shall encounter again connections among anatomical treatises, reports of postmortems supported by pathological cases found in the literature, and medical consultations in which therapy is anchored to anatomy.

4.4 The Brain and the Cerebral Cortex

Malpighi's essay on the cerebral cortex is quite brief and does not appear to have been perceived as especially problematic at the time, despite the fact that his findings were challenged at the end of the century and are now considered an artifact of his investigation techniques.[25] In order to understand the background of his reflections, it is helpful once again to start from Steno. In the section on glands in *De musculis et glandulis* (1664), he argued that identifying the way in which the humors of the brain are secreted required investigation. Steno argued that the dissection of the brain showed it to be full of humors, although the path through which they were evacuated was unclear. In a later letter of June 1669 to Ferdinand II of Tuscany, published in Thomas Bartholin's *Acta*, Steno implicitly supported Malpighi's views on the secretion of a nervous fluid in the brain.[26] Nor were Steno's views unusual. In a passage from the 1664 *Cerebri anatome* summarized by Malpighi, Willis also talked of the animal spirits being first separated from blood in the brain in vessels working like alembics or distillatory organs and then refined with some ferments. All this sounds very much like a modern version of the venerable Galenic doctrine of the *rete mirabile*, the plexus of blood vessels at the base of the brain responsible for the generation of animal spirits. But whereas Willis had trouble finding direct evidence by anatomical, rather than clinical, means that the nerves contained a fluid, Malpighi unambiguously claimed in *De cerebri cortice* that nervous fibers were vessels and that, having cut them, he had collected abundant fluid similar to egg white, which coagulated on the fire.[27]

Sponges figure quite prominently in the literature on the brain and in *De cerebri cortice*. An importance source was the Dutch Cartesian Lambert Velthuysen, a physician in Utrecht and one of the governors of the East India Company. In *De cerebro* Fracassati had contrasted Velthuysen's opinions with his own: both identified spongiform structures in the brain, although Velthuysen relied on chymical principles and identified the sponge as a site of filtration, whereas Fracassati based his claims on visual inspection and the usage of a microscope with subtle lighting. In *De cerebri cortice* Malpighi praised Fracassati while challenging Velthuysen, arguing that the separation of the nervous fluid occurs in glands rather than sponges. We witness here a conflict among mechanists as to the specific device responsible for filtration.[28]

In line with his mechanistic program, Malpighi set out to show that the cerebral cortex is a collection of glands and to refute Wharton's arguments to the contrary in *Adenographia*. Wharton claimed that since the glands are subordinate to the nerves, and the nerves to the brain, "it is not credible that a *servant* should be of the same status and substance as her *mistress*." We witness here the projection of views about social order onto anatomy. Wharton also argued that the brain consists of two parts, the cortex and the marrow, whereas the glands consist of one homogeneous parenchyma. Malpighi answered Wharton's theses one by one in a scholastic fashion, but in a respectful way.[29]

We have seen that Malpighi, like many of his contemporaries, was convinced that a nervous fluid was filtered in the brain and reached out to all the nerves; since glands were the locus where such filtration occurred, glands were expected to be present in the brain. The key issue for an anatomist concerned with visual proof was to locate them. Malpighi argued that the excretory duct was to be found in the white fibers of the brain and cerebellum. He started by boiling a fresh brain, because in this way the substance of the glands becomes larger; then he removed the *pia mater* when it was still warm, added a drop of ink, and removed its excess with cotton. This procedure was similar to that employed in studying the tongue and the skin, except that he now introduced staining. This similarity highlights how difficult it was to evaluate preparation techniques: transferring a technique from one context to another led to results that were later found to be artifacts. This time too, as for the liver, Malpighi found in the literature a case supporting his interpretation. In his treatise on stony concretions in the human body, the Renaissance physician and naturalist Johann Kentmann from Dresden had reported the case of a stone shaped like a blackberry found by Johann Pfeil in a human body. Kentmann's work, *Calculorum qui in corpore ac membris hominum innascuntur, genera XII. depicta descriptaque*, was part of *De omni rerum fossilium genere* (Zurich, 1565), by the Zurich physician and naturalist Conrad Gesner. Kentmann's treatise was one among others of a similar genre mined by Malpighi in his search for stony concretions and other anomalies. Once again, as for the liver, he did not hesitate to interpret this concretion as revealing and preserving the glandular structure of the brain. This time his favorite fruit analogy was with pomegranate seeds and dates, but he also compared the cerebral cortex to a substratum in which nervous fibers grow like plants and compared the ramifications of the blood vessels to the veins and netlike fibers of plants. He also relied on surgical experiences, arguing that when cortical glands become ulcerated, they generate excrescences with the shape of mushrooms, as a result of the extravasation of nervous juice.[30]

In the last section, echoing Borelli's opinion, Malpighi sounded a skeptical note against attempts to locate the sites of various cerebral functions such as imagination and memory. Malpighi also questioned Willis's view of the existence of two sets of fibers carrying sensory impressions upward and stimuli to motion downward, since all fibers seem to go in one direction from the cortex; the absence of any reference to Fracassati's views on nervous transmission based on Magiotti's experiment is striking. In what appears like an allusion to Descartes' views, Malpighi stated that he was happy to leave aside the question whether there is a tension in the nerves due to a fluid in them.[31] Malpighi highlighted the fragility of the cerebral glands and relied on Hippocratic texts, such as *On Sacred Disease* and *On the Glands*, in arguing that the diseases of the brain require longer and more complex cures. It is remarkable to notice his approval of ancient medicine: here Malpighi seems to present himself as the provider of a causal explanation for long-held medical truths, a position found also in portions of his *Risposta* to Lipari.[32]

4.5 The Kidneys: Bellini and Malpighi

When reading a text by Malpighi, it is worth remembering that omissions can be as important as citations. A striking feature of his *De renibus* is the lack of any explicit reference to the work published just a few years earlier by Bellini, *Exercitatio anatomica de structura et usu renum* (1662). There are extensive similarities in the events surrounding the publication of Malpighi's *De pulmonibus* and Bellini's work. In both cases Borelli was the architect promoting publication by the anatomists in his circle and providing an interpretation of their structural findings. His role must have been especially decisive in the case of Bellini, who was only nineteen at the time and had not completed his degree yet. We have seen above that in the second letter on the lungs Malpighi reported Borelli's interpretation of their purpose, virtually quoting from his letter. Borelli's authorial role in Bellini's work was equally explicit, since a large portion of the text was attributed to him in a way that would have involved multiple authorship in our times. But even at the time the arrangement must have looked peculiar, given that in the 1664 Strasbourg edition of the *Exercitatio* the text was partitioned and attributed to the respective authors, Bellini for *De structura renum observatio anatomica* and Borelli for *De illorum usu judicium*.[33] Thus, Malpighi's tactful but uncompromising refutation of Bellini touched Borelli too.

Malpighi thought that the kidneys had a glandular structure, much like the other viscera we have seen so far. In his description of the kidneys' general appearance, he noticed that in bears they look like a bunch of cherries. Stony concretions in the kidneys are very common, but Malpighi did not see them as hardened glands, possibly because they can be dislodged from the kidneys.[34]

A crucial point of the treatise concerns the microscopic investigation of the kidneys' structure; in order to examine this issue, it is helpful to introduce Bellini's *Exercitatio*. Bellini started by reviewing the existing literature from ancient times to Thomas Bartholin and Nathaniel Highmore, but he failed to mention the contributions by the anatomist and physician Bartolomeo Eustachio, as Malpighi was immediately to point out on receiving a copy of Bellini's work from Borelli. Bellini argued that the structure of the kidneys consists not in a parenchyma similar to the heart or liver, as claimed by some, but in a series of fibers or tiny vessels. These are not muscular, since there are no tendons and boiling them caused them to become smaller, whereas muscles increase in volume. The external surface of the kidneys is covered by a large

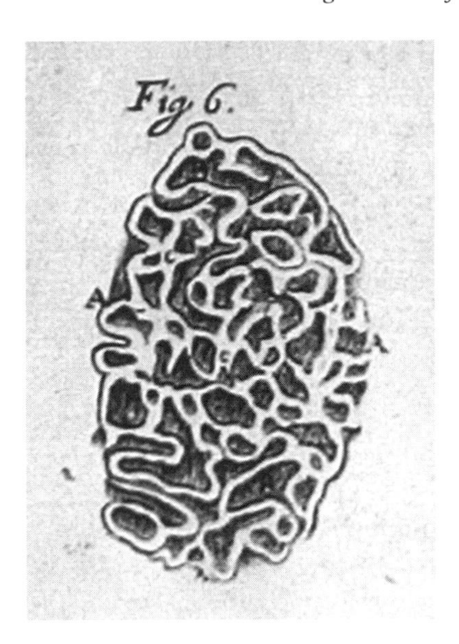

Figure 4.7. Bellini, *De structura et usu renum*: surface of the kidney of a wether

number of *sinuli* or tiny folding vessels where the secretion of urine occurs from their configuration. The *sinuli* are best seen by combining injection of a colored liquid into the blood vessels of the kidneys with microscopy. Figure 4.7 shows the surface of the kidney of a wether—or castrated ram—seen under the microscope after colored injection. Bellini argued that there was no anastomosis between arteries and veins, but rather the microscope revealed the opening of the arteries into spaces marked *ccc* in which arterial blood separated into urine and venous blood. Figure 4.8 shows a portion of a human kidney, where A marks the extremities of the renal ducts, while BB is the area whence urine droplets, marked CC, exude; Bellini ascertained its nature by tasting.[35] Having described the structure of the kidneys, Bellini left to Borelli the explanation of their mode of operation and purpose.

Borelli argued that secretion occurs not by attraction, familiarity, or sympathy, but solely as a result of the vessels' configuration. His account relied on some experiments on capillarity that were being discussed at the Accademia del Cimento. In addition to the study of color change that we encountered in section 1.2, we see here another example of interaction between the experiments of the Medici academy and anatomical research. Borelli drew further analogies with fluids percolating through solid bodies and membranes; for example, mercury penetrates the pores of gold, although air and water do not; some membranes or skins are permeable to water, although not to air. In a similar fashion, arterial blood divides into urine going through the renal small siphons, whereas venous blood goes through the veins. Borelli believed that this process was helped by respiration, since the compression of the abdomen during inspiration would make urine exude from its tubes; he was following in the footsteps of Pecquet, who had argued that chyle moves in its vessels from the compression of the abdomen due to respiration.[36]

Malpighi disagreed with Bellini's findings and Borelli's interpretation. First, he confessed with unconvincing modesty that possibly because of his weak sight or the imperfection of his instruments, he had been unable to replicate Bellini's observation. Malpighi went on to state that upon injecting the renal blood vessels he saw ink stains due to excessive pressure, not the *sinuli*. The *sinuli* or *canaliculi*, as he called them, exist to be sure, but they are urinary ducts.[37] Malpighi went on to challenge Borelli's interpretation too, since it relied on Bellini's structure. He argued instead that by injecting the artery with ink mixed with spirit of wine, and then removing the renal membrane and cutting the kidney longitudinally, one sees the glands hanging like apples from an apple tree: these are the glomerules, a microstructure that had not been previously observed and which pinpointed the site of secretion. Malpighi tried desperately to grasp the connection among the vessels involved, even by means of a vivisection experiment repeated many times. He ligated the renal veins and ureter of a dog, in the hope of increasing the size of the kidney's microstructures: here Malpighi used vivisection as a magnification device to enlarge body parts. The animal survived

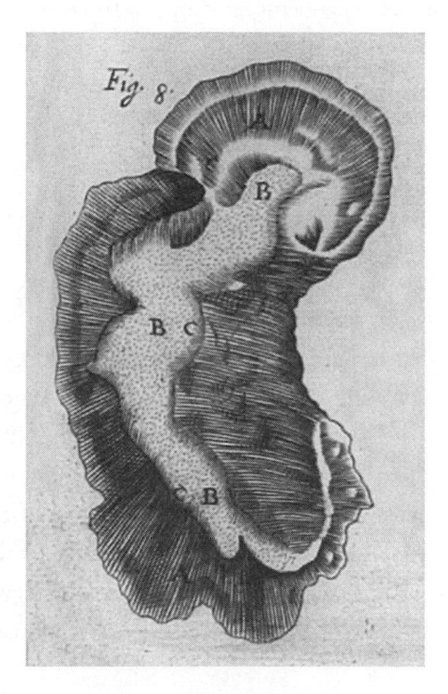

Figure 4.8. Bellini, *De structura et usu renum*: section of a human kidney

a long time, and when the kidney was examined, it was found to be filled with blood. But even this technique failed to reveal the connection Malpighi was after, which had to be inferred by analogy with other organs such as the liver and cerebral cortex. Once again, the localization of the site of filtration was not followed by an understanding of how the process occurred.[38]

There is no question that the purpose of the kidneys is to filter urine, but the precise way in which this operation occurs was most obscure. Malpighi did not question that it occurred mechanically, although it was unclear whether a sponge, sieve, or other structures would offer the best analogy. Relying on pathological observations on patients, he argued that the separation of urine depended not only on the configuration of the kidneys but also on the fermentation of blood, since in certain diseases in which this is abnormal, the secretion of urine is affected; this is another instance in which the normal operations of the body are inferred from pathological states. Malpighi concluded the treatise with some pathological observations, describing a number of cases ranging from hereditary disease to the expulsion of small fleshy particles. While denying any role to the kidneys in the preparation of semen, Malpighi argued that their defective operation may adversely affect the preparation of semen and induce abortions.[39]

Whereas Bellini believed he had uncovered the microstructure of the kidneys to a point sufficient to understand how they operate, Malpighi's investigations had reached a deeper level and revealed a more complex structure, although paradoxically he failed to grasp the connection among the blood vessels and excretory duct and how the kidneys work: the mechanism or mode of operation of the kidneys remained a mystery.

4.6 The Spleen and Its Problems

Whereas in the previous cases Malpighi believed he had been able to identify the glandular microstructures he was looking for, when it came to the spleen he encountered some difficulties that explain why this treatise is the longest in the entire book and involves an especially wide range of investigation techniques. The spleen was considered to be one of the key viscera, as we have seen in Glisson's and Wharton's works, for example, and had attracted considerable attention in the literature; therefore, omitting it would have undermined the completeness of *De viscerum structura*. Thus, Malpighi left no stone unturned to unlock its secret, as shown by the letter of May 1667 to the physician Henry Sampson—Nehemiah Grew's half brother. This is why he was ready to license for the press the final five-treatise version of his book only in March 1668.[40]

Malpighi provided an extended description of the spleen, which is a collection of membranes separated into little chambers, interspersed with a great number of fibers and with a great abundance of lymphatic vessels. In a remarkable passage, he established an identity in nature's operations between healthy and diseased states, as in tumors, arguing in both cases that nature sagaciously or *providè* relies on membranes:[41]

> The generation, growth, and purpose of membranes is so common in nature, that not only single members of the body, which are in whatever way assigned to some purpose, are bestowed with one surrounding them, but also those members, which oppose nature herself and constrain it to deviate from the right way of operating, are sagaciously endowed with those membranes, as occurs especially in tumors; thus we can suspect that nature does not forget the texture of membranes even when it goes wrong.

Malpighi's structural findings challenged Glisson's views on the purpose of the spleen: whereas Glisson believed it to be the receptacle for a watery fluid and saw the fibers and nerves as vessels for this fluid, Malpighi argued that since the fibers are not hollow and the nerve endings are very thin, neither can serve the purpose envisaged by Glisson. Malpighi claimed to have identified a new membrane that had not previously been observed enveloping the blood vessels and nerves, which he named *commune involucrum, seu capsula.*[42]

The spleen consists of membranes forming small cells, as well as cavities. In line with the program of *De viscerum structura*, he identified glands or vesicles in the small cells, similar in size to those of the kidneys, and which he compared to a bunch of grapes. The spleen of a swordfish he had observed at Messina seemed remarkable, but since he could not complete his investigation, he pointed out the matter for future research. As in previous treatises, here too Malpighi found some stony concretions and supported his claim with similar findings reported by the Ferrara physician Ippolito Bosco.[43] As in *De hepate*, Malpighi reported the result of a postmortem he had performed immediately before licensing the manuscript for the press at the S. Maria della Vita Hospital on 5 January 1668. This time he found in the body of an eighteen-year-old woman the spleen filled with glands of varying dimensions, some as large as a chickpea. Also in this case some medical consultations, dealing with a tumor and hardening of the spleen, offer some explanations of the process of the disease and of how to cure it. In *De liene* postmortems were also invoked in ruling out a systematic involvement of the spleen in ascites, or swelling of the abdomen: although in some cases the spleen was involved, generally the cadavers showed that the spleen was not affected by ascites to the point that it could not perform its operations.[44] Once again, these observations show the profound links between anatomy and pathology.

Malpighi considered the viscera as *officinae*—a term rendered as *laboratories* in

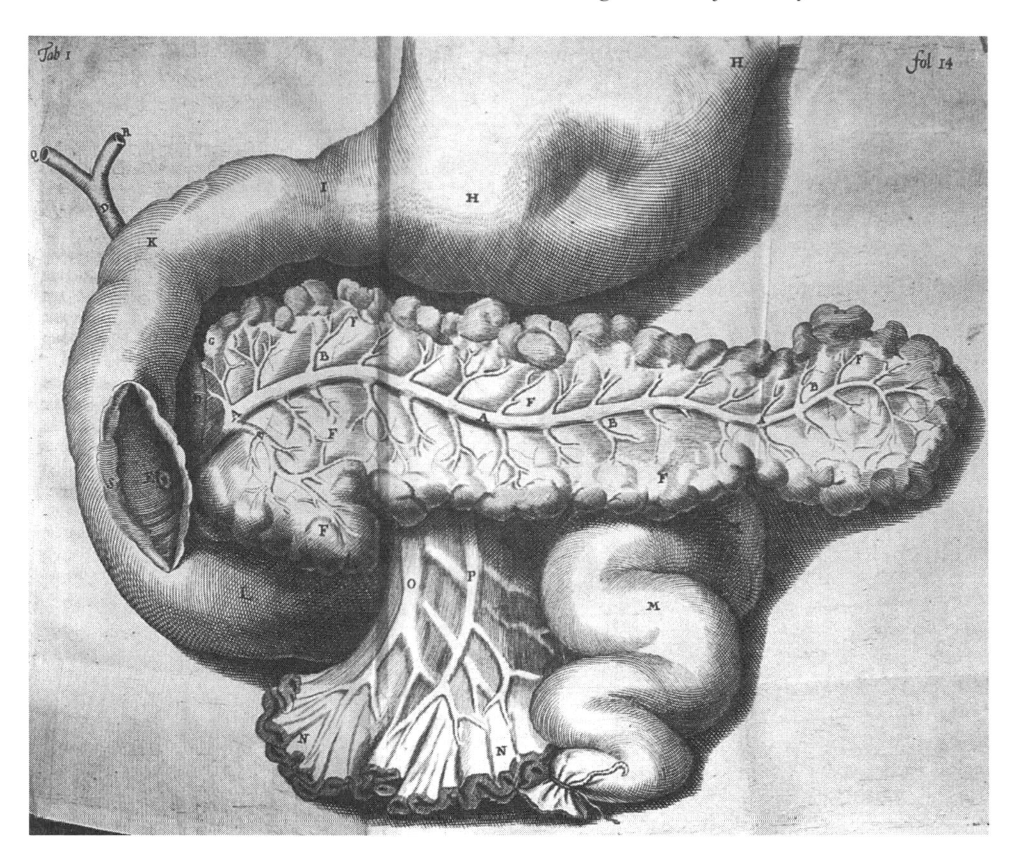

Figure 4.9. De Graaf, *De succo pancreatico*: pancreas with its duct

the contemporary French translation—whose microscopic operations are, as it were, covered by a veil. While providing a detailed description of its structure, Malpighi had to face the problem that the spleen did not seem to have an excretory duct and thus it was unclear what its glands secreted and what their purpose was. He surmised that they secreted a juice directly into the bloodstream—like our modern endocrine glands—although the problem was to provide evidence for this belief by detecting this juice and also to make sense of this whole operation: if glands mainly filter their juice from blood rather than synthesize something new, the question was why they would separate something from blood that was immediately mixed with it again. Malpighi performed a large number of vivisection experiments in the hope of shedding some light on the matter, including ligating the blood vessels feeding the spleen in a dog: he found that the dog lived quite happily, but was especially voracious and urinated a lot, even by dog standards. Dissecting the dog after some time with the help of Fracassati and Bonfiglioli, they found the spleen to be tiny, yet its purpose remained obscure.[45]

Malpighi reported having performed several times an experiment on a sheep; although the experiment failed, it is exceedingly interesting to notice the similarity with the celebrated experiment of the pancreatic fistula performed by the Delft physician Reinier de Graaf in the 1664 *De succo pancreatico*. Both sought to collect the juice produced by the spleen and pancreas, respectively, in order to study its chymical properties. Figure 4.9 shows the pancreas with the insertion of the pancreatic and bile ducts into the duodenum. De Graaf inserted a quill into the pancreatic duct of a dog, leading to a fistula attached to the dog's belly: he collected enough juice to carry out some testing, largely by tasting the juice. Figure 4.10 shows the experimental arrangement of the dog with a fistula attached to the belly and one in the mouth to collect saliva. Malpighi, by contrast, failed to retrieve any juice by means of a *vitrea fistula*, exactly the same term and device employed by de Graaf. Since the spleen lacked an excretory

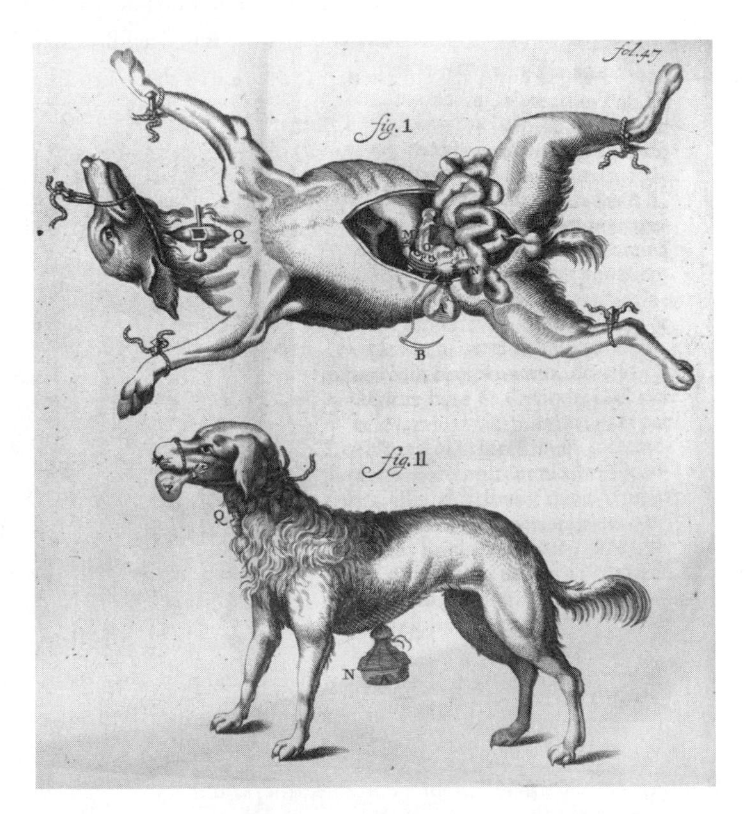

Figure 4.10. De Graaf, *De succo pancreatico*: fistula experiment on a dog

duct, he attached the fistula to the splenic vein, after having ligated the splenic artery. This attempt shows Malpighi seeking to rely on de Graaf's celebrated experiment and attempting to replicate any recent advanced technique of investigation. De Graaf was a close friend of Steno's and even if Malpighi may not have come across his rare work, he may well have heard about the experiment from Steno's mouth at the 1666 dinner in Rome. Further, Malpighi performed chymical tests in the hope of determining the nature of the elusive juice from the spleen. Since all attempts to isolate and collect it had failed, Malpighi proceeded to distill the cells or solid parts of the spleen, mixing the distilled humor with blood. What he found was a liquid with an empyreumatic smell, similar to that obtained by distillation of meat and matters containing sulphur. He mixed the distilled humor with blood recently drawn from an animal, in the hope of shedding light on the role of the hypothetical splenic juice: just like blood gets mixed with the juice secreted by the glands in the spleen, Malpighi mixed blood with the distillate of the cells of the spleen, in the hope of attaining analogous results. The blood turned black, as if it had been cooked. Then he extracted a salt from the spleen, which had the same empyreumatic smell and tasted first acid and then profusely and intensely bitter. He did not know whether the initial acidity he had tasted was a result of the distillation process or whether it was due to the spleen, although he inclined toward the latter hypothesis, because of the acrid smell of the spleen. In order to test the strength of the acidity, he threw the salt in milk, generating such a major turmoil that the milk lost its normal state and its color, becoming fetid. Apparently Malpighi did not taste that milk, although smelling was also a type of chymical test. In conclusion, Malpighi expressed the wish that chymists with more time and greater knowledge would shed light on the matter. In the end he concluded that the splenic juice is mixed with the blood and gives it a more appropriate nature whereby bile can be better separated in the liver and more generally improves the blood and allows the rest of the body to perform its operations more effectively.[46]

The rise of interest in glands in the second half of the seventeenth century changed the anatomical landscape: the years separating Glisson's *Anatomia hepatis* and Wharton's *Adenographia* at one end from Steno's *De glandulis oris* and Malpighi's *De viscerum structura* at the other saw some of the major transformations. Initial attempts by the Pisa group to grasp the minute structure of organs such as the testicles and the kidneys with Auberius and Bellini, respectively, had found a vascular as opposed to a glandular structure. Building on Steno's recent findings about the glands, however, Malpighi believed he had identified common features in the structure of four main body parts discussed here, liver, cerebral cortex, kidneys, and spleen, namely, minute glandules; thus, the localization of glands took center stage in his anatomical project. These transformations led to radical changes in the classification criteria for body parts: the revered notions of their nobility, of sanguineous and spermatic parts, and of parenchyma itself were exploded, thus subverting the traditional taxonomy still employed by Wharton.

Malpighi encountered several problems in the four core treatises of *De viscerum structura*. Possibly because of their microscopic size, Malpighi did not provide figures of glands but relied instead on verbal analogies from the vegetable world to describe them. More crucially, in no instance was he able fully to understand the mechanism of filtration in the glands and to detect their inner structure; at times he had to admit that reason had to intervene where our senses fail us. In addition, he was puzzled by the mode of operation and purpose of the spleen, given that a dog whose splenic blood vessels had been ligated lived quite happily for a long time. Despite these problems, *De viscerum structura* was seen at the time as a remarkably coherent and innovative work in terms of both the key methods of inquiry adopted and results achieved.

Malpighi used a wide range of techniques, such as microscopy, injections, insufflation, vivisection, and chymical analysis; he performed experiments by ligating blood vessels and excretory ducts with a variety of purposes in mind, such as enlarging body parts or cutting off their blood supply, as in excision; and he relied on pathological cases and stony concretions found in animal and human bodies, such as those reported by Möbius, Dodoens, Kentmann, and Bosco, mining the literature for stones supporting his glandular view of the viscera. He also supported his analysis by means of reports of postmortems, which he started compiling systematically on his return to Bologna from Messina in 1666 and which are first mentioned in *De hepate*. Thus, he used disease as a key tool of investigation not simply for pathology but for anatomy as well, in some cases as a fixating agent, in others as a magnifying one. We shall encounter again many of these views and methods in the following chapters.

CHAPTER FIVE

Fat, Blood, and the Body's Organization

5.1 The Necessity of Matter and the Animal's Benefit

Fat is not the first body part to spring to mind when one thinks of mechanistic anatomy and medicine, yet its accumulation and purpose pose philosophical problems relating to the body's organization and goal, as well as to processes such as nutrition and growth, which were originally associated with the role of the faculties. Descartes and Malpighi independently addressed these themes in publications of 1664 and 1665, respectively. In the posthumous *La description du corps humaine*, added by Claude Clerselier to *L'homme*, Descartes discussed nutrition and the nature of fat and blood. By comparing the growth of organs and other body parts to the accumulation of fat, Descartes raised the issues of goal-directness, of the difference between living organisms and dead machines, and of the presence of different levels of organization within the body, denying a role to the faculties in nutrition. Whereas Descartes' discussion occurs in passing in a section on nutrition, Malpighi devoted an entire essay to the nature and role of fat and its anatomical features, with special emphasis on the circumstances and location of its production and the possible existence of appropriate vessels: unlike Descartes, Malpighi was an anatomist who wished to get the anatomical evidence straight and spared no efforts to investigate and experiment, even having Borelli ship body parts of a deer and lion from Pisa to Messina. *De omento, pinguedine, et adiposis ductibus*—an anonymous addition to *Tetras anatomicarum epistolarum*—contains some of Malpighi's most intriguing philosophical reflections on the body's organization and on the tension between the necessity of matter and a purposeful nature.

The expression *necessitas materiae* occurs in the peripatetic and Thomistic tradition in which Malpighi had been trained. In *De principiis naturae*, for example, Thomas Aquinas drew a distinction shaped by the four Aristotelian causes between *necessitas materiae* and *necessitas finis*. The *necessitas materiae* proceeds from the efficient and material causes, which are the prior ones; the *necessitas finis* proceeds from the formal and final causes, which are the posterior ones. Thus, denying that the *necessitas materiae* is at play implies having recourse to final and not mechanical causes. The expression *necessitas materiae* occurs also in Francesco Buonamici's *De motu*, the major work by one of the authors Malpighi identified as one of his formative sources.[1]

The expression "necessity of matter" and its cognates occur in other works by Malpighi besides *De omento*, most notably in *De polypo cordis*, an essay appended in 1668 to *De viscerum structura*. Thus, there are good reasons to discuss *De omento* and *De polypo cordis* together, even though at first glance these topics don't seem to have much to do with each other: both address similar philosophical issues about the body's organization and mode of operation by having recourse to the same key expression. Therefore, this chapter enriches our sense of Malpighi's philosophical itinerary, showing that several years after having left Pisa and Borelli, he was still reflecting and elaborating on the body's organization, teleology, and mechanistic explanations. As we have seen in section 4 of the introduction, the relationships between mechanistic and teleological explanations could be quite subtle. Both *De omento* and *De polypo cordis* contain exceedingly interesting and problematic passages requiring careful exegeses. While accepting a notion of teleology providing a role for a provident artificer, Malpighi was skeptical about another notion of teleology, defending the existence of a principle or motor internal to the organism and acting to its benefit.

De polypo cordis can be approached from a variety of perspectives: an especially interesting one ties it to the medical rage of the time. Malpighi's treatise appeared in the immediate aftermath of the first publications on blood transfusions, some of which were performed in 1667 at Bologna in Cassini's house—probably in Malpighi's presence—immediately prior to the treatise's publication, and it relates to the extensive literature on intravenous injections and blood transfusions; thus, it seems appropriate to review this literature both for its medical and anatomical significance and as valuable background to Malpighi's research. In addition, Malpighi relied more explicitly and extensively than in previous works on diseased states as a tool of investigation. While postmortems had long been used in order to understand the location and progress of disease, Malpighi used disease programmatically to investigate body structures, exploiting it as an enlargement device or a microscope: the heart polyp thus became a tool to investigate the nature and structure of blood. The peculiar status of *De polypo cordis* is underscored by its inclusion in the *Bibliotheca anatomica*, which ostensibly dealt exclusively with anatomy, not pathology: the editors' introduction justified its presence in their collection because of its significance to the study of blood. Lastly, Malpighi performed several chymical experiments with a therapeutic aim, hoping to understand the nature of several ailments and to find a therapy for them. He distrusted intravenous injections, which were then being discussed and had been practiced at Pisa by his friends Fracassati and Bonfiglioli, preferring instead in vitro experiments.[2]

The following two sections are devoted to Descartes' and Malpighi's reflections on fat and its implications on the body's operations, with special emphasis on the tension between the notions of purpose and necessity of matter. Section 5.4 discusses early experiments on blood transfusion, especially at Bologna, thus providing a novel medical and intellectual context for Malpighi's treatise on heart polyps and the nature of blood, which is the subject of section 5.5; I focus especially on his methodological and philosophical concerns.

5.2 Descartes on Fat, Blood, and Nutrition

In the winter of 1647–48 Descartes started writing an anatomical treatise on the human body, *La description du corps humaine*, notably on the motion of the heart and blood, nutrition, generation, and growth. These were topics traditionally related to the faculties that Descartes wished to abolish: he aimed to show that all those operations occurred purely through the organization of matter depending exclusively on its own processes and properties, such as fermentation. Whereas many later mechanists tried to bypass the problem of generation by accepting some form of preformation and then dealing exclusively with nutrition and growth, Descartes adopted what could be described as a mechanistic epigenesis, seeking to explain the process of differentiation of the body parts of the fetus from conception. My aim here is to focus on Descartes' attempt to deal with nutrition and growth in a mechanistic way.[3]

Whereas in *L'homme* Descartes relied on the fiction of the statue, claiming that he was talking of a statue with several features similar to our body, in *La description* he dropped any pretense of dealing with anything but the human body and wished to address healing and prevent disease. Descartes argued that both solid and fluid parts move, the former more slowly than the latter; organs consist of filaments in constant motion that are continuously being replaced. This motion occurs in different ways depending on age: in young bodies, since the filaments are not joined together very strongly and the channels along which they flow are large, more filaments become attached to the organ than are lost; therefore, the organs become longer and stronger, or grow. When the humors flowing through the filaments are abundant and irregularly shaped, in the form of branches, they get stuck and form fat. Descartes pointed out that this process does not occur as nutrition proper does in flesh, but rather it takes place from an accumulation of parts as with dead things. When the humors are less abundant, they flow faster and pick up little by little the parts of fat, and one gets thinner. With age, the filaments of the solid parts become tighter and harder, the body ceases to grow and even to get nourishment, leading to death. This account expands somewhat that offered in *L'homme*, where Descartes had explained growth in terms of the softness of an infant's body parts and the ease with which they can be enlarged; as a result, the particles of blood entering the composition of the solid parts will be larger than the ones they replace, thus engendering growth. Descartes seems to have something akin to elasticity in mind here for his mechanistic explanation of growth; his views on nutrition, growth, and fat can be profitably compared to Aristotle's. In *Parts of Animals*, Aristotle dealt with the formation of fat, which he distinguished between soft like lard and hard like suet, which is more fibrous; he also claimed that fat is produced by blood and associated its excess with aging and even death. Whether an animal has soft or hard fat depends only on its nutritive material or blood and seems unrelated to a distinctive purpose served. Aristotle, however, argued that in moderate quantity fat is beneficial to the animal, contributing to the health and strength of the body, especially in the kidneys, whose natural heat it preserves. Thus, his views appear as a mélange between material necessity—for the type of fat produced—and finality governed by the nutritive faculty of the soul.[4]

By associating growth with fat, Descartes made an interesting gambit against the action of the faculties and against a goal-directness of bodily processes at the same time: it would seem absurd to place somewhere in the body a faculty responsible for getting fat or thin, and it would seem equally absurd to consider getting fat or thin a meaningful goal for an animal. Thus, by leveraging on the accumulation and reduction of fat, Descartes sought to present nutrition in animals in mechanistic terms. At the same time, he did not totally equate growth proper with the accumulation of fat: while providing an explanation of both processes in terms of moving particles of an appropriate nature and shape, he compared the accumulation of fat to that of dead objects, as if to imply that the nutrition of flesh in a living body had different features from the accumulation of dead things. As we have seen in section 4 of the introduction, however, in *Principes de la philosophie*—a work he referred to in *La description*—Descartes equated, in principle, machines made by artisans and living bodies made by nature, claiming that the only difference between them was one of size, and even *La description* treats growth and the accumulation of fat in the same terms. Therefore, it seems reasonable to understand the difference between these two processes in terms of varying levels of organization rather than a difference in kind between dead and living; the accumulation of fat appears to be more random and disorderly than growth. Indeed, La Forge's comment on nutrition and growth explains that, unlike all the other parts of the body, humors, spirits, fat, and possibly some glands do not consist of fibers, thus suggesting a different level of organization.[5]

Descartes further discussed nutrition in relation to the nature of blood. Since he died in 1650, he could not have known Pecquet's *Experimenta* and still believed that the liver was the site where chyle was transformed into blood, a process he compared to the transformation of the clear juice of grapes into red wine, following a tradition going back to Galen. Galen, however, had recourse to specific attraction in his account, something Descartes eschewed. In *L'homme* Descartes argued that blood is a heterogeneous fluid composed of particles of food we have ingested; he attributed a role to taste, arguing that those particles of food that taste too bland or sharp are inappropriate to enter blood or to serve for nutrition, whereas those producing an agreeable taste to the soul will nourish different body parts depending on their position in the body, as well as the size and shape of the particles, rather than the faculties. In *On the Natural Faculties*, for example, Galen had argued that nutrition occurs through a faculty of assimilation, literally rendering that which is ingested similar to the body part that is nourished. By contrast, as Descartes put it:[6] "To suppose that there are in each part of the body faculties that choose and guide the particles of nutrient to where they are appropriate, is to make claims to an account which is both incomprehensible and chimerical, and to attribute more intelligence to these than even our soul has." Much like sieves with holes of different shapes and sizes, the organs can separate the particles from blood that are appropriate to them without a soul.

In the aftermath of *La description*, Malpighi also addressed related issues.

5.3 Malpighi on Fat and Its Philosophical Implications

Malpighi's posthumous *Vita* provides a brief account of the circumstances of composition, publication, and contents of *De omento*. The initial stimulus came while at Messina from examining the fishes' omentum, a fatty membrane covering the abdomen from the stomach and liver to the intestine. His claim is corroborated by his notebook, showing that he started observations on the omentum of a fish on 6 December 1662. He subsequently studied other animals and published the work anonymously because he was uncertain of his conclusions. Malpighi did not seem to appreciate that since the work appeared under the initials "M. M." together with other essays by him, readers could have easily guessed its authorship. When he finally acknowledged it as his own, Malpighi claimed no role in earlier attributions and still expressed doubts about its contents. Yet he claimed that it is certain that fat is accumulated in appropriate places; he also argued that probably its purpose is to entangle and thus neutralize the sharp salty particles present in blood, thus proposing a role for fat within the corpuscular philosophy. While claiming that fat is necessary to animals, Malpighi was uncertain about the existence of adipose vessels or ducts: on the basis of evidence from recently slaughtered animals, however, he inclined toward the existence of hollow adipose structures similar to arteries and veins. As usual, he examined fat and omentum from a large number of animals, including sheep, dogs, rabbits, porcupines, frogs, and also the deer and lion that Borelli managed to ship from Pisa to Messina. In the late spring of 1663 a preliminary version of the essay was sent to Borelli, so that he and Bellini could comment on it.[7]

Having praised the hand above chymistry and mechanics as a tool for anatomical investigation, Malpighi states that he was stimulated to his work by the obscurity of the omentum's purpose. Close to the opening one finds a statement with important philosophical implications on the production and role of fat:[8]

> For I was not satisfied that fat was amassed from oily matter sweating from the blood vessels *from the necessity of nature and the arrangement of the vessels*, in order to keep warm the viscera designated for food, because Nature is not accustomed to accumulate incongruous and useless things in the more internal parts of the body and indeed in the very præcordia, about which she is especially attentive. Further, nature was neither blind in separating and mixing those bodies of which the mass of blood is composed, nor was she wanting in thickening the tunics of the vessels, so that the portion of nourishment continuously carried through them flows out during the journey.

In this passage Malpighi adopts a teleological approach according to which nature does not operate blindly on the basis of physico-mechanical laws, but rather it acts in a goal-directed and purposeful fashion. On its own, this passage may be read in different ways: it could imply that Malpighi denied that the sweating of oily matter from the blood vessels followed the configuration of the vessels and necessary physical laws of nature, but it could also imply that the animal body was so arranged by nature as to prevent incongruous and superfluous accumulations, in a way that does not necessarily deny the existence and role of physico-mechanical laws applying to dead matter and living bodies alike. The second part of the quotation deals with the nature and composition of blood and the problem of separation, the same that lies at the heart of *De viscerum structura*, whose initial three treatises were published only one year after *De omento*: Malpighi argues that nature arranged with foresight the composition of blood and the separation of its parts necessary for nutrition; the reference to blood ties nicely *De omento* to *De polypo cordis*, which we shall examine below.

This passage goes hand in hand with, and is elucidated by, a later one from *De omento* in which Malpighi returns to the same topic:[9]

> Concerning the generation of fat, I cannot agree with those who state that blood, almost fortuitously exuding from the vessels, is transformed into adipose matter; in fact if this happened *from the nature of matter and of the containing vessels*, it would

> occur constantly in all the single parts to which blood vessels are attached; in the lungs, however, in the bladder, in the meninges, and in other membranous parts crowded with tiny slender arteries and veins, we do not observe an accumulation of fat.

Malpighi denies that fat can exude from blood vessels according to physico-mechanical laws of matter and the configuration of the vessels, because it is found in some areas only, not everywhere in conjunction with blood vessels. Here too the passage could be read in two ways: it could imply that there is something else at play besides the nature of matter and the configuration of the vessels, such as a way of operating of nature peculiar to living organisms; but it could also mean that the generation of fat is not fortuitous but has been arranged by nature differently from that of blood exuding from its vessels, although still following physico-mechanical laws that do not differ from those applying to dead matter. Indeed, his solution was not to deny that those laws apply to the living body, but to find a different origin of fat. He proposed three alternatives: the stomach, the spleen, and the glands throughout the body, by which he meant the conglobate glands associated with the lymphatics. The advantage of the third solution is that fat would be produced only in those areas where it is effectively found and it would not be necessary to have recourse to nonmechanical actions to explain its presence at some location rather than elsewhere. We find here an early mechanistic use of glands, enabling the body to work according to the necessity of matter following its structural organization rather than an internal principle; thus, teleology operates at the level of the formation and design of the organism rather than because of a principle internal to the organism, such as the soul.[10]

These are not the only passages in which Malpighi discusses teleological matters and uses expressions related to the *necessitas naturae* and *natura materiae*. There are two further passages relying on cognate notions, drawing a distinction between a mechanistic understanding of the body's operations on the one hand and a finalistic view on the other. The first passage deals with some anatomical features forming a structure resembling a network of vessels that Malpighi had observed connecting the "striae" or grooves of fat:[11]

> Lastly, one can be in doubt whether these bodies are a communion of grooves made by fat, which, melted by heat, moulds some tunnels among the membranes, *by means of the necessity of matter alone, without any intervening end of nature*. This speculation can be refuted if we consider that these bodies, in which one observes only the network of vessels without the membranes, as in the porcupine, extend this network for a long stretch with a sinuous tract, and furthermore with branches propagating laterally, and, if they are observed in an animal still warm, are raised to a considerable height.

Hence, the peculiar structure of an animal—the porcupine in this case—serves to refute a hypothesis, much like the optic nerve of the swordfish served to refute Descartes' theory of vision. Similarly, in the passage immediately following Malpighi argued that ox fat was organized in a structure suggesting more than casual accumulation. In the passage above it is especially interesting to notice the contrast between the necessity of matter alone on the one hand and nature's end or purpose on the other; this contrast helps define the meaning of the notion of the necessity of matter as opposed to finalism in the formation of what seemed to Malpighi like tunnels or vessels. In this case too, as in the previous one in which he had recourse to glands, I suspect that his own refutation of the specific mechanistic account he had proposed does not necessarily imply the endorsement of a form of finalism rejecting physico-mechanical explanations. Arguably he believed that other solutions could be found satisfying anatomical rigor and a mechanistic interpretation at the same time. Incidentally, virtually exactly the same dichotomy between "nullo naturae fine" and "sola materiae & circumambientium necessitate" will be found in *Anatome plantarum*, in which Malpighi was to endorse mechanism. In my interpretation of the passage of *De omento*, Malpighi was seeking to provide and test mechanistic accounts of the formation and accumulation of fat; if the one he had initially thought out did not work, this did not imply that the enterprise as a whole was flawed. Unlike Descartes, Malpighi was an anatomist interested in the anatomical details, who disliked mechanistic explanations when they were only in principle rather than in practice—even though at times he found nothing better to fall back on.[12]

The other passage addressing the purpose of fat is also the last one dealing with the necessity of matter in *De omento*. We have seen above that Malpighi identified that purpose in the entanglement of salty particles in blood, but he also considered other alternatives, notably providing a source of heat for the body. Malpighi was quite skeptical of this interpretation:[13] "Since I hold that one should not attribute to heat as much as it is commonly boasted, and in animals perhaps it is caused by *the necessity of matter alone*; whence we observe heat to be more easily responsible for diseases than the artificer of a tranquil life." This quotation suggests a tension between a purposeful nature working for the animal's benefit and the blind necessity of matter endangering the body's health. Malpighi here associates the necessity of matter with the body's malfunction in the form of diseases due to excessive heat; hence, nature is more intent on removing heat than in generating it. He was quite dismissive of Galen's claim that fat and especially the omentum keep warmth. In *On the Usefulness of the Parts of the Body* Galen had argued that fat is warm, as shown by the sensation of those who use it for massages and the fact that it burns, since cold substances are not easily ignited. He provided further evidence claiming that although a gladiator from whom he had removed almost the entire omentum recovered promptly, he suffered from cold and kept his abdomen wrapped in wool; thus, here Galen relied on the excision of a body part to understand its purpose.[14]

Although at first sight these passages may have suggested a different Malpighi from the strict mechanist we are familiar with, upon careful analysis one finds that they do not necessarily contradict mechanistic anatomy: by denying a specific mechanistic account, Malpighi did not deny mechanism as such, since other mechanistic accounts could be found. His reflections on the body's organization and use of the expression "necessitas materiae" and its cognates are revealing: clearly in *De omento* he was struggling to conceptualize issues about the body's organization to his satisfaction.

Of course, *De omento* was primarily an anatomical treatise, and Malpighi surveyed the literature on the topic and provided a minute description of the omentum and its membranes in many animals. He also talked of the role of fat in the production of milk and semen, since fat is found in milk and castrated animals become fatter. *De omento* contains also an explicit reference to Galileo, who discussed the floating of fishes in *Discorsi intorno a due nuove scienze*. Galileo believed their fleshy parts and some parts close to the bones to be lighter than water, whereas other parts such as the bones are heavier; Malpighi added that fat and oil also play a role in making fish lighter.[15]

Although from the celebrative perspective of the birth of microscopic anatomy and of the discovery of new body parts *De omento* has little to offer, this essay provides rich and rewarding material on Malpighi's philosophical reflections on the body's mechanistic organization.

5.4 Blood Transfusions

The years 1666 to 1668 witnessed the dramatic rise and fall of blood transfusion across Europe. The novelty of the procedure and its medical and anatomical implications made it a major object of debate and controversy. Operations were carried out in Oxford, London, Paris, Rome, Bologna, and Udine. Although the idea of blood transfusion was not entirely new, the timing of the resurgent interest in the matter, in the aftermath of the rise of intravenous injections that we have mentioned in section 3.4, was no accident: the circulation of works by Elsholtz and Major was tied to the rise of transfusions, and in fact both of them engaged in early attempts of blood transfusion on humans. Major first mixed blood to be transfused with sal-ammoniac in a dish, whereas Elsholtz relied on the more traditional direct method. His plate from the 1667 edition of *Clysmatica nova* (fig. 5.1) shows a lamb-to-human and human-to-human transfusion. Elsholtz surveyed the literature on intravenous infusions and blood transfusions, including the cases of Fracassati at Pisa and of others, discussing the therapeutic advantages with regard to several diseases. Memorably, he also sug-

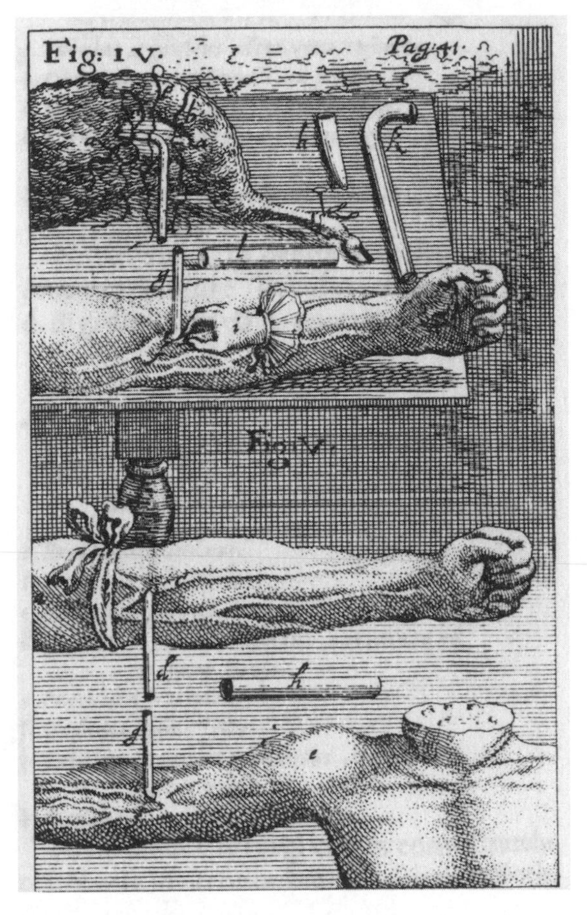

Figure 5.1. Elsholtz, *Clysmatica nova*: examples of blood transfusions to humans

gested that blood transfusions between husbands and wives may result in improved relations in marriage.[16]

Several features are of interest in transfusion experiments. Surgeons were often involved, collaborating with anatomists and physicians, not a surprising aspect in view of the connections with bloodletting: at the Paris *Académie* Claude Perrault relied on the surgeon Louis Gayant to perform the first transfusion among animals in France on 22 January 1667; soon thereafter still in Paris the Royal physician Jean-Baptiste Denis relied on the surgeon Paul Emmerez for several transfusions, including the first transfusion on humans; in London Lower collaborated with the surgeon and physician Edmund King; in Rome the surgeon Ippolito Maguani performed transfusions, while at Udine the physician Giovanni Battista Coris and Geminiano Montanari, professor of mathematics at Bologna, relied on the surgeon Andrea Cerassini.[17] Tech-

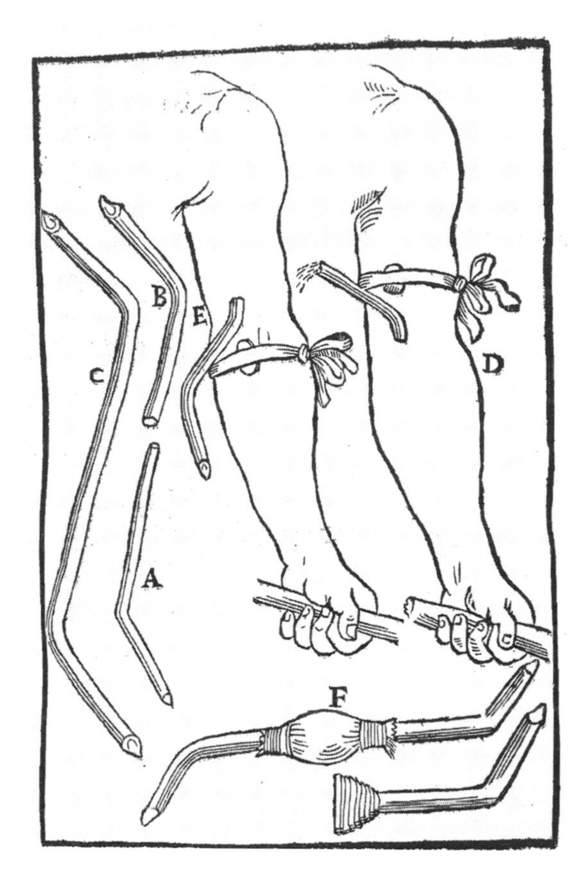

Figure 5.2. [Manolessi], *Relazione*: apparatus for performing a blood transfusion

nical issues involved the instruments and way of performing the operation, from vein to vein or from artery to vein: although the former was also used, the latter proved more effective and became the prevailing method. As to the procedure, Lower provided a detailed description in a letter to Boyle of 6 July 1666, involving ligatures and instrumentation. Baldassarre Coluzzi at Rome suggested a method then perfected by Maguani, involving bent glass pipes joined by a bladder F (fig. 5.2), which allows the surgeon to control blood flow. Although the operation was carried out on animals, the bent pipes were tested on humans for bloodletting, to make sure the animals did not suffer more this way than through standard methods: this is the reason why the figure shows a human arm, although the transfusion was from a wether to a dog. It is unusual for the period that such concerns about animal suffering would arise, and it is even more remarkable that humans were used to test whether the animal would suffer unduly, rather than the other way around.[18] Which animals to use as the donor and recipient also became an issue of debate, grafting being used as a term of comparison by both sides: some argued that the great difference in the blood of different animals meant that blood from one animal could be a poison to another; others claimed that animal blood was best for transfusion to men, because it was purer, and since we eat the meat and drink the milk of different animals, there seems to be no reason why their blood would be pernicious. Blood transfusion was considered together with generation, not a surprising association given that semen was thought to come from blood.[19]

Aside from philosophical interest and basic curiosity, transfusion had a medical role. Several blood transfusions on animals and humans were carried out in the hope of curing the recipient and often involved monetary remuneration. The most spectacular cases involved a patient deemed crazy that allegedly regained his sanity thanks to the transfusion of six ounces of blood from a calf on 19 December 1667 in Paris, and a deaf dog that became more vital and allegedly regained part of its hearing thanks to the transfusion of blood from a lamb in Udine on 20 May 1668. These reports testify to the great medical expectations attached to blood transfusion. As is well known, the first blood transfusion to a human was carried out in Paris on 15 June 1667 by Denis and Emmerez, whereas Lower and King performed one on Arthur Coga in London on 23 November 1667, although both Denis and Lower had performed and discussed transfusions between animals before that time. The death of a patient, Antoine Mauroy, in Paris led to greater caution and a partial moratorium on these early daring attempts. Another patient affected by phthisis died following a blood transfusion by the surgeon Riva in Rome.[20]

Blood transfusions between lambs were also performed at Bologna, starting in May 1667, in the house of the astronomer Giandomenico Cassini. The report in the first issue of the *Giornale de' Letterati* states that the lamb receiving the transfusion died on 5 January 1668. Cassini and Montanari belonged to the same intellectual circles within which Malpighi moved; all three were close personal friends and belonged to the Accademia della Traccia, founded by Montanari in 1665. In a deleted passage from the manuscript of his *Vita*, Malpighi refers to his anatomical collaboration with Cassini, stating that they dissected the eye together and experimented on its parts; this must have happened before 1669, when Cassini moved to Paris. Thus, it would appear as if Malpighi would have been informed of—and probably involved in—the debates and experiments on blood transfusion raging across Europe and in his own town. Clearly the nature of blood and its diseases would have been of central importance in those settings. This account provides a justification and context for a treatise on heart polyps and the nature of blood, as an appendix to a collection on the glandular structure of the viscera. It is to this treatise that we now turn.[21]

5.5 Malpighi on Heart Polyps and the Nature of Blood

From a methodological standpoint *De polypo cordis* is one of Malpighi's richer works, a nodal point in which he developed his previous methods of investigation and outlined a project that would become central to his later research. In *De viscerum structura* we have already encountered instances of his reliance on malformations and diseased states—especially stony concretions from the Renaissance literature and diseased organs from postmortems—to confirm the glandular structure of the viscera. In *De polypo cordis* he chose diseased states as the privileged window into the constituents of blood. Malpighi relied on both descriptions of heart polyps found in the literature, such as the one discussed and reproduced in a work by Thomas Bartholin (fig. 5.3), and his own observations. Here too, in fact, it is possible to identify passages reporting cases taken from his collection of postmortems; one of them mentioned in both *De liene* and *De polypo cordis* refers to an unnamed woman who had died at the Bolognese Ospedale della Vita and is dated 5 January 1668; despite the year 1666 on the title page, *De polypo cordis* appeared two years later.[22]

Luigi Belloni pointed out that later in the century Johann Conrad Brunner aptly called the method of inquiry of moving across different species "microscopium naturae," the microscope of nature, as we have seen in section 1.3. In the same way as the

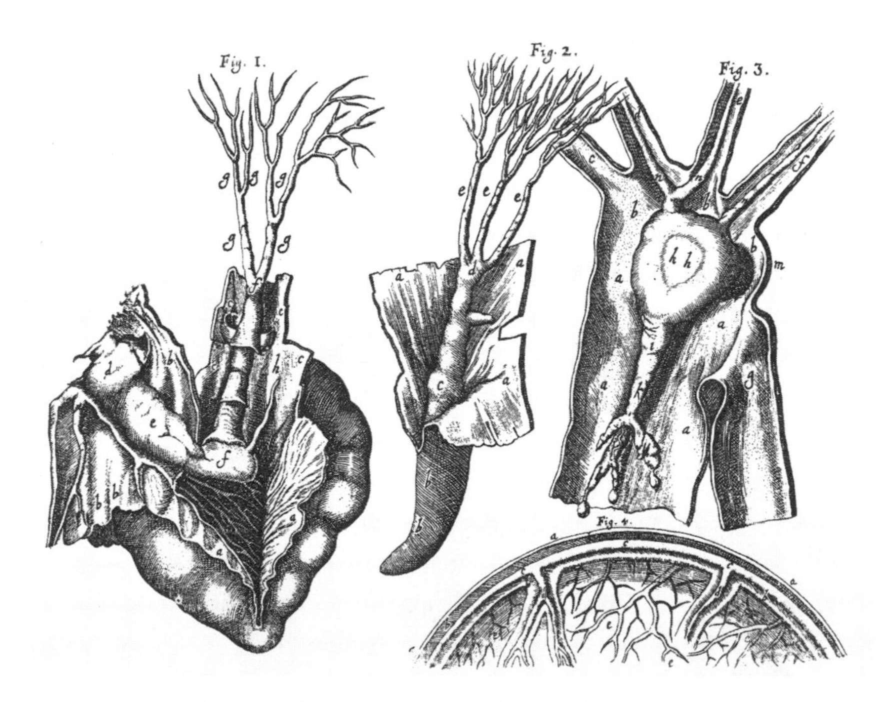

Figure 5.3. Thomas Bartholin, *Centuria*: heart polyp.

optical microscope enlarges and at times reveals body parts that would otherwise be indistinct or invisible, moving across different species whose organs and their constituents vary in size provides the anatomists with a way of accomplishing the same task. In his work Malpighi combined the optical microscope and the microscope of nature. By analogy to Brunner's denomination, I have called the method of inquiry of *De polypo cordis* the "microscope of disease," namely, the use of disease, with due caution and constantly comparing diseased states with normal ones, as a magnifying lens, not as a hindrance but as a tool of inquiry enabling visualization.[23]

The key idea of *De polypo cordis* is to explore the structure of blood by means of heart polyps, structures occasionally found in the hearts of cadavers; the fact that today they are not attributed to a single condition and at least some of them may be formed after death is of no great significance here. Malpighi showed by means of the microscope that blood was heterogeneous and argued that it consisted of more than forty components, including "red atoms"—as he called them—all the fluids secreted by the glands such as saliva and pancreatic juice, and a white reticular structure. On the basis of chymical tests based on fire analysis, Malpighi denied that heart polyps were formed by fat or phlegm—whose existence he doubted anyway—and argued that polyps were formed by the reticular structure. This identification enabled him to pinpoint the culprit of the diseased states in normal ones. Moreover, Malpighi sought to offer a justification based on microscopy for the different strata appearing in congealed blood: the whitish crust at the top consists of the compact reticular structure; below it, the pores of the crust become larger and allow access to the yellowish serum; lastly, at the bottom, the pores become larger still and are filled with red atoms, giving to the lowest part a dark coloration that has led many to mistakenly identify it with melancholia. In fact, a change of position makes them purple.[24]

De polypo cordis opens with a striking praise of the epistemological and heuristic role of imperfection:[25]

> I have always believed that the morbid states, which we see frequently arising in the bodies of animals due to the jokes of Nature or the strength of aberrant disease, shed much light on the investigation of Her true norm and method of operation. In fact those morbid states indicate a *necessity of matter, and determined inclination* revealed in the construction of the animal body. Thus monsters and other mistakes dissipate our ignorance more easily and reliably than the remarkable and polished machines of Nature: hence the present century has learnt more from studying insects, fishes, and the first unformed *warps* in the development of animals, than have all the preceding ages exclusively interested in the bodies of perfect animals.

One feels as if Malpighi had reached a turning point here: he had explored as far as he could the viscera and other organs with increasingly interventionist techniques and powerful microscopes, and now he felt that it was time to focus more systematically on diseased states, lower animals and even plants, and the formation of animals, here introduced with the weaving analogy of the warp.

Erasistratus had already employed weaving analogies in the process of growth: it may be useful to review briefly Galen's views on growth expressed in *On the Natural Faculties*, a work in opposition to Erasistratus's views, which inclined toward mechanism. In the first book Galen defines growth as the faculty of extending that which already exists in all directions. He reports that children in Ionia play with pig bladders by rubbing them against warm ashes; while doing this they recite verses exhorting the bladder to grow and indeed the bladder does grow in volume because of the rubbing and the heat. Galen points out that this is not real growth, however, because while increasing its surface the bladder becomes thinner: real growth cannot be accomplished by art and can only be a property of nature. In an important passage in the second book he criticizes Erasistratus, according to whom growth occurs as in a rug, a rope, a bag, or a basket, by weaving at the end more material similar to what was already there. Galen objects that this process is formation rather than growth: the basket is not yet a basket while it is being made, whereas growth is the increase in all dimensions of all the parts of something that is already complete. As such growth pertains only to animals and plants and results from matter acquired through nutrition. By contrast, Malpighi believed the analogy between weaving and growth to be perfectly legitimate: working on a virtual loom—a mechanical device in its own right—nature operates mechanically with warp and weft. Malpighi will seek to account for the process of growth in *Anatome plantarum*, not only with examples of the vegetable world but also with bones and teeth.[26]

The opening quotation above refers to a notion we are already familiar with: both expressions "necessitas materiae" and the more scholastic "determinata inclinatio" have a philosophical role within a mechanistic framework, emphasizing the regularity of nature and the tendency of body parts to form diseased states, such as obstructions or excessive accumulations. The references to "jokes of nature" and "monsters" have to be interpreted as the outcome of the same necessary laws governing nature's normal course and as crucial tools for uncovering those laws, rather than as the outcome of a playful nature, in Renaissance fashion, whose signs have to be deciphered: although jokes of nature, monsters, diseased states, and lower animals may be unusual or peculiar, they are not qualitatively different from the products of higher animals but can shed light on nature's normal operations by highlighting a particular feature, such as the component of blood responsible for heart polyps. The discussion below and part III will reveal the significance of this program to Malpighi's later investigations.[27]

In a subsequent passage from *De polypo cordis* Malpighi tentatively rules out the finalism implicit in the opening quotation:[28]

> These things will not seem insignificant to anyone who, by assiduously dissecting animals, comprehends the industry of nature equally in morbid tumors and in the creation of the parts' *warps*, because her method of proceeding is nearly identical. Thus I recall noticing that an iron needle bursting out of the fleshy stomach of a hen was covered with a strong double membrane and a coating of fat as well. Moreover, we can consider whether all these things happen by the *sole necessity of matter and motion, without a guiding mover for the animal's benefit.* Similarly, in certain tumors arisen in the lungs, liver, and elsewhere, integuments or multiple bladders are joined, in which the larger encloses the smaller and thus they fit together successively; the conglobation of similar tumors can be regarded as similar in nature to that of polyps, for the matter and mode of production in both cases is presumably the same: in fact, following the usual *law of nature*, from a network of *threads* several layers can be formed, which can remain everywhere separate if what lies between them is not congealable but is the watery fluid which abounds in tumors of this species.

This remarkable passage establishes a near identity between nature's way of operating in diseased and normal states. Notice the reference to the coating of fat, complementing the passages from *De omento* we have seen in section 5.3. Here Malpighi tentatively argues that there is no mover directed to the animal's purpose, no faculty of nature or of the soul, because everything happens according to the same laws of nature. Malpighi adopts a weaving analogy in this passage as well, in one instance employing the very same word—*stamina*—as in the previous quotation. This passage goes hand in hand with that from *De liene* discussed in section 4.6, in which he argued that nature works in the same way in healthy and diseased states, especially in forming membranes in tumors. Growth was one of Galen's natural faculties, together with nutrition and generation. The passage above, however, tentatively accounts for growth in terms of *laws of nature* that are not peculiar to living beings, whether plants or animals. In *De polypo cordis*, however, Malpighi's program appears better defined than in *De omento*. In chapter 8 we will see how mechanists dealt with generation, chapter 9 documents how Malpighi developed these ideas about the mechanism of growth for plants, and in chapter 10 we shall encounter further instances of Malpighi's usage of the expression "necessitas materiae" in his controversy with Mini on the role of the soul in the body's organization.

There is a profound link between anatomy and pathology, since it was through understanding the normal that the pathological could be understood. In his first reply to Jean Riolan the Younger in 1649 William Harvey emphasized this aspect when he stated, "that which is normal is right and serves as a criterion for both itself and the abnormal" and "from pathology the practice and art of therapeusis and opportunities for discovering multiple new remedies, derive," concluding that "one dissection and opening up of a decayed body, or of one dead from chronic disease or poisoning, is of more value to medicine than the anatomies of ten people who have been hanged." Harvey was already relying on postmortems to analyze the diseases of all the organs of the body in his *Prelectiones* to the London College of Physicians. The privileged role attributed to normal states did not prevent him or other anatomists from using pathological states to confirm their own anatomical views. *De motu cordis* established the circulation of the blood on a number of grounds, such as the great abundance of blood pumped by the heart and the role of the valves in the veins in allowing unidirectional flow of blood. However, it also relied on evidence from butchers killing an animal and draining its blood and surgical amputation: in both cases huge hemorrhages can occur, consistently with the notion of circulation, provided that the slaughtered animal is still alive and the heart beating. Harvey also mentioned the swelling on his head about the size of an egg he experienced after a fall, or the speed with which poisons spread throughout the body from wounds, or that medicines applied externally are almost as effective as if taken internally. Harvey also provides embryological evidence; he argued that the liver concocts chyle to make blood, but since in the embryo nutrition is provided by the mother's blood, there is no need for the liver and in fact this organ is among the last ones to be formed.[29]

It is instructive to notice the similarities and shifts with Malpighi's essay and the program it outlined. Also Malpighi relied on diseased states, but he made them the centerpiece of his investigation: rather than corroborating findings established by more direct means, Malpighi's investigation starts from them. At the end of the century, Bernard de Fontenelle, permanent secretary of the Paris *Académie Royale des Sciences*, referred to this method of inquiry in his history of the *Académie* with words closely resembling Malpighi's, emphasizing the methods of both investigating different animals and using monsters as a key to unlock the secrets of normal bodily operations.[30]

In dealing with blood, Malpighi did not miss the chance to revise his own views about the mode of action and purpose of the lungs. Whereas in 1661, following Borelli's advice, he had presented them as performing a purely mechanical operation of mixing the composite components of blood, now he accepted that through filtration they absorb a very subtle vital balsam or salt from the air—called "salt of life," following Tommaso Cornelio's *Progymnasmata*—whose existence is confirmed by the smell and taste of blood. The "salt of life" awakens ("*suscito*") the red portion of blood; thus, contrary to his previous beliefs, Malpighi attributed to the changing color of blood a significant feature in respiration. Malpighi conceived the lungs almost as a reverse gland: in a gland the afferent arterial blood is separated into two efferent fluids, one depending on the gland and venous blood. By contrast, in the lungs two afferent components—venous blood and air—are mixed to form efferent arterial blood. Based on postmortems and the nature of a thick polyp-like crust forming on blood extracted from patients, Malpighi argued that pleurisy and acute fevers occur in conjunction with special atmospheric conditions affecting the filtration of air. On the basis of intravenous injections in a dog, Malpighi questioned the role several authors had attributed to niter in respiration: all he could observe was excessive urination. Despite having recourse to injections in this case, Malpighi expressed doubts about such techniques because they are laborious, ambiguous, and one cannot see what happens inside the body. His alternative was to mix in a container different chemicals with blood when it is still fluid and warm: oil of sulphur and oil of vitriol make blood bloat in the areas of contact and make it look burnt; alum also makes it look burnt. Niter, distilled wine, common salt, sal-gem, sal-ammoniac, sulphur, and hartshorn make blood bright red and seem to prevent coagulation for a while. On the basis of these in vitro experiments, Malpighi conjectured the causes of diseases generating polyps or the coagulation of blood, such as the plague, and on suitable remedies, rendering the blood more fluid. Although he did not refer to blood transfusions, one may wonder what he would have made of the new rage.[31]

In his posthumous *Vita* Malpighi reports that in *Spicilegium anatomicum* the Dutch anatomist Theodor Kerckring objected to *De polypo cordis*, arguing that it is erroneous to believe that heart polyps could be found in live patients; rather, he believed them to be formed after death when blood cools down. Malpighi retorted that their conformation differs from coagulated blood and they are more likely formed not in the short time between the patient's death and the postmortem but during a long preceding illness, so much so that in some cadavers the heart and the nearby vessels, especially the aorta, are vastly dilated and in some cases—as in the case of Cardinal Bonaccorsi—are covered with bony scales. He added that the relevant anatomical specimen was conserved by Silvestro Bonfiglioli, probably in the Aldrovandi Museum, of which he was curator. Although Bonaccorsi's case was the only one in this context to provide permanent physical evidence in the form of a hardened aorta, Malpighi could list two other cases, those of Father Angelo Serafino de Ratta of 1667 and Ascanio Baldeschi of 1677, in which he had personally witnessed heart polyps: both can be easily identified in his manuscript records, since they are listed in a special booklet he compiled on the results of postmortems. Malpighi replied to other objections as well, such that polyps could not be found in living patients because they would obstruct blood circulation and could not be formed while the patient is alive because of the impetus of the blood circulation. To these claims he retorted that since the polyps' formation is slow, blood circulation would not be obstructed at once; moreover, he relied on his observations of aqueducts and water conduits to argue that solid concretions can be formed even when a fluid is flowing, especially since the inside of the heart is not smooth but rather full of asperities and fibers. Malpighi stated, "as happens when stones form in the pelvis, or deposits in water channels," and "this is what happens in aqueducts" or "as occurs in rivers." The analogy with aqueducts is significant in that it draws a parallel with a process unrelated to life. The editors of the *Bibliotheca anatomica* discussed Kerckring's objections to Malpighi, arguing on the basis of postmortems that although some polyps may form after death, others were formed in a slow process during the patient's life. [32]

Cardinal Bonaccorsi's aorta makes one think of the analogy with relics of saints, kept as testimony of their sanctity and of true religion in churches and other holy places. Similarly, mechanistic relics, preserved in museums or in private collections, were believed to provide firm and lasting evidence to the physico-mechanical processes at play in the animal machine. In several instances Malpighi was called as an expert witness to assess whether particular cases were miraculous, as in the case of the incorrupt body of Caterina Vigri, recently studied by Gianna Pomata, and the healing of Sister Serafica Francesca.[33]

An especially interesting passage from Malpighi's *Vita* concerns his attempt to link lesions found in postmortems not only to anatomical investigations of the healthy body but also to symptoms observed in patients. That passage lists a long series of clinical signs, such as feeling of constriction in the præcordia, a vibrating and intermittent pulse, dilatation of the jugular blood vessels, frequent defect of sensibility, numbness and indistinct pulse in the arm, and painful spasm in the sternum. In that passage Malpighi appears as a careful clinician of heart disease. His followers and Morgagni's teachers Albertini and Valsalva made this approach the centerpiece of their careful investigation of aneurisms, a topic to which Albertini contributed a celebrated essay.[34]

The nature and purpose of fat and blood occupy an important position in mechanist reflections on the body's organization and in medical debates on blood and its diseases. Descartes' passing remarks on fat and Malpighi's more extensive investigations call for a broader reflection on this issue from both an anatomical and a philosophical standpoint. Despite Descartes' rather ambiguous comparison of fat to dead objects in *La description du corps humaine* and Malpighi's equally ambiguous references to the necessity of matter and a purposeful nature in *De omento*, ultimately both sought to frame the problem of the accumulation and purpose of fat in mechanistic terms, the former in connection with nutrition, the latter with regard to the body's organization. Malpighi's suggestion that fat would be filtered by glands located at appropriate sites fits well with chapters 4 and 6, linking in surprising ways different portions of his work on glands. Both authors eschewed a teleology immanent to the organism.

In the second part of this chapter blood takes center stage. It is not surprising that at the time of the rise of intravenous injections and especially during the brief window when blood transfusions were performed, the nature of blood and its chymical properties attracted the attention of anatomists and physicians. In *De polypo cordis*, Malpighi's reflections on teleology are enriched by his statements about a purposeful nature acting to the animal's benefit versus the necessity of matter in the formation of heart polyps and in bodily processes in health and disease, including growth. In this regard he employed a weaving analogy as a way to account for growth in mechanistic fashion, an approach that we shall encounter again in *Anatome plantarum*.

In *De polypo cordis* Malpighi extends the range of his techniques of investigation to the study of disease as a way to understand healthy states—the microscope of disease. Ultimately, it is the necessity of matter that provides the crucial philosophical justification for his reliance on the microscope of disease and that underpins the two-way relation between anatomy and pathology, because nature operates in the same way in health and disease. Moreover, he advocates the study of lower animals, plants, monsters, jokes of nature, nature's mistakes, and the process of formation of the animal to gain an entryway into nature's way of operating. As we shall see in part III, Malpighi moved to investigate precisely those areas, such as lower animals, the formation of the chick in the egg, and plants.

CHAPTER SIX

The Structure of Glands and the Problem of Secretion

6.1 Different Perspectives on Glands

Glands were at the same time the central element and one of the most obscure features of mechanistic anatomy. Following Wharton's *Adenographia* of 1656—a text reprinted in 1659, 1664, and 1671—glands became a major topic of investigation in the works by Steno and Malpighi. No chapter-length study could do justice to the size of the literature and complexity of themes in the study of glands from the second half of the seventeenth century to the beginning of the eighteenth century. Moving chronologically and conceptually beyond the period and topics discussed in chapter 4, we will investigate some key aspects of the research on glands around the last quarter of the seventeenth century. Relying on the combined perspectives offered by the *Bibliotheca anatomica* and Malpighi's correspondence and investigations, I discuss the discovery of new glands and renewed studies and debates on their structure and mode of operation.

Several factors affected the understanding of the structure and operations of glands in the last decades of the seventeenth century. Some anatomists conceived glands as a filtration or separation device of something preexisting in blood rather than a site of production or synthesis of something entirely new; others, however, understood the glands' operations in terms of fermentation, involving an elaboration or rearrangement of the constituent parts of a fluid. A major demarcation concerned the role of the soul and its faculties, with some, like Malpighi's Bologna colleagues Mini and Sbaraglia, arguing that structure alone was insufficient to understanding glands because nonliving matter could not effect the required separations. Additional differences concerned the source of the secreted juice: we have seen that Wharton identified it in the nervous fluid, whereas Steno believed arterial blood to be the likely source; Anton Nuck tested those views experimentally. Further, conglobate glands were interspersed in the lymphatic system, but their role was unclear; there was no agreement as to the source of lymph, in terms of both the fluid whence it originated and the location where secretion occurred. Lastly, there were two main positions on the structure of glands: the first, with Malpighi as the main proponent, according to which the blood vessels enclosed a follicle or vesicle that was the actual site of the separation of the fluids; the second—initially defended by Bellini for the kidneys and later by others—according to which glands consisted merely of blood vessels. The vascular view of glands became especially prominent in the Netherlands.

Section 6.2 explores the work on the intestinal glands carried out in Leiden by Pechlin and by the Schaffausen group around Johann Jakob Wepfer, including Johann Conrad Peyer, Johann Jacob Harder, and Johann Conrad Brunner. The discovery of new intestinal glands was not merely another addition to a long list, but had profound implications on recent debates on digestion, the role of the pancreas, and therapy. Brunner developed a novel—if cruel—method of inquiry by excising the pancreas in order to investigate its purpose: his conclusion, much like Pechlin's, was that the intestinal glands served a purpose akin to that of the pancreas by secreting a similar juice; therefore, the pancreas was not indispensable to life. Section 6.3 takes the lead from Malpighi's reply to a request by Bellini about recent studies on the nature of fluids and especially their separation inside the bodies of animals. Bellini mentioned among the authors already familiar to him several mathematicians. By contrast, Malpighi relied on the anatomical, surgical, and medical literature: his reply outlines the state of the field and provides a useful guide to our investigation. Section 6.4 is chiefly devoted to the circumstances of composition of Malpighi's 1689 *De structura glandularum* and an analysis of this treatise, in which he proposed a structural identity between conglobate and conglomerate glands. Once again, his correspondence with Bellini provides valuable insights, showing that Malpighi's work was a response to the attack at a public anatomy by his rival Sbaraglia. Malpighi's correspondence with Bellini contains additional valuable material, such as Bellini's report of his discussion of glands in the anatomical theatre and a synoptic view of glands. Lastly, section 6.5 examines later studies of glands by Nuck, who provided a classification and investigated them in a variety of ways, challenging Malpighi's identification of conglobate and conglomerate glands.

6.2 Intestinal Glands and Their Implications

The 1670s and 1680s witnessed both the discovery of further glands and critical reflections on their mode of operation and purpose, as well as warnings about the new anatomical rage. In the 1670 *Spicilegium anatomicum*, for example, the Amsterdam physician and anatomist Theodor Kerckring sounded an alarm bell against the tendency to see glands uncritically in all parts of the body and the use of the microscope.[1]

In 1672 the Leiden physician Johannes Pechlin published *De purgantium medica-*

mentorum facultatibus, an extensive treatise on purgatives. Its last chapter, however, also dealt with anatomical matters and announced the discovery of some new glands in the intestine. Having ligated a portion of the intestine of a dog in a vivisection experiment, Pechlin saw that it filled with a fluid that could not have come from the liver or pancreas: its taste was salty and resembled pancreatic juice. Figure 6.1 shows Pechlin's experiment, with OOO being the portions of the intestine intercepted by ligatures and P the erupting fluid. The fluid, reasoned Pechlin, must have come from

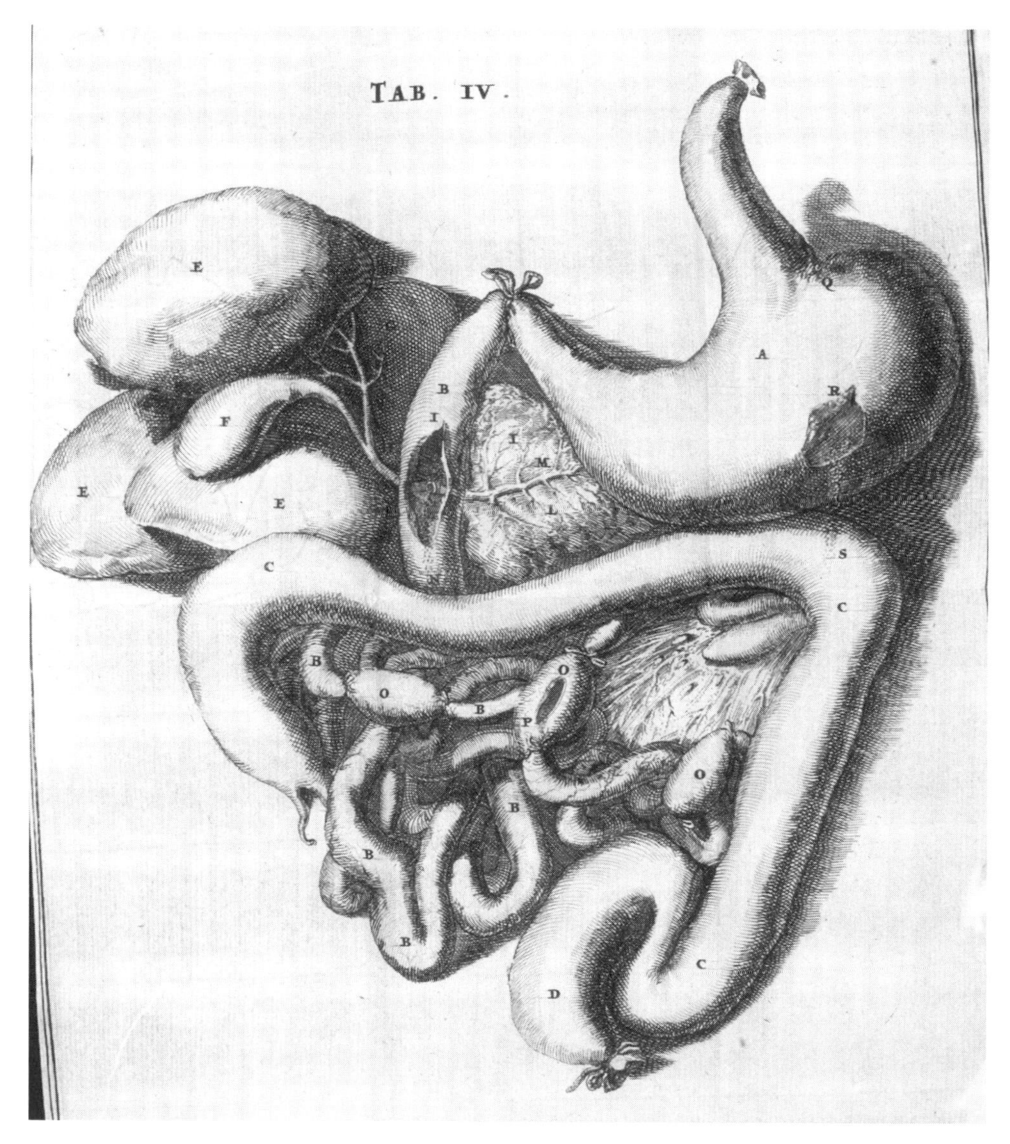

Figure 6.1. Pechlin, *De purgantium medicamentorum facultatibus*: digestive system

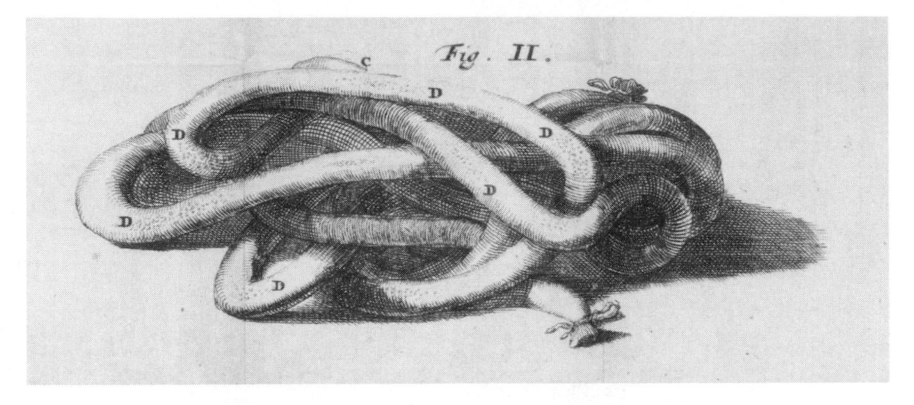

Figure 6.2. Pechlin, *De purgantium medicamentorum facultatibus*: intestinal glands

some glands, and indeed he detected some patches with clusters of glands, which he assumed secreted the fluid he had tasted. At this point he pursued his investigation on the intestine of a pig, which he removed from the animal and cleaned—as we are glad to learn. Inflating it and looking at it against the sun, he detected the glands he was looking for, marked DDDD in figure 6.2. At one point Pechlin stated that he carried out some of his investigations through maceration in warm vinegary water on a piece of intestine provided by Jan Swammerdam, suggesting a connection between them; indeed, one of the dedicatory poems in Swammerdam's 1667 *De repiratione* was authored by Pechlin. About 1672 Swammerdam was in a fierce controversy with de Graaf, and Pechlin himself may have contributed a satirical work under a pseudonym. In the second part of the 1673 *Observationes anatomicae* of the Amsterdam *Collegium privatum*—to which Swammerdam contributed extensively—the pancreatic juice of a sturgeon was found to be not acid but rather bitter, against Sylvius's and de Graaf's opinion. In fact, Pechlin's findings have profound implications against the works and views by Sylvius and de Graaf on the role of pancreatic juice: they believed that alkaline bile mixed with acid pancreatic juice to induce effervescence necessary for digestion. Pechlin's view was that the glands in the entire digestive tract from the mouth and esophagus to the intestine did not secrete fluids with highly specific individual properties, but rather produced an undifferentiated fluid flowing like a river enabling digestion to occur.[2]

In the following years several Swiss anatomists and physicians published works relevant to glands and secretion: in 1677 the Basel anatomist Johann Conrad Peyer, seemingly unaware of Pechlin's work, published *Exercitatio anatomico-medica de glandulis intestinorum*, in which he announced the discovery in the intestine of some new glandular structures secreting a fluid similar to pancreatic juice (fig. 6.3 at D); Peyer judged them to be related to conglomerate glands. It is remarkable that Peyer,

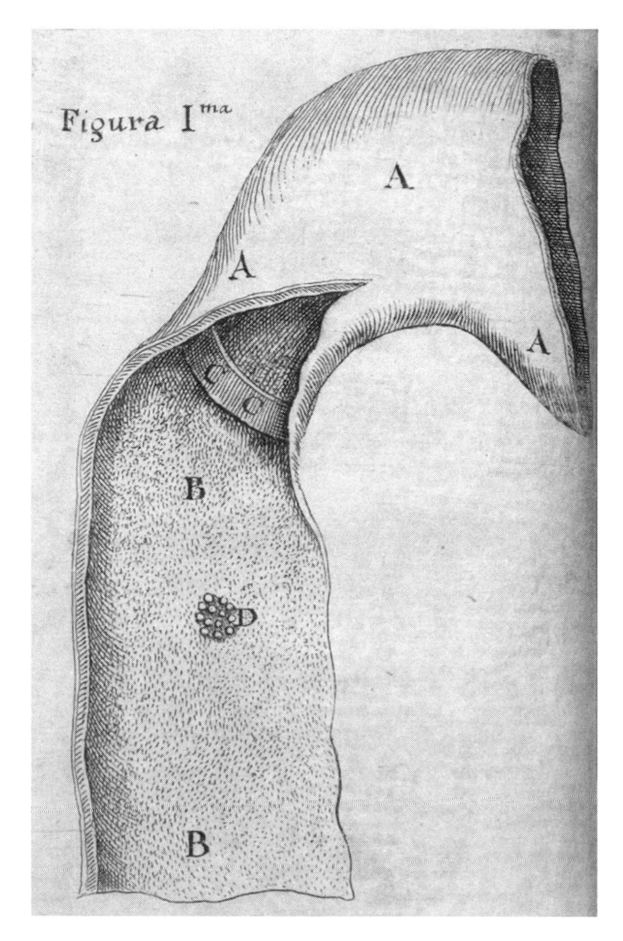

Figure 6.3. Peyer, *Exercitatio de glandulis intestinorum*: intestinal glands

like Pechlin, attributed to these glands a role similar to the pancreas. It was only in 1682 that he acknowledged Pechlin's priority, claiming that previously he had been unaware of his work. It is especially interesting to compare the plates in Pechlin's and Peyer's works: magnification (fig. 6.4) made Peyer's visual presentation more perspicuous. Much like de Graaf had done for the pancreatic juice, Peyer considered the medical implications of his findings: both works bear the words "anatomico-medica" in the title. He argued that his finding had profound implications for a range of diseases and complaints, such as diarrhea, dysentery, and constipation.[3]

Peyer was already aware of Johann Conrad Brunner's experiments on excising the pancreas from dogs, to be published a few years later, and he mentioned them in his work; therefore, it seems that their research proceeded in parallel fashion: whereas Brunner showed that animals could live without the pancreas, Peyer showed that

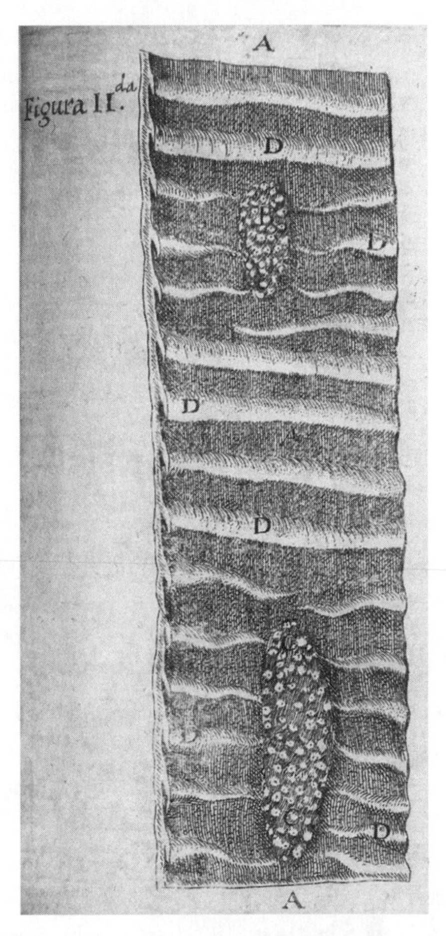

Figure 6.4. Peyer, *Exercitatio de glandulis intestinorum*: glands under magnification

there are other glands that could take on its role with minimal effects on the animal's life. Peyer argued that he expected dogs without the pancreas to be less agile in jumping, because of the resulting lack of salino-sulphureous particles necessary for muscle contraction. Thus, Brunner and Peyer relied on an investigation technique whereby the purpose of a gland was sought by excising it from the body and studying the effects.[4] Paradoxically, Peyer's research did not have the devastating effects on the theory of Sylvius and de Graaf one may expect: by arguing that his glands secreted a fluid similar to pancreatic juice farther down the intestine, Peyer—unlike Pechlin—still accepted a specialized chymical role for pancreatic juice and for the juice produced by his glands, such as inducing fermentation in the intestine, albeit based on different chymical principles. Furthermore, he questioned the role of taste to assay the juice secreted, arguing that the rapid evaporation of the nitro-aerial spirit affected the test. Peyer added to his treatise a study of the chicken stomach, whose glands he claimed resembled an enlarged version of those of the intestine, "velut per microscopium amplificata," an example of what Luigi Belloni, following Brunner, called "the microscope of nature."[5]

In 1679 the Basel anatomist Johann Jacob Harder published *Prodromus physiologicus naturam explicans humorum nutritioni et generationi*, devoted to the humors of the body rather than the problem of secretion and glands in particular. True to its title, Harder focused on the different stages of digestion and the humors associated with generation, such as male semen, the female egg or semen, menstrual blood, and finally milk. Harder reviewed the current literature and compared it to old doctrines going back to Galen, offering a useful synopsis of the field.[6]

In the same year Wepfer published at Basel a remarkable treatise, *Cicutae aquaticae historia*, in which, following the hemlock poisoning of some children, he discussed the effects of poisoning with a series of experiments involving dissections and vivisections of animals. In the same treatise, Wepfer reported the dissection of a woman, Barbara Meyer, executed by decapitation for infanticide, and announced the discovery in the human stomach of new glands, which he had previously detected in the stomach of a beaver. Both findings required extensive preparation techniques and long maceration in water to separate the tunics and membranes of the stomach.[7] In the same treatise Wepfer reported experiments on the pancreas performed in 1677–78 together with Brunner—his son-in-law—and later included in Brunner's treatise on the pancreas.[8]

In the 1683 *Experimenta nova circa pancreas*, Johann Conrad Brunner questioned by means of dog vivisections the role Franciscus Sylvius and de Graaf had given to the pancreas; Brunner ligated and cut the pancreatic duct and removed most of the pancreas, apparently with no major ill effect on the animal. He also tried to make sure that the pancreatic duct did not grow back, as it tended to do because of "Nature's technē," an especially interesting expression suggesting at the same time nature's skill and her artistry in operating mechanically, as in building mechanical devices or mechanisms. Brunner investigated this phenomenon by means of mercury injections into the root of the pancreatic duct, finding that on one occasion the duct had found its way to the intestine again (fig. 6.5): the plate shows the intestine, the portion of pancreas left after a partial excision, the copper pipe used for the mercury injection, and the mercury spurting inside the intestine from an unseen passage.[9] The earliest experiment reported by Brunner dates from 1673, well before Peyer's book was published. There are several motives of interest in Brunner's work. In the dedication to Wepfer, Brunner argued that in applying experiments to medicine he was following Descartes' exhortation, as well as Boyle's and Thomas Bartholin's; Wepfer was presented as Descartes' eager admirer.[10]

Brunner recounts that in 1673, when he was in Paris, the pancreas was at the

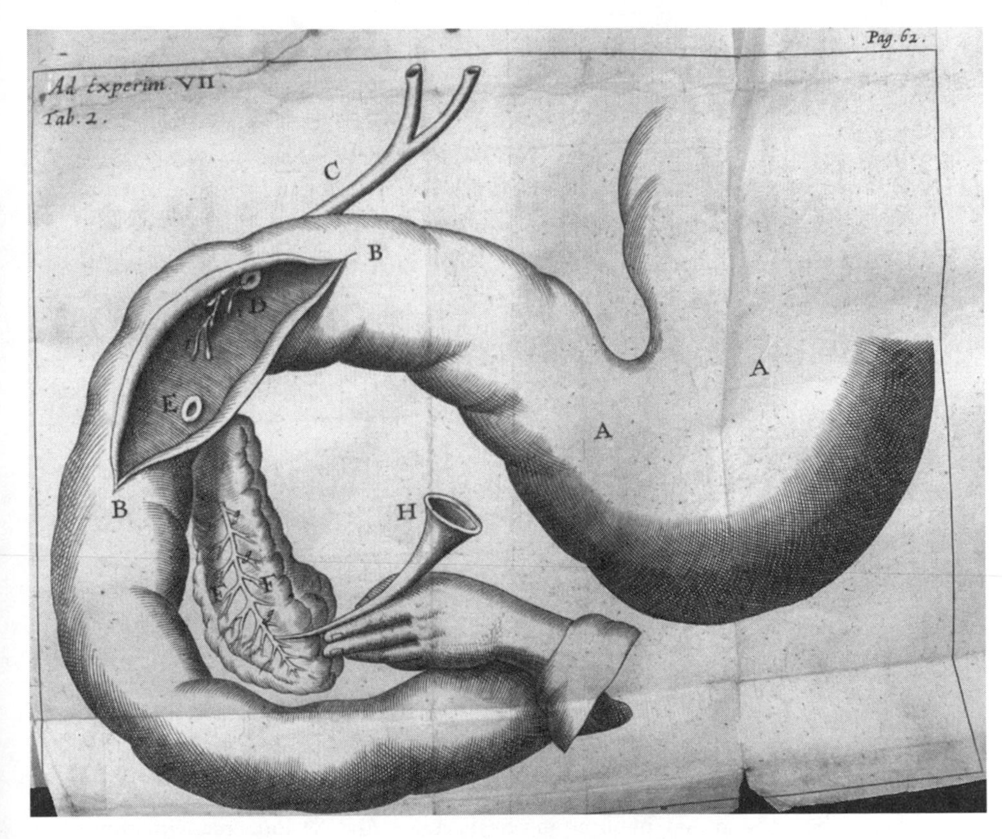

Figure 6.5. Brunner, *Experimenta*: mercury injections reveal pancreatic duct

center of attention, and he even discussed its role in light of his experiments with the royal anatomist Joseph-Guichard Duverney;[11] this interest would have resulted from de Graaf's celebrated work on the pancreatic juice, which had reached its third edition in 1671. At the time splenectomy was not uncommon and was mentioned by de Graaf, for example; Brunner stated that having noticed that the pancreas had no more vascular connections than the spleen, he thought that it was possible to attempt its excision. Brunner learned surgical techniques at public demonstrations in Paris and regularly relied on surgeons for his anatomical experiments on the pancreas and repeatedly acknowledged it, although omitting their names. My sense is that, although many seventeenth-century anatomists, from Malpighi to Steno and from Bellini to Lower, were skillful dissectors, Brunner's reliance on surgeons was not entirely unusual. Moreover, some anatomists were also surgeons: Nuck even wrote a treatise on surgical procedures and experiments, whereas Ruysch lectured on anatomy to surgeons.[12] Brunner also provided extensive details of the actual surgical procedure and its effect on the animal, such as vomiting or the fact that the wound was regularly washed with warm wine. The results of his experiments and reflections were a more radical departure from the views of Franciscus Sylvius and de Graaf than what we have witnessed with Peyer; indeed, Brunner questioned Peyer's conjecture that animals without the pancreas would be less agile.[13]

Brunner started with detailed accounts of seven experiments, showing that the pancreas was not indispensable to life; on one celebrated occasion he noticed that the dog became very thirsty and urinated profusely, although this may have been the result of a splenectomy, since in other instances of excision of the pancreas urination was normal. He framed his new interpretation of the role of the pancreas within a novel view on the role of lymph, arguing that the pancreas, rather than producing a highly specialized juice necessary for digestion, produced a fluid of the same family as other glands such as saliva; the title of his 1715 review of the literature on the subject bears the revealing title *Glandulae duodeni seu pancreas secundarium*. Moreover, pancreatic juice was not acid but rather insipid and mildly salty, similar to lymph.[14] This remark, put forward in passing in the report of the fifth experiment, was to assume a key role later in the treatise: pancreatic juice was indeed considered as belonging to the same category as lymph or saliva, which also helped digestion by loosening and dissolving food much like tepid water does to ice.[15] For understanding digestion Brunner favored elasticity and other mechanical causes rather than chymical processes and questioned van Helmont's archei and the doctrine of effervescence adopted by Franciscus Sylvius and de Graaf.[16]

Thus, the research and publications by Pechlin—and Swammerdam to some extent—and Wepfer's school in Switzerland led to a profound rethinking about digestion and the specificity of the fluids secreted by the relevant glands.

6.3 The Mode of Operation of Glands

The 1674 *De secretione animali cogitata* by the English physician William Cole is a large treatise discussing a mechanical account of glandular secretion and providing a systematic treatment and taxonomy of secretions. True to its title, Cole's work consists of a series of reflections rather than experiments and observations—microscopic or not. Secretions, he argued, could be public, i.e., for multiple uses, or private, i.e., for a well-defined purpose; simple, i.e., involving the production of homogeneous matter, or compound, i.e., involving particles of different types.[17] The presence and role of ferments also played a role in Cole's study: simple secretion involved simple straining or filtering, whereas compound secretion required the presence of a ferment, which Cole defined as consisting of very small particles full of edges and very different among themselves; nervous juice was an example of a ferment. Ferments could also be subdivided into different types, depending on whether they act before or during the process of secretion.[18] Although some historians have found his work rather speculative, Cole's treatise offered a rich picture of the field that was respected by contemporaries.[19]

In 1683 the Middelburg surgeon Antonius de Heide dedicated to the Royal Society *Anatome mytuli*, a distinguished treatise on mussels. He added to his anatomical treatise a *Centuria observationum medicarum*, including material on blood relevant to its composition and the problem of secretion. Two observations from de Heide's *Centuria* concern us here. In the first, de Heide wished to investigate the differences between the blood of healthy and sick patients, intrigued by the fact that they seemed to differ very little. He studied the separation of blood into a serous part and a red one in very thin glass tubes, not much thicker than an animal bristle: he believed that the separation was due to differences in specific gravity and occurred whether the tubes were open from both sides or hermetically sealed, vertical or horizontal. In the second observation de Heide pointed out, as did Malpighi and Fracassati, that the congealed blood from a sheep was bright at the top and dark at the bottom, but the black portion turned red if exposed to the air; he also performed some chymical tests on blood, showing that it turned black and coagulated with *aqua fortis*. Further, he noticed that the blood coming from a dog's artery was much brighter than that coming from a vein. In a series of experiments de Heide examined the behavior of blood drops from a turkey's neck in water: he noticed that whereas in hot water the drops sink to the bottom, forming a nebula, in spirit of wine they form a thin layer at the bottom; in cold water, however, the drops spread on the surface like oil or liquefied fat and tiny portions form threads reaching to the bottom.[20]

In the same year the surgeon and physician Johannes Muys published in Leiden *Praxis chirurgica rationalis*, containing his observation on a portion of blood that is separated in cauterization. Muys attributed the ocular complaint of a young maid of thirteen to acidity of the blood, which, following Franciscus Sylvius, he judged very hard to remove; he referred to an experiment about filtering a mixture of oil and water, which we will encounter below and which he interpreted mechanistically: the experiment consisted in separating oil or water mixed together. A filtering paper treated with oil will filter oil, whereas one made wet with water will filter water. Based on this experiment, Muys thought of opening a fontanel in the maid's arm and keeping it open for many days, reasoning that the humors reaching that incision would stop, turn acid, and attract the acidity of the blood, much like the wet paper filtered water and the oily one filtered oil. By a combination of this remedy and the application of a cataplasm to the eye, he claimed that he effected the cure in one month.[21]

The works on the glands by Peyer, Harder, Brunner, Cole, de Heide, and Muys were mentioned in a letter of 1685 by Malpighi to Bellini, in response to a request for help on the problem of separation of fluids, namely, the process central to the action of glands. Malpighi's survey of the field provides a contemporary account that has helped me structure my work. Although de Heide did not deal directly with glands, Malpighi read his experiments as potentially relevant to understanding the separation of fluids. Malpighi had the good sense not to mention Borelli's *De motu animalium*, a work that would have been on Bellini's desk and that discussed the structure and operation of glands in connection with the notion of fermentation especially for the kidneys and liver: Borelli questioned whether glands contain a ferment, since this may be washed out by the body's fluids, suggesting instead that the glands may be a locus where a ferment from the nerves or other vessels may be constantly supplied.[22] It is noteworthy that in their exchange, neither Bellini nor Malpighi looked strictly at the anatomical literature: in his initial request, Bellini warned Malpighi that he was already familiar with Galileo, Descartes, and Archimedes and was seeking to purchase works by Stevin, namely, mathematicians who had studied the behavior of fluids and especially their equilibrium. Bellini did not mention Borelli's teacher Benedetto Castelli, who had worked on the motion of fluids, probably because in his own view secretion had more to do with hydrostatics, as we shall see. In his reply, apart from stating that he had been unable to find works by Stevin, Malpighi ignored the mathematical literature and focused instead on medicine: the work by de Heide, for example, would have attracted him not just for the anatomy of mussels but also for the additional *Centuria* of remarkable medical observations, since many of Malpighi's publications testify to his interests in the collections of notable cases. These differences in the sources used by Malpighi and Bellini are in line with their broader investigations: when in 1683, for example, Bellini published at Bologna *De urinis et pulsibus*, Malpighi, who had read it for the censors and given it the imprimatur, confessed his inadequacy in dealing with its quantitative style.[23] Their exchanges highlight the fluidity of the boundaries among anatomy, medicine, physico-mathematics, hydrostatics, and surgery.

6.4 Glands in the Theatre: Bellini, Sbaraglia, and Malpighi

In 1689 the Royal Society published Malpighi's *De structura glandularum conglobatarum consimiliumque partium*. The title was not Malpighi's and is somewhat misleading, since the work sought to provide a comprehensive study of glands, both conglobate and conglomerate, starting from the simplest cases and then moving on to more complex ones. Although Malpighi had been interested in the structure of glands from the beginning of his career, the occasion for composing the treatise was an attack by his archrival Sbaraglia at the anatomy demonstration of December 1686. Readers would find it impossible to reconstruct the circumstances of composition from Malpighi's printed text; his correspondence with Bellini, however, provides important details. From a number of humorous letters it appears that Sbaraglia had launched a two-pronged attack, relying, on the one hand, on a few experiments showing that glands do not filter fluids on the basis of the structure and figure of their pores, and arguing, on the other hand, that without an immaterial faculty, structure alone was insufficient to explain the glands' operations. In his letters to Bellini, Malpighi referred sarcastically to Sbaraglia's attack as a prosecutor's charge at a trial against the moderns, including Borelli and Bellini. Curiously, Sbaraglia reported three experiments taken from the *Dissertationes physicae* by the Toulouse Cartesian physician François Bayle, who devoted a brief section of his wide-ranging work to the problem of secretion of different juices from blood. It is worth reviewing his experiments both for their intrinsic interest and for the light they shed on the nature of the exchanges occurring at public anatomies in Bologna.

Although Bayle addressed the problem of secretion in the animal body, he discussed experiments performed with different fluids in vitro. He argued that it was not necessary for the insensible pores of a substance to have a shape suitable to the particles of the fluids to be secreted, or unsuitable to the particles to be retained; the presence of subtle fluids made matters more complex, as his experiments showed. The first experiment involved two small vessels, one of wood and the other of gold, with two small equal circular holes at the bottom: mercury poured in the gold vessel would exit from the hole, much like water poured in the wooden one; however, mercury poured in the wooden vessel would not exit unless it reaches a great height, despite the fact that the holes had the same size. The second experiment was alluded to ironically by Malpighi in the letter to Bellini, when he referred to a "paper smeared with oil, that it aroused nausea in whomever looked at it"; its purpose was to separate a mixture between an oily and a watery substance by means of blotting paper. If the paper is soaked in water, only the watery substance will go through it, while the oily one will be retained; conversely, if the paper is soaked in oil, only the oily substance will go through, while the watery one will be retained. The last experiment was intended to show that the motion of a fluid against its natural propensity—such as gravity, for example—would prevent its going through a sieve whose pores would in normal circumstances allow it passage: to put matters more simply, producing a large crack in a cabbage leaf with a knife, water drops will pass across the crack without falling through. Bayle's experiments offered a more complex picture of secretion or filtration than usual. Although there was a wide gulf between Sbaraglia's philosophical outlook and especially his defense of the faculties and Bayle's Cartesianism, Sbaraglia could tactically exploit Bayle's denial that the structure of pores and particles could explain secretion tout court for his own rather different philosophical purposes: probably Sbaraglia interpreted Bayle's experiments within the framework of like attracts like, for example, blood attracts blood or bile attracts bile, one that was familiar in traditional premechanistic accounts. In addition, Sbaraglia mentioned a further experiment, which Malpighi seems to suggest was only reported rather than performed. Sbaraglia argued that when mixing different fluids and pouring them on a wool cloth, only one type of fluid is separated; moreover, changing the color of the cloth would affect the type of fluid that is separated.[24] Probably Malpighi was referring to Sbaraglia when he discussed the role of color in *De structura glandularum*, as when he argued that whereas according to some the difference of color among glands indicated a difference in nature and therefore purpose, a more careful analysis showed that color was an effect of disease rather than a characteristic feature of the healthy gland. He acknowledged that the color of glands changed in relation to their degree of compression or relaxation, however.[25]

Sbaraglia was not the only one to discuss the role of glands during the annual dissections in anatomical theatres. In yet another letter to Malpighi, Bellini also stated that he had discussed the structure and mode of operation of glands at his public dissections: his letter merits attention for its intellectual implications and for the vivid account he provides of his way of lecturing and proceeding in the Pisa theatre. For simplicity's sake, he stated that he focused on the intestines and lacteals. Moreover, he set on the dissecting table a glass with fresh blood and, while he talked, waited for the serum to separate. He then proceeded step by step, for about two hours, arguing that the glass could be seen as a gland, lacking only some holes whence the fluids can leave. Bellini did not conceal his satisfaction with his own performance, which he believed he carried out following a demonstrative method valid for all glands and also for all types of separations, even those occurring outside the human body. By the last comment he considered processes taking place inside the animal on the same level as those occurring in nonliving organisms, thus erasing the distinction between living and nonliving in that both follow the same physical laws, in line with the mechanistic tradition of his mentor Borelli and much as Bayle had done in his *Dissertationes*.[26] The account of the separation of blood in the glass vessel at the Pisa theatre instantiates his interest in hydrostatics and in the relevant literature.

Malpighi responded to Sbaraglia's oral attack with a treatise published in London. Whereas in previous works Malpighi had chosen as his topic an organ such as the lungs or the liver, or an organism such as the silkworm or plants, in *De structura glandularum* he focused on a constituent of many different body parts, seeking to clarify and systematize our knowledge. In fact, as he stated in a letter to Bellini of August 1687 containing a useful synopsis of the entire treatise, the distinction between conglobate and conglomerate glands then commonly accepted was accidental, that is, it did not concern major differences in their structure and mode of operation: all glands could be treated together starting from the simplest to the more complex ones, following nature's scale of complexity. The organization of Malpighi's treatise is indicative of his way of proceeding from the simplest to the more complex. The simplest gland of all is found in the palate, the esophagus, the intestines, and similar parts: it consists of an oval or lenticular membranous follicle with an excretory duct; the follicle is surrounded by blood vessels and nerves, as well as muscle fibers. According to Malpighi, all other glands would be elaborations of this simplest scheme, which he then proceeds to discuss across different organs and animal species. In the final definition that he provided at the end of his treatise, he stated that the gland consists of a membranous follicle whose shape ranged from round to oblong and that is at times surrounded by a "*lanugo fistulosa*," or woolly collection of pipes, a claim we shall encounter again in section 10.5.[27]

Malpighi found confirmation of his previous identification in *De viscerum structura* of several organs as glandular, such as the liver, kidneys, cerebral cortex, and spleen; in addition, he identified other body parts as glandular, such as the membrane around the heart known as the pericardium, the tunica vaginalis of the testicles, the gastric mucosa, the peritoneum, the pleura, and the testicles themselves. The last case was quite problematic for him, since the testicles consisted of tubules with no obvious follicle; his solution was that the seminiferous tubules would represent a much-elongated version of a glandular follicle.[28]

Malpighi expressed his frustration at the valves in the lymphatics, which prevent ink injections to reach their roots. Probably as a result of this technical difficulty, he could only conjecture that lymph originates from the arteries and is filtered by the smallest glands, whence the lymphatic system originates; other glands filter more lymph so that its amount increases from the periphery toward the center.[29]

In *De structura glandularum* Malpighi did not try to figure out the details of the secretion mechanism, whether it occurs purely mechanically or with the help of some chymical process like fermentation, or the purpose of the juice secreted: unlike Brunner, in this case Malpighi did not rely on vivisection. Neither did he enter the debate on the role of the pancreas and its secreta, stating in the briefest possible way that the glandular nature of the pancreas was universally acknowledged.[30] His aim, true to the title, was the investigation of structures. His commitment to the role of the follicle as the site of separation of fluids did not solve the problem of how filtration or secretion actually occurs: the problem was simply shifted from the gland to its component, as an example of the machine within the machine typical of anatomical investigations of this period. In a letter to Bellini of August 1687 Malpighi stated that he was devoting himself to the study of glands while at his summer house in Corticella, relying on the help of a friendly butcher who provided him with the material he needed: *De structura glandularum* relies extensively on observations of cattle. Moreover, he exploited his expertise with insect anatomy by discussing the shape of the liver and pancreas of the silkworm and cricket. He also described some worms or insect larvae to be found in the stomach of donkeys, including the bent nails with which the larvae adhere to the membrane of the stomach and their internal anatomy.[31]

As to the methods of inquiry, Malpighi had recourse to a range of techniques, such as maceration in water and ink injections. Following a practice he had started in *De hepate*, he relied extensively on the microscope of disease—observing greatly enlarged diseased glands—and postmortems. Quite remarkable was the case of Countess Pentesilea Davia, based on a report by Malpighi's friend Silvestro Bonfiglioli: the structure of her kidney was identical to that found by Malpighi in the left kidney of her brother Antonio and described in the letter to Spon; in both cases the kidney looked like a bunch of grapes in which the *acinus* represented the macroscopic glandular follicle.[32]

In *De structura glandularum* Malpighi had recourse to the microscope of disease in a new fashion: whereas in normal circumstances the fluid secreted by the gland could be insufficient for analysis, in some instances disease allowed him to collect the fluid in a quantity sufficient to enable chymical analysis and to ascertain, for example, whether it originated from blood. In the cadaver of Lorenzo Zagoni, who had died on 23 February 1677, the pericardium contained about eight pounds of fluid, an abnormal amount that enabled testing; in analogous cases Malpighi offered a visual description of the humor and its color, further specifying that it is salty, taste being a standard form of chymical assaying; he also heated it in a pan on the fire, noticing that it evaporated without boiling, leaving a reddish crust smelling of boiled meat. In the 1669 *Tractatus de corde* Lower had followed a similar reasoning and procedures to those later adopted by Malpighi. Lower, however, found that the fluid of the pericardium, once placed on the fire, sets like white jelly; moreover, Lower warned that this procedure could be used only for healthy animals that have died a sudden death, since in those dying of a disease the blood is so altered and lacking of chylous juice that the result would differ. Malpighi also mentioned cases of watery hernia—known today as hydrocele of the testicle—as another instance of disease enabling testing. For example, the fluid collected from the tunica vaginalis exposed to the fire coagulated in a whitish gelatin, whereas mixed with vitriol it coagulates in solid gelatin. Occasionally, however, Malpighi's tests led to different results, as when the fluid exposed to fire evaporated, leaving a thin crust, or when mixed with vitriol it remained fluid; the reader is left to wonder whether the different outcomes were meant to identify different causes for the same symptoms.[33] In other instances the chymical analysis proved easier. For example, when he mixed in a ceramic container the fluid extracted from the scrotal sac with vitriol, he noticed that after a few days the mixture evaporated, leaving a yellow sediment similar to the substance found in diseased glands.[34]

Neither *De viscerum structura* nor *De structura glandularum* included any plates. The existence of a manuscript sheet in Malpighi's hand with about twenty figures of

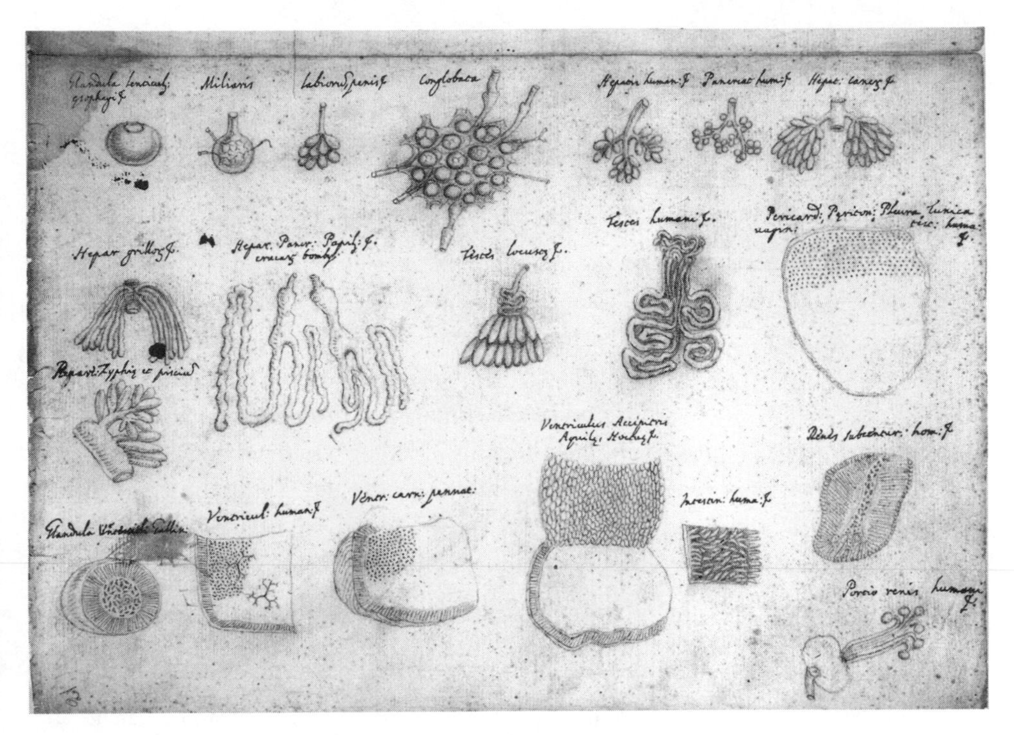

Figure 6.6. Malpighi: manuscript with illustration of glands

glands related to those discussed in *De structura glandularum* is therefore especially fortunate and welcome. In addition, many of the drawings can serve to illustrate passages from *De viscerum structura* as well (fig. 6.6). The order of the glands appears not to be casual, since at the top left we find the simplest gland of the esophagus or the palate, the third drawing from the right in the first row shows the glands of the human liver, and the last figure at bottom right shows the glands of a human kidney. The figures of the glands of the pancreas (first row, second from the right) and of the human testicle (second row, second from the right) are also noteworthy and worth comparing to the figures by Auberius and de Graaf (figs. 1.5 and 8.3). Malpighi's manuscript sheet also includes a page of text with a classification of glands into four species, appearing as a preparatory draft of the more elaborate treatment of *De structura glandularum*.[35]

6.5 Nuck's New Taxonomy of Glands

De structura glandularum was reprinted at Leiden in 1690, thus guaranteeing a Dutch audience to Malpighi's latest views on glands. Moreover, his previous views had already attracted extensive scholarly attention and had established themselves as part of the anatomical canon of the time. In the 1680s, however, some anatomists and physicians such as Pieter Guenellon, Cornelis Bontekoe, Steven Blanckaert, Heidentryk Overcamp, and others were becoming increasingly fascinated with the vascular structure of the body. The doctrine of the vascular secretion was especially prominent in connection with the elaborate injection techniques developed in the Netherlands in the last third of the seventeenth century. There was no strict or automatic correspondence between microscopy and follicles on the one side and injections and vessels on the other; rather, there was a complex interplay between techniques employed and views about secretion, although increasingly sophisticated injections gradually led to the rise of the vascular view of the body and of glands in particular. As Edward Ruestow has argued, "reinforcement between technique and theory may well have been reciprocal," and "to some extent, the consensus regarding secretion could be seen as a decision between the evidence of injection and that of the microscope." Further, from a letter by the Montpellier physician Pierre Chirac of August 1684, Malpighi had learned that in his view glands were merely a collection of vessels, a position Malpighi found unacceptable. It was not until Ruysch's work starting in the mid-1690s, however, that a comprehensive attack on glands was mounted.[36] Other significant research occurred before that time.

In 1685 the physician and from 1687 Leiden professor of anatomy and medicine Anton Nuck started publishing a series of works on different types of glands, notably *De ductu salivali novo*, reissued with additions in 1690 with the title *Sialographia*, and *Adenographia curiosa* of 1691, in which he challenged important notions of Malpighi's *De structura glandularum*. In the *Bibliotheca anatomica* Malpighi's treatise was grouped together with Nuck's works, with a common introduction. Nuck was the heir of the tradition of Franciscus Sylvius and especially Steno, as pointed out by the editors of the *Bibliotheca anatomica*: his first publication dealt primarily with a new salivary duct and the analysis of saliva and involved the chymical analysis of glandular excreta. In line with several other anatomists of his time, Nuck disparaged speculations and abstract theorizing in favor of detailed observations. In *De inventis novis epistola anatomica*, added to *Adenographia curiosa*, he questioned the role attributed to the pineal gland as the site of common sense and of the soul, since it has rather tenuous nervous connections, much like the pituitary. Although the name of Descartes was not mentioned, it is clear that the jibe was addressed at him. On the example of what Thomas Bartholin had done to the liver, Nuck also included an obituary to the pineal gland, which can be read as a criticism of abstract speculation unsupported by anatomical evidence.[37]

There were different ways to approach the study of glands: by structural investigations, involving both the microscope and injections of different fluids; by chymical analysis of their secretions; by vivisection experiments, involving ligatures, for exam-

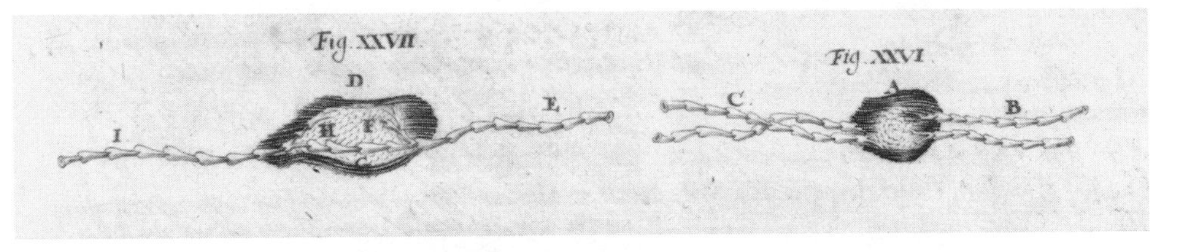

Figure 6.7. Nuck, *Adenographia curiosa*: conglobate glands and lymphatic vessels

ple; and by remarkable case histories and pathology. Nuck relied on a creative combination of all these methods. In his study of saliva, for example, he reported the opinion that secretion occurs by means of a ferment provided by a nerve and proceeded to test the claim by ligating the nerve to the salivary gland: saliva was still secreted, although in smaller quantities; ligating the efferent vein, however, led to increased salivation, thus supporting Steno's view that saliva originates from arterial blood.[38] Curiously, Nuck did not study the effect of saliva on different types of food, but rather compared the effects of different chymical reagents on the saliva of healthy and diseased patients—affected by venereal disease—paying special attention to differences in color and odor. He assayed saliva with urine and spirit of urine, oil of sulphur, and alcohol of wine. He even discussed whether saliva from the higher ducts differs from that in the lower ones, concluding his analysis with an extensive list of case histories from the literature and his own practice.[39]

In *Adenographia curiosa* Nuck provided a list of fifty-three types of glands and argued that conglomerate glands consist of a network of tiny arteries from which arterial blood separates into venous blood and other components; thus, Nuck advocated a vascular and mechanical separation of fluids. He reported without attribution the experiments by Bayle with the oily and wet paper that Sbaraglia had mentioned in the Bologna theatre. Unlike Malpighi, who used ink, Nuck favored mercury injections, which he performed with sufficient pressure to force the fluid from the arteries in the breast to the lactiferous vessels and back from the nipple to the arteries.[40]

Nuck focused on the lymphatic vessels and the conglobate glands, offering a radically different interpretation from Malpighi. It will be recalled that Malpighi believed that the difference between conglobate and conglomerate glands was accidental, in that they were structurally analogous and served an analogous purpose, namely, secretion. By contrast, Nuck saw a radical difference between them: whereas conglomerate glands were truly excretory, conglobate glands were not. He noticed that conglobate glands had afferent and efferent lymphatic vessels attached to them (fig. 6.7); therefore, those vessels could not be excretory ducts. Rather, he argued that the purpose of conglobate glands was to render lymph subtler by breaking it up in smaller components and mixing it with nervous fluid. Whereas Malpighi had argued that conglobate

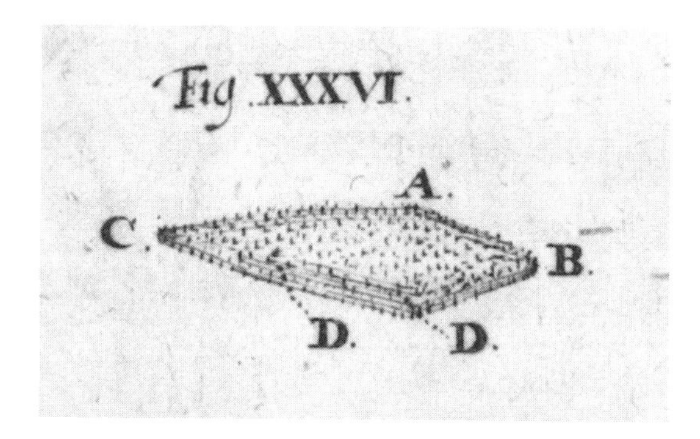

Figure 6.8. Nuck, *Adenographia curiosa*: encrusted globule from a dog's bladder

glands were the source of lymph, Nuck believed lymphatic vessels to originate from arteries; he tried to prove his point by means of insufflation into the blood vessels of the spleen, lungs, and testicles, showing that air passed to the lymphatics.[41]

Adenographia curiosa contains many notable case histories supporting Nuck's views about the therapeutic significance and implications of his findings, such as the role of the lymphatic system in female reproductive organs and its implication for fertility treatments.[42] In addition, Nuck reported a few vivisection experiments only tangentially related to glands but remarkable in their own right. In one of them he ligated the left uterine horn in a bitch three days after copulation; on the twenty-first day after the operation he dissected the dog and noticed two fetuses above the ligature, between the ligature and the ovary. In his view this showed that fecundation occurs through an "aura seminalis" rather than male semen, since he judged that semen could not reach that high. In the other experiment he investigated the origin of bladder stones, suspecting that they formed through the accretion and incrustation of successive layers. He opened surgically the bladder of a dog and inserted a wooden globule in it; the dog survived the operation and lived happily for several weeks, after which he opened the bladder again and found the globule covered with incrustations (fig. 6.8), thus proving his point and also showing the origin of the celebrated bezoar stones, mineral formations in the animal body that were believed to be antidotes to poison.[43] Nuck's challenge to Malpighi concerned mainly conglobate glands, since he treated conglomerate glands much more briefly; other Dutch anatomists would see matters differently, as we shall see in section 10.5.

In the last quarter of the seventeenth century glands were a major focus of anatomical research, since they constituted the key building block of the new anatomy. Many techniques were used: Pechlin, Wepfer, and Malpighi used maceration in water, at times with added vinegar. Anatomists often had recourse to vivisection in conjunction with ligatures for purposes ranging from structural investigations to the study of how glands work, as in the case of Nuck. On the example of early splenectomies, Brunner excised the pancreas of a dog in order to study its role; he also performed mercury injections to test for hidden openings of the pancreatic duct into the intestine. Malpighi also performed injections, especially of ink, in order to trace the path of vessels and their anastomoses; he failed, however, to determine the source of the lymphatics because of the presence of valves in those vessels. Furthermore, chymical assaying of the secreted fluids followed different methods; tasting was the privileged procedure, although occasionally it was questioned not so much for its subjective nature as for the volatility of the fluid or its components, as Peyer did for the secretion of his intestinal glands. In vitro experiments also played an important role: Bellini worked with glass containers, whereas several scholars mentioned experiments with blotting paper to explain how glands work.

Connections with the study of disease are visible in many works, from Peyer's interest in diarrhea, dysentery, and constipation to Malpighi's reliance on the microscope of disease in order to enlarge minute structures. Further, Malpighi used disease to collect enough secretions for chymical testing—a practice questioned by Lower in the case of the pericardium. Thus, disease was both an object and a contested tool of investigation.

Anatomical theatres were important venues for debates and controversies on glands, as we have seen with Malpighi and Sbaraglia at Bologna and with Bellini's teaching at Pisa; Malpighi's *De structura glandularum* originated at one of those functions and defended mechanistic views against Sbaraglia's claim that structure alone could not explain the process of secretion, which depended on the faculties of the soul. Despite this challenge, overall mechanistic views prevailed, although there was no consensus in the mechanist camp about the structure of glands. On the basis of structural considerations, Nuck acutely questioned Malpighi's claim that conglobate glands are the simplest. The different modes of investigation were often interwoven with the outcome of anatomical research. Nuck and other Dutch investigators developed sophisticated injection techniques with different fluids that led to the dissolution of glands or glandular follicles into a network of vessels. While the aim of some injections may have been to preserve body parts, they soon acquired a role in revealing the body's microstructure and challenging the centrality of the glandular follicle, as we shall see in the case of Ruysch.

PART THREE

BETWEEN ANATOMY AND NATURAL HISTORY

Malpighi and the Royal Society

Part III focuses on the last third of the seventeenth century, encompassing three areas at the intersection between anatomy and natural history: the study of insects, generation, and plants. At first this may seem a rather heterogeneous set, but a closer scrutiny reveals profound links and a shared intellectual project among them: although the study of insects and plants involves a descriptive, natural historical component, it also requires tools such as the syringe and the microscope, conceptualizations, and skills associated with anatomy. Although generation seems closer to anatomy, the patient and careful observation and recording of the developing chick recalls methods from natural history.

In 1668 Malpighi received the first letter from the secretary of the Royal Society Henry Oldenburg and broke off his correspondence with Borelli; all his subsequent publications appeared with the imprimatur of the Royal Society of London. Prompted by Oldenburg, Malpighi embarked on the study of the silkworm and other insects, leading to *De bombyce* and his election to the Royal Society in 1669. Subsequently he moved to the formation of the chick in the egg and to the anatomy of plants. I am going to take the lead from his research to explore contemporary investigations of these three fields by a range of scholars. There was something deeper to Malpighi's projects besides their association with the Royal Society, as the frequent cross-references suggest. The study of generation, for example, involved not only the investigation of the chick in the egg but also the challenge to spontaneous generation, the study of the metamorphosis of insects, and the process of germination.

Mechanistic anatomy brings to mind the mechanical explanation of various organs, such as the viscera or the brain. But this is a rather limited version of mechanism, one leaving out fundamental processes such as generation and growth, which, together

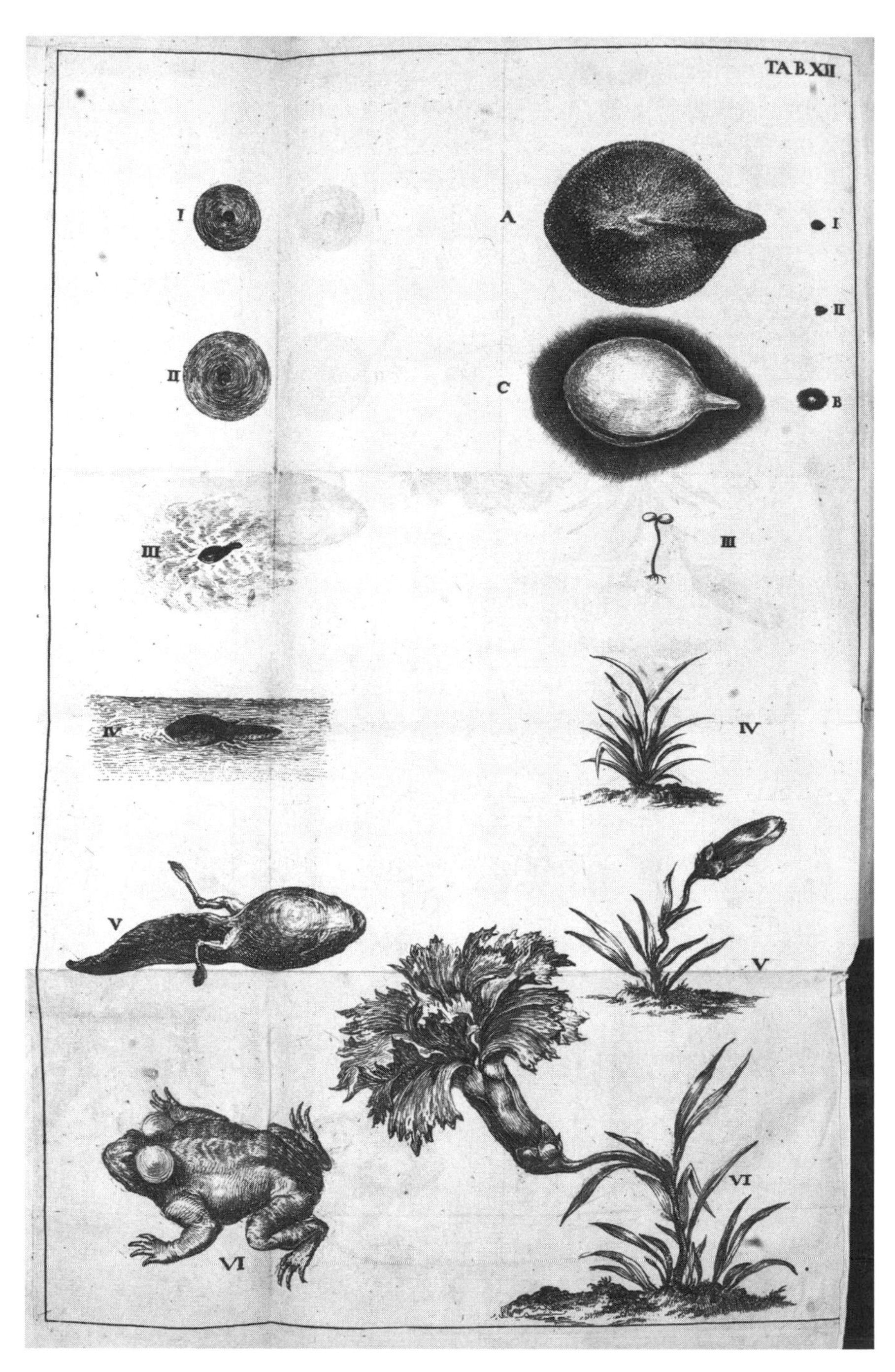

Figure PIII. Swammerdam, *Historia insectorum*: the growth of a frog and carnation

with nutrition, had been considered since antiquity among the main faculties of the vegetative soul or of nature, distinguishing living organisms from inert matter. In *On the Natural Faculties*, Galen had discussed at length their role, arguing that they could not be explained in mechanical terms. Of the three, nutrition seemed the least problematic from a mechanistic standpoint and was accounted for in corpuscular terms, while generation appeared as the most intractable; this is the reason why there was an attempt on the part of anatomists such as Swammerdam and Malpighi to account for generation in terms of growth and growth in terms of nutrition. The following chapters present some of the most challenging mechanistic investigations and reflections on these issues of their time.

The beautiful engraving from Jan Swammerdam's 1669 *Historia insectorum generalis* shows on the same plate the growth from egg and seed of a frog and carnation to maturity, or to the point when they can reproduce. Thus, this plate ties together anatomy and natural history through the study of lower animals, generation, and plants. Malpighi saw this plate soon after its publication, while he was engaged in the study of insects and was about to embark on the study of plants, thus underscoring that several scholars were seeing connections among these fields. In the following three chapters visual representation takes on an especially prominent role and is discussed relying on a comparative approach.

The Challenge of Insects

7.1 Changing Perceptions on Insects

The years 1668–69 can be described as *anni mirabiles* in the study of insects: within a few months readers witnessed the appearance of three major contributions that changed common perceptions about some of their most fundamental features and represented a point of departure for future investigations. In 1668 Redi published in Italian *Esperienze intorno alla generazione degli insetti*, followed in 1669 by Malpighi's *De bombyce* and Swammerdam's *Historia insectorum generalis*, a work written in Dutch despite its Latin title. Within the very short time in which they appeared, Malpighi saw Redi's essay before sending his manuscript to the Royal Society, and Swammerdam saw and cited both the *Esperienze* and *De bombyce* before the printing of his book was complete. In some instances it is possible to trace in the later works responses to specific issues from the earlier ones, such as the effect of oil on the respiration of insects and the generation of galls.[1]

Redi's and Swammerdam's works had an explicit philosophical agenda against Aristotle and Harvey, respectively, and a confrontational style: the former focused on denying spontaneous generation against some seventeenth-century Jesuits and devised sophisticated experiments to that end; the latter denied sudden metamorphosis and observed the entire life cycle, transformation, and classification of insects. By contrast, Malpighi's treatise focused on observation and lacked an explicit philosophical agenda, although of course it did have one. Aristotle and Pliny, for example, the two most quoted classical sources on insects, had denied that they had viscera besides the gut from the mouth to the anus and the stomach: in *Parts of Animals* Aristotle had not gone beyond the claim that the intestine of some insects is not straight but twisted in a spiral. Malpighi's work provided the most resounding yet implicit refutation of this simplistic view about the inner structure of insects.[2] *De bombyce* provided a *historia* of the silkworm including its development, habits, and fine anatomical structure, all described with an extraordinary sensitivity to color in a dramatic departure from Borelli's philosophical tenets.

Starting in 1661, Malpighi's research had opened a new level of microscopic investigations, and other works, especially by Redi, offered refined methodological tools for experimentation. Although the works of 1668–69 were conceived independently and addressed different issues in different ways, that remarkable juncture in the study of insects was part of broader intellectual changes that were coming to fruition at the same time. Redi's essay appeared in Latin translation in Amsterdam in 1671 and 1686, and Swammerdam's appeared in French at Utrecht in 1682 and 1685 and in Latin at Leiden in 1685. Somehow the textile merchant and microscopist Antoni van Leeuwenhoek also became familiar with Redi's work, despite his ignorance of both Italian and Latin. Thus, by the mid-1680s European readers would have had access to all three works of 1668–69.[3]

A closer look at all these texts reveals such a wealth of themes that one may justifiably argue that despite their small size and esoteric nature, insects lay at the intersection of a number of crucial issues in the intellectual world of the third quarter of the seventeenth century; besides generation and metamorphosis, their small size and peculiar modes of reproduction stimulated reflection on refined experimental techniques and analogies with automata and other mechanical contrivances, thus giving insects a role in philosophical debates about mechanism and generation. Further, the need to fix their minute parts and make them visible required refined and novel techniques of investigation and observation under the microscope. Lastly, the illustration of insects and their organs required careful reflection on how to show the arrangements of the parts, the use of white or shaded background, and the use of color. While examining in turn the works by Redi, Malpighi, and Swammerdam in the following three sections, this chapter pays special attention to these themes. It is especially rewarding to look at insects not from the perspective of a single technique of investigation, such as microscopy, for example, but rather from the range of techniques their study stimulated. The mechanistic worldview was an important component of those investigations—especially for the role it attributed to structure as a way to understand bodily operations. Ultimately, however, it was thanks to the novel and sophisticated experimental techniques deployed by anatomists that the results of 1668–69 were made possible. The last section deals with the controversy between Malpighi and Swammerdam and the role played by illustrations.[4]

At the time the definition of insect was rather fluid and included animals having notches on their bodies, such as flies, bees, wasps, and ants, but also spiders and scorpions. It was quite common to apply the notion of insect fairly broadly, if not loosely; I shall follow this habit below. Swammerdam called insects "bloodless animals," meaning that they lack red blood, although he accepted that they had an analogous

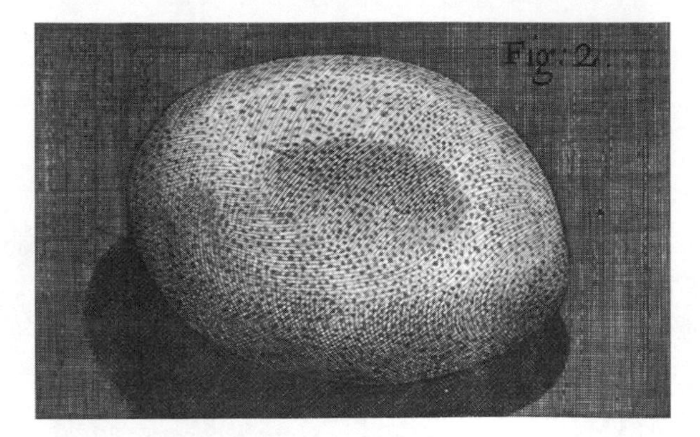

Figure 7.1. Hooke, *Micrographia*: egg of the silkworm

pale fluid. Both Redi and Swammerdam discussed frogs and other animals in their treatises, possibly because they undergo—or seem to undergo—metamorphosis.[5]

In the 1660s insects had attracted the interest of scholars as diverse as Johannes Goedaert—an artist-author of a work on the subject, *Metamorphosis et historia naturalis insectorum* of 1662–68—Athanasius Kircher, who had discussed them in *Mundus subterraneus* of 1665, and Robert Hooke, who discussed them also in 1665 in *Micrographia*.

I have chosen as an entry into the themes of this chapter Hooke's *Micrographia*, a work referred to by Swammerdam. Besides being considered an epoch-making work in microscopy, Hooke's work dealt extensively with insects and touched on all the issues mentioned above. After having described and reproduced his microscope, Hooke discussed the sting of a bee in observation 34 and then devoted observations 37–57 to various features of insects and other small animals and their body parts, such as a snail's teeth. *Micrographia* is lavishly illustrated with figures drawn by Hooke himself. Hooke provided a splendid figure of the silkworm egg: this is his only image of any part of the silkworm, showing the egg in isolation, as if it were a hen egg. He described it as indented on the sides and covered with pits or cavities (fig. 7.1).[6]

The figure of the water gnat is especially interesting because in this case Hooke was able to observe the inside of the body; this was due not to special techniques of microscopic investigation, which he lacked, but to the insect being transparent. William Harvey had exploited the same idea in observing the motion of the heart in some transparent shrimps found in the Thames. Hooke could also detect a beating heart and the gut—the darker portion in the middle of the body extending from the head to the tail—moving with a peristaltic motion with a black substance moving up and down through it. He reported that according to some the gnat is generated by putrefying rainwater; he also observed its metamorphosis into a fly that flew away,

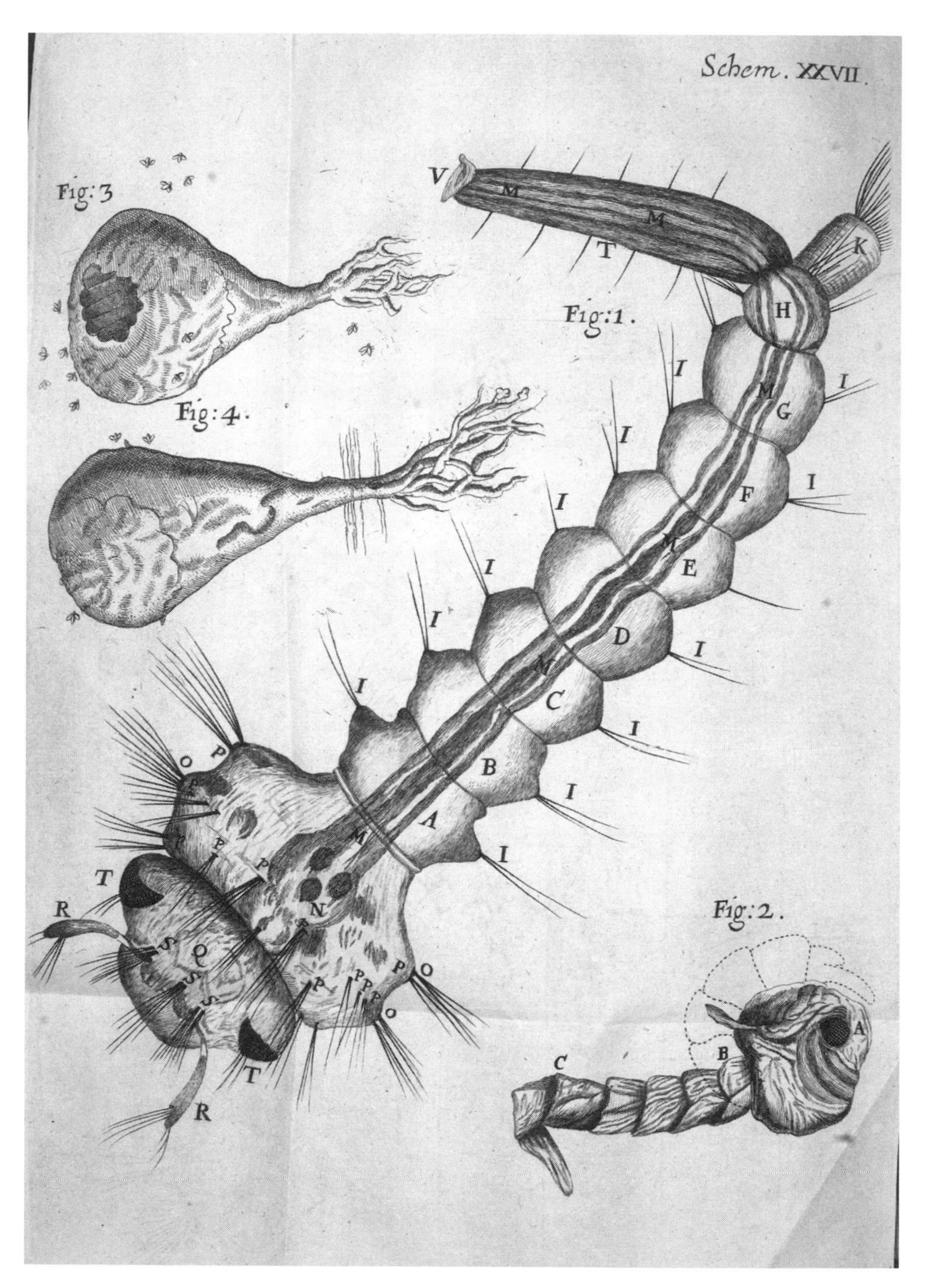

Figure 7.2. Hooke, *Micrographia*: figure of the water gnat

leaving behind the husk floating just under the surface, as shown by his figures 2–4 (fig. 7.2) in his copper plate or "scheme," which is dominated along the diagonal by the larval stage. Moreover, he provided a mechanical explanation for its posture, arguing that since its tail is lighter than water, the larva hangs with its head down. Elsewhere Hooke provided other mechanical analogies, as when he described a peculiar structure—described as a pendulum—under the wings of flies, bees, and other insects, which he initially thought may be related to the wings' motion but later linked to breathing.[7]

Generation of small animals and plants is discussed in observations 19 and 21 with the help of mechanical analogies. Hooke had no reservations about spontaneous generation:[8] "But that *putrifying* animal substances may produce animals of an inferior kind, I see not any so very great a difficulty." Hooke further believed that plants or animals produced spontaneously or casually from putrefying matter may be able to reproduce themselves. The simile he used was that of a set of chimes or bells as part of a clock or a more complex automaton: although the chimes lack parts present in the clock, they work also on their own. Similarly, in talking about the generation of moss, mold, and mushrooms, Hooke put forward the hypothesis that these may be like a clock that has lost some parts and still moves, albeit in a different fashion. In discussing the "tufted gnat," Hooke argued that the same creature may be produced in different ways and in turn may produce different creatures because God could have arranged insects or "automatons" in such a way that in some circumstances they produce one kind of effect or a certain animal and in other circumstances other effects and animals. Likewise, similar automata may be produced by several materials and with different methods. He continued to believe in spontaneous generation for several years. We see here a striking intersection between generation and the mechanical philosophy, one that was not limited to Hooke: both Descartes and his disciple Henricus Regius argued that a living animal could be produced from a drop of fluid, heat, and particles following the laws of motion.[9]

Throughout *Micrographia*, insects contributed significantly to the curiosity and wonder associated with microscopic investigations.

7.2 Redi: Experiments and Generation

Redi's treatise is remarkable for its dazzling display of erudition and for the sophisticated experiments it described. It was written by a member of the Accademia del Cimento and Accademia della Crusca for a philosophical and literary audience at the same time. The issue of spontaneous generation had a long and colorful history stretching to antiquity: from mosquitoes to shellfish and from bees to snails, the range of animals and circumstances of generation were remarkably broad and often surprising to the modern reader. Many of the issues discussed in his *Esperienze*—such as the generation of frogs and the origin of galls in plants—had their roots in investigations and debates at the Cimento. Redi's style was to rely on repeated experiments to debunk claims by previous authors, such as the renowned—although not always respected—Jesuit Athanasius Kircher, and popular credence. For example, Redi challenged Kircher's recipe for generating frogs from earth and also the widely accepted claim that they originate from the rain, either falling from the sky like water drops or emerging from dust during a downpour. In this case Redi's refutation relied on observations and experiments performed at the Cimento: as Borelli had argued in 1657, the frogs allegedly generated by the rain already had food in the stomach and excrements in the intestine. It seems plausible that Borelli was aided in those dissections by Malpighi—his anatomical collaborator at the time—who in this way would have gained some practice in the dissection of frogs that was going to come to fruition in his 1661 investigation of the lungs. Although Redi was far more biting in his attacks against the Jesuits and traditional doctrines than the *Saggi di naturali esperience* of the Cimento, there are analogies between the two for the emphasis on experimentation as arbiter of philosophical matters. As to erudition, however, Redi was in a class of his own: he cited well over one hundred authors, including forty-seven Greeks and four Arabs, often in their original language. Another motif is the fascination with the size and venom of some specimens, especially for those from the Orient: Redi devoted several pages to the scorpions from Tunis he had received from the physician Giovanni Pagni.[10]

Redi relied on several types of microscopes in his investigations, at times giving the impression that he could pick and choose the tool he fancied the most from the Grand Duke's extensive collection; he mentioned microscopes with a single lens, those made by the best maker in England, and one with three lenses made by Eustachio Divini in Rome.[11] Despite the excellent instrumentation, Redi never managed to provide more than a superficial account of the insects he studied. On one occasion he mentioned having observed with the microscope the motions of the fluids in the viscera of diaphanous specimens—the same technique employed by Hooke. On another occasion, when he was in Steno's company, they received several specimens of mantis and embarked on their dissection, finding a canal going from the mouth to the tail surrounded by a confused mass of various threads that they surmised were arteries or veins. They also found several eggs "tied together, or dressed by a thread or a canal that could not be discerned because of its thinness": it seems here that Redi and Steno were unsure whether the eggs were contained in a vessel such as an "oviduct" or were simply tied by a thread. Their study turned to the enumeration of the eggs and soon courtly amusement overtook anatomical curiosity, when they realized that the animals continued to move even without the head or viscera. Moreover, the heads could be reattached to the body because the viscous fluid from the body in drying worked as glue, but those who did not know the trick were startled. Redi added a passage from Ludovico Ariosto's *Orlando Furioso*—one that he no doubt recited to his bemused observers—telling the story of Orrilo, who could reattach his limbs and even his head once they had been cut, as if they were made of wax.[12]

Although in his youth Redi had studied drawing under the draughtsman and landscape painter Remigio Cantagallina, for the illustrations of his treatise he relied on Filizio Pizzichi, a chaplain to the Medici with an artistic talent; his drawings were engraved by Guiduccio Guiducci, whom Redi paid seven lire for each plate. Although I am not aware of copies of Redi's work with the plates colored by hand, Pizzichi's original drawings were colored with tempera; some of the originals were bound in a copy of the *Esperienze* now at the national library in Florence. Most of the plates in the *Esperienze* show enlarged *pollini*—types of lice found on other animals—seen from the top; at times Redi added details of body parts, as in the *pollino* of the swan, shown with a detail of its head seen from the bottom in order to highlight the mouth; occasionally he also showed in the text much smaller life-size drawings of the entire animal. His plates resemble in conception some images produced within the Accademia dei Lincei, such as the weevil of wheat or "*gorgoglione*" from Francesco Stelluti's *Persio tradotto* (1630), which can be usefully compared to Redi's "*punteruolo*" of wheat, from the same family (figs. 7.3 and 7.4). Stelluti's shading adds nothing but rather disturbs the viewer and detracts from the clarity of the image; the cleavage along the length of the rostrum is a result of the lighting, for example, since the rostrum is used for boring, not eating. His figure, however, shows details lacking in Redi's, such as the eyes. In Redi's image the head is curiously partitioned and the overall impression is more decorative than realistic; although both images are to some extent problematic, Redi's shows no clear improvement on Stelluti's with respect to the number of body parts and their structure or general appearance.[13]

Redi mentioned a wide range of experiments testing any claim found in the literature. He immersed the worms growing on a piece of tuna in wine, vinegar, lemon juice, and grape juice, and all survived, but those immersed in oil died, as stated by a number of sources such as the Bologna naturalist Ulisse Aldrovandi, the chymist Andreas Libavius, and the physician Johann Sperling. While those flies immersed in oil could not be revived by exposing them to the sun, those immersed in water could.[14]

Redi's rejection of spontaneous generation relied on his most impressive experiments in which he studied a wide range of rather repulsive putrefying animal meats, from snakes and frogs to pigeons, beef, fish, chicken, bull, deer, and dozens more. A variety of worms devoured them and then turned into flies of different species. He was suspicious of the origin of those worms, having seen on those meats insects

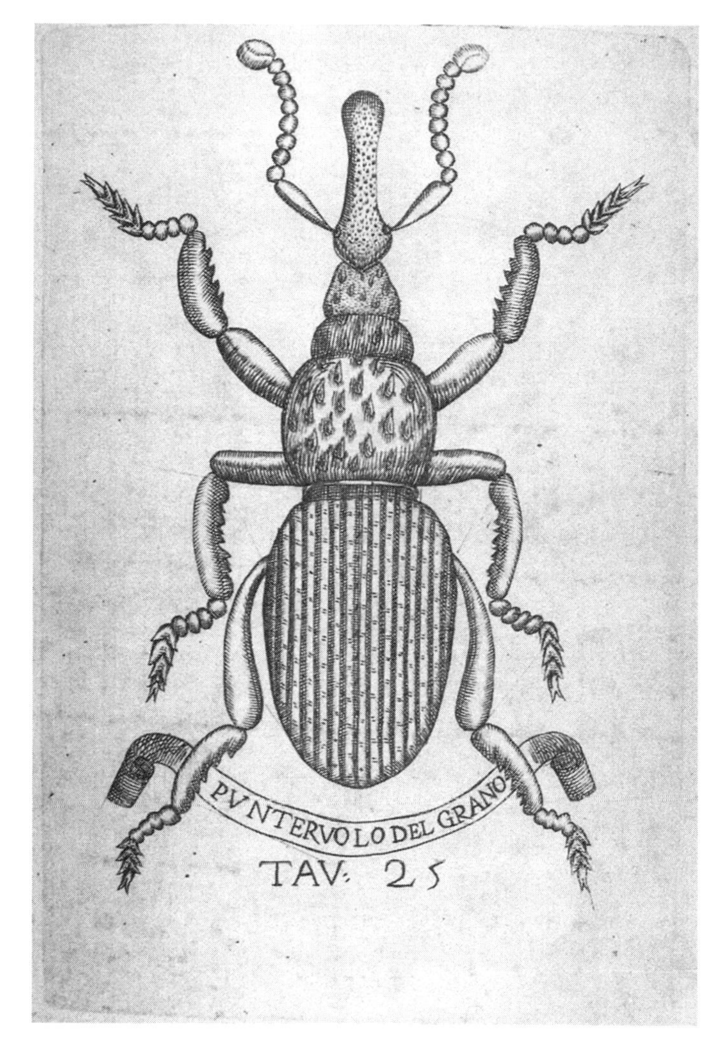

Figure 7.3. Redi, *Esperienze*: "*punteruolo*" of wheat, copper engraving no. 25

of exactly the same type as those later produced. Here Redi relied also on classical sources and everyday observations on the marketplace or the hunting lodge. He mentioned that hunters in the summer, butchers, and women—"*donnicciuole*"—who presumably are in charge of purchasing and storing the meat, keep flies away from it with cloths and nets to preserve it. Moreover, he quoted two passages in Greek from the *Iliad* in which Achilles asked his mother Thetis to preserve the corpse of his friend Patroclus so as to prevent flies from filling his wounds with worms, a request to which Thetis promised to comply. Redi here relied on everyday knowledge and classical sources in conjunction with his experimental research, as if to imply that Homer, butchers, and even market women were wiser than some philosophers. To test his suspicion, Redi employed four pairs of glass containers, filling each pair with different meats, from a snake, freshwater fish, eels from the Arno, and calf. He sealed with paper and thread one from each pair of containers, leaving the others open to the air. Not long afterward he noticed that worms appeared in the open containers, whereas in the sealed ones the meat rotted away but did not generate insects. Redi also buried some pieces of meat and found in this case too that no worm was generated unless the meat had been in contact with flies. Thus, Redi questioned Kircher's recipe described in the twelfth book of his *Mundus subterraneus* for generating flies from their corpses by covering them with honey water and placing them on a copper plate with warm ashes. Considering that fresh air may have played a role in the generation process, Redi refined his experiments by introducing a variant. Instead of sealing the container, he chose a very large one and closed it with a very fine cloth; then, as an

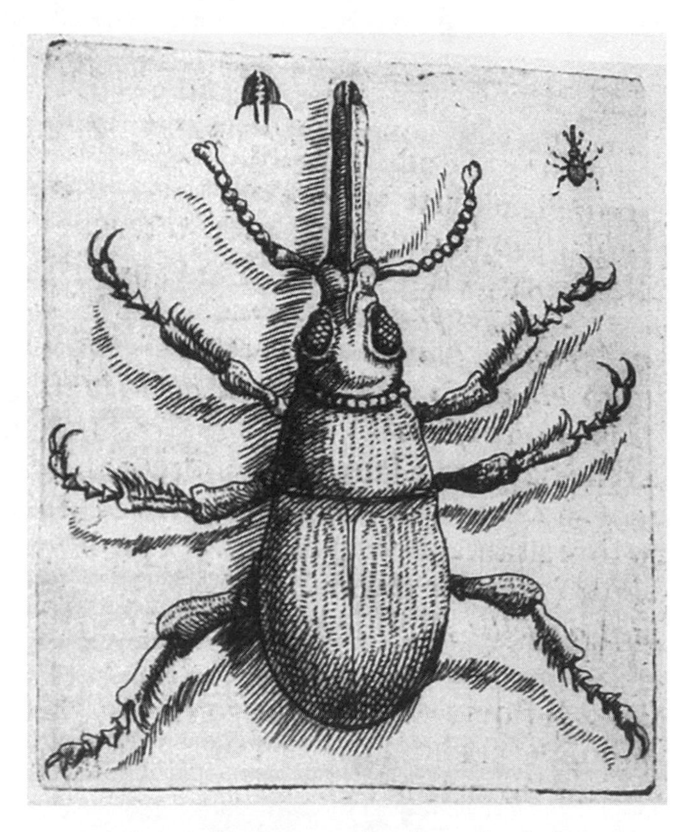

Figure 7.4. Stelluti, *Persio tradotto*: "*gorgoglione*" of wheat

added precaution, he placed the container in a box covered with the same cloth, finding that no worms or insects appeared. In this way he believed he had conclusively refuted spontaneous generation from putrefying meat.[15]

Redi's reliance on two parallel sets of experiments to disprove spontaneous generation shows remarkable sophistication. His method reminds one of other instances from different areas. Torricelli had already compared two differently shaped barometers in order to investigate the role of the seemingly empty space above the mercury column, and the Cimento Academy relied routinely on a rich set of instruments that were contrasted and compared with each other. Moreover, in the celebrated Puy-de-Dôme experiment two barometers were used: while one was carried up the mountain by Blaise Pascal's brother-in-law and court councilor in Clermont, Florin Périer, the other was kept under watch by a monk in the garden of the Minim convent as a "continuous experiment," a significant precaution given the puzzling variability of the height of the mercury column. Based on Pascal's own printed report, it was Périer who thought of using two barometers; Pascal had asked him to perform the experiment several times in the same day using the same apparatus at the top and bottom of a high mountain.[16]

Such methodological precautions must have appealed to Redi, since he employed them again in 1673 in testing the properties of a special styptic water allegedly healing arterial wounds. The methodological significance of Redi's experiments requires a short digression. Initial reports from Paris and London suggested that experiments performed there by the royal physician Jean Baptiste Denis, as well as Walter Needham and others, achieved total success: only in one case did a lamb die because a piece of cloth was inadvertently left in the wound. The other thirty animals survived, as did a horse whose tail was amputated. Some experiments were also performed on human patients, such as mastectomy on a woman with breast cancer and a man with a severe hemorrhage. Similar experiments performed on animals at the Tuscan court by Redi with the assistance of Tillman Trutwin, however, mostly failed. Initially arteries were severed, giving the animals apparently no chance of healing; subsequently the arteries were cut transversally and then longitudinally. In the latter case occasionally a wound was healed, as happened in the case of a sheep. At this point Redi decided to perform experiments using water from a well in addition to the styptic water. Although he did not run perfectly symmetrical runs varying only the type of water used, he was still able to show that some animals treated with water from a well were healed. In an especially interesting and unusually symmetrical case he amputated a wing of eighteen capons and treated six with the styptic water, six with water from a well, and six with nothing at all: they all survived. Since generally the styptic water did not perform better than water from a well, Redi could conclude that the healing properties attributed to the styptic water would be best attributed to nature rather than art. In subsequent experiments on generation, such as those on infusoria, Christiaan Huygens, Antoni van Leeuwenhoek, and the French microscopist Louis Joblot adopted similar precautions and used two phials, one open and the other sealed; at the Paris *Académie* Guillame Homberg also performed experiments on the germination of plants in a vacuum and in air running two parallel trials, a sign that Redi's method was becoming widespread practice.[17]

The last topic I wish to address in this section is the generation of insects in plant galls, an issue addressed in antiquity as well as by modern sources such as Harvey and Gassendi. The roots of this problem for Redi can be found in the debates and experiments of the Cimento Academy as well; oak galls were used as color indicators, a topic we have encountered in section 1.3. The Cimento had discussed the issue already in 1657, when the experimenters had observed insects being generated by eggs in the galls. The issue was also studied at Bologna by the astronomer Giandomenico Cassini, who had debated the matter with Vinvenzo Viviani in 1664 and at the Tuscan court the following year. While rejecting generation from putrefaction, Cassini remained in doubt as to whether the egg was generated by the plant or deposited by the parent insect. By observing with care the structure of galls, especially its regularity and formation process, Antonio Oliva and Borelli came to conclude that the eggs were generated by the plant. This conclusion had an anti-peripatetic significance because it seemed that plants, which had only a vegetative soul and lacked a sensitive one, would be able to generate insects, which have a sensitive soul; in other words, a less noble organism could generate a nobler one, thus transcending the natural order. Redi happily scoffed at this violation—"me ne fo beffe," he wrote—arguing that the notion of more or less noble is unknown to nature; he was using exactly the same argument we have seen Malpighi employ against Wharton in chapter 3. After having examined the thesis according to which the egg is deposited by the parent insect, Redi came to reject it and sided with Oliva. He denied, however, Kircher's and Fabri's claim that oak galls generate spiders, claiming that having opened in the space of a few years more than twenty thousand galls, he had never seen a spider being generated. In this case too the literary and philosophical sources he quoted are revealing: Democritus, Pythagoras, Plato, Anaxagoras, Empedocles, the Manicheans, Plotinus, Johann Vesling, and Tommaso Campanella. While in the case of putrefying meat Redi made his observations directly, in the case of galls he relied on gardeners and other Medici employees who supplied him with huge numbers of galls, as the Leghorn apothecary Diacinto Cestoni was to remark a few years later. By spending more time in the woods, he may have possibly come to observe insects hovering over plant leaves and to ponder on their role, much as he had come to suspect them when they were hovering over rotting meat. In this case patronage resources, far from enabling research, hindered a crucial stage of the investigation. In 1667–68 Redi relied on an artist, possibly Lorenzino Beatrucci, to draw and color a series of tables of galls, while Pizzichi drew the related insects. Redi also purchased twenty-five copper plates to be engraved by Guiducci. Probably he was planning a companion volume to the *Esperienze*, but this project did not reach publication, possibly either because he had already incorporated some material in the *Esperienze* or because he had come to doubt his conclusions.[18] As we will see in section 9.2, Redi's work on galls was to come to Malpighi's attention.

7.3 Malpighi: *Historia* and Anatomy

Despite the huge range of animals he had dissected, insects were not part of Malpighi's standard repertoire. In 1667 he studied a butterfly, a caterpillar, and a silkworm that had been preserved in spirit of wine, but it was not until 1668 that he started systematic dissections of the silkworm in response to a specific request by Henry Oldenburg, who was also interested in the insect for its economic significance. The secretary of the Royal Society had begun a correspondence with Malpighi, impressed by his contribution to the anatomy of sense organs. Malpighi's more than eager response was indicative of his desire for recognition from the Royal Society: he even bred silkworms at home. In his posthumous *Vita* Malpighi was to state that during the summer of 1688 he had worked so hard on the silkworm that he became sick. Despite that, discovering so many "miracles of Nature" had given him a "pleasure no pen could describe."[19]

When he shipped his manuscript to the Society on 15 January 1669, Malpighi was blissfully ignorant of Robert Hooke's stunning *Micrographia*, which had discussed the silkworm and other insects among a multitude of microscopic curiosities. Only in 1671 did Malpighi receive an account of that work by his friend Silvestro Bonfiglioli in Rome, who had Auzout translate the English for him.[20] On receiving Malpighi's manuscript, Hooke and Oldenburg perused it and found it "very curious and elaborate, well worth printing": for once, Hooke did not claim priority. In March 1669 Malpighi was elected a member of the Society in recognition of his goodwill toward it and for the excellence of his contributions: *De bombyce* appeared in the summer of that year. Although Marcello Malpighi's *De bombyce* is generally considered a landmark in insect microscopy, it has been largely ignored by historians of science as to its contents, techniques of investigation, and iconography.[21]

In his epistolary treatise Malpighi combined his skills as a naturalist and an anatomist. He provided an external description of the silkworm and its habits, from its size and the length of the silk thread to the period of incubation and eating preferences. Malpighi moved from the exterior to the interior, from the features most accessible to the senses to progressively inner ones, as if following the progress of his own dissections: the *corium* or "skin" was the transition from the description of the outer appearance, in the tradition of natural history, to an investigation of the inside, in the tradition of anatomy. Previous investigations by Hooke and Redi had shown glimpses of the interior of transparent or diaphanous insects, but even in those particular cases they had been unable to provide any detailed analysis; in other cases, when the inside of the animal was not visible from the outside, they had been unable to penetrate the outer surface in a meaningful way. It is in the dissection of the interior that Malpighi broke new ground, providing an account of the inner structure of the silkworm as it had never been dreamt of. As in all his other publications, he dissected a wide range of other animals, such as locusts, butterflies, crickets, bees, and wasps, as a way to both compare and elucidate specific structures following the microscope of nature.

De bombyce provides a remarkable account of the anatomy of the different stages of development of the silkworm, showing a continuity among them that was to impress Swammerdam, who had made of this claim a key point of his work. Malpighi explored the respiratory system and the nerves, the male and female genitalia, the silk-producing apparatus, and the heart, which consists of a long tube with a series of enlargements that he considered as little hearts. In the butterfly he discovered the periodic reversal in the direction of the heartbeat, from the tail toward the head, then the other way around, obviously in vivisection experiments.[22]

As usual, Malpighi had little to say about his microscopic techniques and his instruments, but several passages in *De bombyce* provide insights into his staining techniques and methods of preparation, showing a resemblance to those he had adopted in the study of higher animals. For example, in describing the yellowish fluid bursting forth on breaking the *corium* or skin, Malpighi stated that it coagulates and forms a crust with the heat of the hand or of fire. He attributed its origin to the digestive organs and argued that it is found in the viscera and the belly of the boiled silkworm and Aurelia, much as in the belly of those affected by dropsy. It is noteworthy here that Malpighi borrowed from medicine investigation techniques and the terminology to describe the silkworm; elsewhere he referred to cachectic silkworms.[23] In describing the fibers attached to the skin, Malpighi used the same methods he had adopted in studying the tongue, the skin, and the cerebral cortex, namely, boiling and staining with ink.[24] In order to study some structures filling the body of the caterpillars, Malpighi studied large specimens—the microscope of nature applied to insects—and suggested extracting those structures from the animal, wetting them, and placing them on a glass surface. Having eventually detected the same structure in the silkworm, he surmised that it consisted of glands, but when he heated it on the fire, it melted and burned like fat; here he was relying on results from *De omento*. Malpighi used similar methods to assay the juice forming silk, noticing that it does not melt if boiled, does not liquefy or burn with fire, but retains a glutinous nature like gum.[25] While studying the stomach, he had recourse to another standard technique, i.e., tasting, finding its honey-like liquid insipid, no doubt a less unpleasant task than tasting the fluid from the belly of the cadaver of the Reverend Father Bolognetti (see sec. 4.3).[26]

Malpighi added twelve plates to *De bombyce*, showing the silkworm's external and especially internal features; it seems plausible that in this, as in most subsequent publications, Malpighi included a large number of illustrations in order to impress the Royal Society. Shading is used to give a sense of volume to the part, although not to give the impression that the object was resting on a support as Stelluti had done with the wheat weevil. As in previous plates, starting from the second *Epistola* on the lungs, Malpighi used letters to identify the parts. In *De bombyce* there is a striking difference between the figures representing the exterior and those showing the interior: the former, including the entire silkworm and details of the head or the back, can be decoded in terms of spatial organization of the parts; the latter show various organs or body parts in isolation, without providing readers with a sense of the proportions of parts that had never been seen before by human eye, or of their organization and relation with each other. Thus, despite Malpighi's efforts to display and clarify, readers could only wonder in amazement without gaining a sense of the interior of the silkworm. Malpighi's images of individual parts look exactly as one would expect from the technique mentioned above of placing the wet part on a glass surface: they are superb individual portraits but lack a sense of the whole, as with the example of the male sexual organs of the silkworm (fig. 7.5), where we have no idea of relative size and of the arrangement of the part with respect to the whole. A related figure of the female genitalia (fig. 7.6) shows the ovary consisting of eight branches terminating in the anus A at the top. Malpighi was at pains to describe the features he had observed and to explain their purpose: comparative anatomist Francis Cole found this portion "the most impressive part" of his treatise. I will not follow either Malpighi or Cole in their description of the parts they identified, but focus primarily on the eggs. Each branch contains sixty or more eggs marked D. While the depression in the silkworm eggs in the ovary is rendered with a spiral line, this technique is not maintained in the isolated egg shown on the right as figure IV, where the indentation is rendered by hatching. The eggs have a shell unlike that of hens because it is diaphanous and flexible. Its external surface is not smooth but rather covered with tiny protuberances like the skin of the fish *squatina*, whereas for Hooke intriguingly they were pitted or covered with cavities. Hooke was extremely interested in the artifacts of microscopy and especially the issue of pits versus warts, which he discussed in the preface of *Micrographia* in relation to the eye of the fly: in the case of the silkworm eggs, however, both accounts appear legitimate because the pits/protuberances are so dense that no univocal description is privileged.[27]

Whereas Hooke provided an individual portrait of an egg to satisfy curiosity and

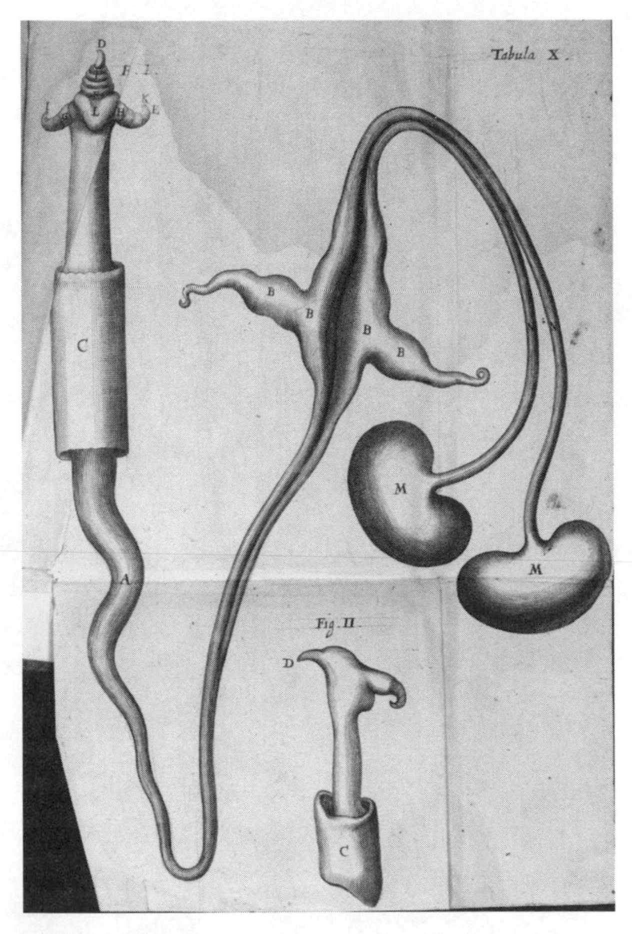

Figure 7.5. Malpighi, *De bombyce*: male genitalia

to show the power of the microscope, Malpighi focused on the silkworm's internal anatomy and provided a much more ambitious figure of the ovary in which the actual shape of the egg visually and conceptually takes second place. The images suggest that Hooke could attain greater magnification, whereas Malpighi could muster greater skill with preparation techniques.

Figure 7.7 shows, on the left, the stomach and digestive apparatus and, on the right, the silk-producing vessels. Since the vessels can be found on the sides of the stomach, it is helpful to see the two organs side by side, even if the reader is left to wonder about their relative size and location. "Reading" the figures is not always straightforward. The silk-producing vessels come in symmetric pairs; the distinguished comparative anatomist Francis Joseph Cole, however, failed to grasp Malpighi's method of representation and charged him with having shown them to be asymmetrical. Malpighi,

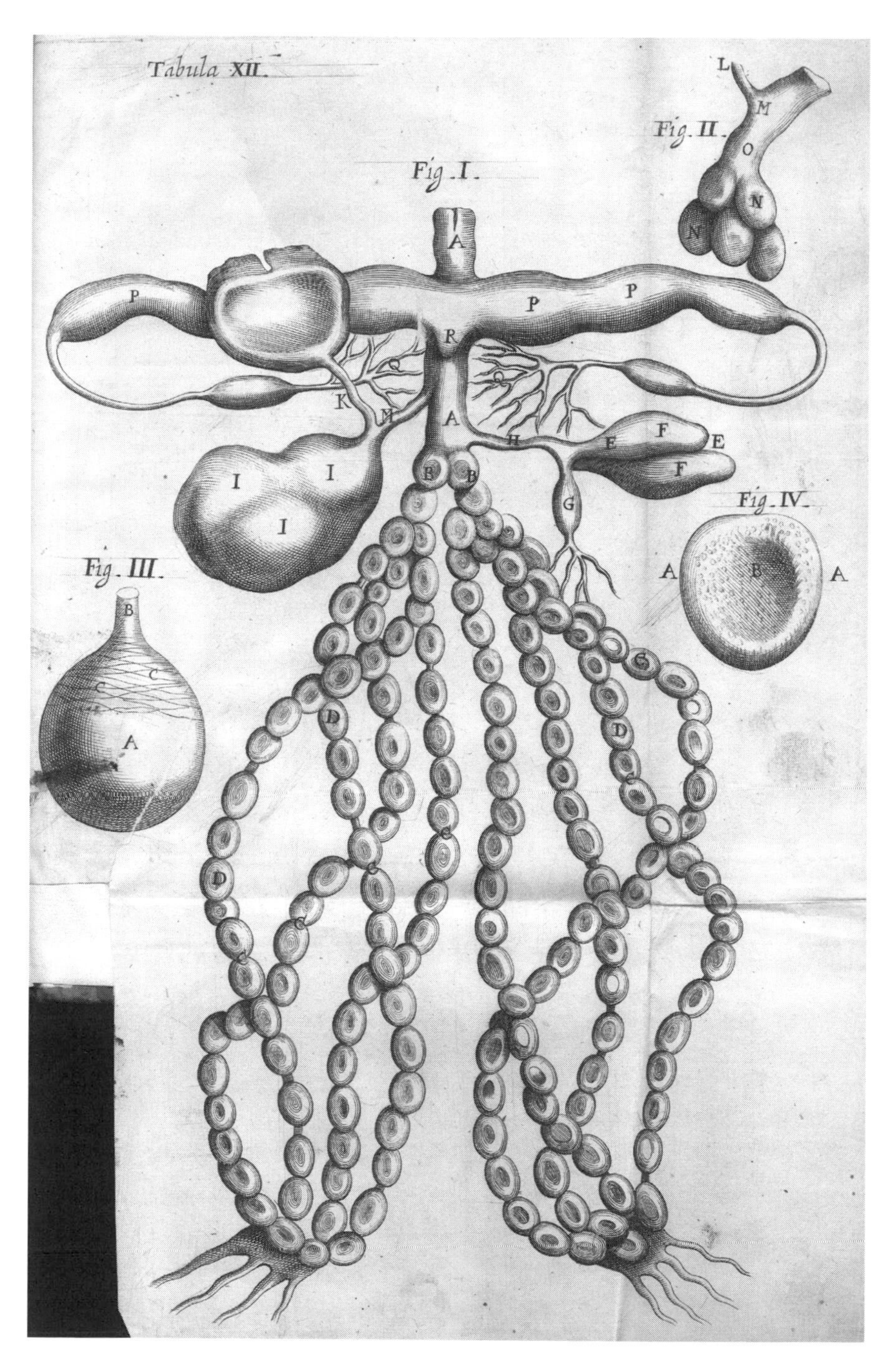

Figure 7.6. Malpighi, *De bombyce*: female genitalia

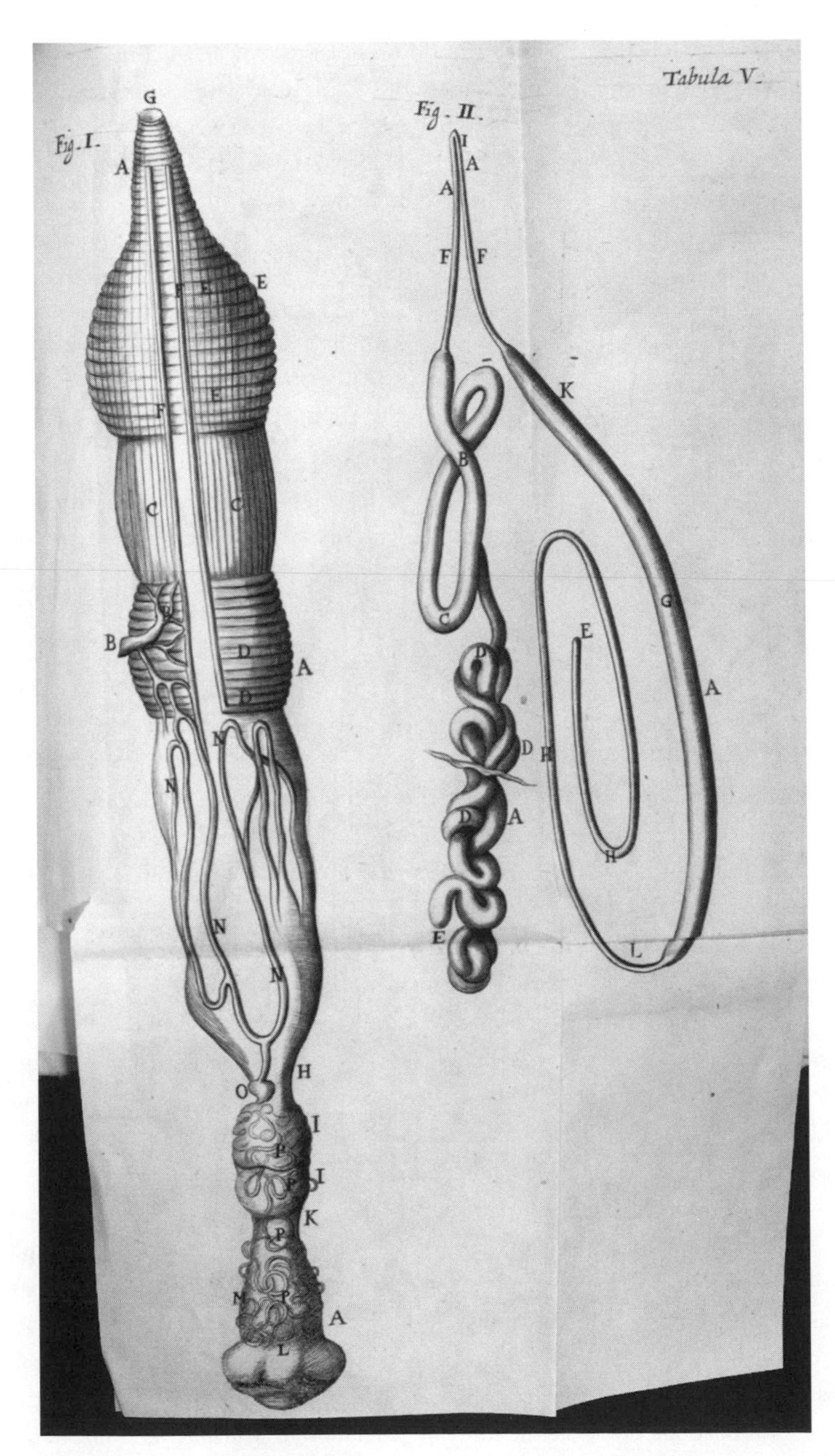

Figure 7.7. Malpighi, *De bombyce*: digestive apparatus and silk-producing vessels

however, had tacitly unraveled the right side in order to show its full length, which he stated to be one Bolognese foot; a manuscript in his hand with a drawing similar to the one in *De bombyce* explains that the right side has been extended.[28]

Malpighi performed impressive experiments on respiration in the silkworm. After having identified structures that he suspected allowed air to enter the body, he sought to probe their role in a series of tests going to a considerably deeper level of understanding than the similar ones just published by Redi, which may have inspired him. First he immersed silkworms in boiling water and observed air coming to the surface. Some air remained trapped between the folds, however, even if the animal had been wet beforehand; therefore, this experiment lacked the evidence Malpighi had wished. If the silkworm is dead, however, no air rises to the surface.[29] Next, Malpighi experimented with oil; by means of a little pointed brush—an unusual tool for an anatomist—he covered the apertures on the sides and noticed that the silkworm went into convulsions. Then he applied oil selectively first to the orifices at the front, then to those at the back, on the right, and on the left, and found that those body parts whose orifices had been obstructed went into convulsions and then became sluggish; the heartbeat became very slow.[30] Applying oil to other body parts while avoiding the orifices, Malpighi noticed that the silkworms showed no sign of discomfort. At this point he changed the substance applied, and used other types of fat, such as butter and lard, which behaved like oil. Nor was it necessary to use fatty or oily substances, because liquefied honey also led to the same results, except that it was easier to wipe honey off. Having experimented with variations to the orifices and the substance applied, Malpighi then changed animals and made similar tests with locusts, grasshoppers, and similar insects, confirming his previous results. He also noticed that those silkworms that had been immersed in water even for hours revived when they were exposed to the sun. Referring to the Society's air pump experiments—"in vestra *Antlia pneumatica*"—he confirmed that insects deprived of air die.[31] In another experiment with a Redi flavor, Malpighi set out to refute the superstition whereby silkworms dislike bad smells: he kept them in boxes full of fetid plants or opium and noticed no ill effect. He confirmed, however, that they are affected by southerly winds, when the air becomes very hot, and also by cold weather.[32]

Lastly, he devoted his efforts to the study of eggs, which he believed were produced by the female and made fertile by male semen, as in chickens. Malpighi attempted artificial insemination by wetting the eggs from a butterfly with semen, but they remained the color of sulphur rather than turning purple, a sign that they had not been fecundated. The change of color was due to some network structures similar to grass or ivy. Given the analogy established above, he tried to rely on larger hens' eggs; thus, these investigations anticipate the study of generation to which he was to turn. Malpighi believed that semen enters the eggs through some apertures that he sought to locate either by placing the eggs near a fire or by boiling them in spirit of wine or oil of sulphur mixed with ink; in this way he found that the ink had entered the membrane under the shell and in the albumen.[33] The detailed study of the organs of generation and eggs of the silkworm and other insects clearly showed the level of organization of the insect body, making the idea of spontaneous generation implausible.

These observations on eggs lead us to a striking feature of *De bombyce*, i.e., its use of color as a general descriptive category, as an analytic tool—to identify fecundated eggs, for example—and as a mode of visual representation, as evidenced by a striking watercolor of a silkworm included in the manuscript version of the treatise preserved at the Royal Society and first studied by Matthew Cobb. Color had become an integral part of Malpighi's descriptions, not only as a significant philosophical feature of the object under investigation, but also as a source of pleasure. For a follower of Borelli who in his first publications on the lungs had nothing to say about the changing color of blood, this seems a remarkable transformation and a so far unrecognized mark of his departure from his erstwhile mentor; it seems plausible that by then Malpighi had come across Boyle's work, in which color and color changes are given a more significant role than in Borelli's reading of the Galilean tradition. Moreover, his training as a physician would have led him to pay attention to color in a range of areas, as we have seen in section 2.3 with regard to jaundice. According to Borelli, colors could easily be changed and were not reliable indicators of the substance at hand; Boyle, by contrast, saw color in relation to the texture or even the roughness of the surface of a substance, one that could be ascertained by touch by a blind man. Thus, he was prepared to attach an ontological status to colors lacking in the Galilean tradition followed by Borelli.[34]

From the first pages of *De bombyce* we read of eggs turning from *violacea* to *caerulea* or light blue, then *sulphurea* and thereafter *cinerea* or ash-colored; nor are Malpighi's identifications of colors approximate: on one single page we find him distinguishing between *cinereus* or ash-colored and *fuliginosus* or soot-colored in describing the color of the just-born silkworm, a color that soon turns into *perlatus* or pearl; the head is *coracinus* or raven-black; the hairs and legs are *ziziphini* or jujube-colored. Elsewhere Malpighi described the color of the silkworm as *achatis* or of agate in those parts free of folds, and *argenteum* or silvery elsewhere. The silk thread is *luteus* or *auratus*, yellow or golden, or also *subalbus* or whitish with *sulphur* speckles. One cannot fail to notice Malpighi's sheer delight in describing and his remarkable sense of color, since only in this way can we explain his extraordinarily nuanced descriptions; we are also reminded of his artistic interests, in which color played a major role. We have an exactly contemporary and relevant letter to the noted Sicilian collector Antonio Ruffo, who had purchased paintings by Rembrandt, including in 1653 *Aristotle with a Bust of Homer* and *Homer*—now at the Metropolitan Museum in New York and the Mauritshuis in The Hague, respectively—dated 1663, thus from the time when Malpighi was at Messina. In the letter to Ruffo Malpighi provided a rich set of artistic news about recent acquisitions and prices. He regretted that a fever—probably the same that he in his *Vita* attributed to excessive work on the silkworm—had prevented him from going to Parma and Correggio to see works by Correggio and Parmigianino; he did go to Mirandola, however, where he saw a nude Venus by Titian with "mezze tinte di Paradiso," or heavenly half tones. These observations on Malpighi's language and artistic interests—especially about color—corroborate Cobb's attribution of the watercolor of the silkworm to *De bombyce*, since the drawing and especially the color range correspond remarkably to those of a development stage described in the text. It is reasonable to surmise either that Malpighi himself was responsible for the watercolor or that the artist who executed it worked directly under his supervision. Here too the letter to Ruffo proves useful, since in it Malpighi states that in the summer of 1668 he had employed a young painter "per dessegnarmi alcune cosette," or to draw for me a few little things, and also to make copies of paintings by one of the Carracci brothers. The young painter executed a few little things for Malpighi exactly at the time of his most intense work on the silkworm; thus, it seems plausible that Malpighi used the same painter to help him draw and color the silkworm and make copies of Carracci. This would not have been the first time Malpighi had relied on a professional artist, since, as we have seen in section 3.3, at Messina he relied on the services of Agostino Scilla; later he also relied on Carlo Cignani, who painted his portrait and for whom Malpighi wrote a consultation.[35]

Malpighi's interest in insects and their generation was not limited to natural history. An exceedingly interesting case occurs in a letter from Stefano Piccoli, a physician from Verona. Having found in an ulcer on the penis of a patient of his affected by the French disease a huge number of worms, Piccoli wavered about spontaneous generation and asked Malpighi for light on the matter. Moreover, after much toil he had found a copy of *Anatome plantarum*, but wondered where he could purchase a copy of *De bombyce*, which he had been unable to find. Malpighi replied that from analogy and the constant way in which nature operates, in his view the juices of plants used to heal the wound may have contained the eggs whence the worms originated, thus rejecting spontaneous generation. We find here a striking and fascinating correlation among practical medicine and the study of generation and insects.[36]

7.4 Swammerdam: Metamorphosis and Classification

Unlike Malpighi, Swammerdam had been fascinated with insects since his student days and had put together a remarkable and imposing collection of over a thousand items, including rare exotic specimens for which he had been offered twelve thousand guilders in 1668. His interest in lower animals can be noticed already from the title page of his early work on respiration—one owned by Malpighi—showing two snails among a wide range of specimens and devices used to experiment on respiration: although the elaborate twisted elements joining them may appear as decoration, in fact they represent their penises during intercourse, since snails are hermaphrodites, a surprising result with significant implications that was announced in his 1669 treatise and that we shall encounter again in the discussion of sexual reproduction in plants (fig. 7.8). This was not Swammerdam's only startling result about sexual reproduction: he also realized that the "king" bee contained egg masses and was therefore a queen.[37]

When in 1669 Swammerdam received from his mentor Thévenot a copy of Malpighi's *De bombyce*, his first concern was to find confirmation of his own views against metamorphosis. Soon, however, he felt challenged to emulate and surpass Malpighi's achievement, discovering or improving upon Malpighi's techniques of investigation; this task occupied him, with significant intermissions due to his religious crises, for the rest of his life. In 1672 he published *Miraculum naturae*, a treatise on the structure of the uterus and generation containing critical passages of *De bombyce*. Three years later he published *Ephemeri vita*, a treatise on the mayfly interspersed with prayers. In the same year he decided to abandon his research and destroyed the text of his work on the silkworm, but somewhat enigmatically he had Steno deliver to Malpighi his drawings, which today can be found at the University Library in Bologna. Some of those drawings highlight inaccuracies in Malpighi's work that he had touched on in *Miraculum naturae*. Lastly, after a period with a religious group on an island off the Danish coast, he returned to research and completed a *magnum opus* on insects, *Bybel der Natur*, which was published posthumously by Boerhaave in the 1730s. Swammerdam had died in 1680.[38]

Unlike *De bombyce*, Swammerdam's *Historia insectorum generalis* is devoted to insects in general. The main thrust of his work can be grasped from the full title, which claims that only gradual changes occur in insects, denies sudden transformation or metamorphosis, and proposes a classification in four orders based on their modes of development and transformations. All these aspects are mentioned in the review in the *Philosophical Transactions*, which further highlights the analogy between plants and animals, and the discovery that snails are both male and female and discharge their excrement from their necks.[39]

Swammerdam did not have kind words for Harvey; he quoted a long passage from *De generatione animalium* in which Harvey had contrasted generation by epigenesis and by metamorphosis: the latter he compared to the activity of a carpenter or a sculp-

Figure 7.8. Swammerdam, *De respiratione*: title page

tor, who fabricates a couch or a statue from preexisting material; the former he compared to a potter, who fabricates a vase by shaping his creation while adding material at the same time. Harvey provided a more precise definition as well, showing the links among the faculties of generation, growth, and nutrition:

> In generation by metamorphosis creatures are fashioned as it were by the imprint of a seal, or cast in a mould, that is the whole of the material being transformed. But an animal which is procreated by epigenesis draws in the material and at the same time prepares and concocts and uses it; at the same time that the material is formed, it grows.

Harvey was relying on Aristotle in claiming that some animals, such as "bees, wasps, butterflies and all creatures whatsoever that are generated by metamorphosis out of a larva, are said to be bred by chance and so are not capable of preserving their own kind"; therefore, they "are born spontaneously, out of material concocted spontaneously or by chance." Aristotle had argued that in the same way as a craftsman can make simple objects by hand but needs more elaborate tools to make more complex ones, so nature makes simple creatures directly but needs tools—the parents—to produce more elaborate animals.[40] By contrast, careful dissections of the different stages of development of insects, especially larvae and chrysalides, revealed to Swammerdam that change is gradual and that the wings of the butterfly, for example, can be seen developing in the different stages of the larva. Further, Swammerdam sided with Redi in denying spontaneous generation, but this aspect was not central to his work. Without mentioning Redi but obviously referring to him, however, he denied that the insects found in galls are generated by the plants, arguing instead that their eggs are deposited by the female.[41]

Swammerdam's fourfold classification largely shaped the structure of his book. The first species of transformation involved insects that hatch from the eggs in their adult form, the only subsequent change being growth. In the remaining three species the insect hatches before having reached its final stage and has therefore to undergo a further transformation through a pupa stage to reach its final form: the study of generation was profoundly intertwined with the taxonomy of insects, and, in a broader sense, a typical anatomical domain had become key to natural history concerns.

Like other scholars engaged in the studies of insects, Swammerdam faced the problem of size and texture. As to size, he relied on single-lens microscopes made by the noted mathematician and at time magistrate and burgomaster of Amsterdam Johannes Hudde, whose skill Swammerdam acknowledged in the text. Swammerdam was also quite generous in imparting knowledge about his techniques: he explained that in order to study the water flea, he placed the insect in tiny half globes of glass specially blown for the purpose, or placed them in drops of water on white or colored paper; we

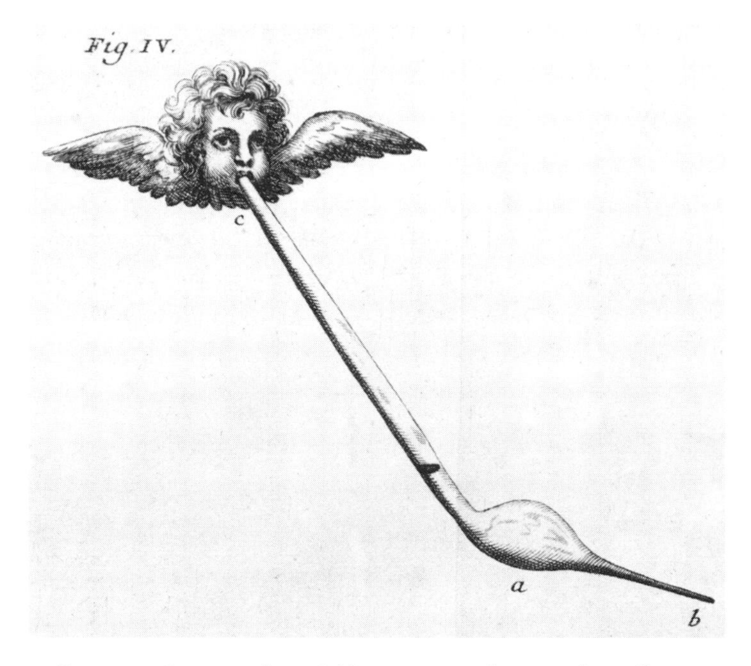

Figure 7.9. Swammerdam, *Biblia naturae*: technique of insufflation

are going to see below that this technique of investigation, designed to highlight the contrast of the insect's contour, had its counterpart in a technique of representation. Swammerdam further stated that he put caterpillars in a bottle filled half with lees of wine and half with vinegar, a combination that, while killing the animal, hardened its parts. Later he was to develop techniques of injection of various substances, staining, drying, using colored glass as a background, and insufflation through a tiny glass pipette, a technique analogous to that used by de Graaf in the study of blood vessels (see figs. 7.9 and 8.3) and that is still in use today by entomologists.[42]

Swammerdam was especially attentive to the issue of graphic representation, as is to be expected of a scholar from a region whose artists were at the forefront in realistic depictions of nature. Artists such as Joris Hoefnagel from Antwerp and Johannes Goedaert from Middleburg were pioneers in the representation of insects. Hoefnagel's works became known through a set of engravings executed by his son Jacob, issued in 1592 and 1630. In 1662 Goedaert published the first volume of *Metamorphosis et historia naturalis insectorum*, followed by a second volume in 1667 and a posthumous one in 1668, a work he claimed was the result of thirty years of research in which he joined a descriptive text with copper engravings that he personally colored by hand. Both works were important sources for Swammerdam, who referred to them extensively in his *Historia.* He criticized Goedaert's use of color, however, arguing that color added confusion to his figures, and claimed instead great accuracy for his own. Overall, Swammerdam was restrained in the use of color, although he colored blood vessels in the engravings of many copies of his 1672 *Miraculum naturae.* In an important passage, he also argued that Goedaert's printing technique was defective because it could not show the white hairs of some insects against the white page. For this reason he adopted a flexible approach, showing dark insects against a white background and white insects against a shaded background, so that all their features and especially hairs stood out, a technique already adopted by Hooke in a number of cases in *Micrographia.* At times Swammerdam relied on both techniques on a single plate showing different stages of development of an insect, as in figure 7.10. In all probability this technique of representation went hand in hand with corresponding observation techniques relying on lighting effects, to highlight white parts against a shaded background. Swammerdam did his own drawings and then had an artist prepare the engravings; whereas Malpighi had no control over this process, which was carried out in London for all his publications starting from *De bombyce,* Swammerdam had his engravers more at hand, yet he was still unhappy with the outcome. In one instance, particularly interesting from our present perspective, Swammerdam showed a water gnat similar to the one we have encountered above from Hooke's *Micrographia.* He referred to Hooke's work, acknowledging him as the first to discover the motion of the inside of the gut and of the different motions of the tail, while criticizing the accuracy of his figure. Swammerdam's own rendering (fig. 7.11) is remarkable in that the insect is shown not only in its different stages of development, life-size and enlarged, but also in its environment, with its posture head-down just below the water surface.[43]

Despite his periodic religious crises that led him to abandon scientific pursuits, Swammerdam had an admiration for the "grand Descartes," whom he praised on several occasions. He was fascinated by the notion of clear and distinct ideas and often referred to it, at times associating it with the use of the microscope, as if to claim that images seen through the microscope allow us to gain clear and distinct ideas of objects and processes, such as the lack of metamorphosis. Perhaps it was not accidental that a late seventeenth-century reader had the Amsterdam 1680 edition of Descartes' *Traitez de l'homme et de la formation du foetus* bound with Swammerdam's 1682 Utrecht edition of *Historie générale des insects,* as to mark the association between both the authors and the topics of natural history and anatomy. Swammerdam also referred to the nearly ubiquitous clock metaphor, but whereas Hooke had compared the organization of insects to that of clocks, Swammerdam used the metaphor differently, applying it to nature as a whole: he argued that an animal's death enables another to live, just as in a clock a weight going up enables the other to descend. His

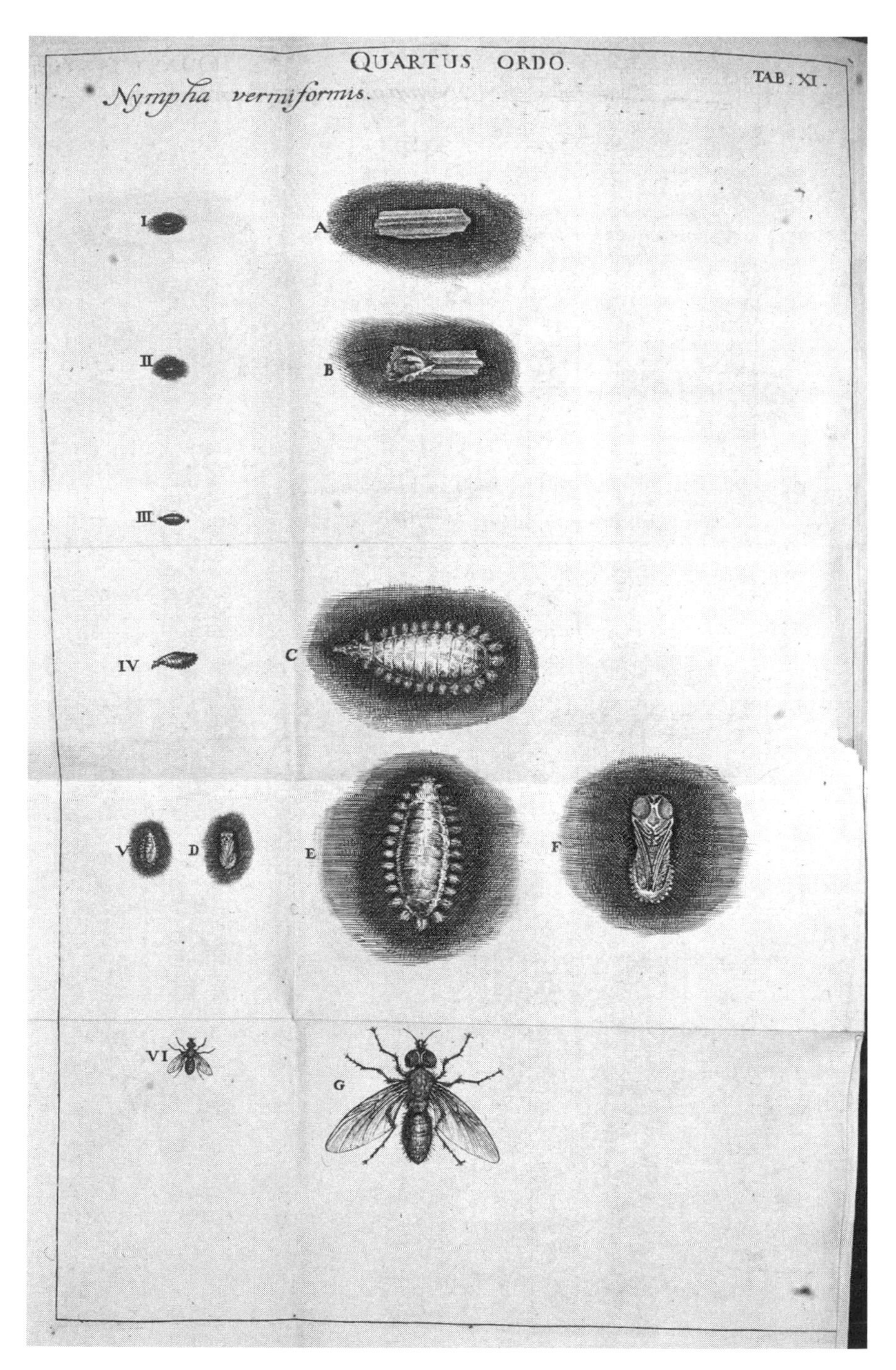

Figure 7.10. Swammerdam, *Historia*: insects against different backgrounds

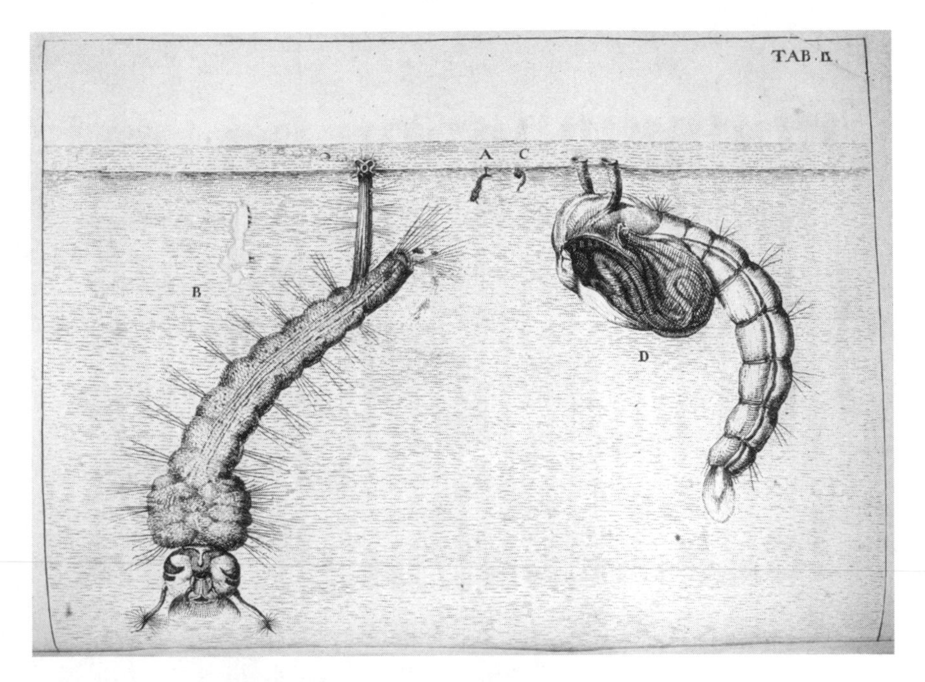

Figure 7.11. Swammerdam, *Historia*: the water gnat

was a striking usage in which nature as a whole is mechanized and in which life and death are rendered as weights going up or down.[44]

7.5 Swammerdam and Malpighi: Microstructure and Iconography

Following the completion of *De bombyce*, Malpighi continued experimenting on the silkworm and especially the issue of generation, becoming convinced that eggs are fecundated only by contact with male semen. Moreover, whereas in *De bombyce* he had claimed that the trachea is strengthened by annular rings, he later recognized that the structure he had detected was a spiral. In 1670 Malpighi became acquainted with Swammerdam's 1669 *Historia*, whose images were the only portion accessible to him in the original Dutch text.[45]

Not long thereafter Malpighi was brought back to the anatomy of the silkworm by an attack from Swammerdam. In the 1672 *Miraculum naturae*, on the structure of the organs of reproduction, the Dutch microscopist challenged Malpighi's work on several grounds. First, Swammerdam criticized Malpighi for his failure to disclose his methods for preparation, either for fixation or for microscopy: indeed, although Malpighi gave some details about his techniques, the information he provided was

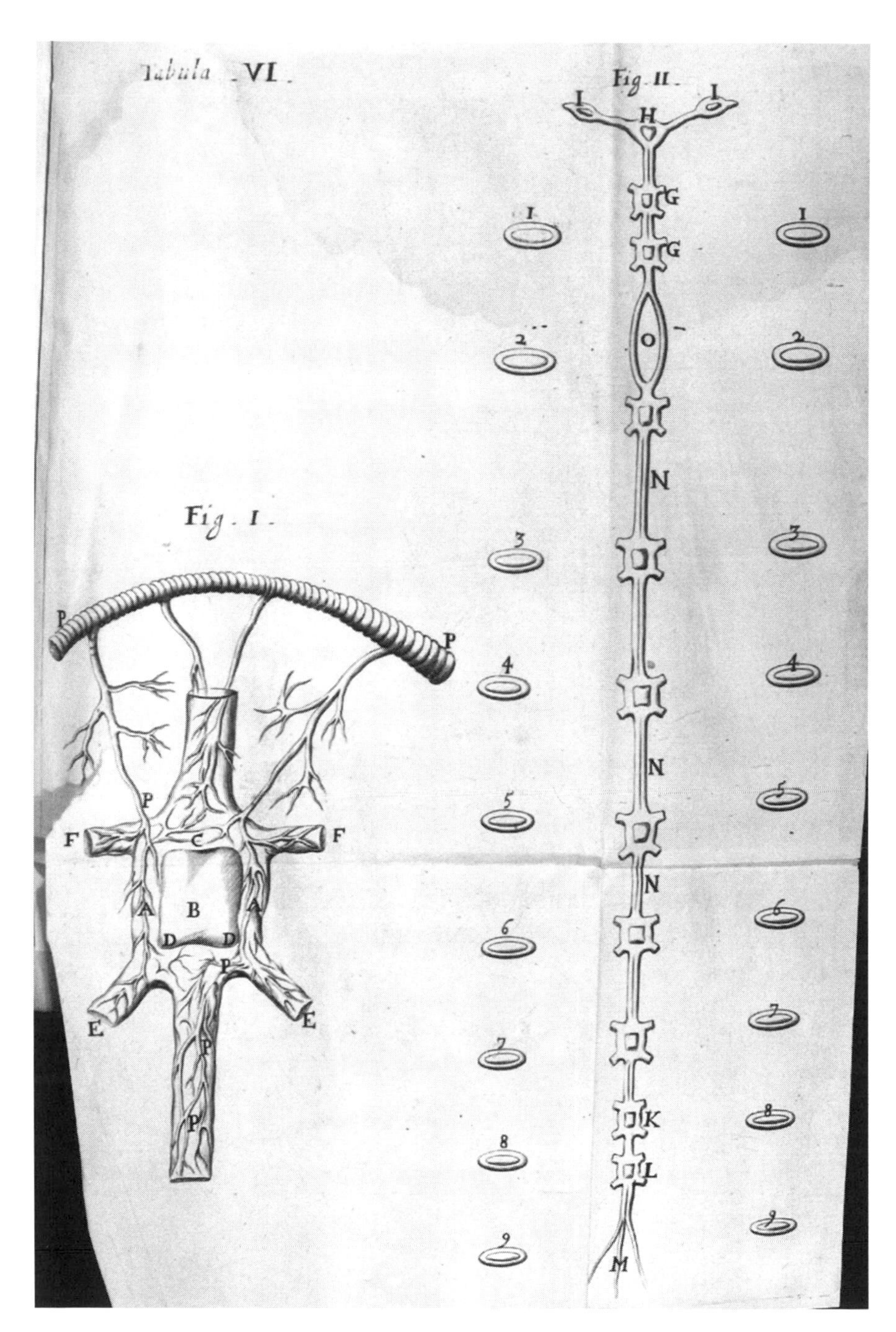

Figure 7.12. Malpighi, *De bombyce*: nervous system of the silkworm

insufficient to allow replication. Then he questioned the accuracy of Malpighi's drawings, especially with regard to the male organs of reproduction and the nervous system: as Swammerdam put it, Malpighi seemed to have conceived the figure with his mind, "*figuram mentem concepisse*," rather than actually seeing them with his eyes. As we have seen above, Malpighi extracted body parts from the silkworm and placed them on a glass surface for microscopic inspection. It was likely this procedure that led Malpighi to change the arrangement of the parts, in that he placed the testicles in an unlikely arrangement, "*modo alieno ponit*" (fig. 7.5), and to overlook the peculiar interconnections between the nervous system and the male genital tract; in fact, the seminal vessels pass through a nerve ring in the final ganglion, as shown at the bottom of Swammerdam's later plate, which can be usefully compared to Malpighi's corresponding one (figs. 7.12 and 7.13). In addition, Swammerdam argued that Malpighi had omitted to show the brain of the caterpillar and had misrepresented the multiple hearts.[46]

In 1675, after having destroyed his manuscript on the silkworm, Swammerdam forwarded to Malpighi via Steno several relevant drawings, some of them in color: this gesture can be seen as a sign of respect for the scholar Swammerdam esteemed most as to the study of insects, but also as a challenge in showing to Malpighi what he had missed and what Swammerdam had seen. As in many other instances, Malpighi waited several decades before answering *Miraculum naturae* and defending the unconventional drawings he had produced. In his posthumous *Vita* he engaged in an elaborate salvaging exercise of his claims and figures, although he acknowledged that Swammerdam had shown the brain of the silkworm more accurately. On balance, there appear to be strong similarities in the art of representation of the microscopic world employed by Malpighi and Swammerdam (figs. 7.12 and 7.13); although both unquestionably produced highly conceptualized images, there were also important differences between them. In order to illustrate my claim, I wish to comment on Swammerdam's figure of the ephemeron or mayfly (fig. 7.14), from *Ephemeri vita*, first published in Dutch in 1675 and translated into English in 1681: of course, Swammerdam also showed individual body parts, but in Malpighi we do not find an overall image of the interior of an insect putting together individual components as a coherent whole, as Swammerdam had done here, showing the medulla spinalis *yyy* with its eleven nodes, the muscle fibers *ddd*, the anus *e* removed from its natural position at the bottom, and air vessels *aaa* like worms running along both sides of the length of the body with their many ramifications, such as *ppp* to the fins, truncated to show the fins' structure. Swammerdam relied on the symmetry of the insect to show different parts, such as the seminal vesicles *fff* of the male, shown on the left in their natural position and size and on the right partly removed from the body and enlarged, thus offering a better view of the muscle structures underneath. This engraving provides

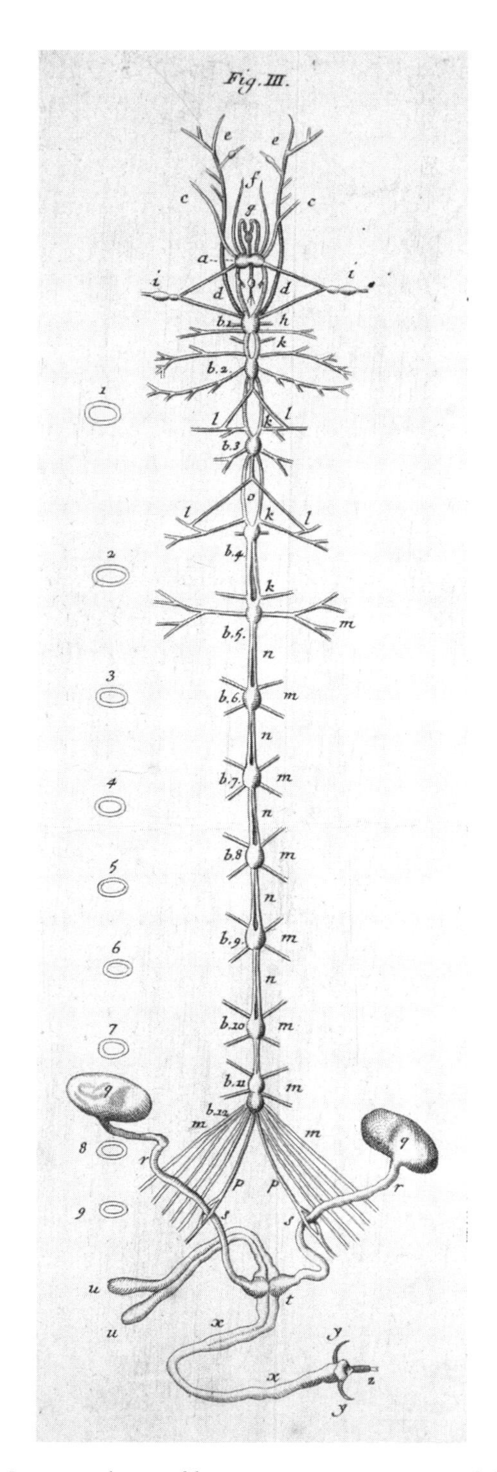

Figure 7.13. Swammerdam, *Biblia naturae*: nervous system of the silkworm

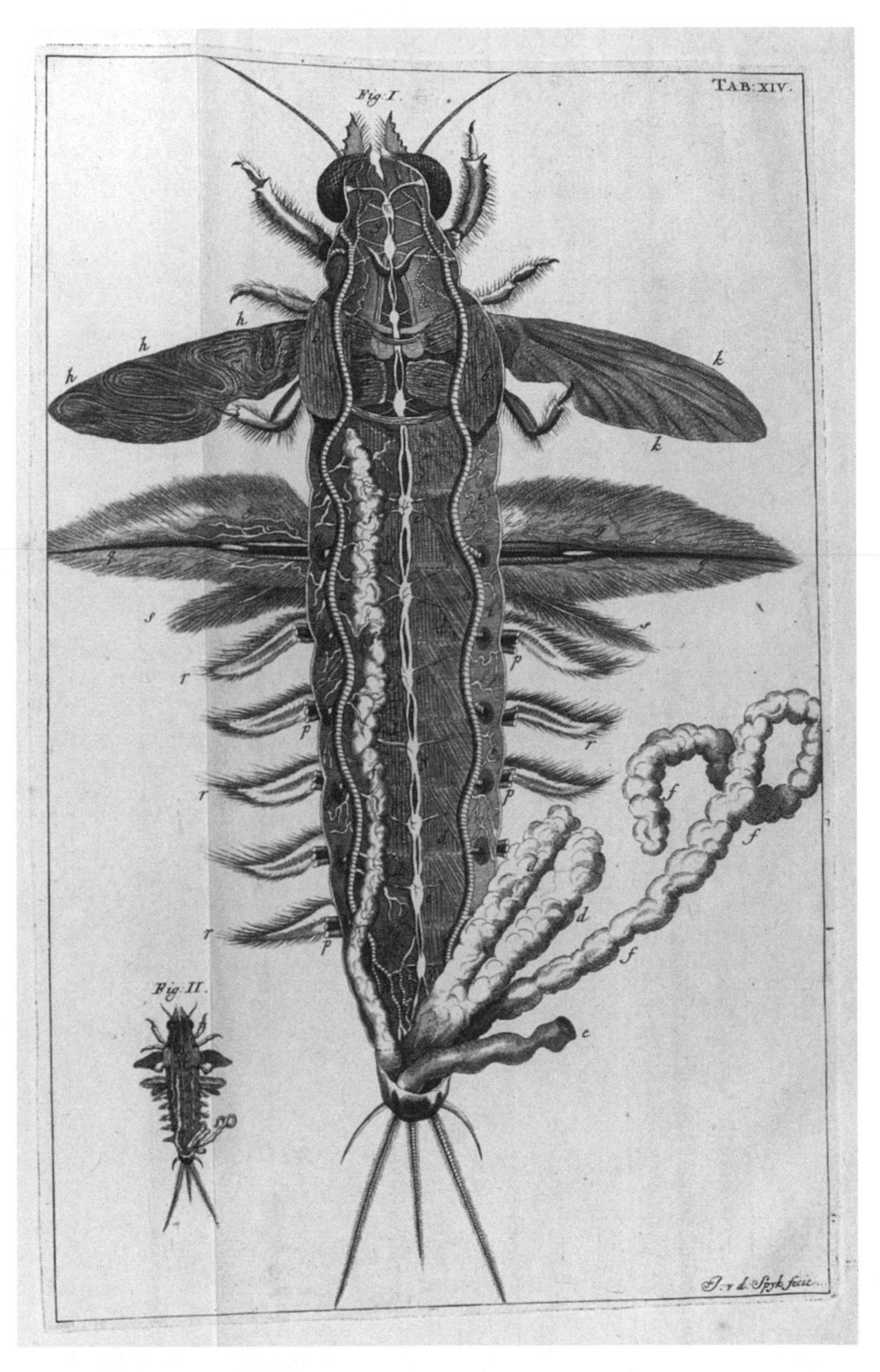

Figure 7.14. Swammerdam, *Biblia naturae*: interior of the ephemeron

an astounding view of the insect, aiming at a more informative representation while not refraining from showing truncated and displaced body parts. This plate highlights Swammerdam's sensitivity to issues like the spatial arrangement, relative size, and mutual disposition of the parts; Malpighi showed individual organs in puzzling isolation, following a tradition going back at least to Vesalius of showing similar parts such as bones, muscles, veins, and arteries together. But whereas this method may have been less problematic for the human body, whose structure was better known, it became exceedingly problematic when it was applied to animals whose internal structure was completely unknown, such as insects. Swammerdam, by contrast, first decomposed the ephemeron—or rather many of them—but then he recomposed the entire insect with all its parts, thus offering to readers an integrated view of its interior, including the size and mutual arrangement of the parts. Despite this significant difference, both Malpighi in his study of the silk-producing glands and Swammerdam relied on symmetry as a tool to display different features of the insect: Malpighi showed the length of the structures and at the same time their appearance in the body of the silkworm, whereas Swammerdam showed the internal configuration of different layers inside the ephemeron's body.[47]

The study of insects posed major technical challenges because of the size and texture of their body parts. Investigators had to develop special techniques to observe insects and their interiors under the microscope, such as fixation and insufflation. Swammerdam found Malpighi's work exceedingly impressive but also frustrating because of Malpighi's insufficient disclosure of his methods: the reader of *De bombyce* had to rediscover the techniques employed by its author. Experimentation played a major role. It is instructive to compare Redi's and Malpighi's experimental confirmation that insects immersed in oil died: whereas Redi merely confirmed a claim found in the literature, Malpighi went further and tied the experiment to the investigation of the insects' anatomy, introducing a number of sophisticated variations showing that the purpose of the apertures on the side of the silkworm and other insects was respiration.

The visual representation of insects and their organs posed new challenges to anatomists, since viewers were asked to make sense of figures showing body parts that had no relation to their experiences. *De bombyce* was the most lavishly illustrated work Malpighi had produced thus far, a probable indication of how determined he was to impress the Royal Society, to which the treatise was dedicated. Anatomists such as Malpighi and Swammerdam developed novel techniques of representations as well, engaging in an iconographic dialogue in which they challenged, and responded to, each other through images. In this regard Swammerdam's sending his drawings of the silkworm to Malpighi is emblematic of the exchange of information, involving also challenges and criticisms, occurring through images. Whereas Malpighi produced individual portraits of many organs and body parts—such as the genitalia or the respiratory system—taken in isolation, Swammerdam strove to develop a visual language showing the relative size and mutual relation among the parts.

Despite its difficulties, the study of insects proved rewarding at many levels and had profound and far-reaching philosophical and theological implications on topics such as the previously unsuspected complexity of their internal organization and their mode of reproduction. The issues of mechanization and generation were crucial at many levels: Swammerdam's striking metaphor of the clock applied to nature as a whole compared life and death with the ascending and descending weights of a clock, whereas Hooke's remarks in *Micrographia* show that mechanistic interpretations did not lead by themselves to the denial of spontaneous generation. Rather, it was the sustained body of observations and experiments on insects that tipped the balance in favor of sexual reproduction, from Redi's sophisticated experiments questioning spontaneous generation to the investigation of the insects' genital parts and the role of eggs and semen in fecundation. It is not surprising that the same circles and at times the very same scholars working on insects engaged in the study of generation; some of the texts discussed here will be mentioned again from a different perspective in the following chapter.

Generation and the Formation of the Chick in the Egg

8.1 Generation and Its Problems

The problem of generation is one of the most demanding—both technically and conceptually—in the entire history of anatomy and over the centuries had attracted the attention of anatomists, philosophers, and historians. Besides the structure of female and male reproductive organs in different animals, several issues are at stake, such as the respective role of female and male parents and the order of formation of the organs in the fetus. The identification of the stages of the formation of the fetus had profound implications on the understanding of the organism: Aristotle, for example, saw the heart as the primary organ and believed that it was formed first, whereas in some writings Galen argued for the priority of the liver over the heart.[1]

Since antiquity, the formation of the chick in the egg had been chosen as a particularly convenient and affordable case for the study of generation, with the chick as a "model organism": the authors of the Hippocratic corpus and Aristotle wrote on it, and so did in modern times the Dutch anatomist and physician Volcher Coiter and his Bologna teacher Ulisse Aldrovandi, as well as Hieronymus Fabricius and his pupil Harvey. Although the formation of the chick in the egg is but a subsection of the problem of generation, this itself is not easily manageable: Howard Adelmann's monumental study of Malpighi's contributions is accompanied by twenty-six excursuses exceeding one thousand folio pages. Adelmann's work is an indispensable tool, yet despite his remarkable technical competence and historical erudition, in this huge amount of material it is often hard to tell the forest from the trees, to characterize the contributions of individual anatomists whose work is presented fragmentarily, and to identify broader themes and problems.[2]

The aim of this chapter is to identify in this vast and at times rugged landscape some key themes and problems associated with generation. Section 8.2 is devoted to William Harvey's 1651 *De generatione animalium*. The problem of generation was a natural area of research for an Aristotelian anatomist and a student of Fabricius; following Aristotle and Fabricius, Harvey studied extensively the generation process of the chick in the egg and of deer, as representative of viviparous animals. I shall discuss the former topic, focusing on the interplay between epigenesis—a term he coined, meaning the formation of the fetus gradually or part after part—and preformation, or the belief that the parts of the animal are preformed and no differentiation occurs but only growth.[3] Harvey's claims about epigenesis and the soul with its faculties, such as the formative faculty or plastic power, were to come under attack by mechanistic anatomists, who found generation especially challenging. By way of contrast, I sketch in a few words the mechanistic explanation adopted by Descartes, who rejected the belief that generation is a faculty of the soul. Section 8.3 investigates changing views on the structure of male and female organs of reproduction and the mystery of conception. The main period I examine corresponds quite closely to that of the previous chapter, from the late 1660s to the early 1670s; indeed, while Redi was investigating the generation of insects in Steno's company, Steno was reaching profound results about the structure of female organs of generation and generation as a whole, with his identification of the so-called female testicles as ovaries. Meanwhile, Reinier de Graaf published treatises on the male and female organs of reproduction in which he expanded on Steno's research. De Graaf did not live to witness the discovery of spermatozoa, whose 1677 announcement to the Royal Society by van Leeuwenhoek was received with skepticism.

Section 8.4 discusses the work on generation by Swammerdam and a group of Amsterdam anatomists, with special emphasis on the philosophical and also theological implications of preformation. With his belief in the encapsulation of all subsequent generations in the first egg, Swammerdam brought preformation to its most extreme version; his and Harvey's views occupy opposite positions with respect to epigenesis and preformation. I also discuss a later correspondence in which Malpighi criticized Swammerdam's positions. In 1673 and 1675 the Royal Society published two works by Malpighi on the formation of the chick in the egg, *De formatione pulli in ovo* and its *Appendix*, which was included in the *Anatome plantarum*; these works are discussed in section 8.5, with special emphasis on their role in Malpighi's *oeuvre* and on iconography. I shall argue that in these works Malpighi wished to address the organization of the organism as a whole, especially the areas related to the natural faculties of generation and growth, as opposed to the operations of individual organs; growth was a motivation for the study of plants discussed in section 9.2. Ultimately, Malpighi hoped to show that both processes result from the necessity of matter and motion and could be explained mechanistically. Yet in this area too, as in his study of glands, one detects a hiatus between his philosophical agenda and the extent to which he judged he had been able to carry it through. I will investigate these themes in other relevant works by Malpighi as well, seeking to disentangle the meaning of puzzling expressions—such as "plastic power"—whose interpretation has eluded several commentators.

8.2 Harvey: Epigenesis and the Role of the Faculties

The publication of William Harvey's *De generatione animalium* (London, 1651) was a major event: within the year, three additional editions appeared in Amsterdam, including one by Ludovicus Elzevier. Harvey's treatise is a starting point for all subsequent studies of generation in the second half of the seventeenth century. It is reproduced in its entirety in the *Bibliotheca anatomica*, together with most of the works discussed in this chapter. While I make no attempt here to provide a comprehensive analysis of *De generatione*, I focus on a few topics at the intersection between philosophical concerns and experimental investigations, such as the role of the soul in the formation process.[4]

As the title page of Harvey's treatise shows, the egg had a privileged role (fig. 8.1): Jupiter, sitting on his throne flanked by the eagle with lightening in its claws, is shown in the act of opening an egg inscribed with the words *ex ovo omnia*, "everything from an egg"; tiny animals escaping from the egg include a bird, a butterfly, a man, a grasshopper, a lizard, a stag, a fish, a lizard, and a snake with a spider hanging from it. The first section of *De generatione* states that all animals are engendered from an egg, even "man himself." By "egg" Harvey understood not exclusively something like the hen's but more broadly a principle of generation common to oviparous and viviparous animals. Quite strikingly, according to him "female testicles"—soon to be renamed "ovaries" and associated with the eggs of viviparous animals—do not "give any indication of being of any use either for coitus or for generation"; thus, he was very removed from views that were to emerge only a few years later. As we have seen in section 7.4, Harvey contrasted epigenesis with metamorphosis and argued that insects and other lower animals are generated spontaneously or by chance out of something damp becoming dry or vice versa. Even in these cases, however, he seemed to believe that they originate from a seed or larva, or something that could be called an egg understood as a principle of generation—hence the privileged role attributed in the book to the most common egg of all, the hen's.[5]

One of the most obvious features springing to the eye in opening Harvey's treatise on generation is the lack of illustrations in the text. This aspect is especially striking if one bears in mind that Harvey's teacher Gerolamo Fabrizi or Fabricius—whom he had chosen as his guide in *De generatione*—had published a lavishly illustrated treatise on the subject, one that Harvey owned and annotated in preparing his own work. Moreover, Fabricius had a remarkable series of colored drawings prepared for his work.[6] In *De motu cordis* Harvey was also very restrained in the use of images, which

Figure 8.1. Harvey, *De generatione*: title page

are limited to his celebrated fourfold sequence illustrating how the valves in the veins of a human arm work. *De motu cordis*, however, presents no new structure; rather, it is a book about actions or motions, which cannot be easily represented through images. *De generatione* is different in that it is devoted to the formation of structures. The process of generation poses considerable problems of description; Harvey was fully aware of those problems and addressed them in the preface to his work. First, in discussing the manner and order of attaining knowledge, he questioned the usefulness of illustrations, which can in no way be a substitute for direct observation. Harvey had rather profound philosophical reasons for mistrusting figures, since he argued that in drawing a man's face, for example, a painter could not look at the model while drawing and therefore would have to rely not only on direct vision but also on imagination and memory, which render things obscure and confused. Hence, a thousand different drawings of the same face, while superficially identical, would in fact all differ from each other and from the original. Moreover, he did not wish readers to rely on images as opposed to reality; as he put it, "those who see foreign countries and towns or the inward parts of the human body only in drawings or paintings . . . make for themselves a false representation of reality." Hence, his strong emphasis on seeing the real thing for oneself made him dispense with images altogether. If images were problematic, so were words for similar reasons. Harvey argued that in describing new things seen in anatomical dissections, neither metaphors using common words nor neologisms would be ideal: "For this very reason Aristotle is in many places thought to be obscure by inexperienced persons, and perhaps it was for this same cause that Fabricius of Aquapendente chose to show the making of the chick in the egg in pictures rather than to explain it in words." Harvey's solution to this conundrum was to adopt the terminology coined and employed by Aristotle and Fabricius, his leader or *dux* and guide or *praemonstrator*, respectively, following the same custom as in geographical discoveries, where the first to explore "new lands and distant shores call them by new names which posterity afterwards accepts"; based on the same principle, at times he also relied on Fabricius's figures.[7]

Harvey divided the formation of the chick in the egg into four stages. The first covers the first four days and involves the formation of blood and its receptacles, such as veins and a simple form of the heart: because of the small size, texture, and transparency of the parts, Harvey was able to provide only a brief account. The second stage involves the division of the fetus into two parts, the head and the body. Here Harvey used some mechanical analogies, as when he compared the body to the keel of a ship; more significantly, he argued that "like a shipwright, [Nature] first lays down the keel for a foundation and then sets up the ribs and the sternum as a deck, and just as he builds a boat so does Nature frame the trunk of the body and the limbs." Curiously, in the preface he had chastised Fabricius for having had recourse to "petty reasonings borrowed from mechanics" in claiming that the inward parts are made before the outward ones. The third stage sees the formation of the viscera, such as the liver, lungs, kidneys, the cone and ventricles of the heart, and the guts, all appearing about the same time. Lastly, in the fourth stage Harvey claimed that "the parts of meaner rank and condition are formed," such as the skin, feathers, and scales.[8]

Harvey's account is based on a careful series of observations framed within an

Aristotelian worldview whereby knowledge is linked to finding causes and the study of generation in different animals is seen as a way to investigate the nature of the vegetative soul. However, Harvey also identified problems and paradoxes with Aristotle: in an important passage, for example, he discussed the relation of the soul to the formation of the chick in the egg, suggesting that the innate heat and the vegetative soul may exist before the chick itself and that they produce the chick as its efficient cause and principle, respectively. Hence, he questioned whether the soul is "the act of an organic body having life *in potentia*," as believed by Aristotle, "for such an act I consider to be the form of the chick," which would thus exist before the chick itself. In the case of animals existing in a succession of forms, such as caterpillar, larva, and butterfly, Harvey argued that the same principle must be present in all, "albeit under diverse forms." Harvey's approach differed from the Stagirite's also because instead of focusing on as broad a spectrum of animals as possible, he chose out of convenience two representative examples from oviparous and viviparous animals, namely, the hen and hind/doe—the females of red and fallow deer, respectively. Harvey, however, agreed with Aristotle in attributing a key role to the heart over the brain, even with regard to sensation, because sensation can be detected before the brain is formed, as when the first rudiments of a body contract when pricked. With regard to the role of the male and female, whereas Aristotle had argued that the egg receives the material from the female and the soul and form from the male, Harvey was prepared to accept that the female—besides the material—does contribute a vegetative soul that is perfected and made fertile by the male. Aristotle's position, however, differed from Galen's, according to whom the role of the parents was more symmetrical in that both contributed a "semen."[9]

Harvey's account stands in stark contrast to Descartes', who sought to provide a mechanistic account of the progressive formation of the fetus without having recourse to the soul and its faculties. While a detailed study of Descartes' account lies beyond my scope here, I wish to sketch it in order to offer a useful term for comparison. Descartes discussed generation mainly in the last few years of his life—before Harvey's work appeared—but his views became known only posthumously with the publication of his treatise on the human body in 1664. His account is unusual in that most later mechanists defended the preformation of the fetus as a way to sideline the faculty of generation and replace it by the notion of growth, which appeared conceptually more manageable from a mechanistic standpoint. By contrast, Descartes thought that he could explain the successive formation of the body's part mechanistically, as if by epigenesis, and provided an account of the formation of the body, including the sense organs, the blood vessels including the *rete mirabile,* and the solid parts. Thus, with respect to the formation of the parts, his views paralleled Harvey's, despite their differences about the role of the soul.[10]

Figure 8.2. De Graaf, *De virorum organis generationi*: title page

8.3 The Organs of Generation and the Problem of Fecundation

We have seen in section 1.2 that Auberius's 1658 study of the testicle was the first investigation of Borelli's Pisa group to reach the press; in view of its rarity, Auberius's single sheet was reissued in the *Philosophical Transactions* for 1668, the same year in which de Graaf—unaware of Auberius's work—published *De virorum organis generationi inservientibus*, a treatise on the reproductive organs in man. On hearing from Oldenburg of de Graaf's findings about testicles consisting entirely of tubules, Malpighi was happy to endorse them by claiming that the same structure could be found in the locust; he also alerted Oldenburg to Auberius's publication. Much as Auberius, de Graaf also relied on comparison with other animals, but the main results of his investigations were due to injections with striking tools prominently displayed on his title page (fig. 8.2), showing at the bottom a syringe circled by a device consisting of the intestine of a small bird with two reeds tied at the extremities; de Graaf used it in his later work on the female organs of generation to display the circulation of the blood by connecting it to the jugular artery and vein of an animal. Other injection devices frame the cartouche with the title: two testicles, one whole and the other sectioned crosswise, are shown at the top, while another testicle is shown in the center just below them. In *De motu animalium* Borelli singled out de Graaf's injection technique with a colored fluid as having revealed that arteries enter the substance of testicles, admitting that de Graaf's work was superior to Auberius's.[11]

De Graaf's treatise contains a large amount of case evidence from patients, and one of the sections of the book is devoted to an instrument he had himself invented so that patients could give themselves enemas, or could have one administered without uncovering their private parts. In his research, de Graaf, who did not hold a university appointment, collaborated with university professors such as his former teacher Franciscus Sylvius, with whom he performed dissections at the Leiden hospitals. His work contains a wide range of experiments and observations, some quite colorful but of dubious anatomical value. For example, he reported that according to Jan Huygen van Linschoten, the Kaffir peoples in Africa are constantly waging war; victorious warriors cut off the penises of the slain or captured enemies and dry them out. After elaborate ceremonies in front of the king, the dried penises end up in necklaces decorating their women, who apparently feel greatly exalted by such ornaments.[12]

Despite his contacts with van Leeuwenhoek, who was also in Delft, de Graaf did not rely on the microscope in any substantive way, preferring other methods of inquiry. In order to investigate the structure of the testicles, for example, he devised an ingenious technique involving a dormouse: he removed the *tunica albuginea*, embracing the body of the testicle, and threw the seminiferous tubules in water; in this way they separate from each other, showing that their structure is not glandular but

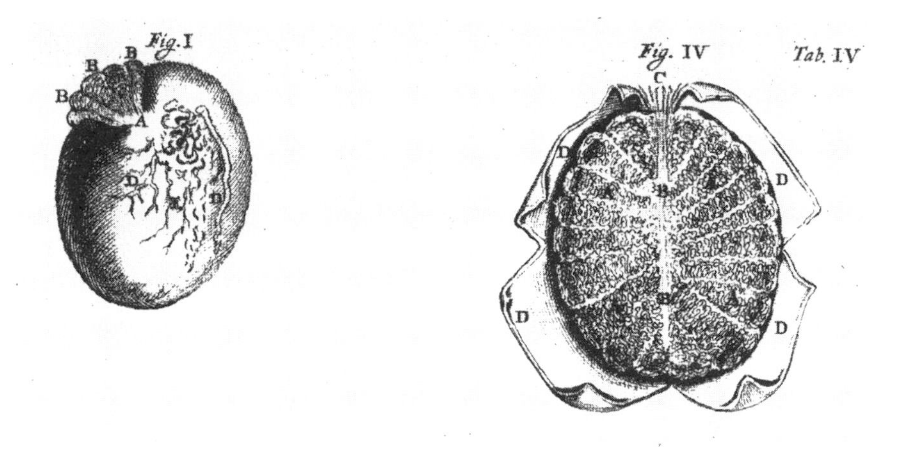

Figure 8.3. De Graaf, *De virorum organis generation*: human testicle, plate IV

consists entirely of tubules for the separation of semen. In 1669 he sent a preparation of the testicle of a dormouse with a drawing to the Royal Society. Figure 8.3 shows a human testicle whole, on the left, and then with the *tunica albuginea* removed, on the right, revealing that its inner structure is not glandular but consists of tubules. De Graaf stated that he drew the figures himself, but of course he relied on an engraver for the plates. In the caption to the figure of the whole human testicle, he stated that the engraver should have shown the attachment of the epididymis—or the tightly coiled vessel connecting the testicle to the vas deferens higher than at Λ. De Graaf must have judged the problem not so serious as to require a new engraving, but serious enough to require a verbal emendation: even drawing the figures himself was not a sure way to accuracy. The structure of the testicles de Graaf had uncovered led him to argue that semen is deposited from blood in the spermatic arteries into the seminiferous tubules just as blood in the hepatic arteries discharges bile into the hepatic ducts. He also relied on insufflation in order to study the structure of the penis; in his later treatise he illustrated the structure of the clitoris in the same way. His home housed a peculiar museum with several anatomical items, such as insufflated and dried seminiferous vesicles and other female and male specimens.[13]

It was de Graaf who introduced the Delft textile merchant and microscopist Antoni van Leeuwenhoek to the Royal Society. In a celebrated letter of 1677 to the Society van Leeuwenhoek announced to the learned world the discovery of *animalcula viva*, later called spermatozoa; tragically, de Graaf's premature death in 1673 prevented him from witnessing those events. The discovery of spermatozoa did not lead to a consensus even among microscopists. The editors of the *Bibliotheca anatomica* questioned specifically the reliability of van Leeuwenhoek's microscopic observations. In a letter to Malpighi of 1686, Bellini expressed doubts about their existence, whereas Malpighi paid no serious attention to them. Possibly following a report by an unknown German friend, he interpreted them as a manifestation associated with gonorrhea; indeed, van Leeuwenhoek's 1677 letter reported that spermatozoa were first detected by Jan Ham of Arnhem in a patient affected by gonorrhea, although subsequent investigations had revealed their existence in healthy individuals and animals as well. In a medical letter discussing whether a woman could become pregnant without losing her virginity, the Roman anatomist and physician Lancisi implied his acceptance of their role in generation. On the other hand, van Leeuwenhoek denied the existence of, and therefore any role to, the female egg in conception, a view the anatomical community rejected.[14]

Four years after his first treatise, de Graaf published a companion piece, *De mulierum organis generationi inservientibus*; this work also had a connection with the Tuscan court, since it was dedicated to Grand Duke Cosimo III de' Medici. The link with Florence, however, was deeper because it was in Florence that Steno—who was about to join de Graaf in the Catholic religion—had given a major contribution to the study of generation in his 1667 anatomy of a shark donated by Grand Duke Ferdinand II. In a passage as brief as it was important, Steno announced on the basis of the structural analogy with oviparous animals that the so-called female testicles of viviparous animals contained eggs and were in fact ovaries; he then renamed the Fallopian tubes of viviparous animals "oviducts," a term he had already used in 1664 for oviparous animals, and studied them by boiling, the same technique de Graaf was to adopt in the study of human and rabbit eggs. According to Swammerdam's friend Justus Schrader, in 1657 the Dordrecht physician Willem Langly had already employed rabbits for his investigations and recognized that the female "testicles" were in fact ovaries. While in Tuscany, Steno carried out further investigations: although they saw the light only in Thomas Bartholin's *Acta Hafniensia* for 1675, he communicated to de Graaf his findings of variously sized eggs in the female "testicles" of "fallow deer, guinea-pigs, badgers, red deer, wolves, asses and even mules." Steno and de Graaf established their results both by comparing the structure of female reproductive organs in oviparous and viviparous animals and by claiming that they had seen eggs in the so-called female testicles; as de Graaf put it, both the oviducts in birds and the Fallopian tubes in viviparous animals terminate in a membranous expansion and the eggs of all animals are fertilized and reach the uterus in the same way. His plates display the structural analogies between oviparous and viviparous animals, such as the cow and the hen, for example (figs. 8.4 and 8.5). Notice in particular the "female testicle" or ovary marked A in both plates: in figure 8.4 B and E are large eggs inside and outside the "testicle," corresponding to the yolks BB and CC in figure 8.5; in figure 8.4 FF is the membranous expansion of the Fallopian tube or oviduct, corresponding to HH

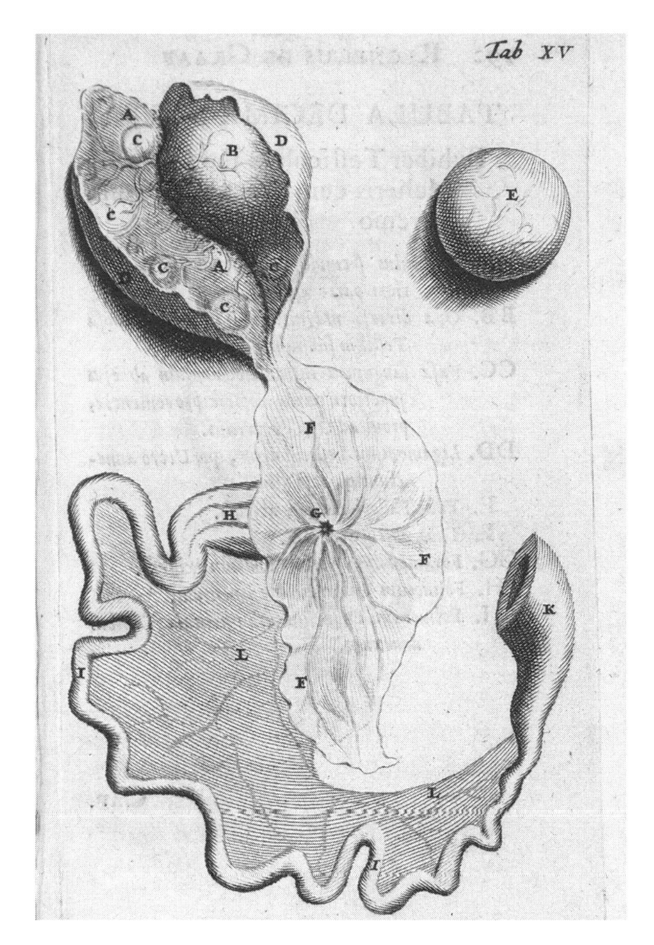

Figure 8.4. De Graaf, *De mulierum organis generationi*: cow's ovary, plate XV

in figure 8.5; II in figure 8.4 is the Fallopian tube, corresponding to the oviduct OO in figure 8.5; K in figure 8.4 and P in figure 8.5 are portions of the uterus. The image on the right in figure 8.5 shows that the orifice of the vagina of a hen that is no longer breeding is so tightly shut that it cannot be injected with air. Another plate (fig. 8.6) displays a uterus with styluses inserted at dd and y showing the urinary vessels, including the ureters and that from the bladder to the urine passage. The same technique of representation with styluses had been adopted by Florentius Schuyl for the blood passageways in the heart in his edition of Descartes' *De homine*.[15]

By 1675, besides Steno and de Graaf, van Horne and Swammerdam had also published their investigations on generation and had started an acrimonious dispute. In the address to the reader de Graaf proudly announced an account of generation based on a new foundation whereby all animals—including humans—are generated from

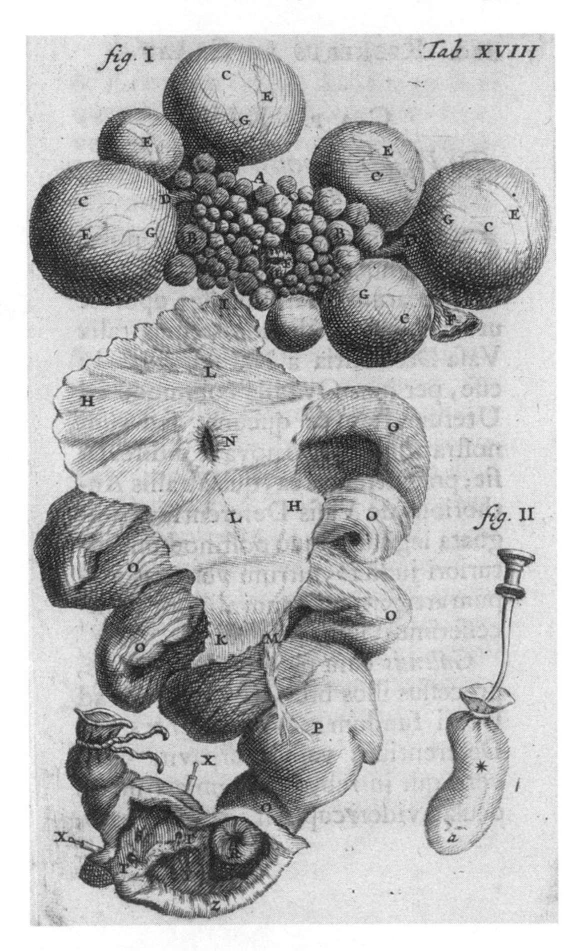

Figure 8.5. De Graaf, *De mulierum organis generationi*: hen's genital parts, plate XVIII

an egg, not one formed in the uterus by semen or its virtue, as intimated by Aristotle and Harvey, but one to be found in females before coition. The main results of de Graaf's investigations can be found in chapters 12–16, in which he laid out his new account relying on a series of observations and experiments on a range of animals, notably rabbits. The great differences among species made generalization problematic; for example, he observed "globulues"—later called *corpora lutea* by Malpighi, who in the letter to Spon believed them to be glands responsible for the generation of the egg—that he argued could be found only after coitus, a claim that is true for rabbits, although not for humans or cows.[16] Besides injections, de Graaf relied extensively on the technique of insufflation, as we have seen: by blowing into the oviduct, he noticed air coming out from the other end, showing it to be pervious. A remarkable plate from *De mulierum organis generationi* (fig. 8.7) shows the female reproduc-

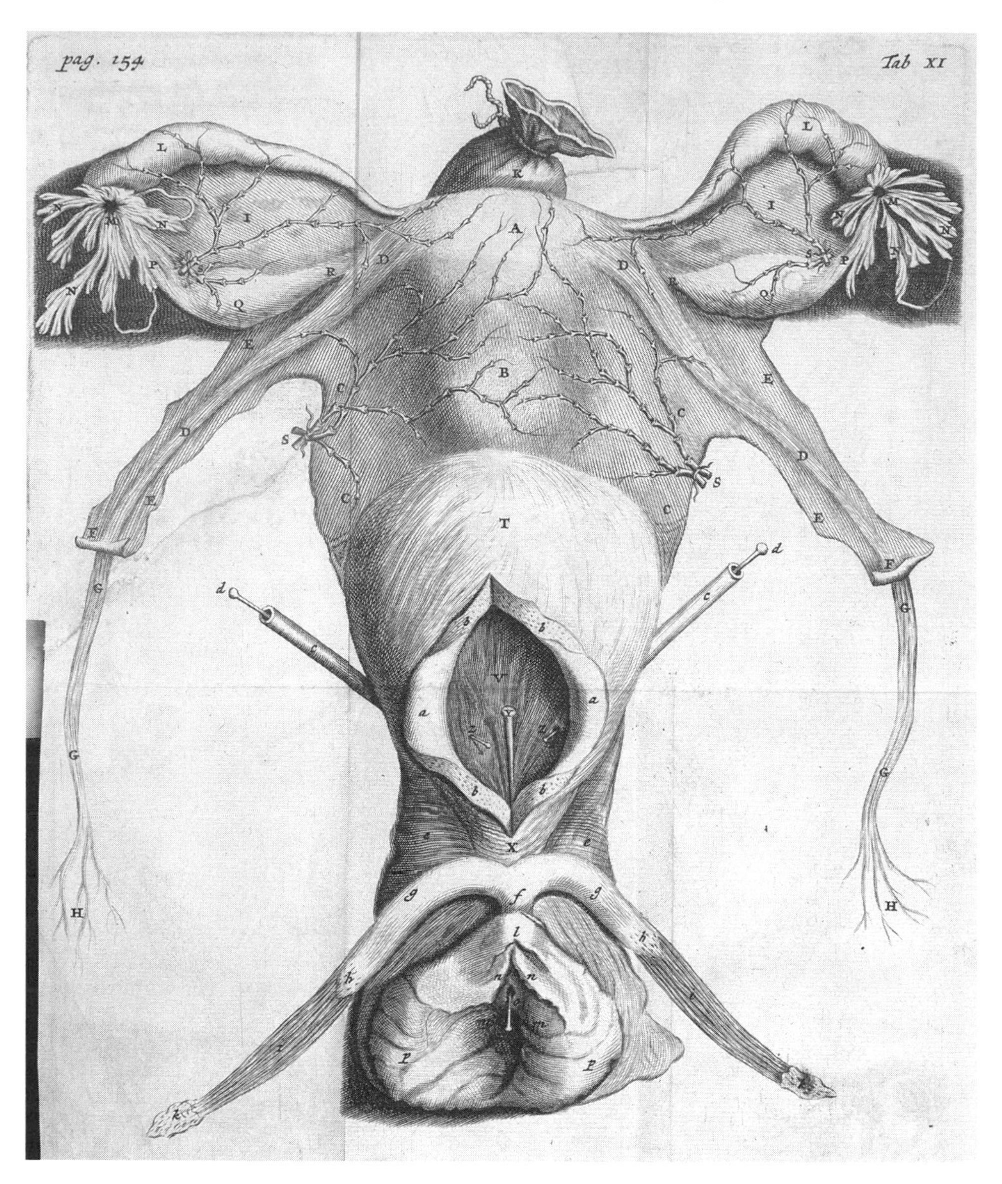

Figure 8.6. De Graaf, *De mulierum organis generationi*: styluses and urine vessels, plate XI

tive apparatus with veins in N highlighted by insufflation; in his study of insects, Swammerdam also had relied on insufflation (fig. 7.8), thus highlighting the fertile transfer of investigation techniques across different fields. After having identified the oviducts and the ovaries, de Graaf faced some problems when it came to identifying the eggs of various animals; he claimed to have identified egg follicles—later named Graafian follicles—that remain in the ovary after the egg is expelled, and the actual

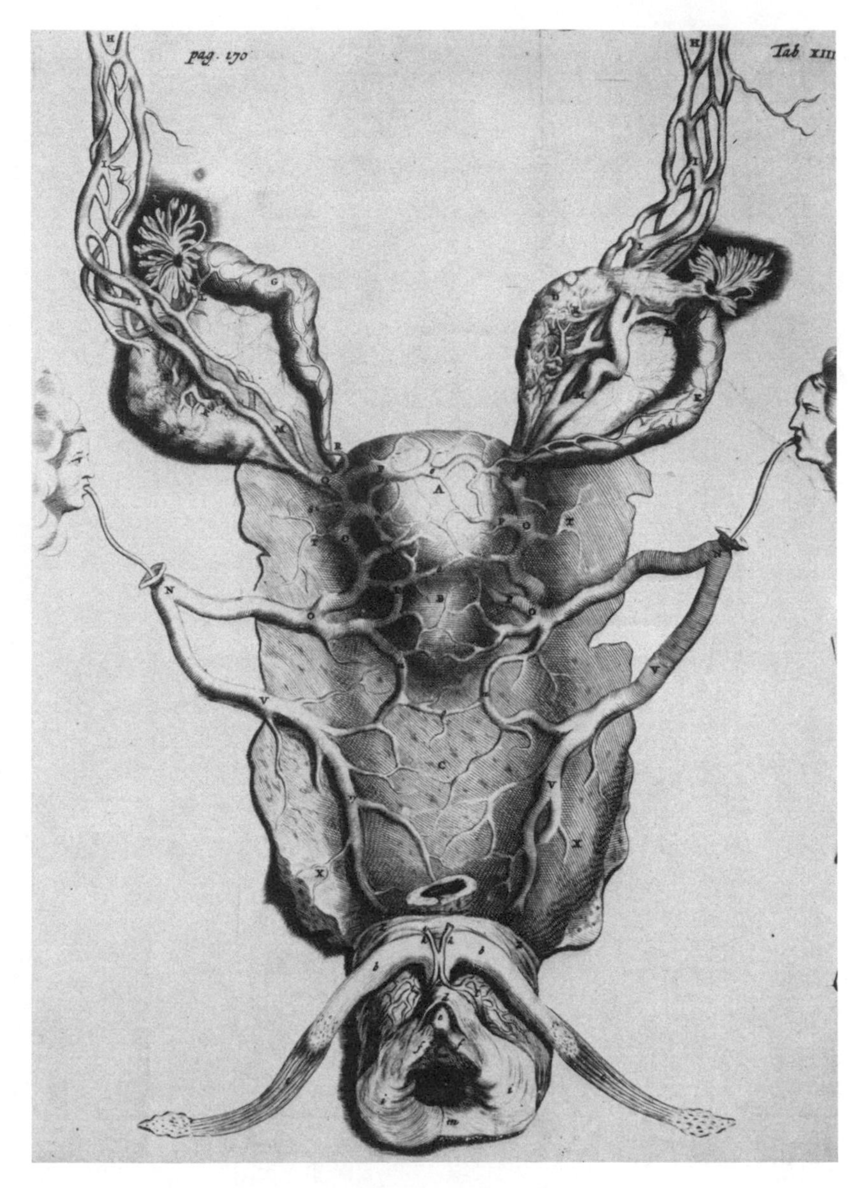

Figure 8.7. De Graaf, *De mulierum organis generationi*: human uterus, plate XIII

vesicles or eggs, which are formed in the follicle. When they are there, de Graaf's vesicles or eggs encompass other matter; he argued that after fertilization in the ovary the eggs diminish in size until they are ten times smaller than before the coitus, and then they are expelled.[17]

It is interesting to compare not only the methods employed but also the images produced by de Graaf and Swammerdam, who followed the mathematician and bur-

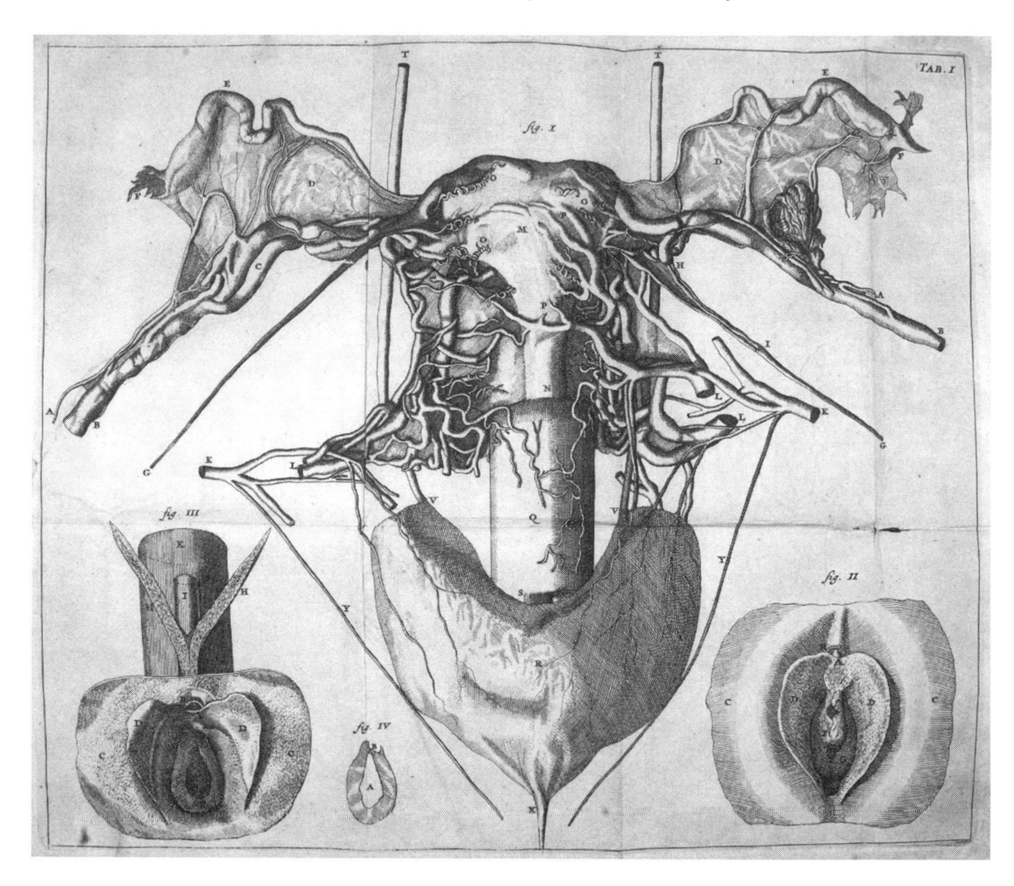

Figure 8.8. Swammerdam, *Miraculum naturae*: human uterus

gomaster of Amsterdam Johannes Hudde in relying on an injection method involving colored wax, thus making vessels visible and at the same time helping to preserve the specimens (fig. 8.8). After de Graaf's dormouse testicle, the Royal Society received for its growing museum of exotica Swammerdam's preparation of a human uterus; moreover, the figures in his *Miraculum naturae* were finely and sparingly hand-colored in red so as to highlight the arteries, joining in this way the object of study with its colored figure, highlighting the preparation method. As we have seen in sections 7.4 and 7.5, Swammerdam was somewhat ambivalent about color: while disliking the blotchy colors found in some books on insects, he did color the images he sent to Malpighi. The color of the arteries in his figure of the uterus was applied with utmost care with a tiny brush and in no way did it diminish the legibility of the figure, but rather enhanced it considerably.[18]

In his treatise de Graaf discussed the manner of fertilization, which he believed occurred in the female testicles or ovaries—rather than in the Fallopian tubes or oviducts. The process of fertilization is an area in which anatomists and physicians as well as philosophers carried out extensive investigations and speculations. De Graaf's views on fecundation remained quite traditional: he argued that since semen, or at least its thicker part, cannot reach the ovaries via the uterus, its action would be like that of a seminal vapor. It is interesting to contrast this position to Harvey's, who compared fecundation to contagion acting by breadth or through a miasma: by seminal vapor de Graaf meant a material agent, whereas Harvey understood something incorporeal, since he found no trace of spermatic fluid in the womb when conception takes place. Harvey also relied on the analogy of magnetization, one that was later taken up in a mechanistic framework in Malpighi's *Anatome plantarum* and Borelli's *De motu animalium*.[19]

Given the widespread circulation of his work and its mechanistic speculations, Borelli's *De motu animalium* is especially relevant here. Borelli was no anatomist and his accounts often verged on the speculative: much as he had done in his collaborations with Malpighi and Bellini in the 1660s, he elaborated on anatomical findings, adding at times fanciful explanations. In his chapter on semen, for example, he did not hesitate to propose that the animal spirits circulate through the body like blood, addressing the lack of evidence by arguing that if blood circulation had remained concealed for so long, the future may lead to the discovery of a circulation of animal spirits too, rushing to the testicles to vivify semen and returning to the brain. In his chapter on animal generation Borelli briefly acknowledged the recent discovery of eggs in viviparous animals, without mentioning any name. His main focus was on the process of fertilization: Borelli's death at the end of 1679 may explain why he did not mention spermatozoa. Relying on the analogy of water transuding through the pores of a baked clay vessel, he argued that semen could reach the ovaries and fertilize the eggs by direct contact rather than through spiritual irradiations or effluvia, in agreement with what Malpighi had shown in the case of insects; in all probability Borelli had in mind the observations on the silkworm published in the *Giornale de' letterati*, where Malpighi argued that eggs are fecundated by male semen that moistens them. Borelli established a profound structural analogy and connection between the brain and testicles: he argued that both are white, soft, have the same taste, and share the same configuration of spongy columns filled with a milky juice. Moreover, in the act of generation animal spirits descend with great impetus from the brain through the nerves to the testicles. Thus, he relied on various methods ranging from structure to consistency: curiously, in this case he relied on color and taste, whose role he had questioned in the past.[20]

Borelli drew an analogy between plants or animals and an automaton or a clock made of cogwheels. He adopted a potentially confusing attitude to plastic powers and virtues; at times he denounced them, only to mention them soon afterward. He argued that semen appears to be an organic body and to possess a "plastic force," by which he meant that it is not homogeneous but has an internal structure and organization that, mixed with the female egg, can form an embryo. In a passage echoing Malpighi's *De polypo cordis* he specified that he understood such organic processes to occur not because of a prudent and intelligent attendant, but through a natural and mechanical necessity arranged by the Divine Architect; thus, he rejected a form of teleology centered on the organism in favor of a preordained divine plan taking place through purely natural and mechanical laws. Borelli relied on other analogies, such as the role of water in the maceration of dry hay or in the germination of plants. But the one he seems to settle on is that of a pendulum as an oscillatory machine: particles of air in the organism work like a pendulum in that they oscillate by being compressed and rebounding activated by heat. Borelli may have had Boyle's views in mind here, who had argued that air consists of spiral particles. In this way he could provide a mechanistic account of generation, applying to living organisms the new mechanical devices that were being discussed in the aftermath of Galileo, Hooke, and Boyle. The language used and the analogies of the egg with a machine and of semen with a powerful ferment endowed with a vital motive faculty suggest that Borelli may have sympathized with some version of preformism, whereby the new organism is contained in the egg and is activated by male semen, similarly to the pendulum that is set in motion by some external action; Borelli, however, stated that male semen not only activates but also completes the structure and organization of the fetus by its plastic power. Ultimately, however, by "plastic power" he seems to understand processes that can be understood in mechanical or perhaps chymical terms.[21]

The seventeenth century was replete with theories of generation ranging from those of scholastic derivation to those with an atomistic flavor, as Karin Ekholm has recently shown.[22] Rather than pursuing an investigation of those theories, I am now going to investigate the more empirically oriented enterprise of the formation of the chick in the egg and the issues of epigenesis versus preformism.

8.4 Swammerdam and the Amsterdam Circle on Preformation

In the 1650s, in the aftermath of Harvey's *De generatione*, the Dordrecht town physician and professor of anatomy Willem Langly studied the generation of rabbits and also the incubation of the chick in the egg, which he illustrated with a number of plates. His work was published posthumously in 1674 by the Amsterdam physician Justus Schrader as *Observationes de generatione animalium*, with a dedication to Matthew Slade and Swammerdam—both members of the *Collegium privatum* of Amsterdam: both were interested in generation and discussed it in their publications with critical references to Harvey, whose work was excerpted in Schrader's book. Between

1666 and 1675 Slade published a number of short tracts that were later reprinted in the *Bibliotheca anatomica*; one of them was explicitly addressed against Harvey. Slade's pieces were communicated by Ruysch, who also sent a short contribution. In the preface to *Observationes* Schrader offered a useful perspective on recent studies of generation. He defended preformation or preexistence, a position he attributed to the prominent Venice physician Giuseppe degli Aromatari, who in 1625 had published as an appendix to a treatise on rabies a brief *Epistola* on the generation of plants in which he had argued that the plant is already contained in the seed much as the chick is already delineated in the egg before incubation. Schrader was one of the first to refer to Malpighi's 1673 work on the chick in the egg, which he linked to Aromatari; Malpighi was aware of Aromatari's work by 1689 at the latest, when he referred to it in a medical consultation on rabies. In *De generatione* Harvey mentioned having met Aromatari in Venice—in 1636—when he had been shown "a tiny leaf formed between the two shells of a bean-pod," something Harvey ascribed to the hot weather; presumably they discussed generation, but no record of their exchanges survives. Since Aromatari's work was quite rare, it was reprinted in the *Philosophical Transactions* in 1694. Thus, around 1670 the three Amsterdam researchers Swammerdam, Slade, and Schrader devoted themselves to the study of generation from an anatomical and philosophical standpoint with a strong anti-Harveyan bend. As we have seen in sections 7.4 and 7.5, Swammerdam had challenged Harvey's views on metamorphosis in *Historia insectorum generalis* and *Miraculum naturae*; his views on generation discussed here complement those we have seen in the previous chapter.[23]

In those works Swammerdam addressed the issue of generation from a theological perspective as well in a most radical fashion: by denying it. As mentioned in the last chapter, in 1669 Swammerdam had denied metamorphosis; his defense of preformation was part of the same agenda. Whereas Descartes had defended a view of generation by successive development, or epigenesis, most mechanists found such views untenable: growth—understood as an unfurling or enlargement—and nutrition seemed treatable from a mechanistic standpoint, but generation did not. It just did not look feasible that matter would create a living organism similar to its parents by simply following the laws of nature. The solution to this problem was to argue that all organisms were contained ab initio in their parents, all humans in "the loins of Adam and Eve" (Eve more so than Adam, presumably), as well as all apple trees in the original seed, like Russian dolls—a view later known as emboîtement. The human race, as well as all other organisms, would end with the death of the individual generated by the last egg. The Oratorian priest Nicholas Malebranche was the leading philosopher defending this radical form of preformation, and it was he who apparently converted Swammerdam to these views, as noticed in a footnote to *Miraculum naturae* by the editors of the *Bibliotheca anatomica*. In this way Swammerdam was able to explain why the offspring of parents without legs or arms could be complete in all their parts, as well as an elusive biblical passage suggesting that Levi was paying tithes before being born. Microscopy, biblical hermeneutics, and mechanicism coalesced in offering support—at times only suggestively—of this view. Karin Ekholm has recently argued that anatomical views about generation were tied to theological issues as well, in the form of the source of the human soul at each generation: whereas Augustine believed that the parents merely pass on the soul to their offspring, Catholic orthodoxy defended the view that God creates a new soul with each generation. Together with Augustine and many Protestant theologians, Malebranche believed that God created all souls ab initio. The occasional resemblance of the offspring to the father posed a serious challenge to these views: it was in this setting that maternal imagination played a role in the transmission of characters from the father and environment to the offspring.[24]

In an exceedingly interesting correspondence with the Modena physician and anatomist Francesco Torti, Malpighi discussed Swammerdam's views about the encapsulation of successive generations in the egg. The occasion arose when the duke's cousin Prince Cesare d'Este ordered Torti to send a monstrous egg to Malpighi as a curiosity. While the owner of the hen that had produced the peculiar egg was concerned whether he could eat the hen, Torti thought the whole matter quite uninteresting, in that he deemed it simply an egg with more yolks. Malpighi, however, was more impressed: he noticed three eggs encased one inside the other. He then proceeded to claim that Torti's egg at first sight seems to support Swammerdam's opinion, but in fact that opinion is so strange and erroneous that it has to be rejected. In a related memorandum first identified and published by Howard Adelmann, Malpighi questioned Swammerdam's views again, arguing that occasional cases supporting it have to be seen as monstrous and therefore one has to wait for another case to discuss the matter with greater care. It is interesting to compare his attitude to monsters in this case to the one we have encountered in *De polypo cordis*, in which he was able to rely on a series of cases. Malpighi thought it more likely that each egg would not contain all fetuses of subsequent generations, but only that of the following generation. I interpret his note to mean that, after the first generation, there would not occur a simple manifestation of preexisting fetuses, but the fetus of each generation would be formed by its parents. Thus, while rejecting Swammerdam's views, Malpighi endorsed preformation only from one generation to the next. Since he did not spell out these views in print, contemporaries remained unaware of them.[25]

8.5 Malpighi and the Formation of the Chick in the Egg

Careful readers of Malpighi's previous works would not be surprised to encounter a treatise on the formation of the chick in the egg, since relevant references could be found in several of his publications starting from the first *Epistola* on the lungs. In the opening of *De polypo cordis* Malpighi had mentioned the formation of the first rudiments of the animal, of insects and fishes, and of malformations as offering the prospect for a deeper understanding of higher animals. In the wake of his work on the silkworm, Malpighi embarked on the chick in the egg; as in Swammerdam's case, there are clear parallels and connections between the study of insects and generation, such as Malpighi's attempt at artificial insemination of the silkworm eggs and the study of their color and properties before and after fertilization. Moreover, there are parallels between the techniques of investigation of the silkworm and the chick in the egg: in both cases he extracted the tiny delicate parts and placed them on a glass plate for microscopic investigation. In some cases he conserved the dried specimens on glass, drying being another noteworthy technique. We know from Lancisi that in order to study the cicatricula, a small scar identified by Fabricius and then considered by Harvey as containing the first rudiments of the chick, Malpighi developed the technique of cutting it from the yolk and placing it on glass for examination, as shown in the accompanying figure displaying different stages of growth (fig. 8.9). *Anatome*

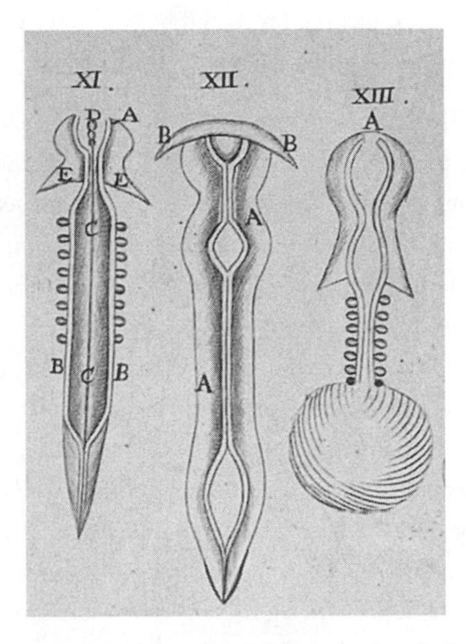

Figure 8.9. Malpighi, *Appendix*: "cicatricula" extracted and placed on a glass, XII

plantarum also contains important passages on generation, which were extracted in the *Bibliotheca anatomica*, as to highlight their relevance to anatomy.[26]

Malpighi was quite elusive about his philosophical agenda, often leaving readers to draw conclusions by themselves. One may even wonder how best to characterize his work: while the term "embryology" dates from the nineteenth century and is therefore anachronistic, "generation" appears as a more appropriate term dating from classical times and used in the titles of seventeenth-century works by Harvey as well as Schrader. Interestingly, however, that term occurs nowhere in *De formatione pulli in ovo*, its *Appendix*, or the letter to Spon, possibly as a way to mark Malpighi's desire to dispose of the philosophical implications of the faculties.

Malpighi was characteristically reluctant to embark on broad speculations unsupported by the evidence as he was eager to rely on visual representations. Therefore, it seems appropriate to start with a crucial passage from the opening of his second dissertation or *Appendix*, in which he presented repeated and expanded observations on the incubated egg, drawing a comparison between artists and investigators of nature:[27]

> It is a well-known custom among painters, most learned colleagues, to refrain from gazing long and fondly at the drawing and the first lineaments of the images to be later completed, to prevent them from growing to adulthood too quickly; but rather they have them put out of sight for a long time until the image of the unfinished offspring has been effaced from the mind of the maker; then after a quick glance, they bring it to completion. I reckon it both advantageous and necessary to follow a similar practice in beholding Nature's first images.

Malpighi continues this self-referential opening by arguing that fantasy is freed from the prejudices by the lapse of time and grasps immediately errors and inaccuracies by means of infrequent but repeated observations. This passage seems especially appropriate as an introduction to a supplement of observations completed eight and a half months after the first series. Moreover, it may well be capturing a profound feature of Malpighi's work, one with such significant differences from Harvey's approach discussed above that we may well reasonably suspect that Malpighi introduced it as a veiled response to Harvey. Whereas Harvey refrained from using illustrations at all, arguing that they relied on imagination and memory and introduced distortions, Malpighi's method seemed to be designed to overcome those difficulties. In a letter to Redi accompanying the 1673 treatise, Malpighi stated that the Royal Society had a supplement of observations—the 1675 *Appendix*—that contained better and more numerous drawings. It is especially significant that Malpighi stated in the conclusion of the second dissertation that the drawings were his own; thus, both his treatises are lavishly illustrated by the artist/investigator of nature in a way that suggests a profound link among observation, art of drawing, and investigation. Armed by these preliminaries, we can now proceed to examine and compare both of Malpighi's treatises.[28]

Malpighi's first treatise, *De formatione pulli in ovo*, opens with a mechanistic reference to the habit of artisans, who fashion in advance the individual components of the machines they want to build in order to see them separately before they are fitted together. Some naturalists, according to Malpighi, hoped that nature would behave in a similar fashion in the formation of the animal so that one could see not the whole organism, but its individual components before they are assembled and its structure could be disentangled, as in a machine:[29]

> In building machines artisans are accustomed to fashion the individual parts in a preparatory stage of the work, so that the components, which must afterwards be assembled, may be viewed first separately. Many of Nature's scholars [*Mystae*] interested in the study of animals, hoped that the same would happen in Her work, because, since it is very difficult to disentangle the complex structure of the body, it was thought helpful to examine the formation of the single parts in their earliest stages, when they are still separate.

Alas, matters turned out to be more complicated because, as Malpighi states, we do not capture the origin of the parts in isolation, but rather the animal appears almost already formed. Malpighi's passage echoes Steno's view that the brain could be understood as a machine, by taking it apart and examining its components one by one and then together.[30] Like Steno's, Malpighi's project relies on a notion of mechanism based on identifying the structure and purpose of individual components before they are assembled. Possibly Malpighi had in mind the assemblage process as well as a way to understand the organism. The mechanistic implications of his opening are quite striking by comparison to what Harvey had written just two decades earlier, for example.

Malpighi argued that the outlines of the "carina" of the chick are already visible before incubation—although after the egg had been fertilized. In all probability this result was due to the fact that he performed his dissections in the heat of summer, which accelerated the development of the chick.[31] At one point in the middle of the first dissertation, Malpighi launched into a speculation based on the analogy with plants, suggesting that the chick with nearly all its parts in the form of small fluid bags may be hiding in the egg and that it may form and then manifest itself thanks to the mixing of those nourishing and fermentative fluids, which generate blood and allow the outlines of the parts to emerge and grow: "For we may surmise that the chick together with the bounding saccules of almost all its parts lies concealed in the egg, floating in the colliquament, and that its nature results from the integration of nutritive and fermentative juices, through the joint action and stirring of which the blood is produced in successive steps and the parts formerly outlined erupt and swell out."[32] Here Malpighi was gesturing toward a chymical account of the formation of the chick after fertilization in terms of the mixing and stirring of fluids, but even this short speculative foray is followed by a retreat: he claimed that since the laboratories of nature are so intricate, they can be deceiving; therefore, he returned to his descriptive analysis. Soon thereafter, however, he claimed quite unambiguously, "It is therefore proper to acknowledge that the first threads of the chicken preexist [praeexistere] in the egg and have a more ancient origin not differently from the eggs of plants." In the second dissertation or *Appendix* he confirmed his views about the preexistence of various parts—such as a fluid filling blood vessels, the vessels themselves, and the heart—and their gradual manifestation as in plants, as to indicate that his mind was toying with such ideas while his eye was unable fully to confirm them. It is in the later letter to Jacob Spon that Malpighi, almost in passing, presents his views somewhat more explicitly: he stated that the cicatricula contains the latent outlines of the parts of the animal. Malpighi often talked of the cicatricula as containing a compendium of the animal, meaning the first outlines of the main parts, which subsequently become visible through growth. The parallels with his research on plants are noteworthy: it seems plausible that given the difficulties involved in detecting the emergence of the first rudiments of the chick in the egg, Malpighi relied on the analogy of plants to the point that some commentators observed that his figures of the developing chick do not correspond entirely to his verbal description and his analysis fit the description of the plant in the seed more than the chick in the egg. Malpighi defended his views to the end of his life, challenging his rival Sbaraglia to provide evidence against preexistence and in favor of metamorphosis.[33]

One of the most striking features of Malpighi's work is the wealth of observations he was able to present of the first four days of incubation. Whereas Harvey's description became more detailed after that point, skillful use of the microscope allowed Malpighi to provide a visual account of development every few hours. For example, he pointed to the formation of glands in the liver, providing new evidence in this regard. In line with his work on the silkworm, here too Malpighi paid attention to color in the formation of several organs; perhaps the most significant case concerns the changing color of blood—a topic ignored in the 1661 *Epistolae* on the lungs—that is here raised on several occasions, as when Malpighi argued that arterial blood was deep red whereas venous blood appeared more yellow.[34] Malpighi explored the formation of the heart (fig. 8.10) and whether it appears before or after blood, remaining agnostic on the matter; he did rule out that the heart produced any form of light, as claimed—possibly metaphorically—by some, by observing it in a dark room.[35] Throughout his works, but especially in the 1675 *Appendix*, Malpighi referred to

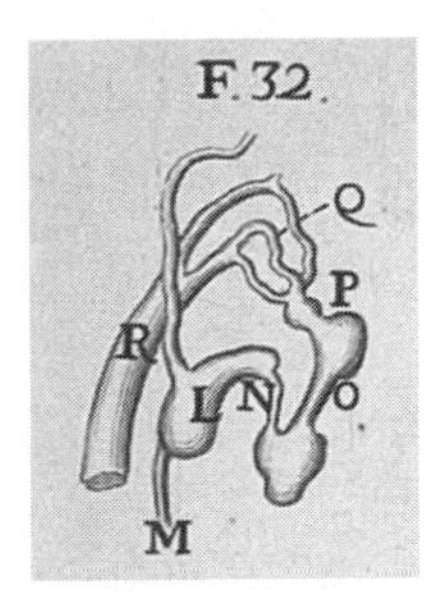

Figure 8.10. Malpighi, *Appendix*: the formation of the heart of the chick, XXXII

fermentation and to the chymical analysis of the relevant fluids by checking whether they coagulated or left a residue when heated on a fire.[36]

In addition to his two treatises on the chick in the egg, Malpighi addressed matters pertaining to generation in his letter to the Lyon physician and antiquarian Jacob Spon, in which he dealt with the fecundation of the egg in animals ranging from cows to butterflies. Malpighi observed that although the male semen cannot reach the uterus directly, it fecundates the female egg by mixing with the serum in the vagina, imparting the principle of motion to it. Thus, he ascribed fecundation not to the direct contact between semen and egg but—mechanistically—to the transmission of an active principle of motion. The letter to Spon contains three apparently enigmatic passages to the "plastic virtue" or "power" involved in the process of generation; another relevant reference to the "plastic spirit" can be found in the posthumous *Vita*. This older notion occurred also in Harvey, for example, who equated it with the "formative faculty." As we have seen, especially in chapter 5, Malpighi was engaged in abolishing such notions from anatomy, replacing them with notions like the necessity of matter. Adelmann explained the occurrence of the expression "plastic power" or "virtue" as "essentially the formative faculty of Galen and other early writers." This would be a truly major reversal of Malpighi's philosophical stance, one that is not supported by other evidence. I believe that the explanation of this apparent paradox can be found in an important passage—ironically reported by Adelmann himself—from Boyle's *The Origine of Forms and Qualitites*, translated into Latin in 1669 as *Origo formarum et qualitatum*. To the objection that the chick is not a mechanical engine but an organism fashioned by means of the plastic power by the soul lodged in the cicatricula, Boyle replied: "For let the plastick principle be what it will, yet still, being a physical agent, it must act after a physical manner; and having no other matter to work upon but the white of the egg, it can work upon that matter but as physical agents, and consequently can but divide the matter into the minute parts of several sizes and shapes, and by local motion variously context them according to the exigency of the animal to be produced."[37] Here Boyle bypasses the problem of the soul by claiming that, whatever that plastic power may be, it must act on matter as physical agents do and has to be treated as such. Malpighi probably found this philosophical gambit quite convincing, since he repeated it—mutatis mutandis—in his *Risposta* to Sbaraglia, when he argued that "the soul is forced to act in conformity with the machine on which it is acting"; therefore, a clock or a mill could be moved by a pendulum, an animal, a man, or even an angel, but it will always be moved in the same way. In conclusion, following Boyle's passage on the formation of the chick, Malpighi's references to a "plastic virtue" or "power" do not signal a return to Galenic doctrines; rather, they suggest an agnostic position with regard to the soul, in that whatever the soul may be, it must act as a physical agent through the material body.[38]

The study of the formation of the chick in the egg was not merely an intellectual recreation for Malpighi; it had a medical dimension as well. In his posthumous *Risposta* to Sbaraglia, Malpighi claimed that the formation of the first stages of the animal led to a deeper understanding of the true cause of abortions, thus offering indications for a therapy and possibly a remedy. As in other instances, Malpighi attributed a chymical cause to disease, but in this case he also referred to Aristotle's belief that excessive or deficient heat hurt the incubation of the egg, and so does thunder.[39]

Generation was among the most philosophically significant and technically demanding problems for seventeenth-century anatomy. Generation was closely tied to the traditional faculties of the soul or of nature, and attempts to disentangle it proved especially challenging: whereas the operations of an organ—such as the filtration of urine in the kidneys—could conceivably be explained mechanistically, the formation of an entire organism from its parents posed huge problems that required a different level of investigation. Anatomists and naturalists had to combine lofty philosophical agendas with the challenging task of providing an account of conception and seeing, describing in words, and representing through images the developing embryo. From the observation and conceptualization of eggs and spermatozoa to the detailed hour-by-hour chronicle of the formation of the chick in the egg and plants, seventeenth-century studies profoundly reshaped the problem of generation and set the agenda for future debates and investigations.

Despite their disagreement about the role of the soul, Harvey and Descartes shared the view of the progressive formation and appearance of the various body parts, a process Harvey called epigenesis. Most mechanistic anatomists, however, did not follow Descartes in this regard and endorsed some form of preformation instead, arguing that the parts of the fetus were present after fertilization. Swammerdam adopted an especially radical version of preformationism, arguing that all later generations of every species were contained in the first female egg. Thus, he sought not simply to explain generation mechanistically but to eliminate it altogether by subsuming it under growth, which appeared more manageable to mechanistic treatment.

Alongside reflections and investigations on the mode of formation of the fetus, seventeenth-century anatomists researched the elusive problem of conception and the structure of male and female organs of reproduction. While male testicles revealed a finer structure than had been anticipated, it was the so-called female testicles of viviparous animals that brought the most profound surprise, when they were identified on the basis of comparative investigations with oviparous animals as ovarics by Steno, de Graaf, and Swammerdam. This finding formed the basis for a profound reconceptualization of the problem of conception and of the respective role of both parents.

Helped by his microscope and microscopic techniques, Malpighi provided an extraordinarily detailed account of the formation of the chick in the egg accompanied by remarkable illustration of the stages of development, especially during the first four days of incubation. Inspired by the example of plant anatomy and specifically the plantlet found in the seed already mentioned by Giuseppe degli Aromatari, in his study of the formation of the chick in the egg Malpighi inclined toward preformation of the fetus before incubation. In a later letter, however, he argued against Swammerdam that each generation was formed by its immediate parents, thus denying that all animals already existed in the first egg. Malpighi conceived the formation of the rudiments of the chick as a chymical process providing the first outline of the animal whence the individual organs erupt and grow. Although his use of the expression "plastic power" has deceived some commentators into thinking that Malpighi lapsed into traditional Galenic doctrines, I have argued that he continued to defend a thoroughly mechanistic program and in all probability he borrowed that expression and its interpretation from Boyle.

The Anatomy of Plants

9.1 Plants between Anatomy and Natural History

Works on plants in the last third of the seventeenth century involved descriptive or natural historical aspects, as well as more interpretive, experimental, and natural philosophical components showing a marked difference from the Renaissance tradition. In the sixteenth and early seventeenth centuries the study of plants was overwhelmingly the domain of apothecaries and humanist physicians; their main concern was the identification, naming, and classification of specimens in relation to classical sources, as well as the study of the medicinal properties of plants. Unlike many previous celebrated herbals, the 1623 *Pinax* by the Basel professor and physician Caspar Bauhin had no illustrations, but with its approximately six thousand plants it was considered the pinnacle of this taxonomic tradition.

In the following decades works by van Helmont and even Harvey, together with the rise of mechanistic and microscopic anatomy, opened new horizons of research on plants with regard to issues such as the motion or circulation of sap and the chymical, mechanical, and experimental investigation of processes such as nutrition, growth, and reproduction.

Some of Malpighi's experiments showed the direction of flow of sap in the trunk, whereas those by Rudolph Jakob Camerarius at Tübingen dealt with sexual reproduction. Alongside a changing understanding of plants went changing methods of inquiry and representation. In the hands of investigators such as Hooke, Malpighi, and Grew, the microscope allowed unprecedented access to minute structures from roots to leaves, allowing them to discover and represent novel features in elaborate plates.[1] As in the case of insects, here I follow contemporary standards in understanding plants in a broad sense, including moulds, for example.

Section 9.2 examines Malpighi's *Anatome plantarum*, a treatise on the structure and "œconomy" of plants that historians of medicine and anatomy would be unwise to ignore as irrelevant to their studies: while systematically comparing plant and animal parts, Malpighi seized the opportunity to expand his reflections on a number of processes such as respiration, growth, and generation. Following the concealed attack by Giovanni Battista Trionfetti in the 1685 *De ortu et vegetatione plantarum*, Malpighi returned to the investigation of plants and in his *Vita* reported several experiments and observations, many of which he had performed himself, whereas others were carried out on his behalf by the Livorno apothecary Diacinto Cestoni and others. This episode is discussed in section 9.3 for the light it sheds on contemporary debates on experimental procedures on generation and the growth of plants, including issues such as spontaneous generation and metamorphosis. Lastly, section 9.4 deals with the works by Grew and Camerarius, focusing on Grew's iconography and study of the "œconomy" of plants and Camerarius's experiments on their sexual reproduction. Their works document how the study of snails opened a new horizon for conceptualizing sexual reproduction in nature, a horizon that soon came to fruition in the study of plants.

As in chapter 7, here too I have chosen Hooke's *Micrographia* as an entry to some themes of this chapter, such as the mechanistic understanding of plant processes, experimentation, and iconography. This choice seems especially appropriate in light of the assistance in microscopy he offered to Grew. Hooke provided a beautiful example of the growing mechanistic understanding in this area with his analysis of the behavior of sensitive plants—such as *mimosa pudica*—a classic topic discussed by several scholars from Descartes and Mersenne to Camerarius. In August 1661, in the presence of several witnesses, including the president of the Royal Society William Brouncker and others, he examined sensitive plants in the garden of Mr. Chissin in St. James's Park. Plants of the mimosa species posed a philosophical problem because their leaves moved by folding when touched, thus challenging a key distinction between plants and animals going back to Aristotle. Besides touching the plant, Hooke dropped *aqua fortis* on it, cut it, rubbed aromatic oil on its leaves, and burnt it with a lens. He noticed that whereas by cutting the plant in its normal state a small amount of fluid was exuded, when the plant had closed no fluid was produced. This led him to propose a mechanical explanation based on the pressure of the fluid moving or circulating in the plant. The action of touching the plant changed the pressure and pushed the fluid back, so that the "sprout hangs and flags."[2]

In another case of a moving plant Hooke realized by means of the microscope that the "beards" of oats consisted of two different filaments twisted together: one of them is spongy and reacts to moisture differently from the other, thus resulting in the movement of the "beard"; he relied on this structure to devise a new instrument, the hygroscope. Although Hooke's celebrated image of the cells of cork has assumed

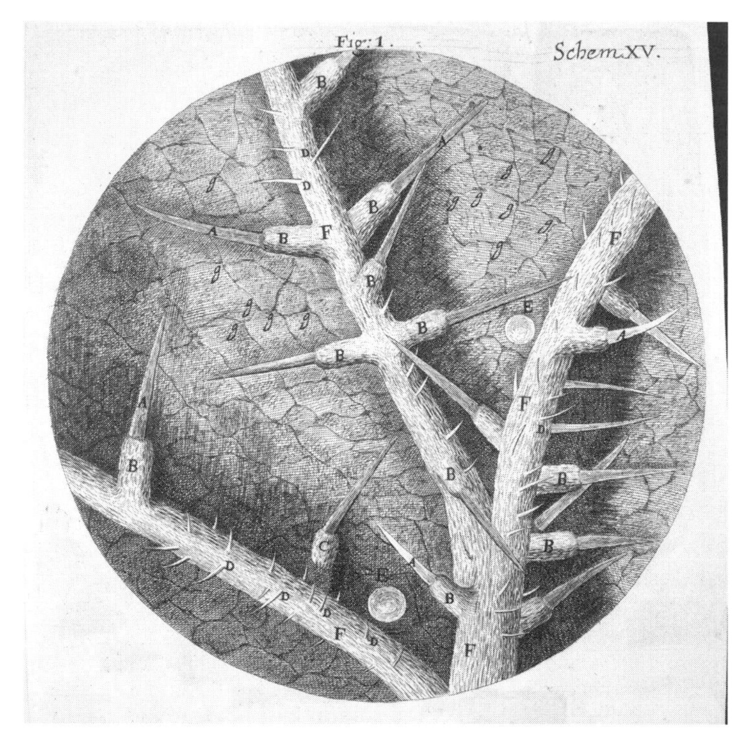

Figure 9.1. Hooke, *Micrographia*: Scheme XV, nettles

iconic status, here I would like to focus on other images from *Micrographia*, also drawn by Hooke, those of stinging nettle and mould. Hooke proudly proclaimed the advantages of the microscope, whereby he could see and understand the reason why nettles sting: they are endowed with a set of "bristles" or spines consisting of a hard hollow aculeus inserted in a bladder that Hooke compared to green leather or the surface of a wild cucumber, working as a syringe. Heroically, Hooke watched under the lens several of those bristles penetrate his skin and release an irritating fluid from the bladder at their base. His striking illustration (fig. 9.1) shows a detailed three-dimensional shaded image giving a powerful sense of the texture of the various parts. The image of the mould growing on the sheepskin of a book (fig. 9.2) is also striking for the way the delicate stalks stand out against the dark background, a technique that would be adopted by Swammerdam in some of his illustrations of insects (fig. 7.9). Lastly, figures 9.3 and 9.4 depict surface leaves: figure 9.3 shows the surface of sage, with a complex structure and some pearl-like globules that the Jesuit Athanasius Kircher believed to be spider eggs, while Hooke identified them as "a kind of gummous exsudation"; in figure 9.4 the microscope shows that the surfaces of some leaves that appeared smooth with the naked eye consist of a network of fibers resembling a

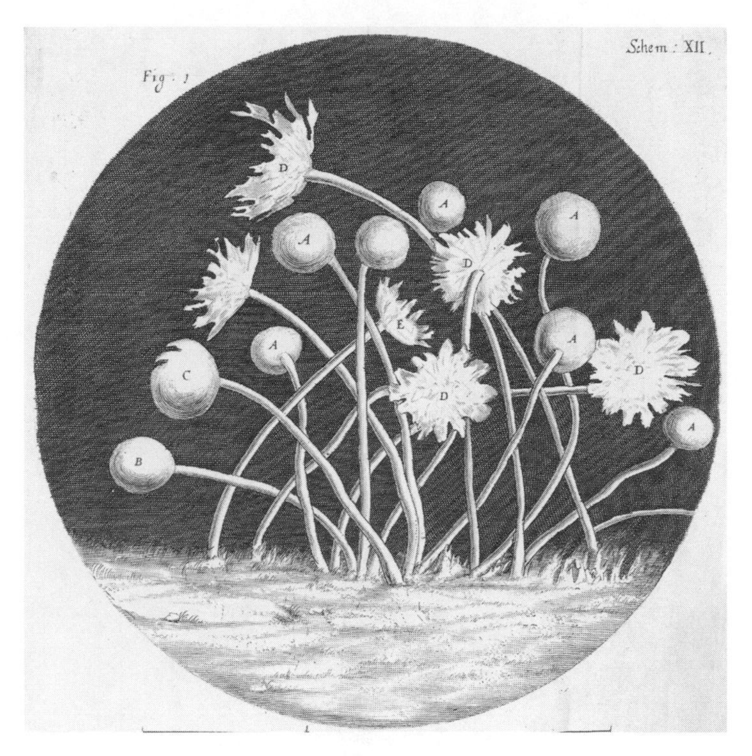

Figure 9.2. Hooke, *Micrographia*: Scheme XII, mould

textile, as if nature were "expressing her Needle-work, or imbroidery," once again a striking weaving analogy.[3]

9.2 Malpighi's Anatomy of Plants: Structure, Iconography, and Experiment

Malpighi's interest in plants seems to date from his Pisa years, when in a lecture he mentioned that the seed contains a fully organized plant that then develops and grows when it is nourished—possibly suggesting his knowledge of Aromatari's work. His first findings date from the early 1660s, when he discovered air vessels in plants that he later called tracheas and that allowed plants to breathe. He tackled this new area in the same way as he studied the anatomy of different animals—including insects—in the hope of uncovering the microstructure of their parts and the processes employed by nature in her operations. His *Anatome plantarum* consists of a programmatic sketch, called by Oldenburg *Anatomes plantarum idea*, and the treatise proper in two parts published in 1675 and 1679. Despite its size and the enormous labor involved, Malpighi was still unhappy: in the preface to the second part he regretted his inability to

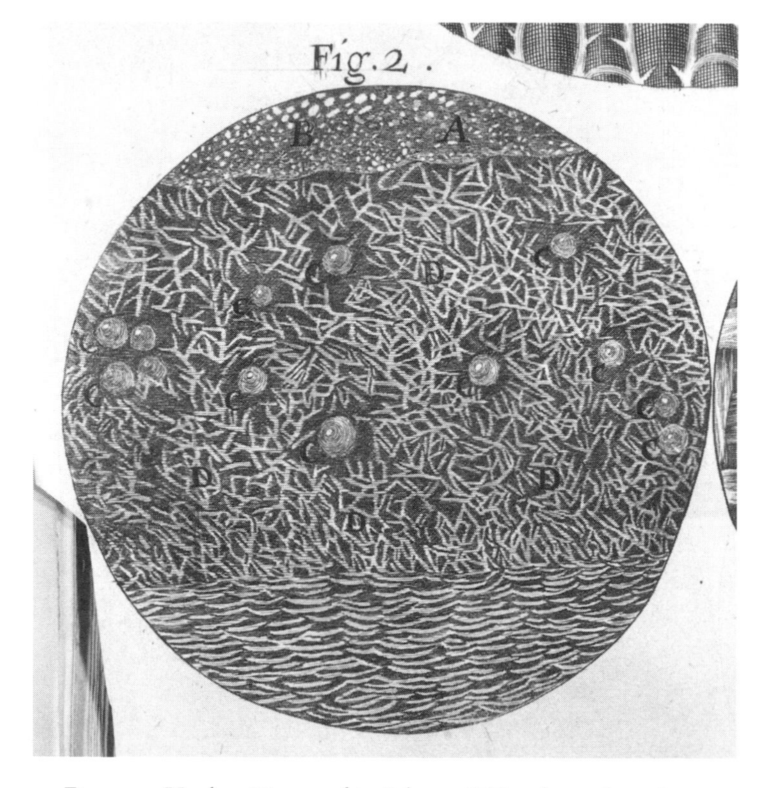

Figure 9.3. Hooke, *Micrographia*: Scheme XIV.2, the surface of sage

study algae, and in letters of 1680 he still bemoaned his inability to visit the coast at Pesaro or Ancona to find a plant with a structure simple enough—one could call it a structural archetype—to allow him truly to grasp the œconomy of plants.[4]

The exchanges between Oldenburg and Malpighi on the subject of plants provide us with valuable information on the composition of *Anatome plantarum* and especially its illustrations. In August 1672 Malpighi revealed that he was not skillful enough to draw with his own hand and complained that he could find no artists or collaborators since those who were expert in such matters found microscopic observations or *micrologiae* tedious. Although in *Anatomes plantarum idea* there are no figures, Malpighi already announced specific ones that he would send. From a letter by the Neapolitan lawyer Francesco d'Andrea to Redi, however, we know that by January 1673 Malpighi had no figures for his book and Count Rinieri Marescotti was offering to pay an artist to draw them. Malpighi's problem with illustrations is surprising, given that in those years he had drawn the figures of his treatises on the chick in the egg; part of the difficulty may be due to the fact that both parts of *Anatome plantarum* are a veritable iconographic tour de force, with hundreds of illustrations of plants and

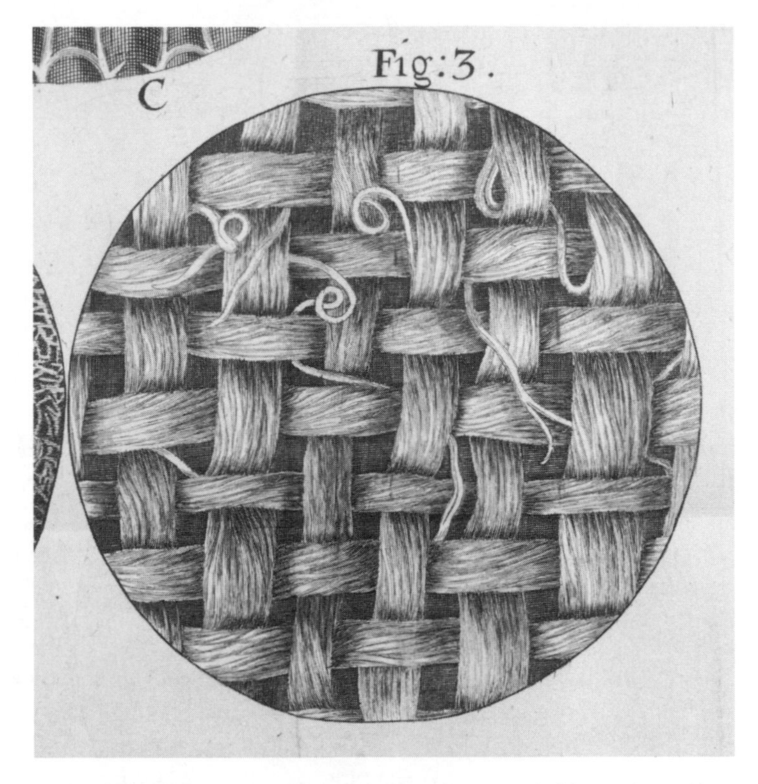

Figure 9.4. Hooke, *Micrographia*: Scheme XIV.3, textile-like surface of leaves

their parts in all stages of development: no viable selective survey could do justice to the dazzling range and often quality of his work.[5]

The preface to the second part of *Anatome plantarum*, a work Malpighi completed and sent to Grew in 1678, offered a rare explanation of the method of representation he had adopted, shedding light on iconography and its problems:[6]

> As to the rest, I offered a series of observations together with figures, so that anyone at all may philosophize relying on any system at will. In like manner, I drew the figures myself, even when they represent the object enlarged by means of the microscope; however, they do not represent distinctly all the parts of the object that are truly there, because by delineating any minute part the figures would become almost measureless. Rather, retaining a certain method analogous to that used by Nature, I delineated only those parts that serve to instruct the reader. In the same way as cosmographers mark out only noteworthy places, neglecting or at least passing by in silence single hovels and trees, so I could sketch out from so many things displayed by Nature only those of which plants frequently consist, in order to make

> more intelligible any type of vessels and their mutual connections. For the same reason I rendered the figures larger and, without neglecting natural integrity, changed the layout, separating the parts. I hope that anyone, also the sternest critics, will kindly grant me this license.

There is a curious tension in this passage: the opening suggests that verbal and visual descriptions were somewhat neutral, in that they could be relied on by readers with different philosophical outlooks; the rest of the passage, however, highlights how problematic those descriptions were and how many choices they involved. Malpighi states that he had drawn the figures himself, including those seen through the microscope, probably in frustration at the reluctance shown by artists to embark on such work. In this striking and rather defensive passage Malpighi admitted having taken many liberties, such as neglecting some portions he deemed uninteresting, changing the layout, and separating the parts. As we have seen in section 7.5, since Swammerdam had challenged his mode of representation of the silkworm by arguing that his figures were conceived with the mind rather than seen with the eyes, it is plausible that here he was obliquely responding to those criticisms by spelling out his conventions. Moreover, in the preliminary publications leading to *Anatomy of Plants*, Grew had drawn figures differing greatly in style from Malpighi's, offering much-enlarged cross sections that showed the relations among the various parts and avoiding the dismemberment typical of many of Malpighi's figures; possibly the passage above was a response to Grew as well. Despite the similarities in visual representation, it seems plausible that in investigating plants Malpighi would have relied on different techniques from those appropriate for insects and the developing chick, as a result of the different consistency of the parts.

In the dedication to the Royal Society of the first part of *Anatome plantarum*, Malpighi wasted no time in dismissing taxonomy as a rather forlorn enterprise. Similarly, in his treatment of flowers, he stated that he wished to leave the task of providing a careful description to botanists. His project, rather, was to study the entire vegetable realm as a single living organism, similarly to what he had done in the study of animals, proceeding from the less to the more obscure. Malpighi's approach to the study of plants was quite novel with its emphasis on their anatomy and especially the systematic comparison with animals, going beyond common analogies dating from Greek times.[7]

In *Anatomes plantarum idea* he mentioned briefly mimosa—the same plant discussed by Descartes and Hooke—several species of which could be seen at the Bologna botanic gardens, as described by its curator Giacomo Zanoni in the 1675 *Istoria botanica*. Although Zanoni is never mentioned in *Anatome plantarum*, it seems unlikely that Malpighi would have failed to rely on his expertise and garden. Malpighi noticed that by cutting a branch or twig of herbs and some trees, especially in winter, portions of the tracheas were left dangling; they showed for some time a peristaltic motion that he surmised may explain the motion of mimosa. Malpighi's views about the peristaltic motion of plants attracted Grew's attention and were discussed by the Royal Society in 1672. Although Malpighi's casual observation does not reveal the same degree of interest devoted by Hooke to the empirical and mechanical side of the motion of mimosa, both addressed the issue seeking some form of mechanistic explanation.[8]

Malpighi's analogy with animals was quite complex and multilayered: on the one hand, he seemed concerned with the overall "œconomy" of the organism. One may surmise that Malpighi wished to study plants in the hope that they would provide a simpler path for uncovering and understanding processes such as growth, as we will see below. Growth was one of the natural faculties according to Galen, together with generation and nutrition. Whereas nutrition could be understood in terms of corpuscular mechanico-chymical processes, whereby particles with the appropriate size and shape are transported by appropriate vessels to body parts specially suited to receiving them, the other two looked extremely obscure. A passage from Malpighi's *Vita* appears especially revealing in this regard. Commenting on the obscurity of the process on growth, Malpighi stated that the only area that had been accounted for was that of bones, a topic he had studied in *Anatome plantarum* in relation to the accretion of the trunk of plants: he relied on a weaving analogy to argue that filaments or fibers form a net that becomes harder with time and is accompanied by other similar structures providing bulk and solidity. Generation also lay at the intersection between the study of plants and animals not only conceptually and terminologically but also methodologically; in both instances Malpighi compared the process of generation in plants and animals and stated that he was keeping a diary of growth, for example, following a natural history tradition. Moreover, Malpighi worked on the two projects together: *Anatomes plantarum idea* is dated 1 November 1671, whereas the two works on generation are dated 1 February and 15 October 1672, and both contributions were published together by the Royal Society.[9] Malpighi also relied on a systematic comparison between plants and animals as a way to understand the role of the various parts, from the bark to the generation of seeds. Often the grammatical structure of his sentences was based on the correlatives "sicut" or "just as" (applied to plants) followed by "ita" or "so" (applied to animals), highlighting the comparative structure of his analysis.

The first part of *Anatome plantarum* explores all the parts of plants starting from the bark, as if Malpighi had a plant on his dissecting table and proceeded to open it as if it were a corpse. Although the bark resembles the skin, Malpighi mentioned this analogy only in passing. Rather, he emphasized its role for the transport and coction

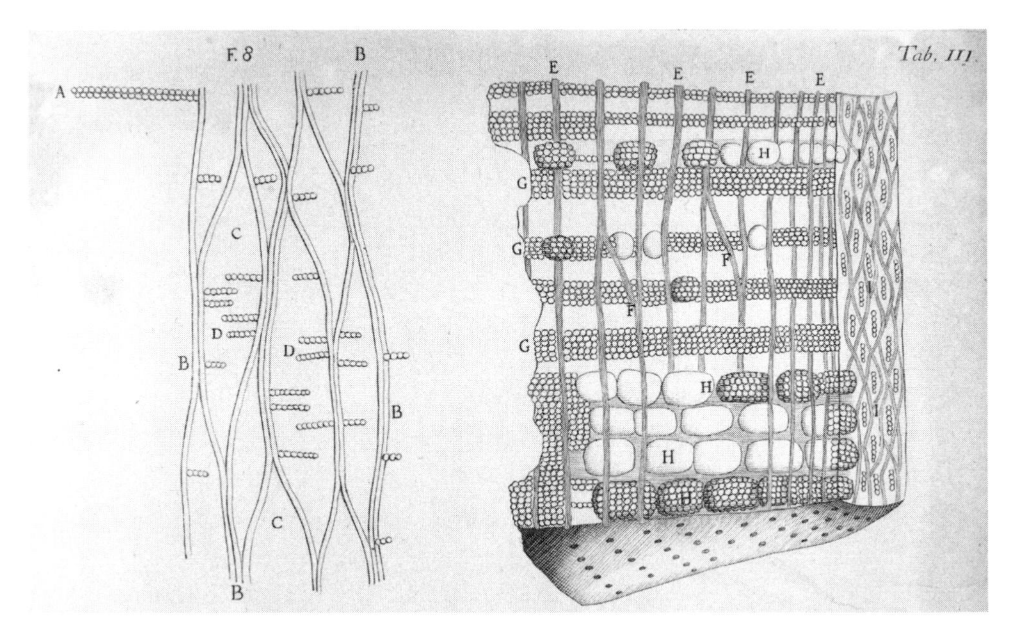

Figure 9.5. Malpighi, *Anatome plantarum*: III.8, oak bark

or elaboration of sap and in the yearly growth. The somewhat stilted engravings (fig. 9.5) show oak bark seen in both longitudinal and radial sections; whereas the former on the left was more common in Malpighi, the latter on the right occurs more frequently in Grew, as we will see below. In the longitudinal section on the left Malpighi identified a band of utricles A, forming the outermost surface, and a network of fibers B that he compared to those of a mantel or cloak; those fibers define areas C occupied by further utricles D. In the radial section on the right Malpighi identified a network of fibers E, communicating through oblique fibers F, which he again compared to a mantle. Weaving analogies are pervasive in his work and, far from being simply a means of describing, assume an epistemological role in explaining the formative process of several structures in which nature appears as a weaver operating mechanically. G represent utricles and H are bodies filled with tartar that can be found mostly in oaks and join the utricles together; Malpighi compared them to tesserae or tiny stones of a mosaic.[10]

Next Malpighi moved to the trunk and stalk, in which he identified spiral vessels that are always open and do not carry any juice, something he considered one of his major discoveries in the vegetable world, lacking from Grew's 1672 foray (fig. 9.6). Despite the important role attributed to them, Malpighi offered a visual representation of them only in one case, and even in that case they do not exactly stand out: readers may want to compare figure 9.6 with Grew's figure 9.21 discussed below.

Here Malpighi shows the longitudinal section of blackberry and vine seen under a microscope with low magnification; the structures K represent the surfaces of the spiral vessels. The vessels themselves are shaped like tubes of unequal sizes and are composed of a lamina L twisted in a spiral. This structure is visible at the bottom right of the figure, where laceration has left the laminae dangling, thus revealing their spiral structure, which would otherwise be hard to detect. Malpighi compared them in their spatial arrangement to the trachea of the silkworm, a comparison he could confirm by studying them under greater magnification, since their microstructures also resembled each other. Given these similarities and the fact that the tubes K are always empty, he attributed the purpose of respiration to them. Longitudinal ligneous fibers M stand alongside the tracheas, whereas transversal utricles N form with them a structure resembling a mat—yet another weaving analogy. Thus, the overall structure of plants resembles that of insects with air vessels surrounded by other vessels carrying nutritive juice.[11]

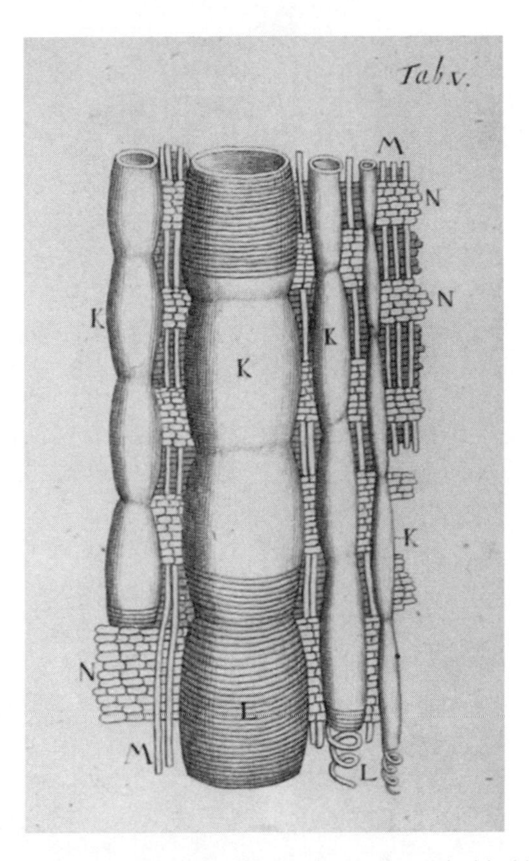

Figure 9.6. Malpighi, *Anatome plantarum*: V.19b, spiral vessels, blackberry and vine

The identification of the purpose of the tubes composed of spiral vessels by analogy with similar structures in insects enabled Malpighi to launch into a brief excursus on the purpose of respiration in plants and animals, in which he expanded on and modified the views he had put forward in *Epistolae de pulmonibus*. Admitting the conjectural nature of his speculations, Malpighi stated that fish and other animals living in water extract from it a matter suitable for breathing and that the lungs of animals separate particles of air that get mixed with blood, in line with what he had stated in *De polypo cordis*. Similarly, he argued that the tracheas of plants separate a portion of air necessary for the fermentation of their fluids. In comparing plants and animals, however, Malpighi mentioned also that air enters the stomach and intestine, claiming that air gets mixed with food and enriches it with its salts—especially niter.[12]

The section on the growth of the stem is one of Malpighi's most significant contributions to mechanistic anatomy. The challenge of mechanistic anatomy was not simply to explain mechanically the operations of one or the other organ but to pro-

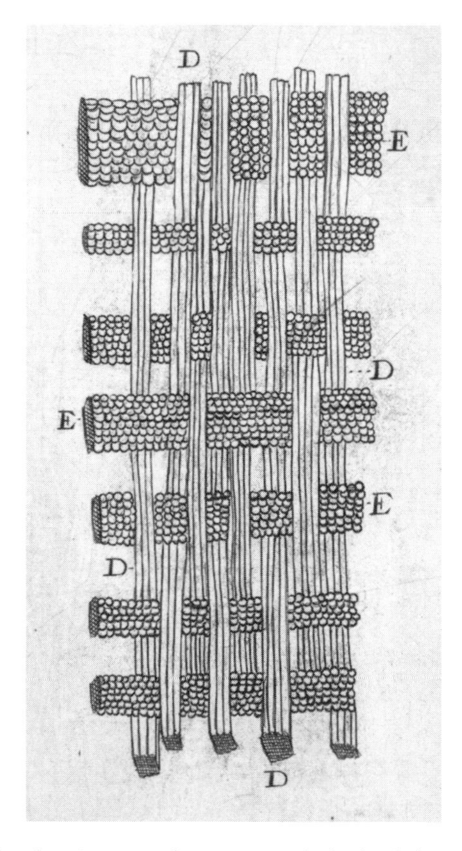

Figure 9.7. Malpighi, *Anatome plantarum*: II.7b, bark of cherry and plum tree

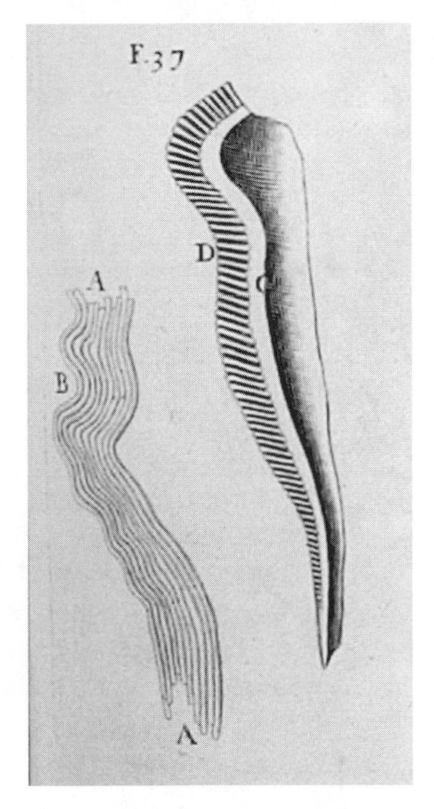

Figure 9.8. Malpighi, *Anatome plantarum*: VIII.37, teeth

vide a mechanistic account of the entire range of the body's operations, including the traditional natural faculties. It is precisely to such issues that Malpighi's work is devoted. Seeking to develop Erasistratean images of weaving, as we have seen in section 5.5, Malpighi sought to document the process of growth in plants, extending his investigations to bones and teeth as well. His illustration of a portion of the bark of cherry and plum trees (fig. 9.7), with its interweaving of fibers D and utricles E, is particularly instructive in this regard; it is useful to compare it to Hooke's plate of the surface of a leaf (fig. 9.4). By studying the growth of fetal bones, especially of the skull, Malpighi was able to detect parallels with the growth of the trunk of chestnuts and oaks. In both the skull and teeth he detected a network of fibers structurally analogous to the liber in plants, which later becomes bony by the effusion of a "petrifucus humor" or "osseous succus," whose existence can be ascertained by the healing of fractures and the ossification of several body parts. Malpighi studied the teeth of oxen, fishes, and the fetuses of other animals: figure 9.8 shows on the left the external plate of a tooth A with its wavy structures B; the longitudinal section on the right

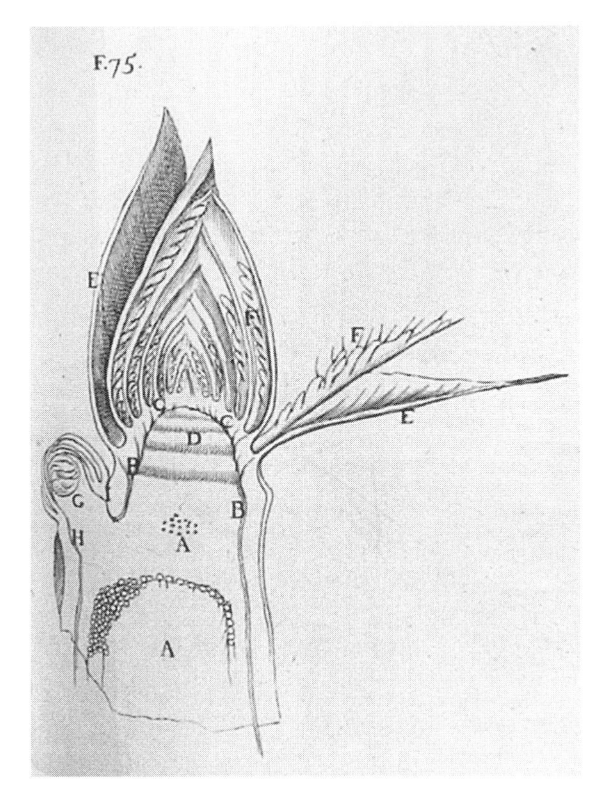

Figure 9.9. Malpighi, *Anatome plantarum*: XIV.75, gem of a fig tree

shows the internal plate C with its covering D, which Malpighi compared to hairs—a result of an optical effect. As in previous instances, Malpighi relied on pathological cases, such as the petrified brain of an ox that he preserved as a witness to his theory, whereby he believed that the gland filtering the bony juice was accidentally stimulated to produce an abnormal amount. In his *Vita* Malpighi returned to the structure of bones and hairs, expanding on the treatment he had provided in *Anatome plantarum*. Malpighi relied extensively on fetal anatomy and monsters to uncover the process of ossification and challenged the view put forward by the Rome surgeon Domenico Gagliardi about some tiny nails holding the various layers together, arguing that they were appendices to the filaments of the bony laminae.[13]

In addition, Malpighi's studies of buds, leaves, and flowers are lavishly illustrated and proceed by systematic comparisons between plants and animals. He considered buds to be analogous to a fetus, in which there is only an increase of existing parts with the addition of new particles to preexisting structures and no new part emerges, with the exception of teeth and horns. In the engraving of the longitudinal section of the bud of a fig tree (fig. 9.9), the encasing of deciduous and permanent leaves E and

F is especially noteworthy; A is the white marrow with some utricles, B is a woody portion terminating with some appendices C ending at the base of the leaves, D are the rudiments of the knots, and G is a future fruit surrounded by woody fibers H and protected by some tiny leaves erupting from the cortex.[14] Malpighi investigated several aspects of leaves, such as their structure and the way they are attached to the stalk; in lemons and oranges, for example, leaves have a peculiar conformation in that they appear to be an extension of the stalk. The bundles of fibers that extend into a leaf emerge after leaf B and after four leaves C terminate in the pedicle D (fig. 9.10). On the surface of leaves Malpighi identified utricles whence a sweaty humor flows forth. Figure 9.11 shows smaller and smaller fibers A, B, and C, ending in small receptacles D. As to the purpose of leaves, Malpighi compared them to the skin, but a skin *sui generis*. In fact, leaves serve a purpose similar to the digestive system as well, whereby sap is refined and concocted. The receptacles we have just seen are compared to glands that allow the expulsion via transpiration of the useless products, analogous

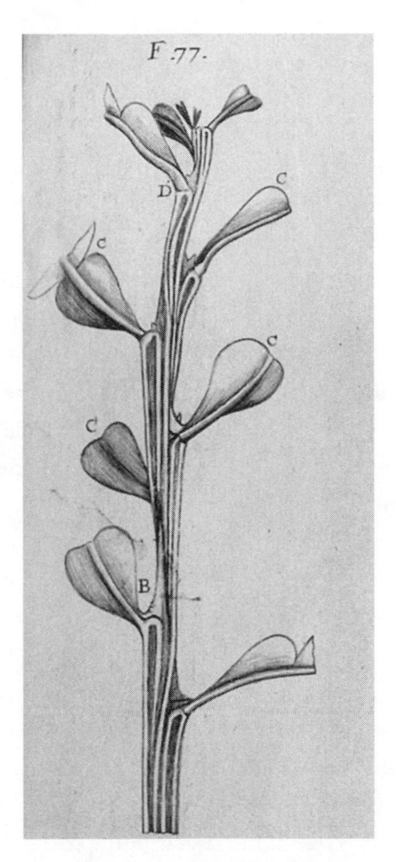

Figure 9.10. Malpighi, *Anatome plantarum*: XV.77, attachment of leaves in lemon

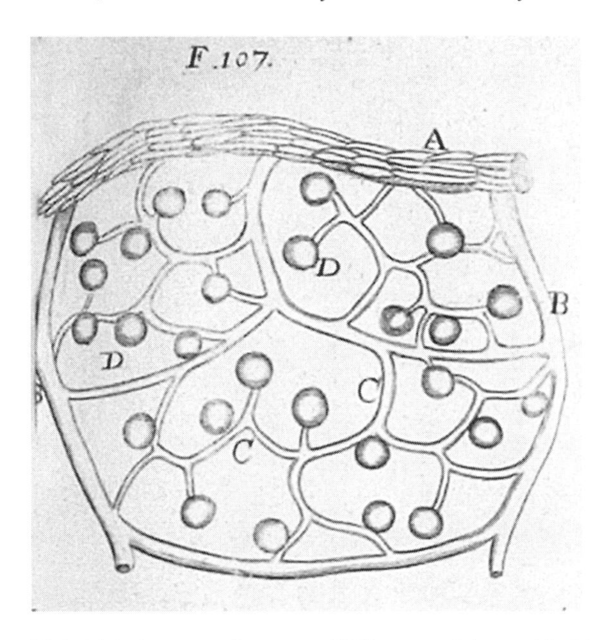

Figure 9.11. Malpighi, *Anatome plantarum*: XXI.107, utricles on blackberry leaves

to sweat: once again it is useful to compare Malpighi's figure 9.11 to Hooke's figure 9.3, showing the magnified surface of sage with globules that Hooke believed to be the plant's exudation. The sap concocted in the leaves then travels "quasi peculiarem circulationem," states Malpighi, to all parts of the plant. Although he seems to have in mind an analogy between sap and blood, one that was not uncommon at the time, there are also significant differences between the circulation of blood and that of sap, as it appears from his cautious language: concocted sap reaches to all parts of the plant but does not return to the leaves. In view of the central role glands occupied in the animal "œconomy," it seems especially significant that Malpighi sought to find their corresponding organ and location in plants.[15]

Malpighi attributed the ascent of fluids through the plant to a range of factors, such as transpiration through the leaves, the anastomoses of vessels throughout the plant, and the expansion and contraction of air due to heat and cold, by analogy with the motion of chyle through the lacteals in animals, which is made possible by breathing. However, he did not refer to the experiments performed at the Cimento Academy in Florence and the Traccia Academy in Bologna on the rise of water in very narrow tubes, even in a vacuum, a topic that had been discussed by Borelli, Montanari, and Rossetti.[16]

The section on flowers is one of the largest in the treatise in terms of both length (seventeen folio pages) and number of illustrations (exceeding one hundred). Of special interest are the illustrations of the fig and chestnut flowers. Malpighi pointed out that whereas normally the pericarp—a seed-vessel or fruit—increases in size convexly in a conical or pear-shaped form, in the fig the exterior portion grows in a concave shape so as to enclose the flower (fig. 9.12). In the chestnut, he distinguished between fertile or "fæcundi" (fruit-bearing) and unfertile or "amentacei" flowers. It was not until 1694 that Camerarius published his celebrated letter *De sexu plantarum*, in which he argued that the former were female and the latter male flowers; although others before him, from Theophrastus to Grew and Ray, had referred to male and female flowers to distinguish fruit-bearing from non-fruit-bearing flowers, the exact meaning of those terms was unclear. Malpighi was far from appreciating the sexual nature of plant reproduction; however, he carefully drew many different types of flowers—an imposing iconographic apparatus on which Camerarius was to rely. In this respect it is intriguing that in 1901 the German botanist and historian Martin Möbius criticized Malpighi for having mixed male and female chestnut flowers (fig. 9.13), as if Malpighi had produced a composite image by putting together features he had observed in different flowers of the same species. As it turns out, however, his drawing was quite

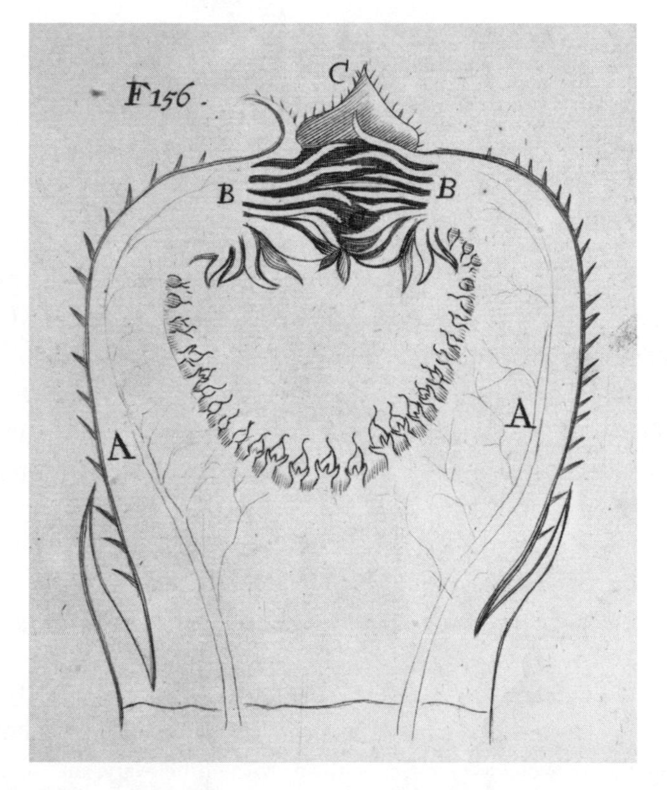

Figure 9.12. Malpighi, *Anatome plantarum*: XXVII.156, fig flower

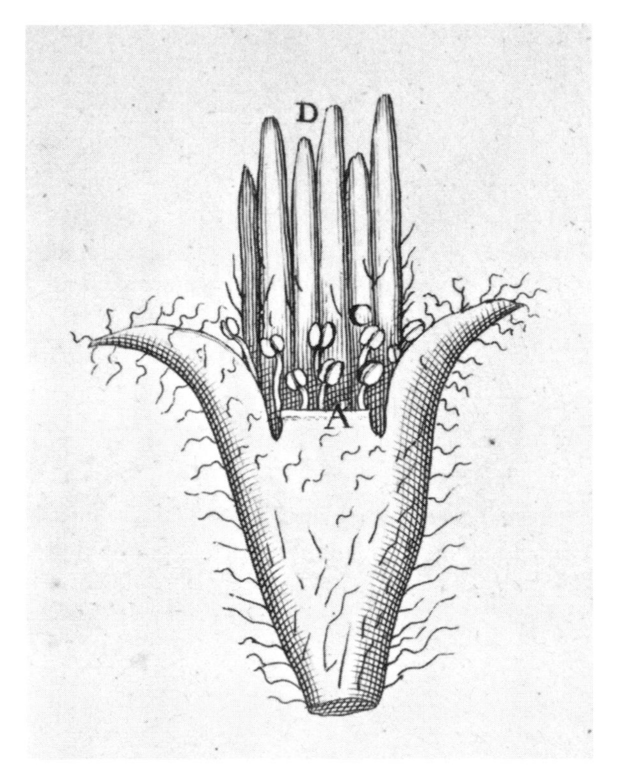

Figure 9.13. Malpighi, *Anatome plantarum*: XXXVII.228, chestnut flower

accurate and relied on a single specimen, because the chestnut flowers he drew—later to be identified as female—often present abortive male stamens, clearly visible in his figure underneath the style-shaped structures D, which he compared to "tubarum uteri." Malpighi considered flowers as the discarded material from the process of seed generation and compared them to the menstrual purgation of women, an analogy seemingly supported etymologically by the use of the term "flower" in the sense of menses. According to Malpighi, the whole process probably occurs according to the "materiae necessitate" without "naturae fine," another occurrence of the expressions discussed in chapter 5. In dealing with the pericarp Malpighi had recourse to similar notions, arguing that its growth occurs for natural causes without any particular purpose of nature; the fact that in some cases it is of pleasant taste to us is entirely accidental.[17]

With the notable exception of the study of growth, so far Malpighi had focused on structures. In the following sections of *Anatome plantarum* his focus shifted toward processes such as the generation of seeds, the growth of the ovary, and the vegetation of seeds. Malpighi drew a sharp distinction between animals and plants with regard to generation, since he believed that whereas animals reproduce sexually, plants do not. Despite this difference, he seized on the opportunity to expand on the theme of sexual reproduction in animals; he implicitly referred to and approved recent anatomical findings by Steno and de Graaf (discussed in sec. 8.3) by stating that the ovaries in viviparous animals are vulgarly called "testicles"; he further argued that whether the female provides the "colliquament" and the egg, the male provides a fluid with an active principle that—following Borelli—can be compared to a magnet. Plants, by contrast, can grow from a seed or a detached branch. In line with his systematic comparison between plants and animals, Malpighi later likened the egg of animals to the seed of the plant.[18]

In *Anatomes plantarum idea* Malpighi had already announced his belief that the insects erupting from galls were not generated by the plant, as believed by several scholars and especially Redi, but rather they were generated by other insects. One of the longest—as well as visually and philosophically most striking—sections of *Anatome plantarum* is devoted to galls. Malpighi provided a thorough description of different types of galls, but a key point was the refutation of Redi's views; unsurprisingly, his name was never mentioned, but Malpighi had already arranged to inform Redi privately through a letter of January 1673 by d'Andrea and subsequently to debate the matter through the same intermediary as well as through the Jesuit Antonio Baldigiani. The climax of the section is Malpighi's report of his observation of the apparatus or "terebra" some insects are endowed with to deposit the egg inside the leaf (clearly visible in fig. 9.14), one that he had initially mistakenly identified as the male organ. Moreover, on one occasion toward the end of June Malpighi was fortunate to observe a fly arching its body, extracting the "terebra" and inserting it into the leaf. He was lucky enough to be able to capture the fly and find in its reproductive organs exactly the same type of tiny diaphanous eggs he had found in the leaf. His plate (fig. 9. 14) shows the oak fly with its "terebra," which he elsewhere showed also in detail.[19]

Besides galls and the vegetation of seeds, the second part of *Anatome plantarum* deals with a number of special topics such as tumors or excrescences, tendrils of vines, plants vegetating in other plants, and roots. Malpighi saw an analogy between plants and animals with regard to hairs and thorns, and when he returned to such matters in his *Vita*, he expanded on this matter. Among plants growing from other plants Malpighi tentatively included mould, especially the variety growing on dairy products (fig. 9.15), which Malpighi found initially white or diaphanous, then green and finally black. It is instructive to compare his rather stylized rendering of different types of mould with Hooke's dramatic and visually arresting plate shown above (fig. 9.2).[20]

Malpighi's study of plants involved a series of ingenious experiments on different aspects of their "œconomy." For example, having discovered different types of vessels,

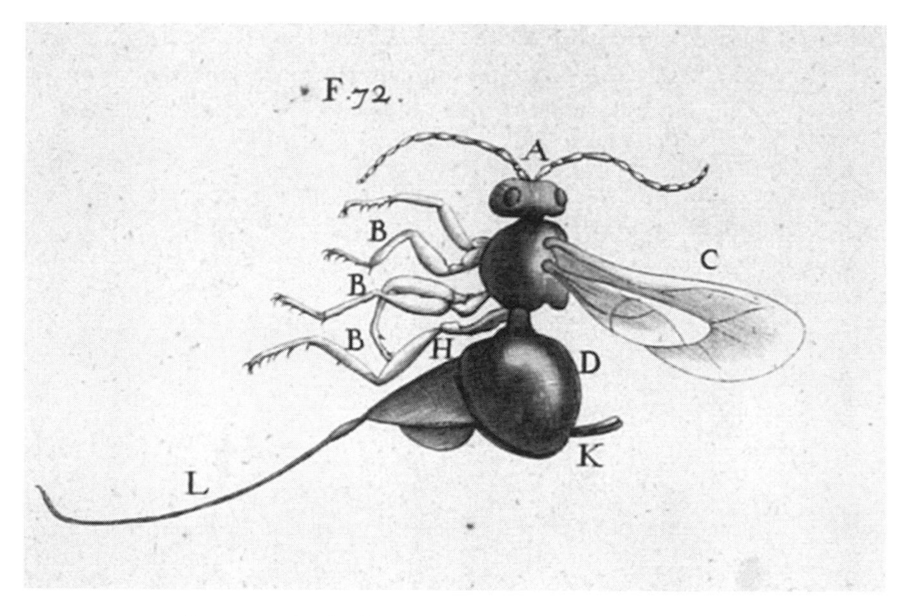

Figure 9.14. Malpighi, *Anatome plantarum*: XX.72, oak fly

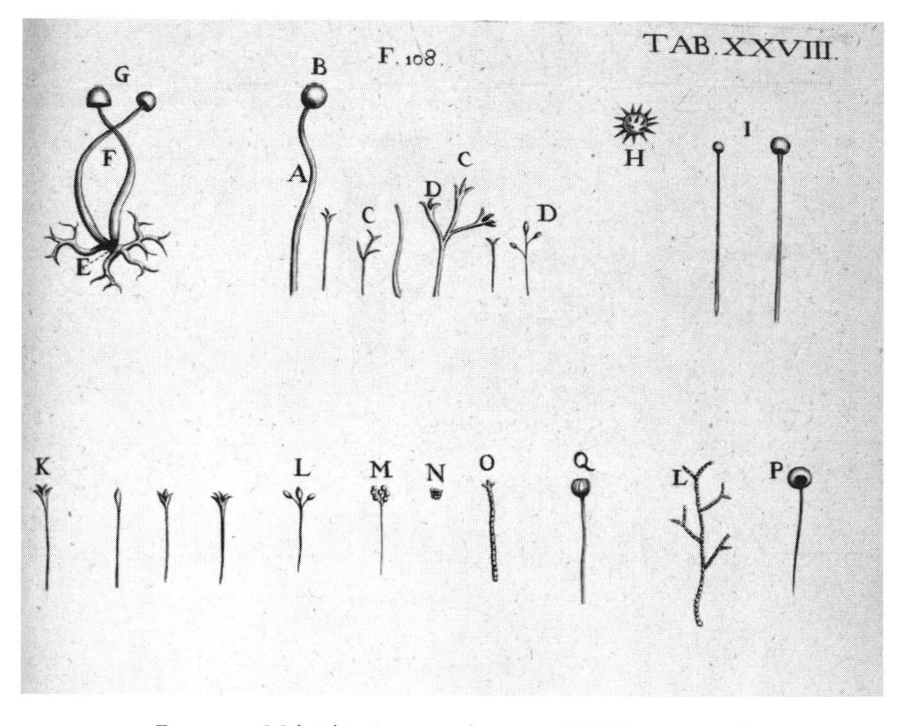

Figure 9.15. Malpighi, *Anatome plantarum*: XXVIII.108, mould

Malpighi showed that some of them contained air by cutting a plant under water and observing bubbles. He planted young twigs of fig, plum tree, and blackberry upside down and succeeded in making them grow, albeit somewhat smaller than normal. This experiment goes hand in hand with his claim that there are no valves in the vessels of plants, contrary to what Hooke had surmised. Malpighi's experiments and concerns seem to be inspired by Harvey's work on circulation and especially the valves in the veins; in this case, however, plants turned out to be different from animals. Yet another "vivisection" experiment concerning the direction of sap flow consisted in removing a horizontal band of bark from the trunk of different trees: Malpighi noticed that it grew back from the top, thus showing that in the external vessels of the trunk sap flowed downward. A similar experiment is reported at the end of the letter to Jacob Spon, in which Malpighi stated that he ligated the trunk of several plants and removed a ring of bark; whereas the upper part became enlarged, the part below the ligature stopped growing. Further, Malpighi explored whether nutrition was carried exclusively along straight vessels or whether an area could be reached through different paths: he lacerated the leaves of gourds and lemons, finding that those areas immediately behind the laceration remained alive, a clear sign that the vessels carrying nutrition form anastomoses, as one observes in networks of vessels in several portions of a plant.[21]

Malpighi performed a number of experiments concerning growth as well. Clearly inspired by Redi's celebrated trials on the generation of insects discussed in section 7.2, he collected earth from well under the surface and placed it in a glass jar covered by several silk veils, to prevent tiny seeds carried by the wind from germinating in the soil while ensuring access to air and water; however, he did not rely on parallel trials. No plant ever grew in that soil.[22] He seems to have been inspired by his experiments on insects to explore the effects of immersing seeds in various liquids; he watched the results of immersing broad beans and beans in water infused with niter, tartar, sulphur, chimney soot, sea salt, vitriol, antimony, rock salt, armenian salt, hartshorn, quicklime, wine, vinegar, lye, human urine, and spirit of wine. Then he observed what happened when he planted the seeds, as well as the effects of sprinkling various fluids, such as human urine, on various seeds, including beet, lettuce, endive, and radish. In another experiment reminiscent of Redi's and his own trials with insects, Malpighi tested whether seeds such as broad beans, wheat, vetch, beans, radish, and lettuce germinate in a jar filled with water and covered with oil. Although he did not explain the reason for covering the water with oil, by analogy with his work on the silkworm one may argue that he wished to seal the container from air. The seeds rotted and the water turned "fœtentissima." Undeterred, Malpighi proceeded with further experiments, this time testing whether by removing the cotyledons or seed leaves from several seeds of broad beans, beans, gourds, melon, and lupine the plants would still grow. Of all the many seeds he planted, only one broad bean grew for a few days. He then repeated the experiment with a larger set of plants, removing the cotyledons once the vegetation had just started; also in this case plants did not fare well. Malpighi concluded that the seed leaves are analogous to the egg albumen, and the earth is the uterus in the vegetation process. Presumably he had his contemporaneous experiments on the generation of the chick in mind here.[23] We shall return to these experiments below while dealing with Trionfetti's challenge to Malpighi.

Among the multitude of topics addressed by Malpighi, here I wish to mention a key notion occurring in the study of the generation of seeds and the growth of roots, that of fermentation. Malpighi argued that the growth of the tiny leaves of the fetal plant is stimulated by the fermentation in the particles of the cotyledons, which he compared to the placenta. As to the operation of roots, he admitted his inability to detect how the roots absorb nutrition and after a suitable fermentation transport it to all parts of the plant. As we will see in section 10.3, in *De motu animalium* Borelli was to seize on Malpighi's account and challenge the notion of fermentation in plants, prompting Malpighi to undertake a systematic series of investigations on this matter.[24]

9.3 Trionfetti, Malpighi, Cestoni, and the Vegetation of Plants

Borelli's work was not the only one to prompt Malpighi to undertake additional investigations of plants. In 1685 the Bolognese Giovanni Battista Trionfetti published *Observationes de ortu, ac vegetatione plantarum*, a treatise on the growth and vegetation of plants in which, while describing and illustrating some new species, he challenged some of Redi's and Malpighi's claims. Thus, Trionfetti's work dealt with both natural history and philosophical aspects of generation. Trionfetti was professor of medicinal simples and prefect of the botanic garden at la Sapienza in Rome and a noted botanist in his own right, who had a genus and several species named after him by Linnaeus and other botanists thanks to his findings in *Observationes.* Malpighi was suspicious of Trionfetti, who was an associate of his rivals Mini and Sbaraglia. In the aftermath of Marsigli's and Sbaraglia's 1689 attacks on Malpighi (see sec. 11.2), Trionfetti was to follow suit with his *Prolusio ad publicas herbarum ostensiones*, also held in 1689 and published the following year, in which he attacked Malpighi with claims echoing Sbaraglia's. A sense of the acrimony of those disputes and of the atmosphere at Bologna in those years can be gained from the actions of Lelio Trionfetti—Giovanni Battista's brother and the professor of medicinal simples at Bologna—who destroyed all the copies of his brother's *Prolusio* he could put his hands on.[25]

Despite the later events and a revealing letter by Bellini providing hints to decode portions of Trionfetti's text, Howard Adelmann was unable fully to appreciate Malpighi's hostile reaction to Trionfetti, attributing it partly to oversensitivity, but wisely concluded, "it is probable that there is more than meets the eye." In fact, Trionfetti's work requires careful handling because it is written in an elusive style typical of the time. A letter by Bellini and a passage in Malpighi's *Vita* explain Trionfetti's baroque rhetorical gambit: Trionfetti did not challenge Redi and Malpighi directly, but rather by mentioning their views as reported by the Danish anatomist Caspar Bartholin and the English naturalist John Ray. Trionfetti challenged by means of experimental trials Redi's statement that "omnia ex ovo producantur." The nature of the intellectual issues involved can be grasped through an unusual detail, the rather long and lavish imprimatur in which Trionfetti was praised not only as a botanist but also as a philosopher, and his book decreed not only as worthy of print but most welcome by all investigators of the secrets of nature. Its author was none other than the Jesuit Filippo Buonanni, who in 1681 had published a study and classification of snails with a defense of the doctrine of their spontaneous generation, *Ricreazione dell'occhio e della mente*. Letters by Malpighi go as far as to state that Trionfetti's book was written at Buonanni's instigation, adding that Trionfetti was responsible for some abusive almanacs against Malpighi. At that time Buonanni was engaged in a controversy over spontaneous generation: his claims were answered not only by Redi in the 1684 *Osservazioni intorno agli animali viventi* but also by Antonio Felice Marsigli, future archdeacon of Bologna University, in a letter addressed to Malpighi published in 1683, *Relazione del ritrovamento dell'uova di chiocciole*. Marsigli challenged Buonanni's belief in generation from putrefying matter and reported his own accidental discovery of the eggs of snails. Marsigli identified four reasons why, following Aristotle, Buonanni had denied that snails reproduce through eggs, such that testacea do not have intercourse and therefore cannot be born of their own seed, and no animal without blood is oviparous. Moreover, Marsigli reported the observations by Ulisse Aldrovandi, Martin Lister, Johann Jacob Harder, Swammerdam—referred to as "Sumerdam"—and John Ray, who witnessed and described their copulation and identified them as androgynous. Indeed, in *De respiratione* Swammerdam had been the first to state—in a brief note and in the title page (fig. 7.8)—that terrestrial snails have both a penis and womb around their neck and therefore share of both sexes, a finding that was to have profound implications for the study of plants.[26]

Buonanni promptly replied to Marsigli in the *Riflessioni sopra la "Relatione,"* in which he argued that Marsigli's finding was not new. Rather, the problem was that if some snails are generated from eggs, one cannot conclude that all are generated in the same way. Thus, Buonanni called for greater caution in defending universal claims and questioned the method of induction, arguing that not all stars are fixed, since planets move, and not all birds fly, as the ostrich never leaves the ground. Malpighi expressed his dislike for Buonanni's views—and later versions of them—which in his opinion corrupt the true method of philosophizing, making everything uncertain and at the same time every bizarre thing plausible. Indeed, Buonanni had challenged a key tenet of Malpighi's philosophy: the uniformity of nature underpinned many of Malpighi's investigations across different species, as we have seen many times. To be sure, Malpighi accepted surprising results, such as the hermaphroditism of snails, but only when they were supported by strong empirical data; as a general tendency he would stick to the uniformity of nature unless he was compelled by overwhelming evidence.[27]

Trionfetti's *Observationes* was dedicated to Marsigli, an indication of Marsigli's shifting allegiances in those years against Malpighi in favor of Sbaraglia and Trionfetti, possibly in connection with Malpighi's opposition to the reform of the Bologna studium discussed in the next chapter. A passage in *Observationes* reveals that Marsigli had witnessed Trionfetti's experiments on the alleged transmutation or metamorphosis of plants, as we shall see below. The main issues at stake were whether plants grow from seed alone or through other means, notably from putrefying matter; the role of the seed leaves or cotyledons in the first stages of germination; and the preformation and metamorphosis of plants. Following the works by Redi and other naturalists, the intellectual world was rapidly shifting away from spontaneous generation in animals of all types, leaving scholars such as Kircher and Buonanni on the defensive; with regard to plants and other organisms such as fungi, however, matters were more complex. Trionfetti questioned whether sea plants, such as *Ceratoides Maritimus* or *Maximus* (fig. 9.16), grow from seed, given that despite his efforts he had been unable ever to find flowers or seeds in sea plants; moreover, he referred to the well-known observation, also reported by Malpighi in *Anatome plantarum*, that a new plant often grows by planting a cut branch.[28]

A related matter arose accidentally in autumn when Trionfetti buried a totally putrefied branch of a spurge known as *Tithymalus Mirsinites*. To his surprise, the plant grew back in the spring, prompting him to undertake systematic experiments to test whether the leaf or juice of putrefying plants may suffice to make the plant grow; he buried rotten specimens from different species and genera of plants, including the juice of *Cyanus major Alpinus incisis folijs*, a new species first described in *Observationes*; figure 9.17 shows the relevant plate from a copy of Trionfetti's book owned by the French botanist Jean-Louis Poiret—an associate of Jean Baptiste Lamarck—showing that the plant was later renamed *Centaurea triumfetti* in his honor. In his trial Trionfetti noticed that several specimens of *Tithymalus*, especially *Mirsinites*, grew back, thus confirming the fortuitous growth he had previously observed. He also reported the opinion of the Montpellier physician Dedu that the seed of plants can be formed in the earth by the aggregation of atoms; analogous views in support of an atomistic view of spontaneous generation had been defended by Edme Mariotte in

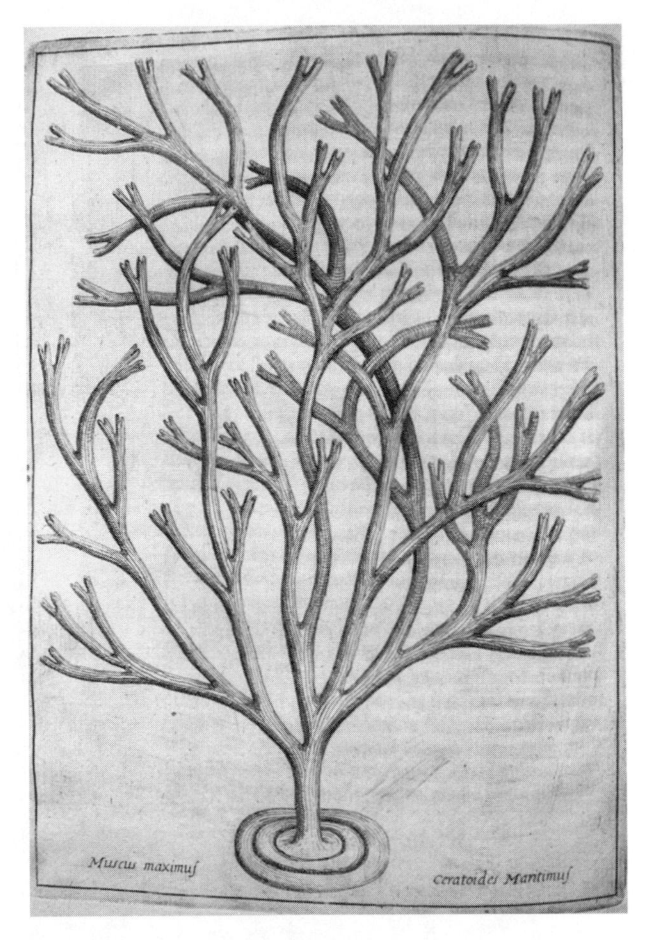

Figure 9.16. Trionfetti, *Observationes*: *Ceratoides Maritimus* or *Maximus*

his 1679 treatise *De la végétation des plantes*. Trionfetti also gestured toward an explanation of the phenomenon along similar lines. He held rather eclectic views, citing authors as diverse as Aristotle, Lucretius, Paracelsus, Cesalpino, and Giovanni Battista della Porta.[29]

In order to test the effect of the sun and air on plants, Trionfetti performed two experiments; in the first he used two pots with *Thlaspi persicus*, one placed in the open air, the other in a closed room in which he lit a fire every day. Whereas the former languished, the latter thrived. In the other experiment Trionfetti placed one hundred seeds of different plant genera in glass jars sealed with paper and then in October stored them in an underground room with minimal air circulation. In May he found that some seeds had become much smaller and were floating in a limpid fluid, others in a resinous one, others were mouldy, while others still had undergone metamorpho-

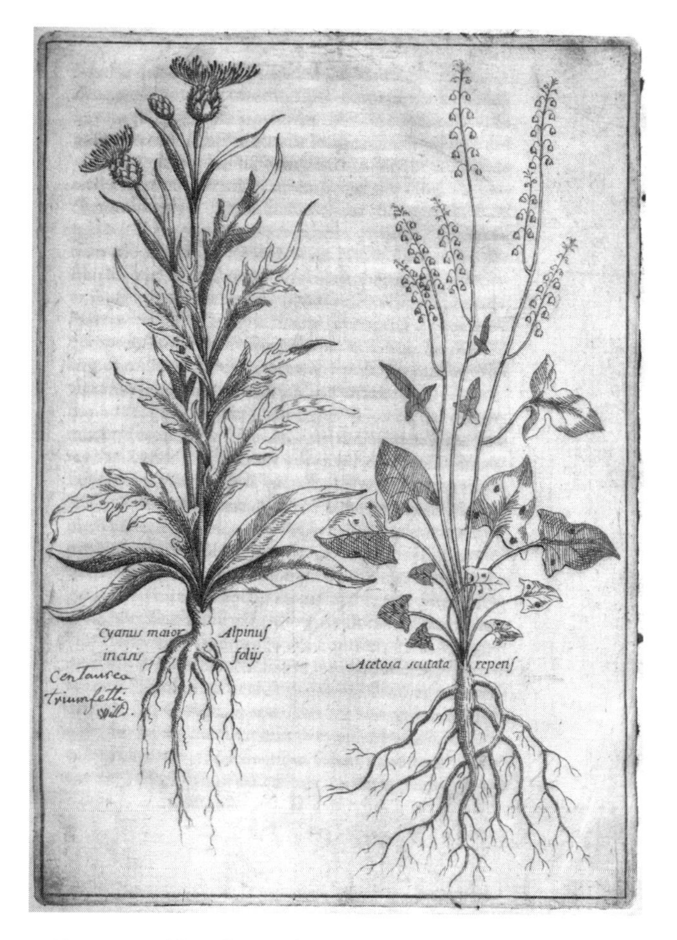

Figure 9.17. Trionfetti, *Observationes*: *Cyanus major Alpinus*

sis, but Trionfetti tantalizingly left this matter for another occasion. Some seeds grew a root, but *Hippolapathus matthioli* grew a root and four leaves.[30]

Trionfetti challenged the view put forward by Malpighi and confirmed by Ray that the cotyledons or seed leaves help the growth of the plant by means of a series of experiments with beans and castor beans. He argued that the cotyledons protect the young plant and prevent the dissipation of the most active particle by the undue entrance of matter from the outside. In order to prove his point, Trionfetti filled the space left after the removal of the cotyledons with earth and wax, showing that the plants grew well.[31]

Lastly, in *Observationes* Trionfetti argued that generation was a much more complex process than "expanding those things that are narrow, increasing those that are small, and unfolding those that are enrolled." It is easy to see in his wording his antimechanistic bend geared against an explanation of generation in terms of growth with "expansion, increase, and unfolding" alone, and in favor of active transformations as well. To this purpose he referred to experiments he had performed in the presence of Marsigli, showing the metamorphosis or transmutation of plants, whereby wheat is transformed into darnel, mint into basil thyme, and other similar examples. Further, Trionfetti argued that a transformation or metamorphosis of *Iacea cinerea* into a degenerate smaller form without flower could be obtained by planting it in a smaller vase (figs. 9.18 and 9.19); by repotting it in a larger vase, he showed that the plant retained its degenerate form, one that Trionfetti went to the trouble of showing and comparing to the primitive one.[32]

On reading Trionfetti's work, Malpighi became deeply concerned: a sense of his intensive work on plants in the 1680s can be gained from his diary, showing extensive experimentation on the role of cotyledons and growth. Malpighi experimented with buried putrefied spurge and was happy to report that he never detected a growing plant. Some of Trionfetti's claims involved sea plants that were beyond his reach, and for this reason he gladly accepted help from the Livorno apothecary Cestoni, a contact he was able to establish thanks to Bellini at Pisa. While acknowledging Trionfetti's *Observationes* as an erudite work, Cestoni challenged his views that plants on earth and in the sea could reproduce without a seed, but he wished to keep mushrooms out of the picture. It was with palpable relief that Malpighi could report in his *Vita* that he had been able to confirm from repeated observations that moss did have seeds, whereas the "sagacious" Cestoni had been able to detect seeds in sea plants. Moreover, echoing a passage from a letter by Cestoni, Malpighi challenged Trionfetti's claims about metamorphosis, arguing that plants can indeed differ considerably in size if moved from the wild to a protected environment and vice versa, although this was not true metamorphosis across species.[33]

Among the most interesting passages in Malpighi's *Vita* are those reporting a series of refined experiments that are of special interest for their methodological implications. Originally Malpighi had removed the cotyledons from some seeds in order to observe the result. Challenged by Trionfetti, in 1685 he repeated the experiment with bean and lupine seeds, with the added precaution of filling the space left by the removal of the cotyledons with wax; no plant grew. Among twelve gourd seeds he planted, however, only one where a portion of one cotyledon had remained attached did grow; its size was only three inches as opposed to more than half a foot for a normal plant. Malpighi does not state whether he had planted whole seeds to make the comparison, or whether he was just giving the reasonable size of a normal plant at that developmental stage. Immediately after this point, however, he stated that he ran two parallel trials, planting a number of mutilated chestnuts and acorns alongside nonmutilated ones; the mutilated seeds did not grow, whereas the whole

Figure 9.18. Trionfetti, *Observationes*: original form of *Iacea cinerea*

ones did. This added precaution of running parallel trials apparently arose at a later stage as a reaction both to Trionfetti's challenge and to the fact that one gourd seed did germinate after all. There is an interesting parallel with Redi's experiments on styptic water that we have seen in section 7.2, in which he had performed two parallel trials in order to weed out those cases when healing is performed by nature. Here too Malpighi needed to weed out—so to speak—those seeds that seemed to be so vigorous as to germinate regardless of whether their cotyledons are removed, as well as to compare the growth of plants with and without cotyledons. His trials did not rely on a rigorous and systematic comparison between the two cases, however, but rather were based on rather loose comparisons, in line with the procedures of the time. In June 1685 Malpighi repeated similar experiments with beans, but this time he let the root germinate for a few days before removing the cotyledons; whereas those plants

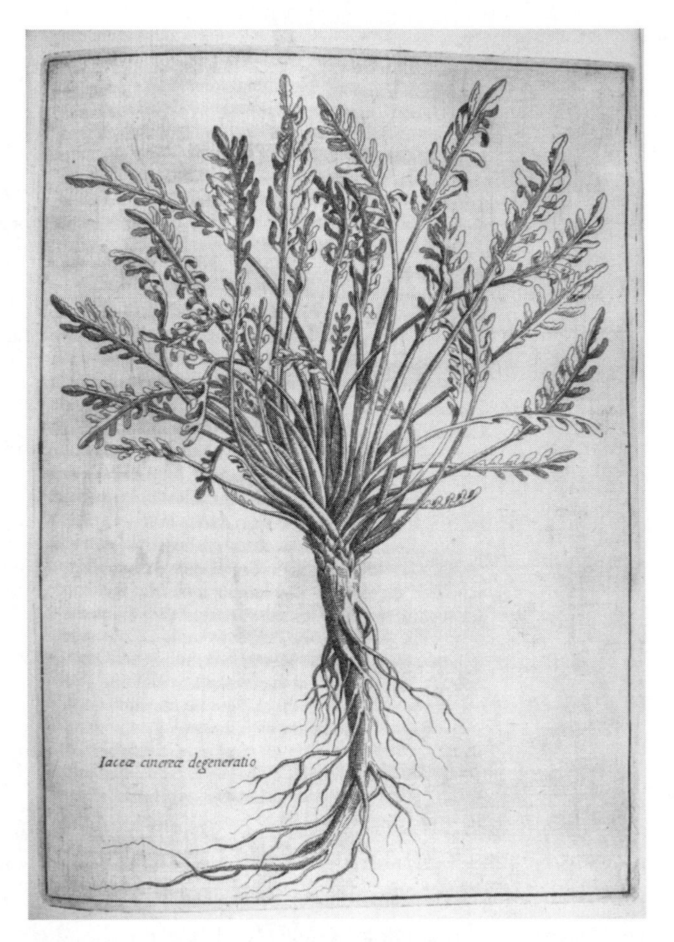

Figure 9.19. Trionfetti, *Observationes*: degenerate form of *Iacea cinerea*

that were left whole germinated, those mutilated on the second and third day soon died. Performing similar experiments with lupines, Malpighi noticed a significant difference in the growth of the stalk and leaves between mutilated and whole plants. Additional variations of the experimental procedures occurred with gourds: Malpighi delayed the mutilation process of twelve plants until they had sprouted and noticed that only two grew but the difference in size with whole plants was in the ratio of one to five. In 1687 he performed additional experiments with gourd seeds by removing only one cotyledon, showing that their growth was hampered. He proceeded relentlessly to experiment with castor bean (*Catapucia major*), arguing that the portions Trionfetti had removed were not the true seed leaves. Malpighi provided an extensive diary and a series of plates detailing its growth.[34]

Malpighi defended his view that the whole plant is contained in the seed against

Trionfetti's attacks, arguing that use of the microscope would have allowed his antagonist to detect parts he had not been able to see with the naked eye; it is easy to see here a significant intersection with issues addressed in the study of the generation of the chick in the egg concerning preformation. Malpighi also challenged Trionfetti's views about the metamorphosis of plants, arguing that he and his friend had been unable to observe the metamorphosis of wheat into darnel. In this case he did not rule it out, but rather argued that if one were to admit the experiment, it would be inappropriate to draw conclusions about the normal course of nature from a monstrous case that could be compared to the malformation of a fetus. Lastly, Malpighi stated that he and his friends had planted spurge and related plants and, although the experiment was difficult to perform because spurge can easily produce buds from the stalk, had been unable to see it grow. He wished to make it clear, however, that he had not stated that all plants grow only from seed but rather had maintained a cautious attitude to the problem. His remark highlights the difficulty of understanding the nature and problem of generation in cases of plants, algae, and fungi.[35]

9.4 Grew and Camerarius: Iconography, "Œconomy," and Sexual Reproduction

It is one of the famous coincidences in the history of science that preliminary versions of Malpighi's and Grew's works on plants arrived almost simultaneously at the Royal Society. The Society informed them of their respective works and encouraged both to bring their projects to completion; although Malpighi was sent Grew's works as they were being printed, in the process Grew had easier access to Malpighi's manuscripts than Malpighi did to his printed books. Grew's work was published first in a short treatise in 1672, *The Anatomy of Vegetables Begun*; other parts appeared subsequently until the volume was completed in 1682 as *The Anatomy of Plants*. Grew's work provides us with the opportunity to contrast his approach with Malpighi's, with special emphasis on the organization of their books and iconography. Moreover, Grew's passing comments on the sexual reproduction of plants are a useful springboard for approaching the experiments by Camerarius in light of the seventeenth-century developments we have been reviewing.[36]

Given the circumstances of composition and the access they had to each other's work, it is difficult to get a clear sense of their respective contributions and of their art of representation over a decade as if they were entirely independent; moreover, often it is hard to detect points of disagreement because the Royal Society actively discouraged disputes. Grew accepted many of Malpighi's findings, such as the spiral vessels and the notion that plants contain air, something he verified by repeating Malpighi's experiment of cutting a branch under water. Grew also adopted many weaving analogies, such as in his description of the roots and trunk. He seemed more optimistic than Malpighi about finding the chymical composition and therefore properties of plants, hoping to explain why some are purgative and others "vomitory" or emetic. Deriving the reasons for the medicinal properties of plants from their composition was a common concern at the time from Redi to the Paris *Académie*. Unlike Malpighi, Grew did not start from the bark, but rather followed "the Method of Nature herself," moving from the seed to the formation of the root, trunk, branches, leaves, flowers, fruit, and back to the seed. Further differences between Grew and Malpighi concern the lack of systematic comparison with animals and iconography. To be sure, Grew did draw significant comparisons with animals, but in no way was this a key feature of his work. Much as Malpighi had done in the preface to the second part of *Anatome plantarum*, Grew spelled out his intentions and aims in the preface to *The Anatomy of Plants* (fig. 9.20):[37]

> In the *Plates*, for the clearer conception of the *Part* described, I have represented it, generally, as entire, as its being magnified to some good degree, would bear. So, for instance, not the *Barque, Wood,* or *Pith* of a *Root* or *Tree* by itself; but at least, some proportion of all three together: Whereby, both their *Texture*, and also their Relation one to another, and the *Fabrick* of the whole, may be observed at one *View*. Yet have I not every where magnify'd the *Part* to the same degree; but more or less, as was necessary to represent what is spoken of it. And very highly, only in some few Examples, as in *Tab.* 40 which may serve to illustrate the rest. Some of the *Plates*, especially those which I did not draw to the *Engravers* hand, are a little hard and stiff: but they are all well enough done, to represent what they intend.

This passage suggests a tension between the representation of single parts and of the whole, or with establishing a visually satisfactory relation among the individual constituents of the plant. It is possible that in this passage of 1682 Grew was criticizing Malpighi, who in the 1675 *Anatome plantarum* had provided pictures of single parts of a plant, such as the bark. By contrast, Grew often preferred transverse or longitudinal sections of a whole trunk. Grew explains that he wanted to show "the Contexture both of the Perpendicular and Horizontal fibers"; thus, he managed not only to combine a cross section with a longitudinal one but also to disassemble the various components while showing how they come together. It is also instructive to compare Malpighi's cryptic figure of spiral vessels with Grew's (figs. 9.6 and 9.21), bearing in mind the importance Malpighi attributed to this finding in his work. Both relied on laceration in order to reveal the spiral structure of the vessels, but Grew made the key element stand out and performed quite a careful operation by sandwiching the elongated spiral vessels between two whole sections of a vine leaf. His figure is quite effective in offering a sense of the spatial arrangement of the parts and their relative sizes by means of the juxtaposition of the different enlargements on the same plate, whereas Malpighi's seems more successful in highlighting the internal structure of the air vessels, without however being able to convey a sense of the proportions of the parts described to the entire plant, much like in the case of the silkworm: in this regard Grew's art of representation can be usefully compared to Swammerdam's plate of the ephemeron (fig. 7.14), in that both share a holistic aim.[38]

Grew argued that the motion of sap occurred in both directions, as could be seen by the bleeding of both ends of a cut root; a similar experiment had been reported by Mariotte. As to the rise of sap, Grew followed the example of contemporary scholars,

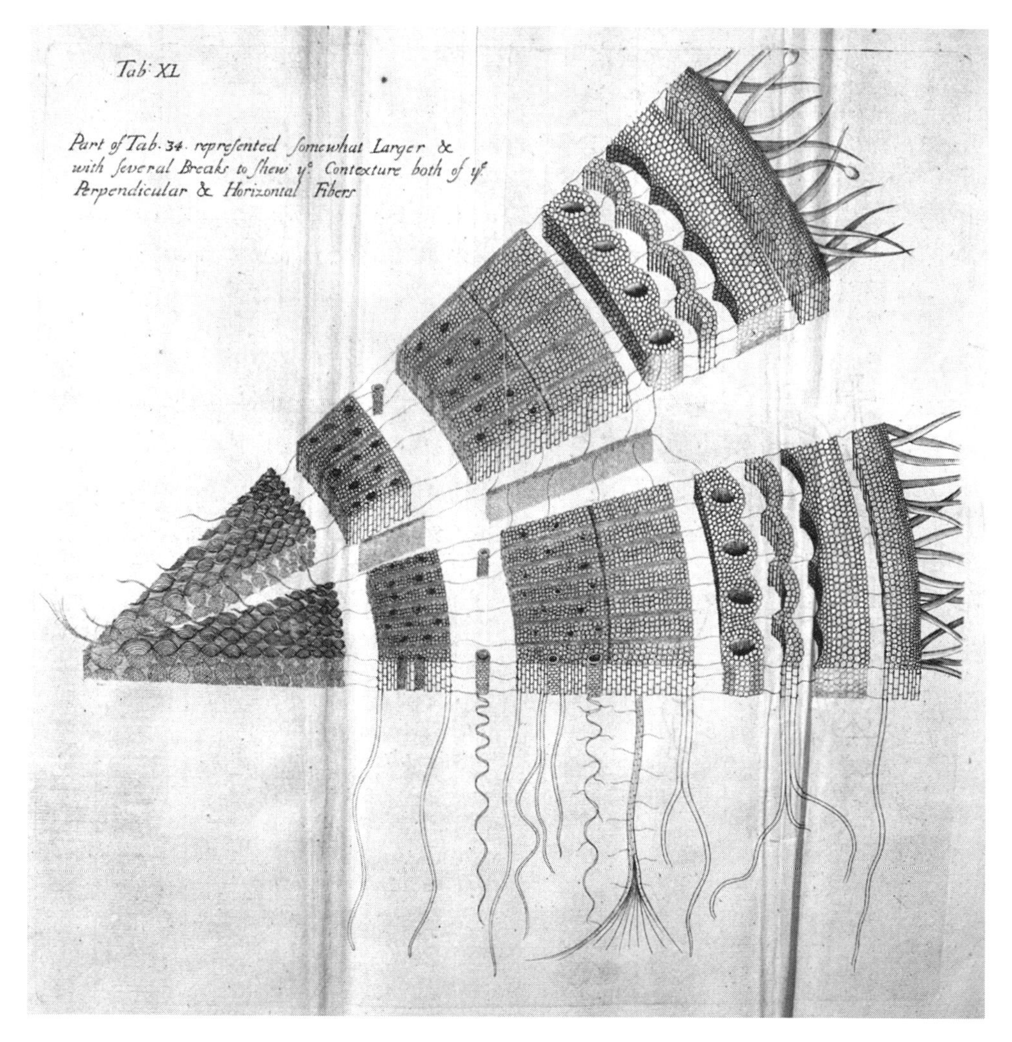

Figure 9.20. Grew, *Anatomy of Plants*: sumach trunk, plate 40

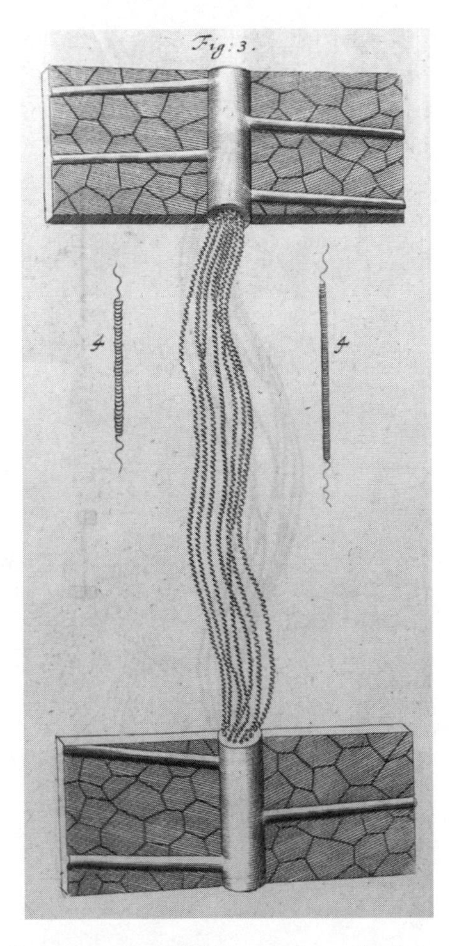

Figure 9.21. Grew, *Anatomy of Plants*: spiral vessels of vine leaf, plate 51

such as Hooke, in drawing a comparison with the rise of water in glass pipes; this he deemed insufficient, however, because water can rise only a few inches in such pipes. Moreover, he attributed also to the parenchyma or pithy part of a tree a role in the ascent of sap, analogously to sponges. However, he deemed this structure to be insufficient as well, since water can rise again only a few inches this way. Following Malpighi against Hooke, Grew denied the existence of valves in the sap vessels of plants. Thus, he had recourse to a combination of the two methods, whereby sap rises first in thin pipes, then the swollen parenchyma exerts a pressure on it and makes it rise further, and so on. His illustration of the process (fig. 9.22) shows different images of a sap vessel surrounded by the parenchyma of the plant; each set of bladders of the parenchyma contributes to the rise of sap to the next level, until sap reaches the top of the plant.[39] This is not the only case of physico-mechanical thinking in Grew's work.

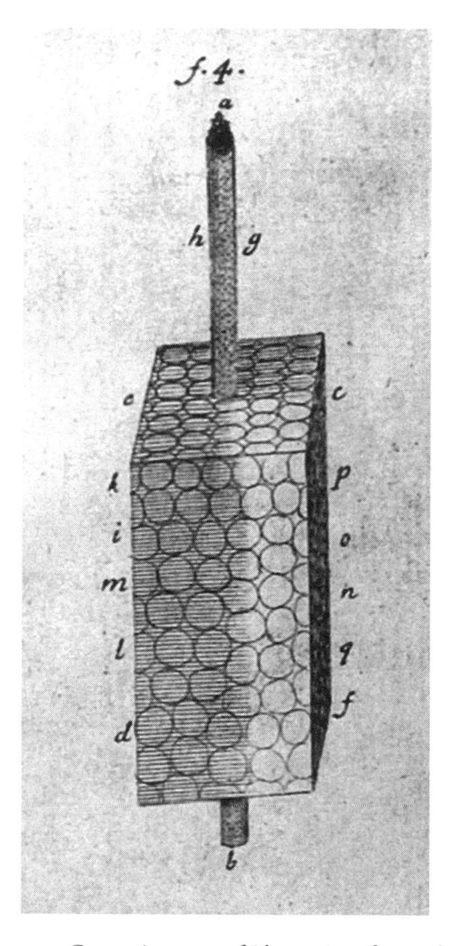

Figure 9.22. Grew, *Anatomy of Plants*: rise of sap, plate 39

One of my favorite examples is the discharge mechanism—a term used by Grew—of the seed of "Coded Arsmart," which is contrived like a bow ready to "ejaculate" the seed from the bottom as in figure 9.23, showing at the top the curled springlike membrane ready for discharge: Grew's curled membrane closely resembles Hooke's image of a spring in *De Potentia restitutiva*, published just a few years earlier in 1678.[40]

One of the most interesting passages in Grew's work concerns plant reproduction. A relevant notion he coined is that of the attire, namely, the parts within the floral leaves or corolla, especially the stamens, or "seminiform attire," and the florets of the disk in composite flowers, or "florid attire." Grew reported that the physician Thomas Millington had told him in conversation that the "Attire doth serve as the Male, for the Generation of the Seed," adding that he thought every plant was both male and female. The explanation he provided for the attire resembles Malpighi's, in that he argued that it was the result of the separation of the redundant part of the sap; Grew also drew the same comparison with the flower of menses of women. But the most intriguing portion of his analysis concerns his comments on the sexual generation of plants. Grew argued that the attire, before it opens, corresponds to the menses in the female, but after it opens it resembles and corresponds to the male, thus suggesting that the same plant could be first female and then male as it develops; his comparison was based on a superficial analogy between penis and style as well as seeds and testicles. In order to justify his views, Grew compared plants to snails, which, as we have seen from the work by Swammerdam and others, were known to be hermaphrodites.[41]

It was precisely this analogy with snails that was seized upon by the Tübingen physician and director of the botanic garden Rudolph Jakob Camerarius in his 1694 *Epistola de sexu plantarum* addressed to his Giessen colleague Michael Bernhard Valentini. From Aristotle and especially Theophrastus onward, several botanists had talked about plants being male and female, although the matter was far from clear and settled: indeed, although it was Theophrastus who had argued that whereas some palms bloom and others fruit, fruits ripen only if one shakes the pollen of the flowers on the bloom, it was Camerarius who provided a careful exegesis of the relevant passages by Theophrastus, highlighting his wavering opinions on the matter. Others argued that Theophrastus talked about the sex of plants only metaphorically.[42] Camerarius

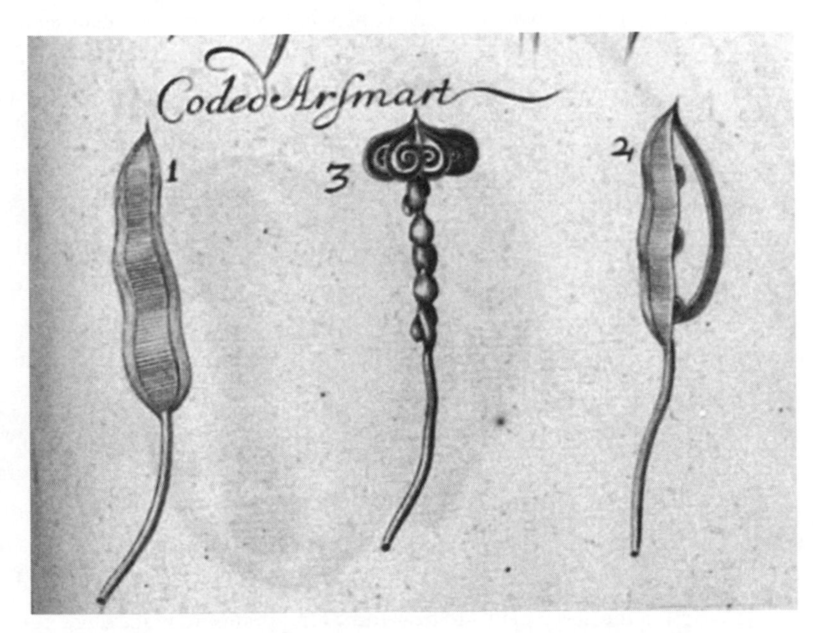

Figure 9.23. Grew's discharge mechanism of the seed of coded arsmart, plate 71.3

chose as the motto for his work the same passage by Theophrastus used by Malpighi, stressing the variety of plants and the difficulty of their study while highlighting their respect for the ancient authority on plants.[43] Camerarius had an intimate knowledge of Malpighi's work: while providing at the outset of his *Epistola* and elsewhere careful verbal descriptions without images of the flowers of different plants, he often referred the reader to Malpighi's figures, using them as the iconographic apparatus for his own work, much as Harvey had relied on Fabricius's figures on generation.[44] The identification of fertile and unfertile flowers enabled Camerarius to offer a preliminary classification based on three different classes of plants depending on their flowers: those that have complete flowers, those with fertile and unfertile flowers on the same plant, and lastly those with fertile and unfertile flowers growing on different plants.[45]

In dealing with the sexual reproduction of plants, Camerarius relied on some features of animals to show that plants were not unique: the sexual mode of reproduction found in most animals, including humans, was not the absolute norm, but rather only one instance found in nature. Whereas in most animal species males and females are separate, in some species they occur together. Referring to works by Swammerdam, Ray, Lister, Harder, and even to that by Marsigli we have seen in the previous section, Camerarius gave the example of snails to highlight that hermaphrodites should not be seen as monsters, but as the common and regular form for some species. Camerarius went back to texts by Aristotle and Theophrastus to argue that in fishes the females lay the eggs and then the males inseminate them: if this can be done in water—he reasoned—it must also be possible in air with pollen in the role of male semen.[46] Once again, an established fact about some animal species served as a springboard to conceptualize plant reproduction. In a later letter to Valentini, Camerarius established a systematic parallel between plants and animals, arguing for example that the strongest argument for sexual reproduction of plants is that both sexes are required, just as in animals.[47]

A distinctive feature of Camerarius's work was its experimental nature. In his *Epistola* he mentioned several experiments designed to test the role of the *apices* or anthers; by physically removing them from castor bean and maize, or isolating the female flowers from them in dog's mercury, mulberry, and spinach, for example, he never obtained seeds, thus highlighting their role in reproduction. A remarkable passage at the end of the *Epistola* reports on problematic cases and failures. First, with plants like club moss, ground pine, and horsetail, he believed that he had identified the male plant with abundant pollen, yet the female flower was missing: the reproduction of those plants remained a mystery. Second, he observed female maize and hemp flowers produce seeds without the intervention of pollen: Camerarius admitted his anger at discovering the hemp plants to bear fruit. His attempt to make sense of these occurrences involved the hypothesis that hemp may be fertilized by the pollen from another species, as he believed may happen with animals, although he raised this as a hypothesis rather than a solution.[48]

Camerarius's work can be seen as the synthesis of three traditions: first, the growing realization—following Swammerdam—that sexuality in nature could take different forms and hermaphrodites were not necessarily pathological or monstrous, a key point that opened a new horizon for conceptualizing plant reproduction; second, the reliance on the systematic analogy between plants and animals, so powerfully exploited by Malpighi, enabled Camerarius to argue that both sexes were required for reproduction; and third, the growing tradition of experimentation on plants pursued by several investigators, including Malpighi, opened new methods of exploration on which Camerarius relied. It is especially suggestive to compare Malpighi's experiments on the role of cotyledons in the early stages of growth to Camerarius's mutilation of flowers.[49]

There is an interesting coda tying some of the topics discussed in this chapter: both Camerarius and Trionfetti referred to each other's works concerning metamorphosis and the sexual reproduction of plants. Camerarius published several short pieces on darnel, all under the same title, *De lolio temulento*, coming eventually to oppose on experimental grounds the claims by Trionfetti and others about metamorphosis. On the other hand, in a 1703 rejoinder to Malpighi's' *Vita* entitled *Vindiciarum veritatis*, Trionfetti mentioned a passage from Camerarius's *Epistola de sexu plantarum* dealing with the changes in the size of flowers from generation to generation. Although Trionfetti was seeking to underpin his own views about metamorphosis, his reference is of broader interest in documenting the early circulation of Camerarius's work in Italy.[50]

In current histories of seventeenth-century science and medicine, the study of plants occupies a marginal position mostly to do with taxonomy and therapy. By contrast, this chapter has documented a rich set of investigations and experiments on the anatomy of plants, highlighting their key role in the study of themes such as growth, the uniformity of nature, iconography, and sexual reproduction.

Malpighi provided an extraordinarily detailed study of plants; although he did not deal with taxonomy, he offered an in-depth description and analysis of all their parts. Malpighi relied on the analogy between plants and animals in both directions: he used animal anatomy as a guide to understanding plants and, conversely, studied processes such as growth in plants in order to tackle the same problem in animals and account for it in mechanistic terms. His investigation of oak galls showed that they were due to the eggs deposited by insects inside the leaf; his work refuted Redi's in dramatic and uncontroversial fashion. Malpighi performed several experiments, including some involving parallel trials. In an intriguing case he noticed that some twigs planted upside down set roots and grew; from this he drew a conclusion not so much about the rooting process, however, but about structures, notably the absence of valves in the vessels of plants. While he was prepared to accept strong empirical evidence supporting surprising findings, in this as in other cases, Malpighi believed in the uniformity of nature as a default option. By contrast, both Buonanni and Trionfetti questioned that belief and highlighted instead nature's diversity. Trionfetti and Cestoni performed several observations and experiments on the metamorphosis and spontaneous generation of plants and algae, issues at the center of contemporary debates.

Iconography occupies an important aspect of the study of plants. Both Hooke and Malpighi highlighted graphically the similarity between plant structures and textiles. As in his work on the silkworm, for the most part Malpighi showed individual parts of the plant in isolation, without giving a sense of the proportions and relations among them. By contrast, Grew devised elaborate forms of representation in which different cross sections of trunks showed the mutual arrangements of the parts and their relations.

Grew hinted at the sexual reproduction of plants by analogy with snails. The growing awareness that hermaphroditism in snails was not a monstrous occurrence but the norm opened new vistas on the reproduction of plants: most plants, like snails, were also hermaphrodites, as Camerarius strove to show with his striking and at the same time problematic experiments. These studies led to a profound reconceptualization of sexual reproduction in nature, abandoning views based exclusively on the study of humans and higher animals and accepting a pluralistic perspective according to which modes of reproduction changed from species to species. This profound reconceptualization of the reproduction of plants contributed to new taxonomic projects in the eighteenth century, such as Carl Linnaeus's sexual system.

PART FOUR

ANATOMY, PATHOLOGY, AND THERAPY

Malpighi's Posthumous Writings

Mechanistic anatomists and physicians were involved in a number of debates and controversies both within their fold and with scholars opposed to their understanding of health, disease, and therapy. Given Malpighi's key role among the mechanists, his works are crucial in exploring those debates and controversies. In 1697 the Royal Society published his *Opera posthuma*, an imposing folio volume including his *Vita* and *Risposte* to Lipari—discussed in chapter 2 for chronological reasons—and to Sbaraglia, which immediately became required reading for anatomists across Europe, as testified by the reprints in Venice and by two publishers in Amsterdam in 1698, in Geneva in the 1699 *Bibliotheca anatomica*, and again in Amsterdam in 1700.

Taking the lead from Malpighi's posthumous writings, including the *Opera posthuma* and his medical consultations that appeared in the eighteenth century, part IV studies anatomy, pathology, and therapy around 1700 and compares his anatomical, pathological, and therapeutic thinking with that of some of his contemporaries. This material can be seen from different perspectives. The works in the *Opera posthuma* share common traits, in that they were engineered by Malpighi to be published after his death and are all confrontational in style. By contrast, the medical consultations were published posthumously against his wishes. With regard to contents, however, the *Vita* is largely devoted to anatomy and specifically to defending from criticism everything Malpighi had put in print throughout his life. The *Vita* provides the historian with a remarkably detailed document on the fortunes of his work over several decades because Malpighi adopted a new, explicit, and aggressive style detailing his qualms and responses to the literature. The *Risposta* to Sbaraglia is mainly devoted to pathology and the connections between medical theory and practice; therefore, it serves as a suitable introduction to the consultations, dealing with medical practice.

Figure PIV. De Graaf, *De succo pancreatico*: title page of the 1671 edition

Thus, part IV looks back at anatomy from the standpoint of the status and fortune of Malpighi's *Vita*, but it also investigates the world of disease that is at the center of the consultations. Literary style emerges as an especially rewarding area of investigation not only for the *Vita* but also for the *Risposta* to Sbaraglia and consultations.

Previous chapters have shown that anatomical research was profoundly interwoven with medical practice: the following chapters are especially devoted to this connection, with a growing attention to the practical side. A celebrated image from the 1671 edition of Reinier de Graaf's *Tractatus anatomico-medicus de succi pancreatici natura et usu* beautifully captures the spirit of that connection: in the foreground a dog with fistulae attached to its belly and throat to collect pancreatic juice and saliva symbolizes vivisection experiments; the dead animals and postmortem scene symbolize anatomical research and the search for the cause and location of disease; finally, the sick patient in bed shows the purpose of the anatomist's labor—not just studying the body's structure and experimenting on it but also understanding and curing disease. The title, highlighting anatomy as well as medicine, is true to the book's contents, since a large portion of it was devoted to understanding the role of pancreatic juice in disease.

The Fortunes of Malpighi's Mechanistic Anatomy

10.1 Mechanistic Anatomy and Malpighi's *Vita*

Toward the end of the seventeenth century mechanistic anatomists became embroiled in controversies on two fronts, externally with those holding different philosophical views, and also among themselves, as a result of different techniques of investigation and the resulting views about the minute constituents of the body. No work is better suited to study those controversies than Malpighi's extensive and often polemical *Vita*.

This chapter uses Malpighi's *Vita* to explore two of the most significant of those controversies and a subsequent attack on Malpighi. Section 10.2 provides a framework for reading his work, relating it to a tradition of writing about the self that inspired him. Section 10.3 moves to Malpighi's extended criticism of Borelli's *De motu animalium*, which was the initial stimulus behind Malpighi's work; his replies form the single most extensive portion of his *Vita*, about eighteen folio pages. Section 10.4 examines the controversy with Paolo Mini. Whereas that with Borelli was an internal affair between mechanists, that with Mini was entirely different, since Mini was a defender of the role of the faculties and an iatrochymist. Malpighi's reply to Mini's wide-ranging objections is the second largest in his *Vita*, about fourteen folio pages. These two controversies provide a vivid picture of a broad set of themes and debates in the history of anatomy and botany in the last two decades of seventeenth-century Italy. Finally, section 10.5 adopts a different perspective: the 1699 edition of Malpighi's *Vita* in the *Bibliotheca anatomica* includes numerous footnotes, mostly supplied to the editors by the Amsterdam anatomist Frederik Ruysch, whose seminal and wide-ranging *Epistolae problematicae* were being published starting from 1696. At that time Ruysch was developing novel injection techniques leading to results that challenged Malpighi's views about the ultimate components of the body and especially the structure of glands. Thus, the most advanced and sophisticated techniques of investigation of the time were leading to striking contradictions rather than mutual support.

The debates between Ruysch and the Leiden professors Govert Bidloo and Herman Boerhaave highlight the stalemate on central issues of mechanistic anatomy. Starting from Ruysch's insidious footnotes, I reconstruct his investigations of those years and his ensuing exchanges.

10.2 Writing about the Self

On 26 and 27 December 1689, during his *Iter italicum*, Leibniz met Malpighi at Bologna: Malpighi's diary and a later letter by Leibniz to Thévenot provide us with valuable insights into Malpighi's projects at the time. Leibniz reports that Malpighi read to him extracts from a work he called "testamento" or will, in which he replied to the objections from his critics and especially Borelli. On his part, Malpighi wrote in his diary that he read (to Leibniz's approval) his responses to Borelli's objections with regard to respiration, the vegetation of plants, and other matters.[1] The word "testamento" indicates a text to be opened after the author's death, thus suggesting that by then Malpighi was already planning to publish a posthumous work in reply to his critics and specifically to Borelli, his erstwhile philosophical mentor to whom he had dedicated his *Epistolae* on the lungs in 1661. Borelli had died on the last day of 1679, thus blunting Malpighi's sword, lest he was to publish a tasteless rebuttal to his former mentor when he could no longer respond. In such circumstances a posthumous reply must have looked like the only viable alternative, especially since he thought death was imminent. A letter to Lorenzo Bellini of October 1689 announced that in the summer of that year he had started writing some memoirs to be left to his heir in defense of his writings, "to remove from my mind the horror of the impending death." Malpighi was suffering from a kidney affliction and could not walk even short distances without urinating blood; other letters from the same time show that he thought his days were numbered. In fact, Malpighi was to live for another five years, carrying out with dogged determination his unusual project in the form of an imposing posthumous volume including not only a one-hundred-folio-page *Vita a seipso scripta* but also the *Risposta* to Lipari and the new *Risposta* to Sbaraglia. In 1691 Malpighi communicated to the secretary of the Royal Society his intention to dedicate his work to the Society, confirming both the intellectual and in some respect polemical scope of his *Vita*, and his desire to have the text delivered after his own death.[2]

As it happens, other reasons may have strengthened Malpighi's resolve and coalesced in his mind into a coherent plan; in 1689 Malpighi came under the combined attack of Marsigli, chancellor to the Bologna studium, Mini, and Sbaraglia. Marsigli informed Malpighi of his support for his enemies. This warning was followed by Mini's attack on Malpighi at the anatomical conclusions defended on January 13 in front of Sbaraglia, Marsigli, and others; the public anatomy in April of the same year;

and Sbaraglia's simultaneous attack, which will be discussed in the following chapter. Under such circumstances a posthumous rebuttal to Mini's theses in his *Vita* and a *Risposta* to Sbaraglia's attack published as a supplement to his *Vita*, all under the auspices of the Royal Society in London, looked like the safest option. Thus, the year 1689 turned out to be crucial for Malpighi, since at about the same time his Bologna enemies teamed up against him, with the support of Marsigli, and he conceived a plan for his next publication, one whose style was to differ dramatically from his previous ones. Gone was the polite and formal Malpighi: the author of the *Opera posthuma* was a different man; blood may no longer have been flowing through his veins, but his quill seems to have been dipped in poison rather than ink. Being dead allowed Malpighi to avoid the ethical charge associated with self-praise, especially against antagonists who had long been dead. This picture is confirmed by the preliminary fragment of the autobiography, first published from Malpighi's manuscripts at the beginning of our century. In a passage present only in the manuscript fragment, Malpighi wrote that up to that point he had refrained from speaking in defense of his works "for sheer modesty, tolerating the insults of my enemies." The reason for writing his *Vita* was to show to his "*posteri*, to whom alone I leave these memoirs, so that they will learn to how many vexations is exposed one wishing to live in freedom, philosophizing for truth's sake alone."[3] This passage sheds light on the character and style of Malpighi's *Vita* and of his *Opera posthuma*. While alive Malpighi wanted to present himself as a modest scholar seeking truth in a dialogue with nature: modesty, civility, moderation, dislike for controversies, respect for other scholars, and disinterest in personal advantage were crucial ingredients of the scholarly ethos of the second half of the seventeenth century. By publishing after his own death, Malpighi thought he could free himself from a number of constraints that had prevented him from writing on certain subjects and in a direct style, stressing his role of victim in controversies with his rivals, and plunging into the midst of scholarly debates he had been following in silence for decades; rarely do we have such a detailed author's perspective on virtually the complete critical fortune of his entire production.[4]

It may be useful to contrast Malpighi's *Vita* with that of one of his predecessors in a Bologna medical chair, Gerolamo Cardano. His *De vita propria liber* was written in the last few months of his life, possibly with the intent of showing to the pope that despite his errors, he was not a heretic. It was first published in Paris in 1643 by Gabriel Naudè with an influential preface or *iudicium* in which he portrayed Cardano in critical terms from a moral standpoint. Cardano starts with a reference to the *Meditationes* of Emperor Marcus Aurelius, which is chosen as a model—although later he claimed that he was not portraying himself as he ought to have been, as the Roman emperor had done, but rather as he was, including vices and errors. Cardano's *De vita propria* follows also Svetonius, *De vita Caesarum*, in adopting not a chronological but

Figure 10.1. Malpighi, *Opera posthuma*: frontispiece

rather a thematic order, including chapters on physical exercises, personal character, family affairs, manner of walking, and thinking. Besides Marcus Aurelius, Cardano refers to other classical models such as Sulla, Caius Caesar, Augustus, and Galen, who had written *De libris propriis* to assert authorship over his own works while distancing himself from works falsely attributed to him. As Arnaldo Momigliano put it, "In the Hellenistic age kings and politicians seem to have monopolized autobiographical writing as an instrument of self-assertion and self-defence. Roman politicians borrowed autobiography from the Greeks in the second century B.C. for the same purpose." Although Malpighi, following a humanist tradition initiated by Petrarch, referred generally to classical models at the outset of his *Vita*, he also emphasized intellectual issues and said little about his character and family. I would argue that his source was not a specific text, but rather knowledge that classical authors had used autobiographical writing when dealing with political, military, and literary affairs as a weapon of self-defense. Malpighi did not miss a chance to express his recriminations and indulged in listing the offenses he had to endure, even over decades-old minor events. These considerations reinforce the impression that Malpighi's text can be characterized as an *Apologia pro vita sua*, in which large portions follow an old-fashioned style of academic disputes avoided by Malpighi in his lifetime publications.[5]

Malpighi's *Vita* is a composite work, much like his letter to Spon, which was framed as a reply to Spon's inquiry as to whether Malpighi had anything to add to his works in view of a new edition of them. Curiously, it was Jacob Spon's father Charles who had edited the ten volumes of Cardano's *Opera* in 1683. Over a few pages Malpighi dealt with themes as diverse as the growth of horns, the structure of kidneys, and generation.[6] Likewise, extensive portions of the *Vita* resemble treatises on respiration, the structure of the viscera, bones and teeth, plant anatomy, hairs, feathers, and nails.[7] As a general rule, however, they appear as refutations of his rivals' opinions, this representing a major difference from the letter to Spon. When occasionally Malpighi was not stimulated by controversy, he stated that his research had not been challenged.[8] The narrative of Malpighi's life provided the connective tissue offering some degree of cohesion to this wide-ranging collection of anatomical research. If the letter to Spon could be characterized as a series of *additions*, the *Vita* is a framework for a series of *defenses*. In this regard the frontispiece of *Opera posthuma* (fig. 10.1) is of considerable interest: it shows a stormy sky with three panthers asleep under a tree with the inscription "in portu dormiunt." Silvestro Bonfiglioli, who was responsible for preparing for publication and transmitting Malpighi's last work to the Royal Society, explained the significance of the allegory: the tree represents the Royal Society, under whose protection Malpighi's works can rest sheltered from the impending storm. The prestige and geographical distance of the Society gave Malpighi's *Opera posthuma*, so embroiled with strictly local and personal controversies, a much-needed aura of being *super partes*. In all probability the allegory was sketched by Malpighi himself for an unknown artist, who produced a more refined version, on the example of the related frontispiece in *Anatome plantarum*. It is therefore clear that Malpighi's claim in his letter to his nephew that he wished his posthumous works to be kept privately among the Royal Society papers is unconvincing, since the society had published all his previous works and private manuscripts not intended for publication are seldom accompanied by elaborate frontispieces. Bonfiglioli's uncertainties in the identification of the animals under the tree, described as dogs, panthers, or tigers, suggest that he was not responsible for the illustration, and that this can therefore be reasonably attributed to Malpighi, on the example of the frontispiece of *Anatome plantarum*.[9]

Although Malpighi's *Vita* has been extensively mined by historians for information on dates and events in his life, it is useful to provide a critical analysis of the work as a whole, amounting to over one hundred folio pages, taking into account circumstances of composition, genre, contemporary conventions and styles, and the status of posthumous works. I wish to preclude a misunderstanding of my argument about the difference in style between Malpighi's *Vita* and his previous works. I am not claiming that in his lifetime Malpighi was constrained by a number of literary canons of modesty and civility, whereas in his posthumous works he was free. The elaborate construction of his autobiography and the variants among the manuscript fragment, the manuscript draft, and the published version show that posthumous works also followed rules and constraints, but they differed in some crucial respects, because Malpighi could no longer be accused of seeking personal advantages, pecuniary or in status.

10.3 Levels of Mechanical Explanation in Borelli and Malpighi

In his *Vita* Malpighi acknowledged his debt to Borelli in the briefest possible way, stating that Borelli had introduced him to the free way of philosophizing when they were colleagues at Pisa. In the manuscript fragment Malpighi stated that Borelli introduced him to free and "Democritean" philosophy, an adjective lacking in the 1697 version; reference to an openly atomistic philosophy by the pontifical archiater would have been singularly inappropriate, especially at a time when such views were under close scrutiny. This brief tribute was immediately followed by the accusation that it was not Borelli, as claimed in *De motu animalium*, but Malpighi who first observed the spiral structure of the heart muscle.[10] Being dead was a virtual necessity in replying to Borelli because Borelli had always been at least formally very polite toward Malpighi, Malpighi had praised him in his letters to the Royal Society and had dedicated several of his early works to him, and Borelli could no longer defend himself. In the extensive rebuttal of his formal mentor, some topics are treated in detail over many pages, but at times it appears that Malpighi could not refrain from ridiculing or denigrating him. On page 2, for example, we read that Borelli acted like an oracle, *velut oraculum*. On page 25 Malpighi tells us that he communicated to Borelli his discovery of cavities of air in plants, to which Borelli replied, "I have made it myself, but my sight does not help me." On the same page Malpighi tells us that on learning of his discovery of the structure of the optic nerve of the swordfish, Borelli replied that he would find something similar in other large animals, such as oxen, as it happens an erroneous claim.[11] Other contentious areas are treated more extensively and reveal different forms of mechanistic anatomy and medicine advocated by the two contenders. Applying mechanical notions to anatomy proved less problematic in explaining the motion of arms and legs in terms of levers, as in the first part of *De motu animalium*, than in accounting for microscopic structures.

It is worth reflecting here on a reading strategy for *De motu animalium*, a work, unlike Malpighi's *Vita*, not intended to appear posthumously—Borelli just happened to die before his work appeared in print. Often the book looks unfinished and contradictory, leading one to suspect that Borelli was unable to complete his project. *De motu animalium* poses further problems to the modern reader, not simply because of its technical contents and awkward Latin, but also because of the conventions adopted, typical of Borelli and his time. In the "Proemium ad Lectorem" in the 1670 *De motionibus naturalibus*, Borelli laid out his rules for mentioning living authors, based on "modestia" and "moderatio." Although Borelli was not always consistent in avoiding mentioning living authors with whom he disagreed, his works contain a number of allusions and coded messages requiring considerable interpretative efforts; often they are of great interest because Borelli's work relates to some of the major debates of his time. The issue of references to contemporary debates and controversies is one of the most challenging for the historian because actors follow self-consciously elaborate styles and conventions depending on the period, local traditions, and personal inclinations. Indexing criteria also require attention: the *editio princeps* of *De motu animalium* contains indices of chapters and propositions. If the modern editor then produces a name index, as in *On the Movement of Animals*, the result can be misleading: the modern reader checking for references to Malpighi, for example, will find a string of praises and compliments. Yet when Malpighi read Borelli's work, he must have thought differently, since in his *Vita* we read, "I had broken off scholarly correspondence with him, and he consequently became so indignant and inflamed against me and my works that he took occasion in the things he wrote in his extreme old age, such as his *De motu animalium*, to invalidate my work."[12] Four areas in particular proved contentious: respiration, diuresis, the cause of fevers, and the growth of plants. I explore them in turn.

Differences between Borelli and Malpighi on respiration date from 1661. Borelli's disagreement with Malpighi's interpretation of its purpose led to a shift between the first and second *Epistola*. Whereas initially Malpighi had talked of the fermentation of blood, later he dropped it and endorsed Borelli's interpretation quoting *verbatim* a passage from his letter (see sec. 1.4). Although Borelli had urged Malpighi to publish and Malpighi had accepted his mentor's views, in *De motu animalium* Borelli still challenged Malpighi. Refraining from mentioning proper names, he argued that "famous anatomists" believed that respiration was instituted to mix in the lungs the principal constituents of blood (namely, the serous, red, and chyleous matter with lymph) so that the smallest part of all constituents would touch each other. This task would be achieved by the periodic compression by the alveoli inflated with air. Whereas in 1661 Borelli had proposed this very view, he later found it questionable. There can be no doubt as to which proper names Borelli was omitting, since in his *Vita* Malpighi identified himself as the target of propositions 107 and following of Book II of *De motu animalium*. Borelli's display of civility, rather than appeasing Malpighi, irritated him further. This way of proceeding, however, was not uncommon, and Malpighi himself had frequently adopted it, as in *De renibus* while attacking Bellini and Borelli.[13]

Borelli refined his understanding of respiration between 1661 and the end of his life. In 1661, at the time of Malpighi's discoveries, he expressed his views about grafting jasmine and vine into a lemon tree in a letter to Malpighi, who excerpted it in his second *Epistola*. In the second part of *De motu animalium* Borelli argued for three purposes of respiration. The example of grafting was still referred to at the very end of proposition 127, but this time it occupied an ancillary position in his account. First, respiration serves the purpose of dividing blood into its smallest elements, in order to undo improper mixing and to render curdled glutinous portions of the blood fluid again. This function is explained by arguing that all liquids are composed of minimal hard granules and by analogy with containers filled with white and black seeds.[14] Secondly, Borelli sought a more satisfactory role for air in respiration. While arguing that air alone cannot directly penetrate the lungs and enter the blood, he claimed that air mixed with the serous watery juices in the recesses of the lungs does mix with the blood. The role of air particles in blood is crucial to Borelli's atomistic and mechanical account. In proposition 115 he argued that the smallest particles of air are spiral machines that can be compressed by an external force and then spontaneously rebound like a bow. In the following proposition we read: "An automaton presents a certain vague similarity with animals, since both are self-moving organic bodies employing the laws of mechanics, and both are moved by natural faculties," among which he included gravity. Borelli's notion of machine was more closely tied to motion than Malpighi's, and he often had recourse to the notion of automaton, as when he discussed the motion of the heart.[15]

Vital operations in animals can be compared to a pendulum clock oscillating with a given speed, because they also depend on motions that have to take place with well-determined "speeds, rhythms and periods." Thus, animals need air, whose elementary elastic components activate the particles of blood with their oscillations, as in pendulums. In the same way as a clock in which the pendulum does not move almost suffocates, "veluti suffocatur," animals in which the oscillatory motion of blood is obstructed die. Thirdly, respiration serves to make blood. At the end of its circuit through the whole body blood is "altered and unfit for nutrition." By mixing with chyle, refashioning its components in an appropriate way, and interspersing them with springlike air particles, blood is reformed and ready to resume its function. We witness here the use of elasticity, whose role in anatomy was pioneered by Pecquet. Also here Borelli had recourse to an atomistic analogy, that of mosaics forming elaborate, orderly representations consisting of elementary particles. In his explanation of this third function of respiration, Borelli relied on the netlike structure put forward by Thomas Willis in *Pharmaceutice rationalis*. In his account Borelli excluded chymical notions and especially fermentation; besides arguing that fermentation cannot occur in the narrow alveoli of the lungs, he also argued that they could not contain a ferment, because they are continuously cleansed by the flowing blood—an argument he often deployed.[16]

In his *Vita* Malpighi quoted Borelli's letter of 1661 to emphasize how astonishingly unfair Borelli had been. But he went further, providing a nearly six-folio-page rebuttal of Borelli's views, old and new. Malpighi defended the existence of fermentation in blood, despite the pulmonary capillary network being very thin, on the grounds that blood particles are considerably smaller than the sections of the veins and arteries. He also pointed out that at the beginning of the second part of *De motu animalium* Borelli himself had introduced the notion of fermentation in order to explain the contraction of muscular fibers. Thus, the objection against the use of fermentation in lungs—that fermentation does not occur in narrow tubes—contradicted Borelli's own account of muscular contraction, which relied on fermentation in even narrower spaces. Further, Malpighi challenged the soundness of Borelli's anatomical representation of lungs, arguing that the capillary vessels were considerably more complex than his figures and arguments suggested, so much so that even his source, Willis, had omitted to describe them. In fact, Borelli did not provide anatomical but rather physico-mechanical drawings illustrating his views about mixing fluids; he discussed, for example, the mixing and unmixing of black and white seeds or fluids going through the configuration of figure XVIII.10 and 11 (fig. 10.2, *bottom right*).[17]

Malpighi confirmed and refined the views he had put forward in *De polypo cordis*, in which he had suggested that only a portion of air enters the blood in the lungs. He expressed doubts on its identification with niter on the basis of his experiments with

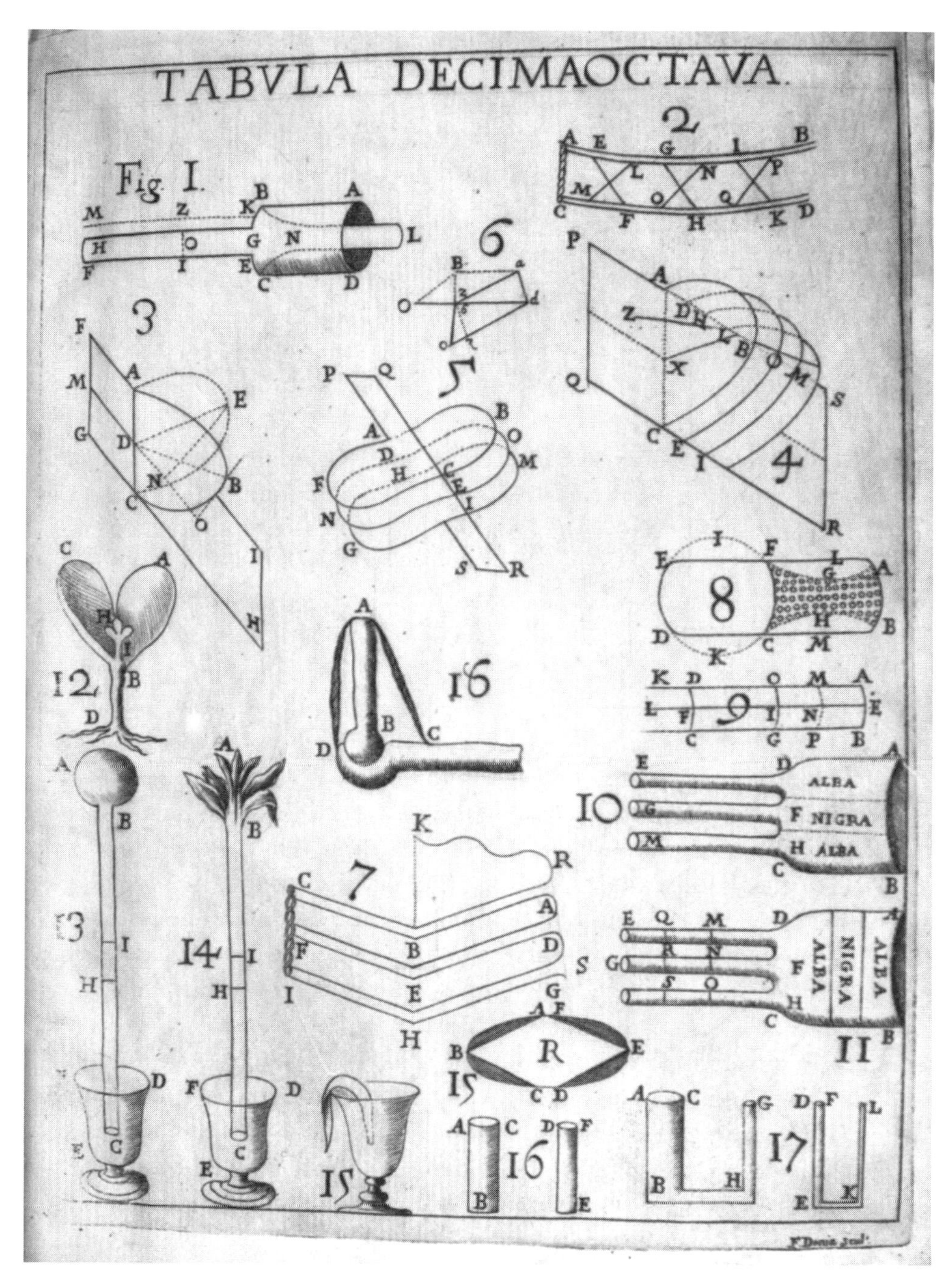

Figure 10.2. Borelli, *De motu animalium*: the lungs, XVIII, 10–11

blood injections in a dog: apparently six ounces of niter produced no sensible mutations. It is revealing that in 1673, explaining his views on respiration to the Neapolitan lawyer Francesco d'Andrea, Malpighi had referred to *De viscerum structura*, containing *De polypo cordis*, rather than *Epistolae de pulmonibus*. Similar views about the separation of a portion of air in the lungs can be found in his *Risposta* to Sbaraglia.[18] In his *Vita* Malpighi was more specific and called the portion of air penetrating the lungs *principium fermentativum*, namely, an extremely active and mobile principle similar to light, possibly identifiable with phosphorous. Whereas in his 1689 encounter with Leibniz Malpighi had heard similar views with skepticism, later he must have found them more convincing, since he stated that although Borelli had claimed that the smallest spiral particles of air enter the blood, it is more probable that something hidden in air, and also in water, extremely mobile and active, similar to the nature of light, is separated. In fact, a luminous and inflammable body is extracted from urine, which is a portion of the blood, as it appears in Boyle's *Phosphorous*. Thus, for Malpighi respiration was primarily chymical rather than mechanical.[19]

The second area of contention concerned the kidneys, a topic that brought to light new differences between Borelli and Malpighi. In 1662 Bellini had put forward his views on the structure of kidneys and at the end of his *Exercitatio* had left the pen to Borelli for the explanation of purpose. Borelli argued that the kidneys work mechanically based on their structure, i.e., a junction between arteries, veins, and renal ducts. The separation of urine depended exclusively on the shape and size of the tubes and was aided both by blood pressure and by respiration, which would supplement arterial pressure.[20]

In *De renibus* Malpighi put forward a more elaborate glandular structure and modified Borelli's interpretation by using the notion of fermentation together with simple filtration. Moreover, Malpighi adopted a much more cautious style, emphasizing the difficulty of the problem and inadequacy of our understanding. Both the introduction of fermentation and the skeptical tone attracted Borelli's attention. In *De motu animalium* Borelli's explanation of the anatomical structure of the kidneys was left rather vague and no picture was provided. Rather, he chose to focus on methodological issues. In proposition 142, in the chapter "On cleansing of the blood in the kidneys," he discussed the "objections written by a very famous anatomist against our opinion," quoting several passages from Malpighi's *De renibus*, without mentioning the author by name. Borelli reports Malpighi's passage:[21]

> [The most famous anatomist] confidently states: "the artifice by which urine is separated in the kidneys is very obscure. Although (he says) this whole work of the glands (i.e. the kidneys) conforms to reason, since that smallest (namely elementary) and simple structure of the canals of the glands is hidden, we can only ponder

> on this problem and solve it in terms of probability. This machine must carry out the separation by its internal configuration. Actually it is doubtful whether it corresponds to those which we use to human ends and which we imagine to be very similar. Although the sponge, riddle, tubes, and sieves appear to be of analogous structure, it is very difficult to assimilate the texture of the kidneys to those structures whatever the similarities. Further, since Nature is very fecund in her ways of operating, machines unknown to us will be found, which may even be beyond our grasp."

Borelli asked rhetorically, "Can the internal configuration of the kidneys be anything else than the shape of the canals?" claiming that it was irrelevant whether the canals were round-shaped or angular. In the same proposition Borelli admonished: "Do not say that the luxury of Nature appears in the multiplicity of flowers and plants. You are rather mistaken or deceived since Nature always acts seriously." Both passages highlight the tension between Malpighi's bewilderment at the fecundity and complexity of nature and Borelli's emphasis on understanding it on the basis of rigid mechanistic presuppositions: Malpighi wanted empirical proof of what Borelli already believed a priori. Further, whereas Malpighi believed that fermentation plays a role in the separation of urine, Borelli questioned whether fermentation could occur in narrow tubes and in a short time. In his view such a separation could be replaced mechanically by forcing the blood at high speed into the capillary tubes of the emulgent artery. Later Borelli defined fermentation as "an internal movement of the elements of a composite body which are activated by their own motive force or by that of another additional body."[22] In *De motu animalium*, however, he dropped the views he had previously held that respiration takes a role in diuresis.

Malpighi treated the dispute on structure and function of kidneys very briefly, arguing that Borelli was "*sinistre interpretando*" his own writings. While agreeing with the principle of a mechanical separation of urine put forward by Bellini and Borelli, Malpighi claimed only to have expressed doubts on the type of mechanism employed in the separation, not on its mechanical nature.[23]

In his *Vita* Malpighi inserted extensive extracts of his correspondence with Borelli on fevers and the cause of the 1661 epidemics at Pisa. Those passages shed light on the problematic links between their philosophico-anatomical program and practical medicine. Also the search for causes of diseases within the novel anatomical framework is discussed in the correspondence reported by Malpighi. It is not immediately clear why Malpighi devoted about three folio pages to such matters, illustrating in far greater depth Borelli's opinions rather than his own. A plausible explanation is that both the extract of a letter of September 1659 and the extensive correspondence of 1661 report Borelli's opinions on the presence of a ferment in the blood, an opinion later denied in the last chapter of *De motu animalium*. There Borelli argued instead for a nervous origin of fevers, conjecturing that the alteration of nervous juices irritates the nerves, is transmitted to the heart, and provokes fevers. Borelli relied on Fracassati's and Bonfiglioli's experiments on intravenous injections to question whether mixing blood with salts or sulphur induces fevers and argued that phlebotomy had no effect on fevers, since patients in France and Spain, where phlebotomy is practiced very frequently, fare neither better nor worse than in Italy, where it is practiced far less frequently. Possibly Malpighi wished to divulge views by Borelli contradicting those found in *De motu animalium*, as well as Borelli's doubts that jaundice was associated with bile and his sincere admission at the end of his letter that medicine was not his profession.[24]

We move now to the last topic of contention, on plants. In *De motu animalium* Borelli had several passages from Malpighi's first part of *Anatome plantarum* in mind, although it is doubtful whether he saw the second part, which appeared in the year of his death in 1679. In proposition 178, for example, Borelli attacked Malpighi's comparison between seeds and the eggs of oviparous animals because "eggs contain nutritive juices for feeding, unlike the seeds of plants." For example, Borelli claimed that the root does not grow from some internal juice in the seed. Borelli observed in the forest the long, black, ligneous roots of bay berries, looking like threads. While some reached the length of half a foot, their cortex was intact and the substance of the seeds white and hard with the same taste, shape, and size as that of other berries without roots. Thus, the growth of the root resulted not from the outflow of the substance internal to the seeds, but from an addition from the outside, since the bulk of that long thread was almost the same as the bulk of the berry, which was not diminished at all.[25]

Malpighi expressed views contrary to Borelli especially in the second part of *Anatome plantarum*, but also in the first part one finds a brief treatment of this matter.[26] Moreover, Trionfetti had raised objections similar to Borelli's, so that Malpighi's experimental refutation of those claims applied to both. There can be no doubt about whom Borelli had in mind in proposition 179, stating, "It does not seem possible that aqueous juice can be transformed into a plant by ferment present in it or in its seed," as if that juice that is sucked by the roots of the plants were rough and unsuited for feeding and it must be fermented and digested by the very powerful juice that is kept in the pores of the follicles and in the plant. According to Borelli, plants grow on water alone and their functions can be explained in mechanical, as opposed to chymical, principles. He further argued that the duty of the philosopher is to avoid being deceived by the metaphoric terms "fermentation, digestion, maturation, exaltation and plastic virtue." Malpighi was one of those who had employed those very notions, such as plastic virtues, whose legitimacy was challenged by Borelli. It is ironic, however, that Borelli himself did not hesitate to talk about plastic virtues immediately after having decried them.[27]

The last topic taken up by Malpighi in his attack on Borelli is also the most extensive, amounting to more than seven folio pages and four plates with dozens of illustrations. Malpighi reported that in *De motu animalium* Borelli had challenged his views on the function of *folia seminalia*, denying that they nourish the plant and objecting to the use of fermentation in plants. Malpighi wasted no time in taking up Borelli's challenge. The address to the reader in the second part of *De motu animalium* is dated 22 December 1681. In 1682, thus during the first available season, Malpighi had already started experimenting with bay seeds, the same mentioned by Borelli in his attack, as well as palm and date. Whereas Borelli appears to have performed extemporaneous observations in woods, Malpighi emphasized that his own account was not based on observations *fortuito facta*, but on a systematic study over several months, rigorously recording in a diary the growth of the plants. No criticism, whether on major or minor aspects, was spared. Borelli had stated that no excretory vessels exist in plants, but this was false because Malpighi had observed excretory orifices on leaves, analogous to sweat pores in the skin. Borelli had stated that plants grow on water alone, but this too was false and indeed Borelli himself had referred to a variety of bitter, sweet, acid, or oily juices absorbed by the roots of plants. This time as well the notion of fermentation was at the center of the dispute and was identified by Malpighi as a major concern for Borelli, who, while questioning it in places, had used it freely elsewhere in order to solve his own interpretative problems.[28]

Despite the disparate subjects under investigation, some common elements can be identified. Whether he was dealing with lungs, kidneys, fevers, or plants, Borelli never challenged Malpighi's structural findings, but rather took issue with their interpretation and purposes. Whereas Malpighi relied repeatedly on the notion of fermentation to bridge the gap between his anatomical findings and interpretative program, Borelli tried—or pretended—to dispense with chymical notions and preferred purely mechanical ones. Fermentation and other chymical notions were at least in principle acceptable to Borelli and his colleagues because they were ultimately reducible to the motions, separations, and combinations of particles. In many cases, however, they retained a dubious status not entirely dissimilar from that of the Aristotelian notions the mechanists wished to displace. In a revealing passage echoing one on plants quoted above, in which fermentation, digestion, maturation, exaltation, and plastic virtue are called "deceiving metaphoric words," Borelli challenged the abuse of terms such as "fusion and precipitation" by the chymists, who do not care about the mechanism ("artificium mechanicum") by which such fusion and precipitation are achieved and therefore ignore whether they can be applied to the matter in discussion.[29] Chymical notions had to be anchored to a mechanical basis, but the proliferation of the notion of fermentation signaled that simple mechanisms were invisible. This of course does not mean that Borelli was opposed either to chymistry or to chymical explanations in anatomy and medicine, as we have seen in section 2.2, and in his correspondence with Malpighi, he referred a few times to chymical remedies. The second part of *De motu animalium* considers chymical notions in the fabric of the animal body, including fermentation, for such fundamental functions as digestion. After van Helmont, it was relatively common to refer to fermentation, especially in the mechanical sense favored by Descartes and Willis. Borelli's concern was with the uncontrolled proliferation of chymical explanations introduced as ad hoc remedies to the problems of mechanical interpretations. This is also the view of the editors of *Bibliotheca anatomica*, who emphasized precisely this aspect in the preface to *De motu animalium* and shared his concerns.[30]

Borelli's and Malpighi's approaches differed in a number of ways dependent on their disciplinary affiliations and expertise. Their controversies over the operations of the lungs, kidneys, muscles, and seeds, as well as the causes and cures of diseases, show that despite the wealth of original findings by Malpighi and others, mechanical explanations proved problematic and controversial. The lengthy negotiations at the time of the *Epistolae* on the lungs and their different views thereafter testify to the hiatus among anatomical dissections, complex microscopic analyses, and an account of use or purpose. Whereas Borelli remained until the end of his life fully confident and optimistic in his approach, Malpighi, without abandoning the mechanistic program, at times doubted our ability to grasp the depth and complexity of nature; by using a broader set of explanatory tools, he recognized the tensions between a strictly mechanistic program and anatomical investigations that were proving more and more like soundings of the abyss.

The dispute over fermentation was not simply a private affair between Borelli and Malpighi, but had a broader European import. Besides the editors of the *Bibliotheca anatomica*, also Walter Charleton, William Cole, Archibald Pitcairn, and William Cockburn across the Channel attacked the notion of fermentation in different contexts.[31] Those controversies around 1700 raise the more general issue of the status of chymical explanations and remedies in anatomy and medicine.

10.4 Paolo Mini and the Soul-Body Problem

Whereas the dispute with Borelli was an internal affair to mechanistic medicine, that with the Bologna professor of medicine Paolo Mini touched on the assumptions of mechanicism and posed problems of a rather different nature. In the 1678 *Medicus igne, non cultro necessario anatomicus* Mini attacked Malpighi—without ever mentioning his name—and mechanistic medicine while defending a chymical understanding of bodily operations curiously tied to Galenic faculties. His belief in the key role of the faculties led him to disparage anatomy, microanatomy, and postmortems, which he judged useless to determine the causes of disease. Paradoxically, Mini taught anatomy at Bologna and was in charge of several public dissections at the anatomical theatre, missing no chance to attack Malpighi. The views he put forward on those occasions can be reconstructed from a variety of sources and provide a vivid picture of the intellectual atmosphere at Bologna in those years. Several topics were touched on in his treatise and at public functions, notably the operations of the lungs, liver, kidneys, and spleen and the broader philosophical question of the relationship between structure and *usus* or purpose.[32] Mini died in the summer of 1693, over one year before Malpighi; therefore, Malpighi found himself, as in Borelli's case, in the position of having to defend himself from a dead opponent. Although he must have composed the bulk of his *Vita* by the time Mini died, he probably felt relieved that his reply was going to appear posthumously.

The theses put forward by Mini in January 1689 and discussed by Malpighi's enemies in the library of a Bologna monastery deny that microscopic anatomy and the anatomy of insects and plants were of any use to physicians. More crucially, Mini denied that the separation of humors occurred mechanically as in a sieve: given the central role of sievelike separation in mechanistic anatomy, it is easy to see how radically opposed Mini's and Malpighi's systems were. His emphasis on a God-created immaterial soul in the first thesis posed significant philosophical issues for understanding the body's operations.[33] By analogy with the mind-body problem of the interaction between a nonextended thinking soul and an extended body in Cartesian philosophy, we can call the interaction between the immortal soul and the mortal body the "soul-body problem": although this issue was not new in the second half of the seventeenth century, it did take new shape and significance with the rise of mechanistic anatomy and the attempt to explain the operations of the body in terms of the conformation and motion of its organs.

Having been a medical student and friend of Malpighi's, later in life Mini turned against his erstwhile mentor, thus violating a Hippocratic precept; in his *Vita* Malpighi hastened to point this out in the very first reference to Mini, as if to disqualify his antagonist from the very outset on moral grounds.

Mini had been able to infuriate Malpighi by claiming implicitly that his findings on the lungs, liver, and kidneys were truly due to the Padua anatomist Giulio Casserio, Johann Jacob Wepfer, and the Rome anatomist and physician Bartolomeo Eustachio, respectively. In addition, Mini argued that those anatomical discoveries were of no use to medicine. The latter claim was more extensively addressed in Malpighi's *Risposta* to Sbaraglia; the former Malpighi took so seriously as to embark in extensive exegeses of the relevant passages to defend the novelty of his own findings.

Malpighi was an exceedingly learned and tenacious opponent who had once conceived and started the project of a history of Italian anatomy. Concerning the lungs, he drew a distinction about their substance either consisting of thin membranes, as he had claimed, or being spongy, as claimed following tradition by Casserio. He also argued that the structure of the lungs was inaccessible without the microscope. As to the liver, Wepfer had claimed in a letter to the Danish anatomist Jacok Henrik Paulli that the cooked liver of a swine looked *quasi*-glandular, yet according to Malpighi the restriction to one animal, the word *quasi*, and again the lack of the use of the microscope—without which the termination of blood vessels and bile ducts in the tiny glands could not be detected—show that Wepfer had not really seen glands. As to the kidneys, several commentators, including Sbaraglia and Mini, attributed Bellini's and Malpighi's findings to Eustachio. Malpighi took it upon himself to defend not only his work but also that by his "charissimus amicus" Bellini. This time he stated only implicitly that without a microscope, Eustachio could not have seen the tiny miliary glands in the outer circumference of the kidneys. Moreover, he argued that, according to Eustachio, the kidneys are a continuous body with a series of furrows, as if incised in wax by a style, a structure altogether different from that put forward by Bellini and himself.[34]

In dealing with the lungs, Mini put forward a wide range of objections to Malpighi's *Epistolae*, a text implicitly referred to many times although never mentioned explicitly. While accepting the structure put forward by Malpighi, Mini challenged his and Borelli's problematic attempt to explain the purpose of respiration purely in mechanical terms, as mixing blood. The most dramatic objection relied on a vivisection experiment performed at a public anatomy at the Bologna theatre. Mini cut the trachea of a dog and tightly ligated a bladder to the portion attached to the lungs; he noticed that the bladder filled and emptied when the animal breathed, but after five or six times the animal died. This experiment seriously undermined Malpighi's and Borelli's early mechanical interpretation of the lungs' operations, since air preserved its mechanical properties—as evidenced by the fact that the bladder kept inflating—but could no longer sustain life. Mini argued on this basis that air acts in a way different from that surmised by recent anatomists. Rather, he stated that air consists of a dense part and a thinner one that enters in contact with blood because of the *virtus attrahens* of the lungs in the live animal; when the animal is dead, the separation of the thinner part fails to take place. It is surprising that Mini quoted Lower's work, since in *De corde* he had reported the experiment performed with Hooke whereby venous blood injected in the inflated lungs of a dead dog emerges bright red or arterial. Following Lower, however, Mini could argue that blood changes color because of air, recounting the observation of the caked blood exposed to air that had been performed by Malpighi at Pisa and that was published by Fracassati in 1665. Another criticism concerned disease and fermentation. Mini pointed out that in fevers blood would be mixed by the increased heat and—according to the moderns—its enhanced fermentation; therefore, respiration would be superfluous. Yet respiration does not stop or slow down in fevers; according to Mini, its purpose is rather to decrease fermentation.[35] At the end of the first *Epistola* on the lungs Malpighi had compared them to the gills of fishes and the placenta in women. Mini objected on the basis of a claim by Basilius Valentinus that even when the water surface is entirely frozen, fishes are surrounded by fluid water, yet they die—an evident sign that the role of respiration is not mechanical; he also objected that in the placenta there is no pressure from the outside air and therefore its way of operating must differ from the lungs. Quite characteristically, Mini objected to Malpighi's idea that the serous or white part of blood renders it fluid, something in his opinion that was due to a vital spirit.[36]

We now know that Malpighi had changed his mind on the operations of the lungs and in *De polypo cordis* and *Anatome plantarum* had put forward the view that a portion of air does enter the blood through the lungs. In his *Vita* he pointed out that both publications predated Mini's *Medicus* by several years, yet Mini had ignored them. Malpighi dismissed the observation of fevers as too uncertain to draw any conclusions from them, but the main thrust of his rebuttal was to point out that Mini's objections were not original: the experiment with the bladder attached to the trachea of the dog had been performed by Swammerdam; the realization that blood changes color when it is exposed to air—although published by Fracassati—was due to Malpighi himself, and so was the observation that fish breathed in air, something he had published in *Anatome plantarum*; in his *Vita*, however, he attributed the fishes' discomfort and death when the water surface is frozen to the abnormal inflation of their air bladder rather than to the lack of separation of an active principle from water.[37]

With regard to the liver, Mini attacked Malpighi explicitly and implicitly on several grounds. Mini still defended its role in sanguification and offered a number of observations on jaundice and the nature of bile, arguing that since bile is secreted in the ears too, the liver is not the only site of its production. He went on to ask about its mode of separation according to the Neoterics. Malpighi relied on Joseph-Guichard Duverney's work in claiming that earwax is secreted by glands in the ear, which he inferred were analogous to those in the liver. Malpighi questioned the identity of bile and earwax, pointing out that although both are yellow and bitter, their taste is different; once again, taste was a key tool for chymical assaying. Overall, he replied only briefly, referring for a more complete rebuttal of Mini's views on the liver to the work by Johannes Bohn.[38]

The most extensive and philosophically significant exchange concerned the role of the kidneys; it is here that Malpighi provided a refutation of the role of the faculties and an analysis of the operations of the body with regard to the soul. Mini argued that a sieve does not separate fluids and that, adopting the sieve analogy with wheat and oats, for example, the oats would block the holes of the sieve and therefore no separation would occur. Moreover, in intense fevers, the fierce ebullition of blood would prevent the separation of urine, as boiling water in a container does not exit from a small aperture until ebullition stops. Lastly, from the supposed structure of the kidneys, all fluid excrements should be separated in them, but experience shows that only urine is. Later he also brought an argument from the subtlety of the humors: urine being finer than bile, it should pass through the pores whence bile is secreted, since it passes more easily through a cloth. Thus, Mini concludes, separation in the glands occurs because of the regimen of the faculties rather than the conformation of the parts, as if the faculties had a discerning power going beyond mere structures: this was a crucial point in the dispute. Indeed, in the opening of *Medicus* Mini had argued that the lack of vital operations in a corpse did not descend from a defect of the organic instrument but in its vital powers. Analogously, understanding guitar playing involves not only the motion of the strings but also of the player acting, as it were, as the faculty does on our body.[39]

Malpighi sought to rebut Mini's points one by one, arguing, for example, that since fluids contain solid particles, they can be separated by a sieve. As to pathology, Malpighi questioned whether Mini's experiment was appropriate: he argued instead that the separation of urine is affected not so much by temperature as by the vigor of the heart's contractions, which make blood particles move so forcefully that they pass from artery to vein without encountering the pores of the vessels; when the motion of the heart becomes less vigorous, the particles of blood no longer move in a straight line along blood vessels but get in contact with their sides and urine is separated. Malpighi offered a typical mechanistic explanation to Mini's objection concerning the obstruction of the holes in the sieve: he argued that not only the shape but also the disposition or situation of the parts affect separation or filtration; he gave the example of sweating, which he related to the motion of the heart, similarly to what he had stated above for the kidneys. Malpighi further pointed out that bile turns bitter and sluggish in the gall bladder but is fine, sweet, and diaphanous when it is secreted from blood; in fact, in cases of jaundice bile is secreted by the kidneys, a clear sign not only of its subtlety but also of the fact that directive faculties are unnecessary.[40]

Malpighi returned to the problem of the necessity of matter and the body's organization in an important discussion on the last point raised by Mini, that the animal could not subsist without the faculties. Malpighi sought to answer Mini's objections with detailed accounts of how mechanistic operations work or may work in principle. In his *Vita* Malpighi reported Steno's account of the structure of the gland, with the nerves functioning like reins in compressing the veins and thus facilitating the secretion of humors from arterial blood, for example. Further, he provided a remarkable philosophical account of his mechanistic understanding of the body. He argued that the soul must necessarily act in conformity with the body's structure; therefore, the faculty in action is unhelpful as an explanatory device—it is a mere name. A passage from his contemporary *Risposta* to Sbaraglia echoes and clarifies his thought: Malpighi argued that a clock or a mill can be moved by a pendulum with a lead or stone as bob, an animal, a man, or even an angel, but if it is broken one has to repair its gears regardless of who or what is moving them.[41] In the important passage from his *Vita* Malpighi states: "The industry of the supreme artificer fashioned the animal and molded it with very small particles, some of which remain firm in mutual adaptation, the others perpetually flow to perform their functions; therefore at the time determined by Nature, the whole remains firm, decreases, and dissolves, so that we gradually die."[42] The editors of the *Bibliotheca anatomica* added an extended footnote at this point, arguing that they had thought the same for a long while, on the example of a ship going around the globe in which wood planks are replaced bit by bit so that on her return her shape would be unchanged but her constituents would not; this statement appears as a reformulation of a classic problem of identity reported by Plutarch in his life of Theseus, known as the ship of Theseus, a topic discussed by ancient philosophers in relation to growth. Later, however, the editors state that they came to accept that a large portion of the body remains the same, as testified by the bones, which last for centuries, and also other softer parts such as muscles, although not fat. A justification for this is that young people eat more because they grow, while older people eat less; however, they would have to eat more than they do if a large portion of their body were replaced.[43]

Malpighi's passage continues:

> *All these things occur because of the necessity of matter and motion.* This structure and order of motion is maintained in the living by the principle of life, whose cause—although it is unknown to us—probably derives in animals from the perennial motion of blood and from the propagation of the nervous fluid, with the help of air and other factors external to the body, whereby from this gentle and natural motion the particles seeking to fly away are constrained and are replaced by others; *and this is the source of the faculties, or at least the instrument that the soul employs in us.* Otherwise if the faculty flowing from the soul had (as it is maintained) the power to overcome the activity of the spirits and particles seeking to fly away, its force and energy would not be overcome in a given time, but would remain the same and therefore animals would last forever. But if in the faculty a reaction is admitted, it will be necessary to admit in it a gentle activity to guide and constrain motion, which I cannot conceive in something incorporeal, since in us the soul is the source of mortality. Wherefore in the same way as the cause of mortality comes from the

> joining together, the motion, and figure of the parts, which cannot be maintained such perpetually by the motion of blood and of the spirits, so the conservation of life in brutes comes from the same factors; in us, however, it comes from a rational soul, which has its own appropriate operations independently from the body, and the remaining ones bound to it.

Malpighi is quite cautious about the cause of the life principle, arguing that it is unknown to us, but he is adamant in claiming that everything happens following exclusively physico-mechanical rules, following the necessity of matter and motion; hence, the role of the faculties is reduced to a mere shadow. We notice here the same expression we have encountered in *De omento*, *Anatome plantarum*, and especially *De polypo cordis*, in which Malpighi had tentatively denied the existence of a goal-directed mover acting to the animal's benefit. In his *Vita* he mentioned the expression *necessitas materiae* again while discussing generation and comparing the formation of eggs in plants and stones. Further down in the quotation, Malpighi draws a distinction between animals and humans. The reason why the faculties flowing from the soul play no role in animals is that if their activities were governed by the faculties, the animals would not be mortal. In addition, Malpighi finds it inconceivable to attribute to an incorporeal faculty the ability to affect motion: life and death follow from the same material principles. In humans, however, the conservation of life comes from the rational soul, although the pontifical archiater does not elaborate on this vexing philosophical and theological problem.[44]

Lastly, Mini challenged Malpighi's painstaking investigation of the spleen. In *De viscerum structura* and in later works Malpighi had identified glandular structures that he believed to be crucial to understanding the spleen's operations; however, this remained a mysterious organ with glands without an excretory duct and that did not seem to filter anything. Malpighi surmised that the glands in the spleen separate a portion of blood and prepare it for the separation of bile in the liver. Mini objected that the existence of glands does not necessarily imply separation and that nature can effect this process without them. Moreover, he questioned the purpose of separating some fluid that is then mixed again with blood. Malpighi's line of defense was to rely on induction in claiming that the purpose of glands is always to separate humors, and that nature offers other examples of fluids that are first separated and then mixed with blood again, such as lymph and saliva. He then compared Mini's purely confrontational style with Bohn's constructive criticisms, taking the lead from Bohn's words to put forward his last thoughts on the spleen. Bohn had noticed that splenetic blood is intermediate between arterial and venous; recalling his earlier experiments on blood whereby its properties are altered by the addition of salts and other substances, Malpighi argued that blood in the spleen is slowed down and is infused with a humor that facilitates the separation of bile. Although he could not detect excretory vessels, he argued that their existence could be evinced by reason.[45] Here we see Mini raising a fundamental objection about the body's operations and Malpighi relying on reason when the senses failed him.

Although Mini was a less than formidable opponent and his book did not have a broad impact, some of his views remained a thorn in the side of mechanistic explanations of the body. Other problems were to come from more savvy and sophisticated opponents.

10.5 Ruysch's Challenge and Boerhaave

Readers of the *Bibliotheca anatomica* will find a number of extensive and exceedingly interesting notes by a number of authors on many subjects of Malpighi's *Vita*. Those drawn from, or directly provided by, the Amsterdam lecturer in anatomy and surgery Ruysch are especially interesting in that they capture a crucial state of transition in his own techniques and views and in the anatomy of the period; it is to those notes and Ruysch's changing attitudes that we now turn.

Referring to *Observationum anatomico-chirurgicarum centuria* (Amsterdam, 1691), the editors reported Ruysch's confirmation of Malpighi's views on the glandular struc-

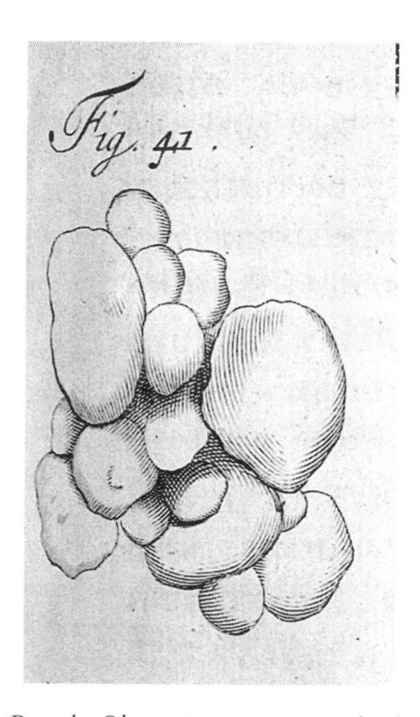

Figure 10.3. Ruysch, *Observationum centuria*: glands in the liver

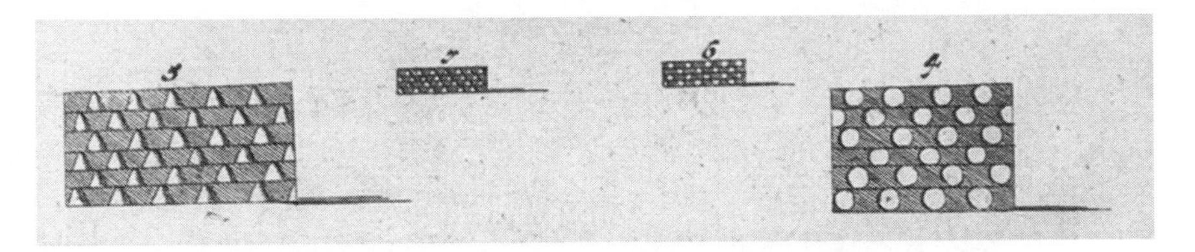

Figure 10.4. Ruysch, *Epistola problematica prima*: microstructure of the skin

ture of the liver: at that stage Ruysch accepted as unproblematic both Malpighi's results and methods. In fact, his *Observationum centuria* reports several cases of a hardened liver in which the glands have become visible. In the case reported in the *Bibliotheca anatomica* Ruysch claimed that some glands were the size of a pinhead, others of a lentil. In an earlier observation on the body of an old woman who had died of dropsy, Ruysch found the liver hardened and its surface uneven because the glands had hardened and had become enlarged, as shown in the accompanying plate (fig. 10.3). The editors of the *Bibliotheca* confirmed the same opinion from their own observation of a woman of sixty-five whose liver consisted of white tubercules. Thus, in this case both Ruysch and the editors relied on disease to confirm the existence of glands.[46]

The editors reported from the *Epistola problematica prima* (Amsterdam, 1696) Ruysch's clarifications on the preparation to make the microstructure of the skin visible. Such *Epistolae* consisted in a letter addressed to Ruysch by a student or colleague raising anatomical questions, accompanied by Ruysch's reply. Ruysch found this format congenial and published sixteen such *Epistolae*, in which he put forward what he took to be his most significant findings. In this instance he clarified that the netlike structure and pyramidal papillae cannot be seen without maceration in spirit of wine, a remark revealing his attention to preparation techniques. In the accompanying plate (see fig. 10.4), figures 4 and 5 show the netlike body and pyramidal papillae as seen with the microscope under great magnification, while figures 6 and 7 show the same structures only doubly enlarged, thus giving a sense of their diminutive size.[47]

Malpighi fared less well with the spleen; the editors added long excerpts from Ruysch's *Epistola problematica quarta* (Amsterdam, 1696), on the conflict between Bidloo's atlas, showing glands in the human spleen (fig. 10.5, at E), and Ruysch's lectures and preparations, showing that there were no such glands. Bidloo was a former student of Ruysch's who, after publishing a huge and lavishly illustrated *Anatomia humani corporis* in 1685, became professor of anatomy and medicine at Leiden in 1694. Ruysch argued that there are indeed no glands in the spleen and the transverse fibers are to be found in calves, not in humans. In the spleen Ruysch found arteries, veins, lymphatic vessels, and nerves; he noticed that some extremities of the blood

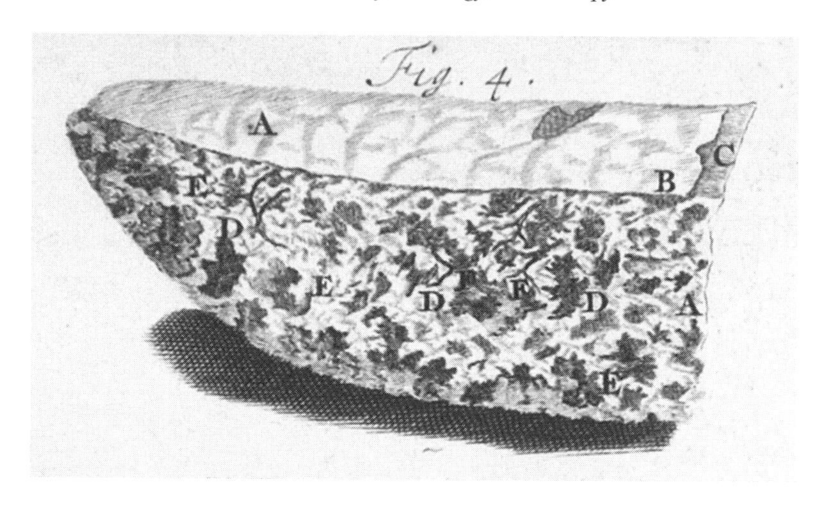

Figure 10.5. Bidloo, *Anatomia*: human spleen, XXXVI.4

vessels were enlarged, although these were not glands. In his plate (fig. 10.6) he compared a human spleen in figures 1, 2, and 4 with a spleen of a calf in figure 3, dissected so as to show the transverse fibers: both the plate and the text are based on Ruysch's novel injection technique, "nostro novo artificio." In a rejoinder, Bidloo endorsed Malpighi's views again, on both a priori and observational grounds, leaving the editors of the *Bibliotheca* unable to adjudicate the matter. Up to that point Malpighi's views constituted anatomical orthodoxy. Bidloo defended himself in his *Vindiciae*—in response to Ruysch's first ten *Epistolae problematicae*—in which he claimed that in the 1691 *Observationes* Ruysch himself had stated that the spleen is glandular, in agreement with Malpighi. Ruysch's swift reply accused Bidloo of having shown a spleen from a calf rather than a human, and he frankly admitted that he had changed his own views on the spleen. Ruysch explained the origin of the misleading idea of glands in the spleen, whose existence he repeatedly denied: he argued that the extremities of the arteries degenerate into a fluid substance that collects in small lumps, giving the false impression of a gland.[48]

As a result of his new injection techniques, Ruysch started to question Malpighi's results and his techniques of investigation. Ink staining had revealed to Malpighi glands in the cerebral cortex; monstrous states or petrified specimens reported in the literature—by Ruysch among others—or preserved in the Aldrovandi Museum had confirmed the glandular structure of the liver, cerebral cortex, and spleen. Malpighi had sought to confirm the structure he had found in the brain by means of petrified remains and monstrous states observed in postmortems and reported in the literature by Johann Kentmann in 1565 and Johann Jakob Wepfer in 1681, adopting a technique I have named "mining for stones." Moreover, diseased states reported in the

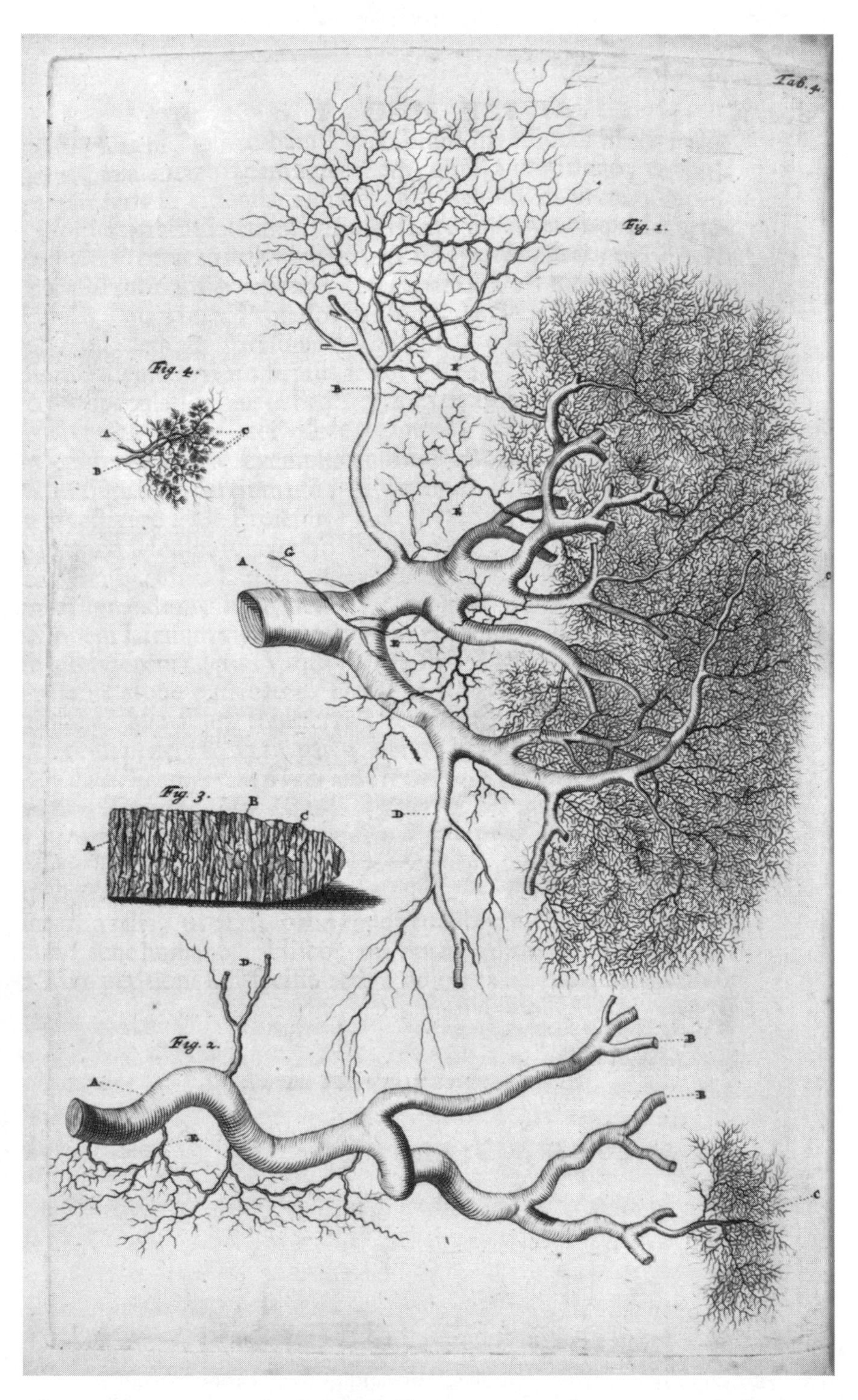

Figure 10.6. Ruysch, *Epistola problematica quarta*: spleen in man and calf

many *centuriae* of notable cases published throughout Europe or observed in animal dissections or postmortems had enlarged glands and made them visible throughout the body; finally the microscope of nature had helped uncover glandular structures in many organs by moving across different species. Against this robust background, a single technique of investigation led to a major challenge to Malpighi's views on glands in general and especially the cerebral cortex, whose softness and rapid decay had made it exceedingly difficult to investigate. Ruysch argued that monstrous and petrified states had to be treated with greater care; his injections managed at the same time to preserve the specimens and to show that no glands were to be found: injections reduced globular structures to a network of vessels, and the glands found by Malpighi thus appeared to be an artifact of his preparation techniques. Injections were already known to several anatomists, including Malpighi, who in 1661 had used mercury to highlight the vessels in the lungs. Dutch anatomists, however, developed a range of new techniques, from devising new mixtures including oil of turpentine or terebinth, to preserve cadavers in a surprisingly fresh state, to new injection devices, as we have seen in de Graaf's study of the male and female genital organs or Swammerdam's study of insects (figs. 7.9, 8.2, 8.7). No other method besides injections could reveal the microstructure of organs, according to Ruysch.[49]

In the crucial response to the *Epistola problematica duodecima* on *De cerebri corticali substantia*, by Michael Ernst Ettmüller, soon to become professor of medicine in Leipzig, Ruysch announced the vascular structure of the cerebral cortex. Ruysch had found a method to inject the arteries in the cerebral cortex in 1698, the year before the *Epistola* was published: although his name is best remembered for his discovery of the valves in the lymphatic vessels and his museum based on injection preparations, he later stated that he considered the successful injection of the cerebral cortex and the disproof of its glandular structure as the major finding in his forty-year career as an anatomist.[50] Govert Bidloo's 1685 engraving of the glands of the cerebral cortex (fig. 10.7) and Ruysch's 1699 engraving of its vascular structure (fig. 10.8) highlight striking differences: the former shows a lump of globular structures with excretory vessels at the bottom resembling an ivory comb, following Malpighi's description of the fibers of the brain (see sec. 3.2); the latter shows an intricate web of blood vessels kept in a fluid and observed without magnification, with no glands. Ruysch had Bidloo's atlas in mind when he claimed that seeing in anatomical books the cerebral cortex depicted as consisting of globular follicles made men nauseous. He often emphasized that not only was he making claims about what he had observed, but he was also able to preserve and display his preparations in his museum as permanent witnesses to his views—in larger number and more dramatic fashion, one may add, than petrified or bony parts kept in the Aldrovandi Museum as permanent witnesses to Malpighi's views.[51]

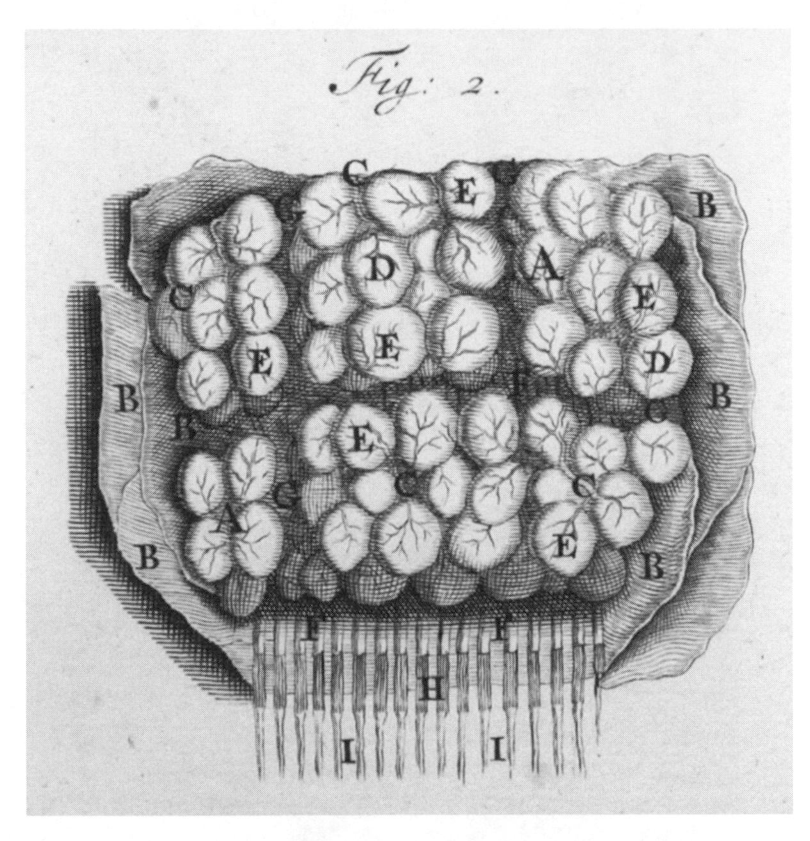

Figure 10.7. Bidloo, *Anatomia*: glands in the cerebral cortex, X.2

The ten-volume series of Ruysch's *Thesaurus anatomicus*, published in Amsterdam between 1701 and 1716 in both Latin and Dutch, was conceived as a guide to his museum and helps document the development of his views over the years. The "public" nature of the volumes can be judged by the address to the reader in the second volume, in which Ruysch explains that he had been forbidden to translate into Dutch some portions—dealing with the organs of reproduction—lest younger or female readers be offended.[52] Thus, display of the anatomical preparations took the place of replication: the secrecy surrounding Ruysch's injection techniques and the monetary value attached to his specimens—Ruysch sold his collection to Peter the Great for thirty thousand guilders—meant that those who wished to see for themselves could visit the museum for a fee or look at the plates in his books, although not perform the injections.[53]

The impact of Malpighi's views on glands and of Ruysch's challenge to them can be measured by the exchange between Boerhaave and Ruysch published in 1722: their letters, printed together in *Opusculum anatomicum de fabrica glandularum in corpore*

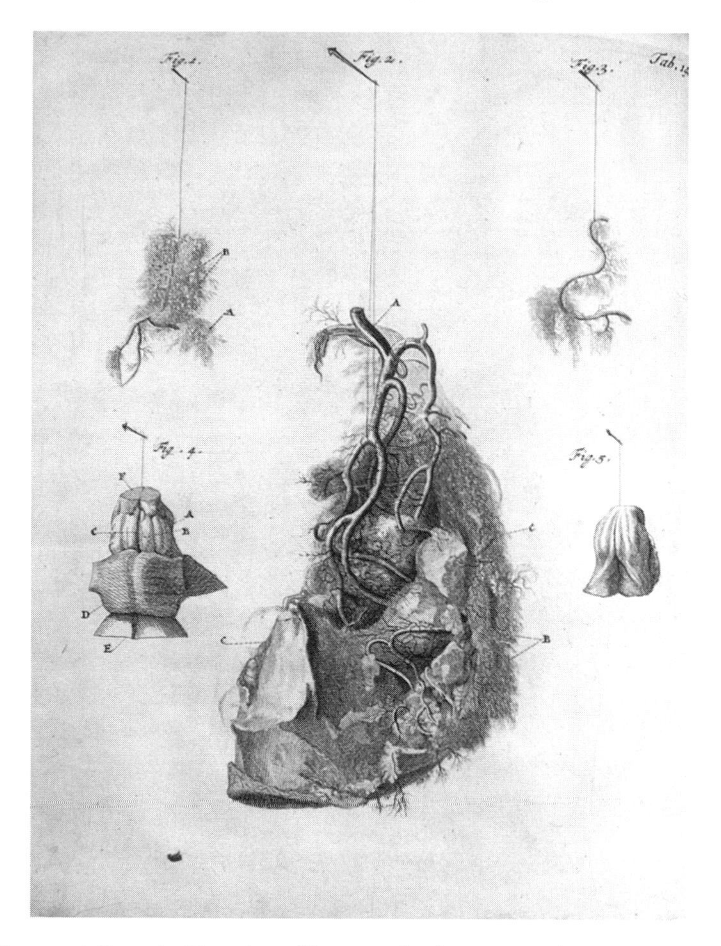

Figure 10.8. Ruysch, *Epistola problematica duodecima*: vascular cerebral cortex

humano, hinge on Malpighi's work, especially *De viscerum structura* and *De structura glandularum*. While Ruysch was largely opposed to Malpighi's views, Boerhaave sought a compromise; for example, Boerhaave tried to identify the "*lanugo fistulosa*" Malpighi had seen surrounding the follicle with a network of arteries, in line with Ruysch's vascular anatomy, an identification Ruysch rejected. There was no acrimony in their exchange, however: the aging authors were friends, and Boerhaave often spent several weeks of his vacations at Ruysch's, where they reportedly performed observations together.[54] Their exchange is an extraordinarily rich document on the state of the field: their differences were tied to new methods of investigation that had been developing since the seventeenth century, notably microscopy and new injection techniques, and their debate recapitulates some of the methodological themes we have encountered so far. Their figures provide a useful way to reconsider their exchange: Boerhaave's plate (fig. 10.9) shows a composite gland, with a, a, a representing the simplest follicles, b, b, b their excretory vessels, and c the common excretory vessel. It would be more appropriate to use the word "diagram" than "figure" because Boerhaave had not observed such a structure: his was a conceptual representation rather than a real one. As Ruysch put it, the figure was only an abstraction that he had been unable to see: he would concede defeat to anyone who would show him such a structure in a dissection. His own plate (fig. 10.10) showed a portion of the human mesentery prepared with his own injection method and observed through the microscope. Ruysch claimed that glands and glandular *acini* were nothing else but a network of vessels; in fact, he argued paradoxically that his figure was also an abstraction because the vessels were too minute to be drawn unless the entire plate became a black spot. With all his art, however, Ruysch—much like Malpighi—had to admit defeat about the actual process of secretion, arguing that his vessels acted because of a certain inexplicable force introduced by God at the moment of creation.[55]

In much the same way that the microscope helped to uncover previously unknown and invisible structures, the new injection techniques also made visible body parts that had previously escaped the human eye because of their transparency, size, or ephemeral nature. Much as the microscope was liable to create illusions and aberrations due to a variety of factors, however, injections were also liable to create distortions due to excessive pressure, and even extravasations of the injected fluid. As Boerhaave put it, echoing the judgment by Jean Jacques Manget's imposing *Theatrum anatomicum*, the pressure of injected wax could distend some membranes and make them invisible or in some cases even exit from the smallest arteries, confounding matters.[56] Other methodological aspects also came under scrutiny: was it legitimate to draw conclusions about the structure of the liver of higher animals from the structure of the liver

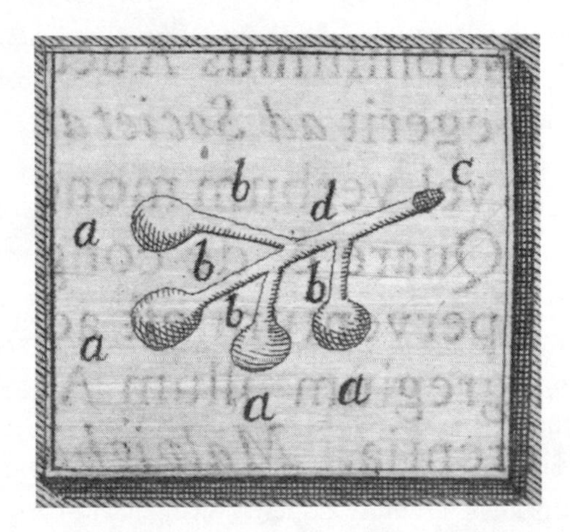

Figure 10.9. Boerhaave, *Opusculum*: diagram of an ideal gland

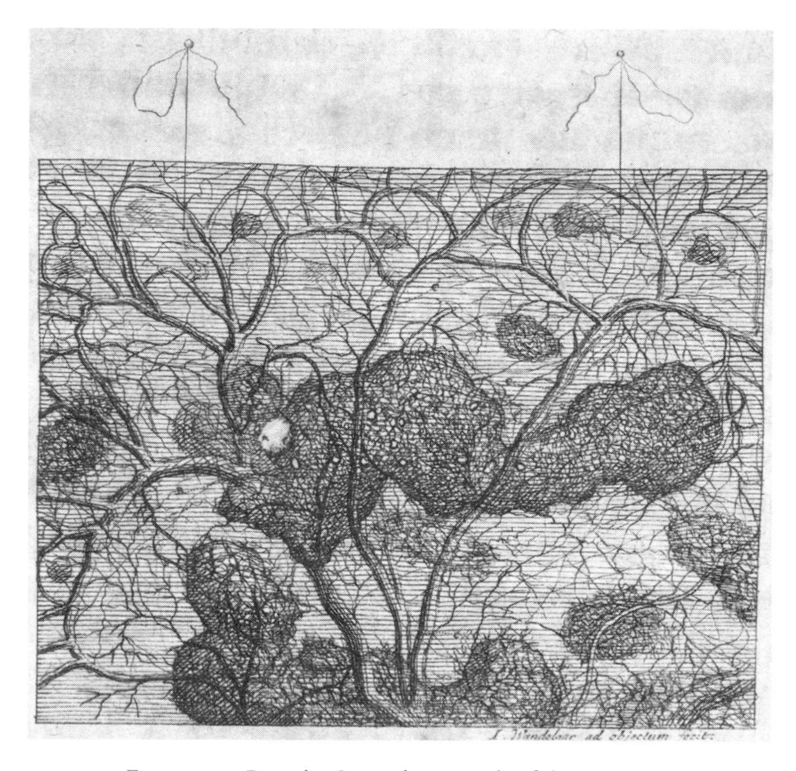

Figure 10.10. Ruysch, *Opusculum*: vessels of the mesentery

of lower animals, including snails, lizards, crickets, and silkworms? Further, as we have seen above for the cerebral cortex, Ruysch argued that it was not legitimate to infer the structure of glands from pathology: besides questioning the microscope of nature, he questioned the microscope of disease as well, since obstructions to blood vessels could create other abnormal or preternatural states, but these are no indication of the existence of glands, least of all do they illuminate their structure. This left him with only one tool, preventing the triangulation among different methods that would have been available using a wider range of techniques.[57] In some cases Ruysch flatly denied the existence of glands: besides the cerebral cortex, he denied their existence in the peritoneum, pericardium, and other membranes. Against his own earlier views, he challenged Malpighi's claim that the liver was a gland, but he was more defensive on the kidneys, on whose structure his opinion had also changed over time; although he could not deny altogether the existence of follicles or vesicles, he questioned whether they were truly glands. In the *Thesaurus anatomicus tertius* he drew a distinction between human and animal kidneys, where glandular structures are more pronounced. He also included a plate of a human kidney showing in the cortical area minuscule black dots that he argued Malpighi had misidentified as glands. I was able to detect them in the original plate only under magnification. In the *Thesaurus anatomicus decimus* Ruysch stated that the so-called glands are only a ball-shaped collection of arteries.[58]

At the end of his response to Boerhaave, he summarized his views in four points: (1) whereas Malpighi thought that glands produce the appropriate juice, Ruysch argued that the juice is made in the smallest arteries and only deposited in the follicle; (2) rather than using the term "gland," he argued that "*crypta*" would seem more appropriate; (3) preternatural tumors cannot be taken as revealing the existence and structure of glands; and (4) whereas Malpighi took cutaneous tubercles to be glands, he argued that they were nervous papillae for the sense of touch.[59]

The peculiar circumstances of composition and style of Malpighi's posthumous *Vita* enabled him to adopt a different style than in his previous works. Whereas in the past he had sought to be polite and modest, often omitting the names of his antagonists, in his *Vita* he adopted a much more explicit and aggressive style, mentioning works and page numbers while responding point by point to the objections and criticisms he had received throughout his life.

Three major challenges to Malpighi's views came from Borelli, Mini, and Ruysch. Borelli questioned the notion of fermentation and its role in mechanistic explanations of a number of key processes in animals and plants; both Borelli and Malpighi, however, shared a belief in mechanistic anatomy and wished to dispense with the faculties of the soul. By contrast, the controversy with Mini focused on whether mechanistic explanations were at all adequate to explain the operations of the body, such as secretion. Generally, Malpighi relied on observation and experiment, but his response to Mini also spelled out his philosophical views about the soul, life, and mortality.

While Malpighi's *Opera posthuma* was being readied for publication, Ruysch's new injection techniques were starting to challenge the glandular structure Malpighi had attributed to a host of organs. Malpighi's *Vita* in the 1699 Geneva edition of the *Bibliotheca anatomica* appeared with a number of footnotes questioning his views. While this work was being printed, Ruysch proceeded to challenge Malpighi's positions even further in his *Epistolae problematicae*. Following de Graaf's and Swammerdam's early death, Ruysch was left as the chief expert on sophisticated injections. The contradictory results attained by microscopy under high magnification with elaborate preparations and by highly interventionist injections proved especially troubling: the former had led to a glandular structure of the body, the latter to a vascular one. The controversy between Bidloo and Ruysch captures the far-reaching implications of the techniques Ruysch was developing at the time and involved a striking visual dimension as well. It is especially significant that in the course of the 1690s Ruysch changed his mind on the glandular nature of the body: whereas at the beginning of the decade he followed the canon established by Malpighi, by the end he had reached opposite conclusions on a number of points.

Boerhaave's debate with Ruysch beautifully documents the anatomists' growing awareness that their techniques of investigation were not neutral tools of inquiry but a problematic part of their research. These types of reflections, however, were hampered by the secrecy with which research was being carried out. Ruysch did not teach or divulge his techniques, which survived largely through his preparations rather than know-how or replication, thus making it difficult for anatomists accurately to reproduce them and assess their merits and problems. Moreover, Ruysch charged an entry fee to his museum and sold his collection, whereas the monetary aspect was absent in Malpighi. The dichotomies replication/display and openness/secrecy are interwoven with the development of Ruysch's *anatomia subtilis*. The exchange between Boerhaave and Ruysch was also revealing of striking additional limitations, highlighted by Ruysch's inability to explain the key operations of filtration and secretion in his vessels. Thus, despite their increasing sophistication, investigations of fine microstructures had reached a startling double impasse, with regard to both the actual structure of many organs and the mechanical explanation of their operations. Yet more problems were forthcoming from the perspective of practical medicine, as we will see in the following chapter.

From the New Anatomy to Pathology and Therapy

11.1 A Bologna Controversy and Its Wider Implications

The increasingly sophisticated investigations carried out by anatomists in the second half of the seventeenth century left many physicians skeptical. Research on lower animals such as insects, or even plants, had no medical implications in sight. Moreover, as London physician Gideon Harvey pointed out in 1683, techniques of inquiry such as microscopy and the preparation methods associated with it had become so invasive that one could question their reliability. In the 1665 *Micrographia* Hooke had praised microscopy, enabling the investigator to study nature undisturbed, over vivisection, with its brutal interventionist methods; less than twenty years later, microscopy had become the emblematic interventionist technique:[1]

> The necessary point of Anatomy consists chiefly in the temperament, Figure, Situation, connexion, action, and use of the parts; and not in superfluous, incertain, and probably false, and indemonstrable niceties, practiced by those, that flea Dogs and Cats, dry, roast, bake, parboil, steep in Vinegar, Lime-water, or *aqua fortis*, Livers, Lungs, Kidneys, Calves brains, or any other entrails, and afterwards gaze on little particles of them through a microscope, and whatever false appearances are glanced into their eyes, these to obtrude to the World in Print, to no other end, than to beget a belief in people, that they who have so profoundly dived into the bottomless pores of the parts, must undeniably be skilled in curing their distempers; whereas those pretended Anatomical Physitians, who have so belabour'd and tortur'd the particular parts, are generally the least knowing in the whole body of Anatomy.

Even more radically, the very role of anatomy of higher animals and humans was questioned. Although these were Europe-wide concerns, Bologna offers an especially fruitful perspective from which to explore them through the acrimonious dispute among Malpighi, Sbaraglia, and their supporters between the end of the seventeenth and the beginning of the eighteenth centuries. Some of Sbaraglia's criticisms against invasive techniques of investigation echo the passage by Gideon Harvey just quoted.

A sense of the acrimony and pettiness of the dispute can be gained from a *Historia medica* dedicated by the Bolognese physician Giovanni Battista Giraldi to Sbaraglia, who is referred to as primary professor of medicine, a title allegedly not existent at Bologna. The publication, more like a large miniature book than a small pamphlet, consists of twenty pages in diciottesimo; the only copy I am aware of survives among Malpighi's manuscripts. Malpighi included a copy with the manuscript of his *Opera posthuma* purely to ridicule and embarrass its dedicatee by showing the rhetorical flourishes and glowing tone employed by Giraldi in addressing Sbaraglia; it was probably by accident that the Royal Society published it. Despite its local roots and occasional pettiness, however, the dispute gained broad European resonance for its profound medical significance: from Leipzig, for example, Johannes Bohn wrote a rebuttal of Sbaraglia's work that circulated in manuscript form until Sbaraglia reprinted it with a response. Moreover, the relevant texts from Malpighi's *Opera posthuma*—including Giraldi's tiny work!—were given pride of place in the opening section of the 1699 edition of the *Bibliotheca anatomica*, thus gaining the most visible place in the most authoritative anatomical publication of its time. Thus, the dispute both reflected and contributed to broader European developments.[2]

Sbaraglia's attack on Malpighi appeared in the wake of the onslaught by Antonio Felice Marsigli on Malpighi for his opposition to university reform and his lack of commitment to the practice of medicine, whose chair he held. Thus, besides the problems of the glandular or vascular microstructure of many organs and the unresolved puzzles about their mode of operation that we have seen in the previous chapter, Sbaraglia put his finger on another issue: the pathological and therapeutic relevance of the new anatomy. As a result, Malpighi was pressed to spell out the pathological and therapeutic implications of his own research and of those by other anatomists, providing contemporaries and later historians with an invaluable document on the *status quaestionis* by one of the leading scholars in the field. Sbaraglia'a attack resonates with the shifting concerns by the European medical community toward the study and classification of disease and the emphasis on therapy around the turn of the century in the wake of Thomas Sydenham, whom he mentioned in his 1704 response to Malpighi.

The controversy involved several printed texts and manuscripts; among the publications, some had a European circulation, whereas others were rare ephemeral pamphlets. Rather than brave this bibliographic maze, I shall focus on the main texts between 1689 and the early eighteenth century. The controversy has attracted considerable scholarly attention, especially on Malpighi's extensive and profound *Risposta*; as a result, our picture of the whole affair is colored by Malpighi's own perspective. In my study I have sought to study the texts chronologically as they appeared in print, resisting the temptation to focus exclusively on Malpighi's and ignoring Sbaraglia's. I examine Sbaraglia's opening salvo in the next section, Malpighi's extensive *Risposta* in section 11.3, Sbaraglia's even more extensive rebuttal in section 11.4, and young Giovanni Battista Morgagni's defense of Malpighi in the last section. As a student of Malpighi's pupils Valsalva and Albertini, Morgagni considered himself as his intellectual grandchild. It is especially helpful to study Morgagni's defense of Malpighi in relation to his own contemporary contributions to anatomy, such as the 1706 *Adversaria anatomica prima*.[3]

11.2 Sbaraglia's Challenge to Malpighi's Research

Much as in the dispute with Lipari, this case also involved a connection between intellectual affairs and administrative matters. The chancellor to the Bologna studium and archdeacon of the Bologna Cathedral Antonio Felice Marsigli was engaged in a reform process aimed at raising the profile of the institution. As we have seen in section 9.3, Marsigli had been initially sympathetic to Malpighi, addressing to him a printed letter against Buonanni on the eggs of snails; subsequently, however, he had shifted his allegiances, siding with Giovanni Battista Trionfetti on the issue of the metamorphosis of plants. Briefly put, Marsigli wished to decrease the number of lecturers and attract more prestigious ones by increasing their salaries, a needed move given that at the time any Bolognese citizen taking a degree from the university was entitled to a university post; this plan was opposed by Malpighi, who feared that under the new scheme the choice of professors would not be the best. Malpighi's opposition enraged Marsigli. On 5 January 1689 Malpighi received an ominous verbal warning via Domenico Guglielmini—a pupil and follower of Malpighi who at a later stage was to support him in his inaugural lecture as professor of the theory of medicine at Padua and in some anonymous pamphlets—that the archdeacon had become the protector of his enemies. Malpighi was warned that Marsigli was ready to meet the publication expenses of anything written against him; the archdeacon issued an extraordinary series of charges against Malpighi, such that he had opposed a salary increase for Sbaraglia, Mini, and Lelio Trionfetti so that they would stop working and the studium would decline with Malpighi's death, that he had never taught or wanted to have students, and that he was stealing his salary. This stinging attack was followed by Mini's challenges in January and April 1689 and by the appearance of Sbaraglia's pamphlet *De recentiorum medicorum studio*, which Malpighi dated to the spring of 1689. In the summer of the same year Marsigli objected on procedural grounds to the standard exemption from defending conclusions, thus preventing for a while Malpighi's student Albertini from obtaining his degree.[4]

Sbaraglia and Malpighi shared an enmity dating back at least to 1659, when Malpighi's brother Bartolommeo had killed Giovanni Girolamo Sbaraglia's elder brother Tommaso for reasons unrelated to academic affairs. As a condition for the ensuing pacification, both Marcello and Giovanni Girolamo had to swear friendship to each other. Sbaraglia was thirteen years Malpighi's junior, having been born in 1641; at the time of their dispute he was forty-eight, whereas Malpighi was sixty-one. As we have seen in section 6.3, Sbaraglia was in charge of several public anatomies and seized on those opportunities to attack the moderns and Malpighi with them. However, it was not until 1689 that the confrontation reached a climax with the appearance of *De recentiorum medicorum studio*. Given Marsigli's threat to finance all publications against Malpighi, it seems likely that he was putting his money where his mouth was, showing his determination to challenge and intimidate Malpighi.[5]

Sbargalia's work is a rather thin and disappointing first publication for a professor of medicine and a physician in his late forties. Despite the fact that the text is entirely focused on Malpighi, his name is never mentioned, in line with the customs of the time. The kernel of Sbaraglia's attack is the claim that the anatomy of small parts, the anatomy of plants, and comparative anatomy are useless or at best largely irrelevant to the art of healing and the practice of medicine. Given the academic context of the dispute, Sbaraglia's reference to the practice of medicine is of interest, in that Malpighi held precisely that chair; hence, Sbaraglia's attack implied that, despite his international recognition as an anatomist, Malpighi was neglecting that portion of medicine most linked to his academic role, in effect a dereliction of duty. As to confirm this interpretation, in his *Risposta* Malpighi pointed out that Sbaraglia's aim was to "take away from me the credit of practical physician."[6] It is not difficult to see in this attack a link with Marsigli's views. Moreover, most naturalists, whether they were interested in insects or plants, were physicians; therefore, Sbaraglia's attack had broader implications.

Sbaraglia underpinned his attack with a series of quotations from ancient sources, mainly Hippocrates, Galen, and the Roman physician Celsus, and some of the moderns, including Paracelsus. He argued that anatomy is useful for finding the connections among the parts and the purpose of those parts; modern anatomists, however, have become obsessed with the minute ending of a tiny vein, for example, and have been unable to point to new purposes of the parts, merely confirming what was already known about the main organs. He argued specifically concerning the pancreas, the spleen, the brain, the lungs, the liver, and their diseases. Sbaraglia questioned the role of comparative anatomy between animals and humans and also the study of lesions in cadavers. He challenged the uniformity of nature in the process of generation, arguing that some animals reproduce through eggs, others by generating the animal itself, and some through metamorphosis, which undermines the preexistence of the animal in the egg. We notice here a revealing connection with the views put forward by Buonanni in his reply to Marsigli (see sec. 9.3). Although Sbaraglia's pamphlet was mainly critical, it did contain a rare positive recommendation as to which research line to pursue: he seemed to advocate a comparison of the humors of healthy and sick people, notably their blood. As to remedies, Sbaraglia advocated experience as the only arbiter, in the form of the study of and comparison among diseases.[7] Of course, his was not an isolated voice on the European scene, and it resonated with other claims; in England, for example, Thomas Sydenham advocated a new medical empiricism through a focus on disease and its classification. Sbaraglia, however, did not quote Sydenham at this stage, and it is unclear whether he was aware of his work, as he would be in subsequent years.[8]

In an especially interesting passage on the diseases of the solid parts of the body, Sbaraglia outlined his therapeutic strategy and understanding of the notion of indication. He argued that in those cases the indication is drawn from the disease and from its cause, which most of the times is universal, not from the differences in the individual body parts; thus, ulcers are cured in the same way wherever they occur, regardless of the minute composition of the parts affected. In the case of the cornea, one needs especially pleasing remedies, but these depend on the purpose of the part rather than on its microstructure. Sbaraglia added that special remedies are found through experience and comparison rather than the examination of the smallest components.[9] We shall see below how Malpighi conceptualized the notion of indication and the ensuing therapeutic strategy.

11.3 Malpighi: The Medical Significance of the New Anatomy

Soon after Sbaraglia's onslaught, Malpighi's call to Rome as pontifical archiater blunted the insidious attack by removing him from a university position and placing him close to the center of power in the Papal States: Innocent XII, a former cardinal legate in Bologna and friend of Malpighi's, greeted him in person at the gates of the eternal city, where they embraced in tears.[10] It is possible that Malpighi's call to Rome may have been received with both relief and consternation by his enemies, since Malpighi had been removed from the university, although sadly and embarrassingly for Sbaraglia and Masigli, he had become not the pope's anatomist but his personal physician.

Malpighi claimed to have penned his learned *Risposta* to Lipari in barely two weeks. Although it is unclear how long it would have taken him to compose his related and extensive *Risposta* to Sbaraglia, his is certainly a polished and elaborate piece, an exegetical tour de force, a learned and passionate defense of rational medicine and the importance of anatomy to the art of healing through the search for structural causes of diseases. Malpighi's rebuttal is especially valuable to the historian for his profound reflection on the status of anatomy and medicine in his own time and for his extensive review of the field and of what he considered the main achievements of mechanistic anatomy, pathology, and therapy. Coming at the end of his career and of decades of investigations in which he had played such a major part, it is entirely legitimate to view his *Risposta* as marking a major stage in the anatomical and medical thinking of its time.

Much as he had done in the *Risposta* to Lipari, here too Malpighi compared himself to Galileo: the opening, the structure, and to some extent the contents of his work are modeled on the *Assayer*, following Borelli's advice to learn from Galileo's text in controversies. The opening is especially striking in that Malpighi, much like Galileo, provided a litany of vexations and injustices to which he had been subjected during his career. By adopting such an openly Galilean style, Malpighi was once again presenting himself as the Galileo of medicine, the investigator who had opened new horizons to anatomy by means of the microscope, much as Galileo had opened new horizons to astronomy with the telescope. In addition to using Galileo's *Assayer* as a template for his own reply to Sbaraglia, Malpighi relied on some of Galileo's most distinctive doctrines expressed in that work and applied them to medical matters. Sbaraglia had quoted *De decretis Hippocratis et Platonis*, in which Galen, comparing Plato and Hippocrates, had stated that whether the parts of fire have a pyramidal figure or those of the earth a cubical one is of no use to the art of healing. The text alluded to was *Timaeus*, one that was well known to Malpighi and Galileo, in which the four elements were given geometric shapes appropriate to their properties. Quite remarkably for a physician/anatomist involved in a medical dispute, Malpighi sided with Plato against both Hippocrates and Galen in claiming that the elements do not act because of their being hot, cold, moist, or dry, but because of their motion and shape. This remarkable passage continues:[11]

> Therefore, that heating in the fire is an effect of its figure and motion, and is, among the many effects fires produces, perhaps the least significant. Rather it has no existence but in the sensory organs of touch of the animals, being outside the passions of the animal, motion and insinuation. And since in our body fire, air, water and earth operate not with their sole qualities (which are also controversial), but with their figure and motion, Plato, by assigning those properties, does not bring a superfluous notion to the physician, but rather Hippocrates is in defect in neglecting them.

It is not difficult to identify in this passage an application to medicine of Galileo's so-called doctrine of primary and secondary qualities of the *Assayer*, one derived from Plato as well as from the atomist tradition. Much like Galileo, Malpighi also distinguished between properties such as figure and motion, existing independently of our senses, and qualities that reside only in our senses, such as hot and dry. Air, for example, does not act because it is moist and hot, but rather because it consists of spiral particles whose motion and compression cause healthy and diseased states. Here Malpighi was expanding on ancient doctrines on the basis of recent developments: in line with Borelli's *Delle cagioni* and his own *Risposta* to Lipari, Malpighi's approach subverted traditional medicine and therapeutics by rejecting the doctrine of the four qualities. Moreover, it is remarkable that Malpighi, while pontifical archiater, defended atomism at a time of a religious backlash against such doctrines. It is plausible that this was an additional reason why his *Risposta* appeared posthumously.[12]

Malpighi argued forcefully that the moderns had reformed medical practice as a result of their anatomical investigations. Of course, he often had his own microscopic investigations in mind, although he did not believe that observing was a simple matter even with microscopes: witness the failure in this field by the mathematics professor Geminiano Montanari, who also made microscopes, and a group of most learned scholars—a possible reference to the Congresso Medico Romano—to attain any results. He explained carefully and in detail how a host of therapies of many organs, such as the spleen, lungs, liver, kidneys, eyes, and heart, had been affected. Our sense of these changes of seventeenth-century pathology and therapeutics is quite vague; therefore, Malpighi's account is of extraordinary interest for the light it sheds on this area and is thus worth examining in some detail in at least some instances. For example, the transversal fibers of the spleen—the same whose existence was challenged by Ruysch—constrain it and force blood from a larger area in its chambers to a narrower one in the splenic vein, so it does not coagulate; Malpighi seems to rely on Benedetto Castelli's mathematical theorem on flowing water here. From this structure revealed by the microscope it emerges that the appropriate remedies in the diseases of the spleen are those that give the right tension to those fibers rather than remedies making them too soft. A similar reasoning would apply to conglobate glands, whose structure resembles that of the spleen.[13]

In response to Sbaraglia's sweeping attacks, Malpighi attempted to offer detailed analyses of the achievements of mechanistic and microscopic anatomy and medicine. For example, at one point Sbaraglia had argued that with all their findings, the moderns had not changed the understanding of the operations of the body, in that the kidneys separate urine and the liver bile. Malpighi, by contrast, highlighted that whereas according to the ancients the kidneys attract urine because of sympathetic attraction and the liver makes blood because of a natural faculty, according to the moderns the kidneys and the liver are aggregates of glands separating urine and bile, respectively, because they have apertures similar to the particles they are filtering. Bile is no longer an excrement but a ferment useful to digestion. From here Malpighi proceeded to explain that the rejection of sanguification in the liver altered established therapy. His account of the disease of the kidneys known as spurious diabetes, leading to excessive urination, is one of my favorites for the links it establishes between the new corpuscular philosophy and therapeutics. According to the ancients, this disease was due to the excessive heat of the kidneys, whereby they attract an immoderate amount of serum. Others also explained the disease by the existence of an acrid humor that impregnates the kidneys' flesh and enhances their expulsive faculty, a notion that mechanists like Malpighi wished to abolish altogether. The standard therapy in these cases was to use astringents and to cool the kidneys, abstaining from providing fluids. Relying on reasoning inspired by new anatomical knowledge to which he had given major contributions, Malpighi argued that empty spaces are intermixed with those water particles, and often salts are to be found in those spaces. Those salts that are reduced to minute parts by the fermentation occurring in the kidneys can form small wedge-shaped particles that force with some violence the apertures of the glands in the kidneys, leading to excessive urination. Hence, the therapy derived a priori from knowledge of the causes of the disease is to administer fluid and watery remedies that can dilute those salty wedges so that they can be expelled without damaging the kidneys. What is especially interesting here is the explanation of disease in mechanistic terms, the key notions being no longer those of excessive heat and attraction but rather filtration and the shapes and motions of particles. Malpighi also defended the microscope by using it directly to observe a layer appearing on the surface of urine. Traditionally that layer was thought to consist of fat, leading to a therapeutic strategy, but the microscope showed it to be an aggregate of sandy particles, which changed the therapy: this is a rare example of direct application of the microscope to therapeutics. Malpighi referred to the application of the microscope to therapeutics in another instance, although in this case his idea was presented merely as a project: identifying the figures of different salts by placing them over a glass and observing them with a microscope would enable the physician to identify those salts in the blood of healthy patients. By comparing that blood to that of patients affected by pleurisy, dropsy, cachexy, and other diseases, the physician would discover which salt is in excess and which is lacking, thus devising a therapeutic strategy based on microscopy. Since examining the inner structure of glands was beyond Malpighi's ability in healthy and diseased states alike, blood was the next likely candidate, and we know from *De polypo cordis* that he examined it; possibly he tried to investigate the blood of diseased patients without finding anything of notice.[14]

In conclusion, therapy can be best tackled by adopting a layered analysis starting from new anatomical findings and then moving to pathology or the understanding of the causes and locations of diseases, the rationale for the cure, the pharmacopœia, and the circumstances of application of specific procedures and *medicamenta*. Malpighi strongly believed that the new anatomical findings had led to a new understanding of disease; therefore, the rationales for the therapies provided by mechanistic anatomists were very different from traditional ones. Although mechanistic anatomy did not alter the available pharmacopœia, at times the circumstances of application of traditional procedures and *medicamenta* did change, however: standard remedies and procedures such as bloodletting were prescribed—or avoided—in circumstances different from traditional ones.

In a justly celebrated passage Malpighi argued that since nature operates by an ever-constant necessity, we can grasp her "artifici," or, we could say, mechanisms. He then listed a number of mechanical devices such as threads, beams, levers, cloths, fluids, cisterns, canals, and filters, forming the machines of our body. Hence, nature's uniformity provides the philosophical underpinning for this approach. Following this general list, Malpighi proceeded to provide concrete examples of "models," to use his own word, starting from the camera obscura for the eye, canals—or bladders, one may add—filled with fluids representing arteries, the articulations of the bones with threads attached to them, a machine simulating the thorax that expands and contracts filling and emptying with air, and the *statua humana circulatoria* by the archiater to the duke of Württemberg Salomon Reisel, displaying blood circulation, the chymical and mechanical processes associated with digestion, and the filtration of blood in the kidneys.[15]

As it appears from his 1693 essay in *Miscellanea curiosa*, Reisel did not actually build the *Statua*, although he discussed the most appropriate materials for its construction, such as the skeleton of wood or iron, the blood vessels near the heart of elder tree, and, as Luigi Belloni has emphasized, the pineal gland of conical glass or crystal held by a silk thread inside a spherical glass globe (see 6 in fig. 11.1 and fig. 11.2). Reisel's remarkable plates show the bladder (9), sphincter (10), penis (11), and larynx (14).[16] Reisel's primary aim was to convince the skeptics of the truth of the circulation of the blood and included images of blood squirting out of openings in the foot and hand.

It would be erroneous to take Malpighi's examples as curiosities or instantiations of anatomical structures. I do not think it has been sufficiently appreciated that—contrary to Reisel—Malpighi framed his examples as part of a general reflection on pathology and attached a specifically pathological significance to nearly all of them, thus extending the notion of nature's uniformity not only from machines to organs but also from healthy to diseased states. Moreover, Malpighi relied for the most part on devices that had been actually built in order to investigate disease. For example, in a medical consultation of 1687 for a case of gout, Malpighi sought to explain what happens inside the body, the cause of the disease being an excess of acids in the chyle. He also stated, "All this can be seen in proportion also mechanically mixing spirit of vitriol, or another acid that is especially austere, with different fluids." Here Malpighi used the term "mechanically" rather broadly, in conjunction with a chymical operation reproducing in vitro processes occurring inside the body in order to investigate disease and to devise suitable therapies.[17]

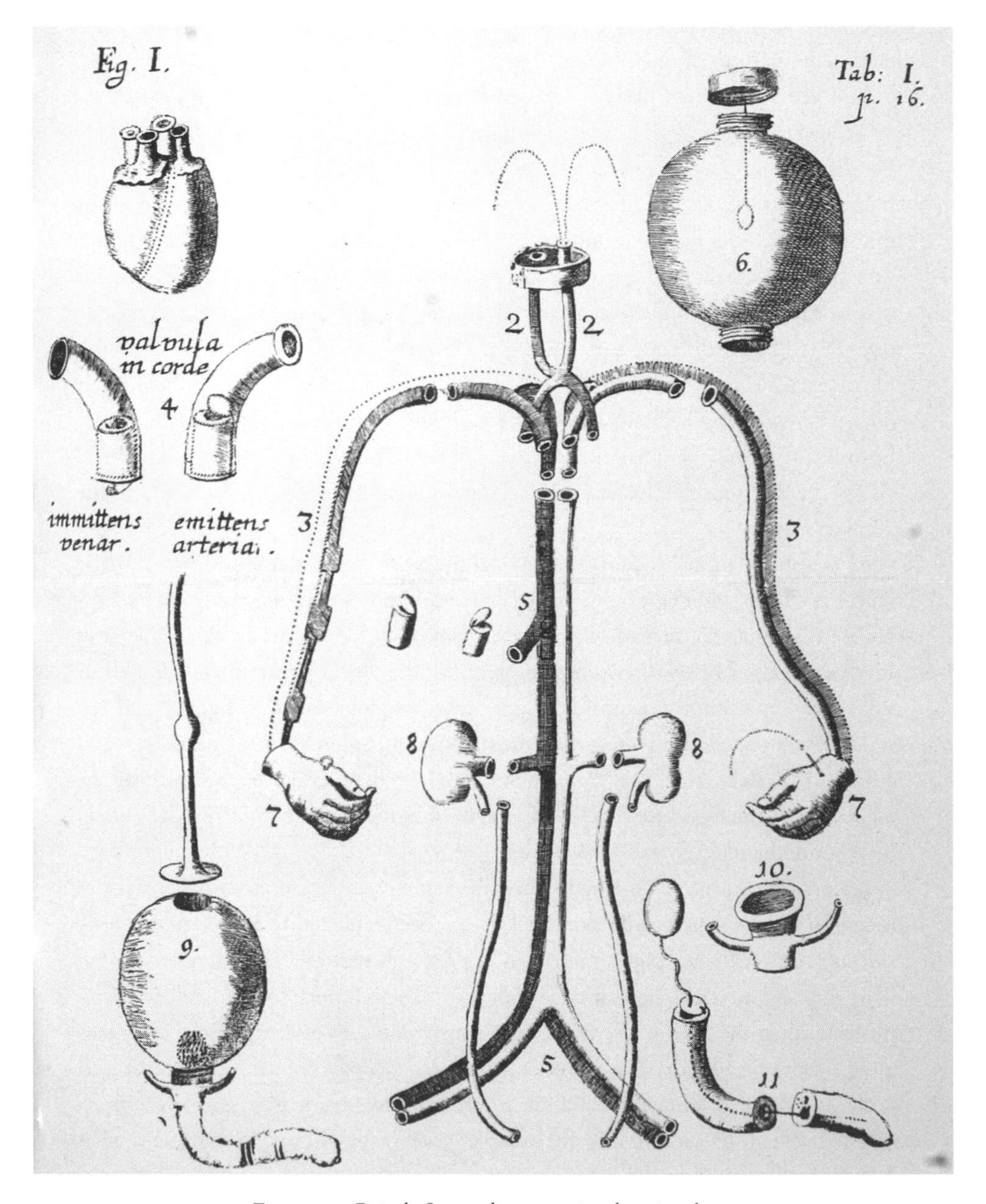

Figure 11.1. Reisel, *Statua humana circulatoria*: plate 1

In the *Risposta* Malpighi mentioned that the camera obscura could serve to understand sight and its lesions. One wonders also whether he was familiar with the model of the eye constructed and described by the Venice maker Giovanni Battista Verle in collaboration with the Padua anatomist Antonio Molinetti for Ferdinand II of Tus-

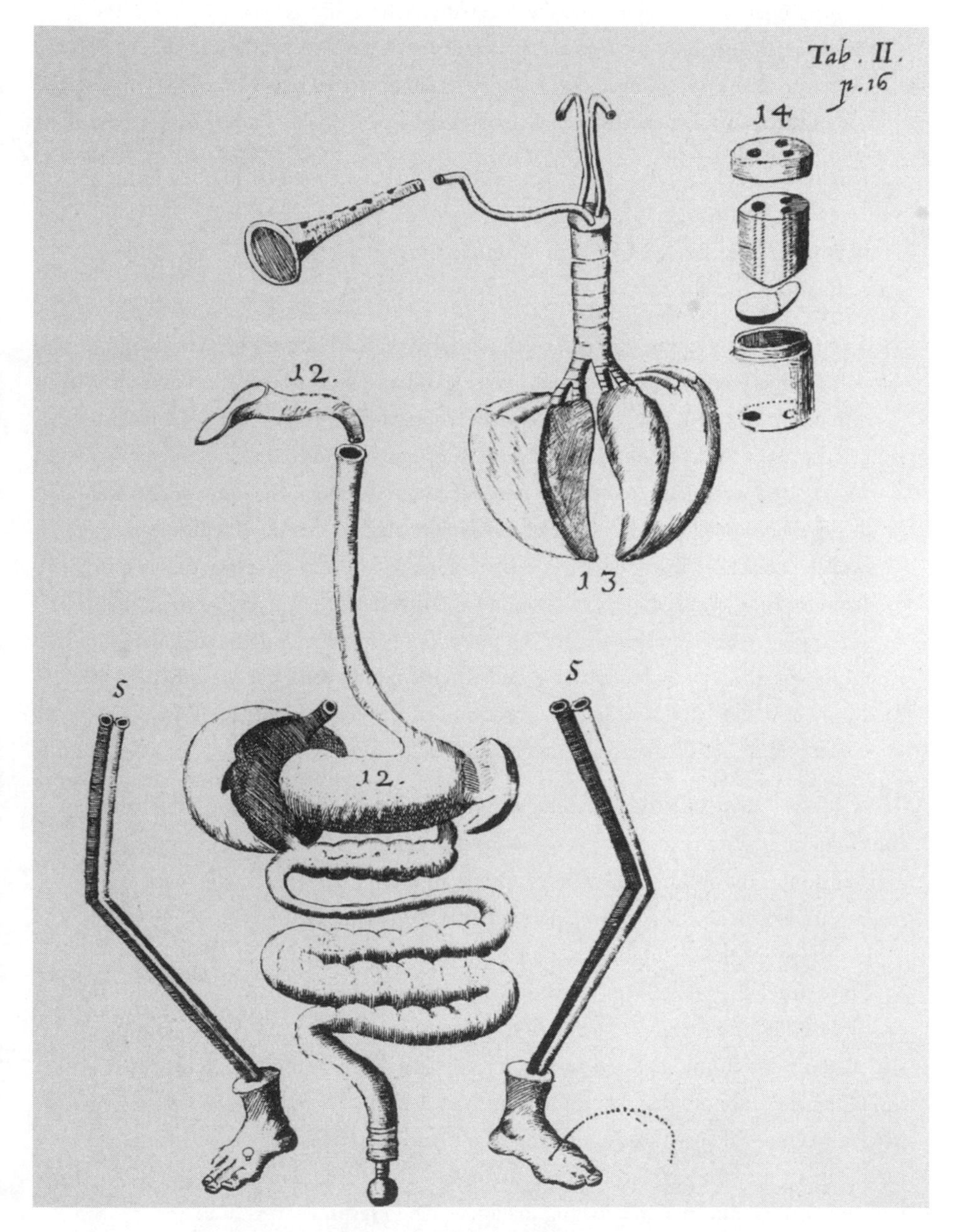

Figure 11.2. Reisel, *Statua humana circulatoria*: plate 2

cany. However, we know that he experimented on the eye and the properties of its parts with his colleague Giandomenico Cassini. As Malpighi put it:[18]

> An evident proof of this is the camera obscura, in which the mathematician produces all those effects that are observed in the sight in healthy and diseased states of the animal, displaying *a priori* the necessity of those effects that occur from the variety of the figures of the lens and the excessive distance or nearness of the parts. Therefore the way of seeing and its lesions are demonstrated by means of the cognition of the man-made machine analogous to the eye.

The model of the artery, closely resembling one mentioned by Harvey in the second reply to Jean Riolan, *De circulatione sanguinis*, would enable us to study blood circulation and its diseases, by which Malpighi presumably meant aneurysms. Harvey states:[19]

> If you take what length you will of the inflated and dried intestines of a dog or wolf (such a preparation as you find in an apothecary's shop), cut if off and fill it with water, and tie it at both ends to make a sort of sausage, you will be able with a finger-tap to strike one end of it and set it a-tremble, and by applying fingers (in the way that we usually feel the pulse over the wrist artery) at the other end to feel clearly every knock and difference of movement. And in this way (as also in every swollen vein in the living or dead body) anyone will be able to teach students, by demonstration and verbal instruction, all the differences occurring in the amplitude, rate, strength, and rhythm of the pulse. For just as in a long full bladder and an oblong drum every blow to one end is felt simultaneously at the other, so in dropsy of the belly, as also in every abscess filled with liquid matter, we are accustomed to distinguish anasarca from tympanites.

Thus, Harvey suggests a use of the intestine sausage going even beyond the diseases of the circulatory system: anasarca is a swelling up of the entire body, and tympanites is a distension of the abdomen. It is appropriate to recall here several references to accumulation of mineral deposits in aqueducts mentioned in *De polypo cordis* to explain other diseases of the circulatory system.

The artificial thorax serves to study what happens when the lungs fill with fluid or solid bodies and therefore "helps to uncover a priori nature's way of operating and the phenomena in the diseased states of respiration"; probably Malpighi was thinking along the lines of what Swammerdam had done in *De respiratione*, when he had used a mechanical apparatus consisting of a bladder attached to a tube inside a glass phial (fig. 11.3) to understand the punctured thorax; Swamerdam's example also had a pathological significance in showing instances when respiration is hindered.[20]

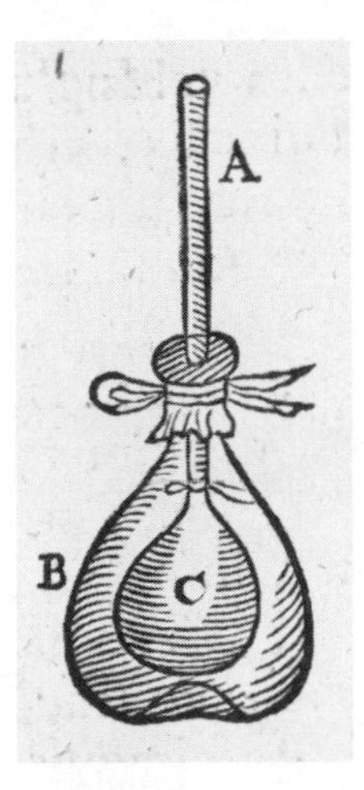

Figure 11.3. Swammerdam, *De respiratione*: figure 1

The passage on the articulation of the bones, especially the reference to walking, swimming, and flying, seems to be a reference to Malpighi's philosophical mentor Giovanni Alfonso Borelli, who in the first part of *De motu animalium* (1680) dealt precisely with these topics in relation to mechanical devices. Although Malpighi did not refer to pathology with regard to the mechanical devices for representing motion, it seems that such machines would have had immediate surgical applications. The role of tendons in moving muscles was well known since antiquity and had been singled out by Vesalius at the 1540 Bologna anatomical demonstration, when he warned barbers of the dangers of accidentally damaging the sinews or tendons during venesection.[21]

It is in this context that Malpighi provides the often-quoted simile of the mill: in the operations of vegetation (including growth, nutrition, and generation), sense perception, and motion, the soul has to act in accordance with the machine to which it is tied, and a clock or a mill is moved in the same way by an attached weight or stone, an animal, a man, or even an angel. His conclusion is that if the mill is broken, one has to repair its wheels, whose structure is known, rather than the moving angel or the faculties of the soul, whose mode of operation is unknown—a remark especially suited to rebut Mini's views. Malpighi adopts here the same strategy inspired by Boyle that he had adopted in his essays on generation, professing agnosticism with regard to the soul and its faculties, whose existence is irrelevant to anatomy, pathology, and therapy. His remark on repairing the mill has a pathological relevance in line with that of the preceding passages and with the whole *Risposta* to Sbaraglia. Although Malpighi stated that the study of plants had a role in philosophy and natural history rather than medicine, at one point he even stressed the pathological and therapeutic significance of his work on plants, especially for surgery, when dealing with generation and the cure of tumors.[22]

One expression occurring repeatedly in the *Risposta* to Sbaraglia and especially in the passage just discussed is "a priori"; over half a dozen times Malpighi emphasizes that it is possible to establish medicine a priori, by which he meant from the cognition of the causes and from the mechanical operations of nature when it is not hindered, or when it is altered, that is, in disease. Whereas in his *Risposta* to Lipari he had emphasized the uncertainty of medicine and its conjectural nature, in the *Risposta* to Sbaraglia he adopted a much more confident tone, stressing a strict causal link tying anatomy, pathology, and therapy; even in those cases when the eye cannot reach the minutest parts of an organ, the mind can grasp their configurations and modes of operation, thus enabling the physician to understand the nature of disease and to devise a therapeutic strategy a priori.[23]

The models and machines mentioned so far concern mostly the solid parts; in response to Sbaraglia's complaint that comparative anatomy ought to concern the humors of the body in health and disease, Malpighi provided another remarkable list of investigations by the moderns, such as Franciscus Sylvius and de Graaf for pancreatic juice; Bohn for bile; Nuck for the humor of the eye, lymph, and saliva; and Boyle, van Leeuwenhoek, and de Heide for blood. In addition, Malpighi mentioned that he too had contributed to this area of study, as we have seen in *De polypo cordis* and *De structura glandularum*. He then proceeded to list some of his recent findings, such as assaying the serum found in the thorax of a woman who had died of dropsy of the chest or hydrothorax. Compared to his standard postmortem reports, it appears that in this case Malpighi devoted special care to chymical tests: he must have thought it especially noteworthy that, after cooking, a networklike structure formed on its surface, while the rest congealed somewhat. He then mixed the serum with spirit of salt, with which it curdled; with spirit of vitriol, with which it remained for many days like frozen; with spirit of niter, which turned it like ricotta; and with the spirit of sal-ammoniac, with which it retained its nature and fluidity. He then proceeded to list other forms of assaying he had performed, ranging from tasting lymph, cooking the diseased serum found under the meninges, and cooking—although not tasting this time—the serum from the vagina of cows.[24]

Malpighi offered a detailed analysis of the advantages of comparative anatomy, another key point of Sbaraglia's attack. Some of his remarks are predictable and in line with his previous research, such as that the brain of fishes has manifest fibers and that the diaphanous blood vessels of frogs are suited to display circulation. In addition, he argued that hares have manifest nervous fibers as well; the posterior portion of the cornea in the eye of the owl shows that the image on the retina is inverted; the role of fat is made clear in frogs and the silkworm, a notable role for insect anatomy; the motion of bile toward the intestine is made clear in snakes and fishes, where the gallbladder is close to the intestine and far from the liver; and the valves in the colon can be best seen in fishes, as he had observed in 1668 with his friend Bonfiglioli, a result then published by Steno.[25]

It is not easy to offer an intellectual characterization of the controversy in simple terms. It would be misleading, for example, to frame it in terms of Sbaraglia's Galenism versus its rejection by Malpighi. As we have seen in chapter 2, Galenism is a complex label involving several themes. If in one respect Malpighi was unquestionably strongly opposed to Galen, namely, with respect to the role of the faculties of the soul, in other respects he was a staunch defender of rational medicine, of the search for causes, and of the relevance of philosophy to medicine so strongly advocated by Galen. The careful reader of the *Risposta* will notice the central role of a *terminus technicus*, namely, *indicationes*, one Malpighi repeated dozens of times, often joined with the qualification "a priori." The *indicationes* provide the therapeutic strategy based on the knowledge of the cause of the disease; by contrast, Malpighi argued that followers of medical empiricism such as Sbaraglia would administer remedies without *indicationes* and without method. In an especially explicit passage of his *Risposta*, Malpighi stated that medicine, even practical medicine, does not consist in finding a remedy, but in the study of the signs, the search for causes and *indicationes* to be found a priori; this requires knowledge of the animal œconomy and an exact cognition of the mechanical structure of the solid parts and of the nature of the fluids elaborated in the viscera in the healthy body. By highlighting the a priori nature of *indicationes*, Malpighi stressed the crucial role of anatomy in pathology and therapy.[26] As we will see in section 12.3, he routinely provided *indicationes* in his medical consultations, where they appear as a key constitutive element of his narrative and medical thinking.

11.4 Sbaraglia's Empiricism and Methodological Concerns

In at least one respect Malpighi must have been able to embarrass Sbaraglia, whose *Dissertatio epistolaris* ended with the words "Scribebam raptim Gottingae idibus septembris 1687." Malpighi pointed out that the place and date were fictitious, since the letter echoed the conclusions defended by Mini on 13 January 1689 and his public anatomy of April 12 of that year; therefore, it dated from shortly thereafter. Malpighi added that matters of such importance ought not to have been discussed "raptim" or hastily. In the 1701 Bologna edition of the *Dissertatio* in *Exercitationes physico-anatomicae*, following Malpighi's *Opera posthuma*, Sbaraglia simply wrote "Vale idibus septembris 1687," thus excising the embarrassing "raptim" and the place of publication.[27]

This, however, was not Sbaraglia's only concern. In 1693 he printed Bohn's *Prae-*

lectiones therapeuticae publicae—a defense of Malpighi's dated from 1691 that had circulated in manuscript form—with a rebuttal entitled *Dissertatio epistolaris secunda*; in the 1701 *Exercitationes* Sbaraglia republished both texts, adding this time an *Appendix* with a response to the *Bibliotheca anatomica*, one to the preface by the Calvinist expatriate Pierre Regis to the Amsterdam 1698 editions of Malpighi's *Opera posthuma*, and even a short qualification to the 1698 Venice edition by Faustino Gavinelli, who had attributed the *Dissertatio epitolaris* to Mini rather than Sbaraglia.

Bohn presented a list of instances in which the new anatomy had provided a novel understanding of pathology and therapy: saliva and the humor in the eye, for example, were understood to come from specific glands rather than the brain; blood circulation had affected the understanding of several processes in the body; and several types of hernias required the understanding of the structure of the abdomen and processes in the intestine to be treated.[28] Sbaraglia's reply highlighted some weaknesses in Bohn's *Praelectiones*: Bohn had apparently omitted to defend microscopic anatomy and the anatomy of plants, as if he had conceded the point; despite the novelty of the finding about saliva and the eye humor, no progress had been made in the therapy for relevant diseases; moreover, the circulation of the blood was irrelevant to Sbaraglia's attack on the three anatomies. In conclusion, Sbaraglia labeled Bohn's reply ineffective and beside the point since the issue was not etiology but the nature of disease and *materia medica*.[29]

In an *Appendix* Sbaraglia briefly dismissed the way the matter was framed in the *Bibliotheca anatomica*, arguing that the editors' standpoint was irrelevant: the formation of the chick in the egg and the vegetation of horns, studied by Malpighi and included in the *Bibliotheca*, pertained to clinical medicine like the satellites of Jupiter, he claimed—an intriguing Galilean allusion.[30] Sbaraglia focused on Regis's preface, which had presented Malpighi as a victim of obscurantism and compared him to William Harvey, Gassendi, Descartes, and especially Galileo for the vicious attacks he had been subjected to in his own town—a clear reference to Sbaraglia. Fortunately, according to Regis, Malpighi's views were universally accepted and only Spaniards, Lusitanians, and Muscovites, who still lie in darkness, give credit to contrary opinions. In his reply, Sbaraglia questioned whether the findings Malpighi had presented as conjectures were universally accepted as proofs and stressed that Regis had referred to Malpighi's contribution to natural history rather than medicine. Obviously piqued by Regis's statement, Sbaraglia went as far as to defend the Portuguese, Russians, and especially the Spaniards, notably for their legal and theological studies. Moreover, he sought to rebut Regis's attack on the notion of faculty, arguing that according to Galen that term should be used when the essence and mode of operation of an organ are unknown; since they are generally still unknown, the notion of faculty is still entirely legitimate and is used by Willis, Glisson, Franciscus Sylvius, Charleton, and

Ettmüller. It would be especially interesting to establish whether Sbaraglia's positions exerted any impact on the emergence of vitalism at centers such as Halle and Montpellier in the eighteenth century.[31]

The 1701 *Exercitationes* included an addition to Sbaraglia's work on generation, *De vivipara generatione scepsis* (Vienna, 1696), under the title *De vivipara generatione altera scepsis*, in which he questioned the existence of eggs in viviparous animals, doubted that in those animals that lay eggs those eggs contain the whole animal as claimed by Swammerdam, and defended the existence of metamorphosis.[32]

Sbaraglia reserved by far his most extensive rebuttal to Malpighi's views for his 1704 *Oculorum et mentis vigiliae ad distinguendum studium anatomicum, et ad praxin medicam dirigendam*, an imposing 750-page tome attacking not simply Malpighi's *Risposta*—discussed mainly in the *Pars secunda apologetica*—but his entire oeuvre from the letters on the lungs to the posthumous *Vita*. Sbaraglia's opening noticed such a difference *in genere morali* between his antagonist's *Opera posthuma* and the works published when he was alive that he was led to question their authorship; thus, he recognized, as we have done in the previous chapter, that Malpighi's style had changed dramatically. *In genere scientifico*, however, Sbaraglia detected no difference; therefore, he felt justified in proceeding with his rebuttal, in which Malpighi is never mentioned by name. While Malpighi had attacked him, violating their oath of friendship, argued Sbaraglia, he had scrupulously followed it, a claim he felt the need to support with a quotation from Lancelotto Corradi's *Commentaria de duello* (Milan, 1553).[33]

It is not easy to offer a characterization of Sbaraglia's *Vigiliae*, a tome ranging from an extensive—if not always up-to-date—exegesis of the literature, to a criticism of Malpighi's techniques of investigation, to occasional dissections and experiments. As we have seen in the previous chapter, Ruysch challenged Malpighi's glandular anatomy by relying on injections and proposed an alternative structure instead. By contrast, Sbaraglia proposed no new method of inquiry and identified no common denominator for the microstructure of body parts, but rather sounded cautionary notes based on a variety of concerns. For example, in the address to the reader he argued that anatomical structures could not be understood without understanding their purpose, as in the spleen, whose alleged glands—which Malpighi revealingly at times refers to as "vesicles," as Sbaraglia pointed out—separate nothing. Furthermore, he questioned the methods used by Malpighi, such as maceration in water, which was required to reveal the glands in the spleen. At one point Sbaraglia mentioned Ruysch but did not refer to his criticism of Malpighi's structure of the spleen in the 1696 *Epistola problematica quarta*, whose contents were excerpted in the *Bibliotheca anatomica*. In the address to the reader Sbaraglia proceeded to caution anatomists about boiling body parts in order to study their structure and discussed the effects of boiling on several organs: although this procedure had been applied in the past, as he documented with a valuable list of cases including Iacopo Berengario da Carpi, Andreas Vesalius, Gabriele Falloppio, and in the case of the tongue Niccolò Massa, its uncritical use led to problems. Sbaraglia also questioned the ease with which Malpighi had enlisted in support of his belief in the glandular structure of the cerebral cortex the occurrence of stones and other concretions, such as the blackberry-shaped gray stone found by Johann Pfeil in a brain. For example, argued Sbaraglia, there was no evidence that the stone originated in the cerebral cortex, nor was its color proof that it originated from the gray matter of the brain, since gray stones are found in the kidneys as well. He questioned the role Malpighi attributed to his hypothetical glands in the cerebral cortex, since the injection of ink in the carotid artery did not apparently reach the cortex, which remained uncolored; ink staining, however, could also be problematic. He then proceeded to challenge the evidence derived from Wepfer's case of a newborn whose skull was filled with vesicles, reported by Malpighi in *De structura glandularum*; Sbaraglia argued that the finding did not support the glandular structure of the cerebral cortex. In this case too he was apparently unaware of Ruysch's 1699 *Epistola problematica duodecima*, which challenged Malpighi's views precisely on this point.[34]

Sbaraglia sought to find contradictions at many levels in Malpighi's extensive works: with regard to the uniformity of nature, for example, he questioned whether generally it was licit to extend conclusions from a diseased state to a healthy one. Commenting on the passage in which Malpighi had compared the mover of a mill to a dead weight, a brute, a man, or even an angel, he pointed out that at the outset Malpighi had stated that the way the soul works in us is ineffable, yet he then proceeded to state that its mode of operation is mechanical. At one point Sbaraglia made use of the very term "mechanismus," meaning more mechanistic explanations in general than a specific device; it is ironic that it was Sbaraglia, rather than Malpighi, who apparently made use of this key term.[35]

"Bone Deus!" exclaimed Sbaraglia at Malpighi's attempt to frame the use of *china-china* to cure fevers in rationalist terms: the remedy was introduced in 1650, he argued, without any a priori cognition of the nature of the disease. The only reasoning about the use of *china-china* concerned the investigation of its application, not its nature, least of all that of the minute parts. He also questioned Malpighi's understanding of Sbaraglia's own views: whereas Malpighi seemed to believe that a comparison of the fluids of healthy and diseased bodies by means of distillation, filtration, and other procedures would yield knowledge of physiology and pathology, Sbaraglia argued that those procedures would tell us something about the defect or excess of this or that component, not their nature.[36] A rather curious twist occurred in Sbaraglia's discussion of Malpighi's project to determine microscopically the excess or defect of different salts in blood in order to study the nature of the disease and its cure. Since the figure of salts can change, argued Sbaraglia, Malpighi's project would be rather dubious; moreover, even admitting for discussion's sake that their figure remains constant, their figure is unrelated to their properties, as shown by several authors such as Galileo, whose *Discorso* on bodies in water argued precisely this point against the Aristotelians, namely, that buoyancy depends on the specific gravity rather than on the shape of the floating body. Even Malpighi's single application of the microscope to therapeutics was challenged: Sbaraglia argued that the microscope merely confirmed that the layer found on top of urine was not oily, as several physicians had already observed through other means.[37]

Vigiliae ends with two supplements, one on *indicationes* and the other on microscopy. The first highlights once again the importance of this key term in Malpighi's work and in the controversy. Quoting Thomas Willis and Franciscus Sylvius, Sbaraglia defended the empirical nature of *indicationes*, which would follow, not precede, the remedy; he further referred to an interpretative tradition distinguishing *indicationes a morbo* from *indicationes a parte*, differing like genus and species: only the latter rely on anatomy, hence Malpighi's view that *indicationes* are based on anatomical knowledge and his insistence on a priori medicine would be misplaced. Indeed, medicine is so full of obscure parts that it is mistaken to think that it can be derived a priori. Sbaraglia referred approvingly to Thomas Sydenham's *Opuscula*, whose emphasis on the study and classification of disease and skepticism over the role of anatomy went hand in hand with his own views. Sbaraglia, however, appears to have been a rather bookish defender of empiricism, more interested in its intellectual defense and in discussing passages from Hippocrates and Celsus than in applying it to nosology, or the classification of disease.[38]

The portion on microscopy evaluates the problems encountered with this instrument. Although Sbaraglia could not be called a microscopist, he asked his colleague in the chair of mathematics—in all probability Geminiano Montanari, who fabricated microscopes—to make two microscopes for him, one more and the other less acute, in order to carry out some observations.[39] Sbaraglia referred to a few authors in the literature on microscopy, from Buonanni to Kerckring and especially Johann Franz Griendel, author of *Micrographia nova*, which seems to have been his main source on the subject. He expressed reservations about very acute lenses and compound microscopes, warning about likely defects due to the lack of transparency and smoothness of glass. Besides problems with magnification, Sbaraglia discussed issues related to color, figure, and number of parts observed, as in the number of components of the composite eye of the fly. An especially interesting aside is devoted to Galileo and the telescope; Malpighi had modeled his *Risposta* on the *Assayer* and, more generally, had presented himself as the Galileo of medicine. Although it is unclear whether Sbaraglia appreciated and decoded the literary and philosophical implications of Malpighi's

Figure 11.4. Sbaraglia's medal

work, he associated the microscope with the telescope, arguing that both ought to be treated with care.[40]

Together with the criticisms of Malpighi's investigative technique, such as "mining for stones," Sbaraglia's tome highlights a growing skepticism and methodological awareness about the problems of innovative and interventionist methods of inquiry. Ultimately, his disagreement with Malpighi involved different philosophical perspectives with a long history and a following reaching to our own times. Although Sbaraglia's attempts to enlist Galen on the empiricist side were as forlorn as Malpighi's efforts to frame the use of *china-china* in rationalist terms, theirs was a profound exchange touching on crucial themes in the history of medicine. Despite their fierce disagreement and profound mutual dislike, Sbaraglia did imitate Malpighi in one respect: he also had a medal (fig. 11.4) cast by the same artist as Malpighi's, Ferdinand de Saint Urbain, showing a sickle cutting a vine hindering the growth of a tree with the words "INUTILES AMPUTANS." Presumably the tree represents medicine, the vine those investigations that are of no use to therapy, and the sickle Sbaraglia's own iconoclastic empiricism.

11.5 Young Morgagni's Covert Intervention

Sbaraglia's *Vigiliae* was followed within months by a rebuttal and a defense of Malpighi in the form of a double epistolary treatise by two unknown authors, Horatio de Florianis and Luca Terranova, published at Rome. Contemporaries suspected that those names were pseudonyms hiding one of Malpighi's followers, but at the time and for two and a half centuries thereafter the name of the common author remained everyone's guess. It was only with the publication of Morgagni's manuscript autobiographies in the mid-twentieth century that the dramatic background of this event has become known. Morgagni states that only three people at Bologna were aware of his authorship: in all probability they were Valsalva, professor of anatomy and surgeon at the Incurables Hospital, the professor of anatomy and surgery Giacomo Sandri, and Albertini, who held Malpighi's chair of the practice of medicine. In privately claiming authorship for both works, Morgagni revealed that he had drafted de Florianis's *Epistola* in less aggressive terms than Terranova's, so that he could claim its authorship in case his identity was discovered, while pretending ignorance about Terranova's *Epistola*. As soon as the manuscript was smuggled from Bologna to the Rome physician and anatomist Antonio Pacchioni, who saw it through the press, twenty-three-year-old Morgagni dramatically drafted his will, obviously fearing for his life. Both *Epistolae* are addressed to a certain Antonio, a student of Malpighi's—possibly Pacchioni himself, who was associated with Malpighi and for whom Malpighi had written a testimonial. In a passage probably inserted by Pacchioni, Florianis's *Epistola* announces the publication of Pacchioni's discovery of conglobate glands *circum cerebrum*, a reference to *De glandulis conglobatis durae meningis humanae*, issued in the same year—1705—by the same publisher.[41]

The contents of the *Epistolae* range from a logical and even grammatical critique of Sbaraglia's work to a more philosophical and medical analysis. Morgagni quibbled with Sbaraglia's use of Latin and defended Malpighi's *Opera posthuma* by arguing that, far from being a polished work, it was put together from loose sheets that the Latin translation and London edition had made even worse. On other occasions Morgagni raised more substantial and damaging points, as when he challenged Sbaraglia's claim about the discovery of glands in the ear, arguing that Steno had already mentioned them several years earlier. Morgagni defended Malpighi's notion of a priori medicine, arguing in a refined philological analysis that those words had to be interpreted not as a deduction from first principles, but rather as a way of proceeding based on a rational assessment of the causes, in line with Hippocrates and Cardano. Concerning the role of the anatomy of minute parts, Morgagni mentioned the structure and direction of muscular fibers of the heart and other muscles, highlighting their significance for surgery. He further defended Malpighi's disputed claim about the glands of the cerebral cortex, challenging Sbaraglia's injection experiment by arguing that body parts decay rapidly after death, thus vivisection would have been required. Lastly, Morgagni defended Malpighi's use of the microscope, which provides reliable results if it is used properly and with caution, accusing Sbaraglia of ignorance about the principles of dioptrics.[42]

In 1706 Morgagni published the first of six *Adversaria anatomica*, a project eventually completed in 1719; the last five, dated 1717 to 1719, are animadversions on Jean

Jacques Manget, *Theatrum anatomicum* (Geneva, 1717), a huge digest of the recent anatomical literature. The first part, however, is unrelated to the others: in it Morgagni presented a broad analysis of the field consisting in new findings, discoveries to be found in the older literature that seemed to have been neglected or forgotten, and findings related to recent controversies.[43] In his *Vita* of Morgagni, Giuseppe Mosca stated that the anatomist from Forlí feared the deceits from too sharp microscopes, injections, and other similar ways of observing, wishing to observe nature in freedom rather than constrained by artificial techniques. Indeed, only occasionally did he rely on such techniques; for example, one plate in the *Adversaria prima* suggests the use of limited magnification, whereas the *Adversaria tertia* contains a reference to injection of colored wax. The *Adversaria* bear witness to Mosca's claims, in that some plates are likely to have been drawn with the help of limited magnification. Whereas Lancisi reported that Malpighi had argued that if Eustachio had relied not only on the knife but also on microscopy and on injections—which he had used only for the kidneys—he would have relieved the rest of humanity from the task of anatomical investigations, Malpighi's intellectual grandchild eschewed precisely those techniques as suspect. Mosca continued that although Morgagni studied comparative anatomy, he abhorred transferring knowledge from animals to humans.[44]

In light of these comments, Belloni's claim that Morgagni was Malpighi's closest follower requires qualifications. Malpighi's practice and legacy consisted of a vast body of views, findings, and techniques; both Malpighi and Morgagni believed in the crucial role of anatomy for pathology and medicine, but whereas Malpighi sought to clinch the mechanical mode of operation of body parts by pushing a wide range of techniques to the limit—and at times beyond—Morgagni focused on exact description and accurate classification for its own sake, with a weaker and less explicit philosophical agenda. The project of studying the location and causes of diseases through anatomy can already be found in Malpighi, but it is in Morgagni that it takes center stage in his celebrated *De sedibus, et causis morborum per anatomen indagatis* of 1761.

Thus, despite Morgagni's defense of Malpighi, it is striking to notice a dichotomy between Malpighi's own line of research and that of his immediate followers, including Morgagni himself. Albertini, for example, whose sister married Malpighi's brother Bartolommeo, was entrusted with Malpighi's literary remains, yet his medical interests were as a clinician rather than an anatomist; nobody at Bologna pursued microscopic anatomy after Malpighi's departure for Rome and subsequent death. Antonio Vallisneri, who was a student of both Sbaraglia and Malpighi, practiced microscopy at Padua later in life but does not appear to have learned it from Malpighi. Giorgio Baglivi was Malpighi's protégé in Rome and dissected his body after death; he was an influential author who pursued anatomical investigations, including microscopy, but his premature death in 1707 deprived the medical world of one of the most active and visible investigators. Although Lancisi was informed of Malpighi's technique to investigate the cicatricula of the chick in the egg, his main contributions were not to microscopic anatomy. More broadly, neither Hooke nor van Leeuwenhoek—who lacked a formal education—pursued anatomy proper, but rather a wide range of research stimulated primarily by curiosity and occasionally by a mechanist agenda. Swammerdam's religious crises and death prevented him from pursuing his investigations further, whereas Grew restricted his microscopic investigations to plants. Thus, at the end of the century Malpighi was the preeminent microscopic anatomist and the only practitioner to hold a relevant university position with students he could train; yet he did not establish a *Coro microscopico* to instruct pupils and colleagues in microscopy. These reflections point to the need for a study of the role of training in techniques of investigation among the generation of anatomists around 1700, as well as of the role of the dispute between Sbaraglia and Malpighi in the reorientation of anatomical and medical research in the eighteenth century.[45]

Despite its Bolognese roots and at times pettiness, the controversy between Sbaraglia and Malpighi touches on key themes in the medical world at the turn of the century, notably the problematic relationships between the role of anatomy and especially the new microscopic findings on the one hand and the art of healing and the study of disease on the other. The controversy had European implications: the involvement of Bohn at Leipzig and Regis in Amsterdam, the inclusion of key texts in the *Bibliotheca anatomica*, and the links with contemporary debates about disease and therapy testify to its broader significance. Rather than focusing exclusively on Malpighi's *Risposta*, I have sought to analyze both sides of the controversy, including little-studied texts, from Sbaraglia's *Vigiliae* to Morgagni's *Epistolae*.

Malpighi's *Risposta* is an extraordinarily rich text, providing detailed organ-by-organ analyses of the perceived success of mechanistic anatomy and of its impact on pathology and therapy. We have gained a deeper understanding of this key text by showing the philosophical relevance of Malpighi's literary style based on a Galilean template: both Galileo and Malpighi relied on optical devices, defended atomism, and suffered vexations for their research. Malpighi was a staunch defender of the search for causes and rational or even a priori explanations and therapies. He relied on material devices to investigate disease, such as the camera obscura, filled bladders, the artificial thorax, and chymical in vitro experiments.

By contrast, Sbaraglia put his empiricist finger on the growing dichotomy he perceived between increasingly elaborate and problematic anatomical investigations and the art of healing: in his view no amount of a priori reasoning could lead to effective therapies. His cry resonates with similar ones heard elsewhere in Europe, especially in England with Sydenham. However, whereas the English Hippocrates investigated and classified disease as ways to devise effective therapies, Sbaraglia's approach remained at a more bookish level.

A preliminary sketch of the generation of anatomists after Malpighi has revealed an analogy in the development of microscopy and injections, problematic and highly interventionist techniques that led to contradictory results; neither was uncontroversial, and, in both cases, the main practitioners formed an exceedingly small community. Much like Ruysch with injections, Malpighi did not train a generation of microscopists and did not discuss extensively his techniques of investigation. Moreover, by 1700 microscopy on its own would have been insufficient; rather, it would have been necessary to assess the limits and features of a broad range of techniques of anatomical investigation, microscopy and injections above all.

The growing number of publications of Malpighi's anatomical works at the end of the century reached an abrupt halt in 1700: except for the partial translations in the *Bibliotheca anatomica, medica, chirurgica, etc.*, one has to wait until 1743 for the subsequent edition of his anatomical writings or *Opera*. By contrast, in the first half of the eighteenth century several editions of his medical consultations saw the light, emphasizing a shift of interest toward disease, as if Malpighi's followers wished to continue defending their master from Sbaraglia's attack: it is to an analysis of those texts that we now turn.

Medical Consultations

12.1 Between Theory and Practice, Carnival and Lent

Throughout this work we have seen that anatomical research was closely interwoven with medical practice: anatomists were routinely engaged with pathology and therapy in so many ways that often these areas of medicine cannot be disentangled. In this chapter we turn to medical practice from the perspective of an underestimated and understudied literary genre, medical consultations.

A consultation is a letter usually written by renowned physicians in response to a request from patients, their relatives, or their attending physicians, known as a letter of referral. Patients tended to belong to urban elites, such as nobles or ecclesiastics. Consultations date back to thirteenth-century Bologna and probably originated from legal examples; they often involved a form of compensation and were more common in Italy. Compensation may not have always come in monetary form: the apostolic protonotary at Castrocaro, for example, promised to remember Morgagni at the holy altar. Unlike case histories, consultations often lacked information on the entire course of the disease and the outcome of the therapy; at times, however, consultations were not isolated documents but formed a series illustrating the course of the disease, while the outcome was later added to the manuscript with a brief annotation or a cross. Whereas case histories may be retrospectively colored by the outcome of the therapy, consultations were written while the outcome was not yet known and therefore retain a tentative nature. Most features remained constant over time, although variations occur depending on when and where they were composed and the personal style of the physician. Consultations followed obvious constraints: notably, they were written *in absentia* and responded to cases that had been at least partly conceptualized by the attending physician or the patient; therefore, they can legitimately be seen as a form of commentary or reflections on a text rather than nature. Despite such constraints, they provide valuable evidence and rare insights into the way of thinking of a physician and practical medicine, if not downright bedside practice.[1]

There are additional reasons why consultations are such important historical documents. Some historians have drawn a sharp and anachronistic distinction between anatomical research and medical practice in the seventeenth century: the former would be tied to the emergence of modern science, the latter would be a remnant of a bygone era. Bruno Basile, for example, argued that the nature of medical consultations, especially their being at a distance rather than based on a physical inspection of the patient, forced authors to rely on ancient methods. Basile then shifted his attack on medical practice as a whole, arguing that the Medici archiater Francesco Redi was an experimentalist and a Galilean as a naturalist, but he was pre-Galilean or even Aristotelian as a physician. More recently, others have dismissed in the same vein Malpighi's remedies as resembling the potions of Macbeth's witches.[2] It is easy to see how such thoughts may have originated, given that the seventeenth-century pharmacopœia on which Redi, Malpighi, and their colleagues relied included the use of human urine, ground human skull (used against mental disorders), spirit of human blood, viper's meat, crab eyes, pulverized red coral, broths with frogs and tails of river crabs, hartshorn jelly, and the application of warm animal viscera. As Jack Pressman has powerfully reminded us in the case of lobotomy in recent times, however, the notion of therapeutic success has to be seen in a historically sensitive fashion.[3]

Despite all this, it is profoundly misleading to separate consultations and medical practice more broadly from anatomical investigations. Historians who dismiss Malpighi's or Redi's medical practice implicitly compare it with modern remedies rather than those available at the time. It is implausible to see in Redi and Malpighi cases of "intellectual schizophrenia"—as implied by Basile—but when we notice similar forms of reasoning and remedies used in the eighteenth century by Giuseppe del Papa, Albertini, Morgagni, and many others, one realizes that it would be necessary to issue a similar charge against at least a long medical century. Many of the preparations recommended by Malpighi could still be found in the official pharmacopœia of 1770, such as the *Antidotarium* of the Bologna College of Physicians.[4] Thus, medical practice, including medical consultations, is a powerful antidote against a double anachronism that makes the very same protagonists of seventeenth-century medicine look like nineteenth-century laboratory scientists when they carry out anatomical or physiological investigations and hopelessly old-fashioned pre-Galilean charlatans when they try to cure their patients.

Medical consultations have generally been considered to be standardized texts, virtually indistinguishable one from the other. All those who believe this to be the case need only compare Redi's and Malpighi's: their best examples differ like carnival and lent. Whereas Malpighi—in line with the majority of physicians—invariably saw disease and the medical consultation associated with it as an occasion to be serious and somber, Redi at times subverted this convention and used his pen as a therapeutic tool producing hilarious linguistic fireworks that retain their freshness and iconoclastic power to the present day; thus, the carnivalesque element of his texts was part of the therapy. Redi's consultations vary considerably, but when he detected patients' obsessions and manias about diseases that in his view were not really there, he unleashed his biting wit and displayed remarkable psychological talents. This observation highlights an important difference between their texts: although both wrote to patients and physicians, Redi's most memorable consultations were either written for patients or conceived with the patient as the intended reader in mind. He often stated that he was writing as a friend rather than a doctor and seemed to take off his medical robes, take the patients under his arm, and talk to them as if he was not the physician. This style made patients feel as if Redi was their ally against the alien medical sect, thus subverting social and intellectual conventions alike. By contrast, Malpighi's texts read like capsule treatises displaying a profound knowledge of anatomy and pathology. Although his consultations are not brimming with jokes, they are not necessarily devoid of psychological subtlety. "Non abbiamo rimedi," replied Malpighi to Antonio Vallisneri's father Lorenzo, who had suggested during a visit to the sick anatomist that his superior knowledge would guide him to the appropriate remedies. Malpighi was rarely so abrupt in his consultations, and one may wonder whether the lengthy anatomical etiology of the case may have served a psychological as well as a theoretical need in reassuring the patient of the physician's expertise. Often Malpighi located therapeutic deficiencies in the lack of remedies rather than of anatomical knowledge.[5]

In line with his position and temperament, the overwhelming majority of Redi's consultations were in Italian, as one would expect from the court physician to the Medici—who promoted the use of Tuscan—a noted poet and literary man, and a prominent member of the Accademia della Crusca, an elite group devoted to Italian literary studies. In fact, arguably Redi's courtly position not only affected his choice of language but also enabled him to adopt his extraordinary carnivalesque style throughout the year. In his study on the history of laughter, Mikhail Bakhtin has identified a medical tradition dating back to Hippocrates and including François Rabelais, whereby laughter is used to heal and regenerate. It is not surprising that Bakhtin's work on carnival and Rabelais is proving useful here, since the author of *Gargantua and Pantagruel*, just like Redi, was a literary man and a physician: both shared an interest in medical texts, patients, laughter, and the power of words.[6]

I set out my inquiry in the following section by discussing next the publication history of Malpighi's consultations and the problems associated with it. These reflections lead to some critical reflections on Adelmann's edition of Malpighi's *Correspondence* and provide at the same time a more precise characterization and a brief historical account of this medical literary genre. Section 12.3 is devoted to an exegesis of the structure and contents of Malpighi's consultations, paying special attention to the connections with anatomical investigations. Section 12.4 establishes a comparison with the consultations by Redi, a friend and correspondent of Malpighi and one of the most visible Italian physicians at the time. In many respects Malpighi's and Redi's consultations fall at the opposite end of a spectrum with regard to the emphasis on etiology and the role of experience and observation; most texts by contemporary physicians fall somewhere in between, although hardly anyone was as subversive as Redi in his use of laughter. Lastly, section 12.5 broadens the horizon by sketching some features of medical consultations by eighteenth-century anatomists and physicians who were followers of Malpighi and Redi, notably Vallisneri and Morgagni. Thus, besides analyzing and comparing their consultations and the history of their transmission, this chapter provides useful benchmarks to explore further these difficult texts. It is my contention that medical consultations from Malpighi to Morgagni constituted crucial documents in the anatomical investigation of the causes and locations of disease.

12.2 Publishing Malpighi's Consultations

In 1713 readers gained access to another new side of Malpighi's work, besides the *Opera posthuma*, in the form of a *Centuria* of medical consultations on a wide range of diseases. The collection of Malpighi's consultations had allegedly been put together for the benefit of clinicians by Gerolamo Gaspari and Giovanni Francesco Mauroceno, students of Antonio Vallisneri, Padua professor of medicine and a former student of Malpighi; in fact, in a private letter Vallisneri acknowledged it as his own.[7] The event was a fiasco. At Bologna Malpighi's followers, and especially Albertini, were incensed, but also at Padua Morgagni expressed his negative views on the operation, while at Rome Lancisi's initial enthusiasm soon turned into bitter disappointment. Albertini argued that Malpighi had always opposed the publication of his consultations—as indeed was common practice among physicians, including Redi.[8] Some of those that had been included appeared like short private notes unworthy of publication; moreover, the authorship of some of them was dubious. This publication represents a curious fate for someone like Malpighi, who had so carefully engineered his posthumous legacy, and for his followers, always concerned with his fame and reputation in connection with the long-standing controversy with Sbaraglia and his followers.[9] Whereas Malpighi's anatomical writings were in his possession at the time of his death and were entrusted to his followers with precise instructions, by their own nature medical consultations were scattered among a wide range of patients and physicians over whom little or no control could be exerted. Albertini's concern for Malpighi's reputation took the form of a *Monitum literarium*, an ephemeral publication of errata that is still occasionally found among some of the copies of the *Consultationum centuria*; relying on Malpighi's manuscripts, of which he was the depositary, he pointed out a number of embarrassing mistakes, my favorite being "fermentatione" instead of "ferruminatione" and especially "lacte et vino" [!] instead of "lacte vaccino." Apparently the editors relied for the transcriptions on someone unversed in medical matters, a most unwise choice for such technical texts. As we are going to see, medical consultations were often copied and assembled to form an informal clinical manual; therefore, it is hard to tell at which stage errors had been introduced. In response to Albertini's charges, Vallisneri argued that even short notes by Malpighi were worth publishing; he compared them to the sketches of a great master who reveals his artistry in just a few lines. In any event, the embarrassment must have been acute because in the same year the publisher issued a second corrected printing taking into account Albertini's *Monitum*.[10]

Whatever Albertini and his Bolognese associates may have thought at the time, Malpighi's consultations are an extraordinarily important document for the historian in many respects. From 1691 to his death in 1694, Malpighi was archiater to Pope Innocent XII, whom he had also attended to while still a cardinal several years before. The large number of his surviving consultations and of observations relevant to medical practice in his correspondence testifies to the importance of therapeutics to his professional life. Moreover, while Malpighi's anatomical publications contain a surprisingly large amount of information about pathology and therapeutics, his consultations are emphatically not lists of diseases and therapies but often include highly theoretical sections relying on recent anatomical findings in the best tradition of rational medicine. Disease was explained in relation to the normal operations of the body, but it could also offer a window onto those operations—the microscope of disease. Thus, Malpighi's consultations, far from constituting a peculiar isolated corpus following rules of its own, form a continuum with his other works and often illuminate them.[11] Luigi Belloni, a perceptive scholar of Malpighi's work, included twenty-three consultations in his edition of Malpighi's *Opere scelte*, successfully weaving them with Malpighi's other writings. There is no question that Malpighi's consultations are worthy of serious investigation on par with his other writings and that they add precious insights to our picture of his activities. Later editions of Malpighi's consultations followed in 1743–45, with a series of collective volumes, and in 1747, together with consultations by Lancisi.[12]

Medical consultations were a tricky genre for Howard Adelmann. The editor of Malpighi's correspondence missed no chance to highlight all references to Malpighi's dislike of medical practice in his correspondence, a dozen or so, all cross-referenced. Possibly respect for his author's preferences inspired him to carry out a dubious selection, one with far-reaching consequences for the image of the Bologna professor of the practice of medicine in his daily life and activities. Adelmann chose to include letters dealing with medical matters but to exclude consultations. His criterion of demarcation was "to regard as a letter any communication provided with a personal salutation, even though in content the letter amounts essentially to a *consulto*." Adelmann concluded by expressing the wish that "someone with the proper qualifications will edit and publish Malpighi's *consulti* with suitable medical annotations."[13] Anyone familiar with medical consultations can sympathize with Adelmann, since they often lack the date and addressee and are a veritable nightmare for the editor of a correspondence. Yet it is regrettable that Adelmann was so modest, because the recent laudable edition of Malpighi's consultations falls short of his high standards in editorial practices and scholarly apparatus, in omitting systematic mention of whether a text had already been published, for example.

First of all, it is not difficult to find Malpighi's consultations in the three manuscript volumes in Morgagni's hand preserved at the Accademia delle Scienze dell'Istituto di Bologna, studied by Adelmann, which he did not include in the edition of Malpighi's correspondence despite the presence of forms of salutations.[14]

But the issue goes beyond one of simple neglect of some sources. A slightly more sophisticated survey of the sources throws further doubts on Adelmann's decision. Consultations may survive through different channels. First, there are the original documents that were dispatched either to the patient or to the attending physician. Second, we often have drafts retained by the author for his own records. Third, there may be copies generally made by other physicians either of the dispatched documents or of the copies in the author's possession. Many of Malpighi's consultations survive in copies, such as those in the three volumes at the Bologna Accademia. The process of copying is relevant to my analysis, because epistolary consultations tended to be collected to form a manual where date and personal references, such as forms of salutation and the names of the attending physician and the patients, were omitted. The consultations were then arranged according to illness of the relevant body parts; for this reason it is difficult to order consultations chronologically. Besides reasons of convenience or ease of reference in reorganizing consultations, there were also privacy reasons involved. A consultation by del Papa is accompanied by a letter from the patient authorizing the publication of his name, suggesting that the publication of names of live patients was not commonly accepted. It is therefore clear that either Morgagni or someone else before him deliberately omitted forms of salutation and other material from the original documents, adding a suitable heading for the appropriate diseases and their locations. In an exchange between Lancisi and Vallisneri on the publication of a second *centuria* of Malpighi's manuscript, the former proposed to arrange them for identification purposes under the standard headings *de morbis capitis, de morbis pectoris, de morbis infimi ventris*, etc.[15]

This reconstruction of the fortunes of the original epistolary consultations is not just a philological exercise, but bears important consequences for my argument. It is not difficult to find cases in which the original document and a copy of the consultation survive, showing evidence of these transformations. The letter by Bernardino Ramazzini to Malpighi of 4 September 1689 starts with the salutation, "Illustrissimo Signore Signore Padrone Colendissimo," whereas in the copy in Morgagni's hand this salutation is omitted and the letter bears the title "Difficilis Respiratio." The salutation in Malpighi's reply to Ramazzini is also omitted. This instance shows that even if one were to adopt Adelmann's criterion, this ought to be applied only to dispatched documents or faithful and complete transcriptions, not copies intended to form a medical collection. This example explains Adelmann's inclusion of consultations such as those written to Antonio Ferrarini and Ramazzini at Modena and Angelo Modio at Rome, for which the dispatched documents survive, and the exclusion of others surviving in modified copies. The loss of forms of salutations can be quite significant for a variety of reasons, such as a sign of the social standing and level of familiarity between the correspondents, for example.

Here I wish to argue an even more radical stance, namely, that to my knowledge all consultations by Malpighi are epistolary exchanges that should have been included in his correspondence. Even if some consultations lack forms of salutations, we have to suppose that a copyist omitted them, or that forms of salutations were enclosed in an accompanying sheet. In both cases the presence or absence of forms of salutations must be seen as an accident resulting from the way the documents have been transmitted rather than as a classification criterion. I wish to qualify my statement by arguing that not all consultations by any author can necessarily be classed as letters. There are instances of texts called *consilia* that were not written for a specific patient, but were rather general treatises on a disease and its therapy; however, this is not our case.[16] On the other hand, one may argue that some texts by Malpighi and others, such as Redi, known as consultations should more properly be called medical letters. In some cases they offered medical advice to people they knew very well or whom they cured regularly, such as Bologna patients to whom Malpighi wrote while out of town in his villa at Corticella, or friends of Redi's like Carlo Dati and Vincenzo Viviani, not to mention family members. In some cases Redi's letters were reports written from the bedside to others, such as the Grand Duke. These observations show that consultations are part of a seamless body of medical letters and call for a fine-grained investigation of this important genre.

The consequences of Adelmann's decision to exclude the consultations from Malpighi's correspondence are considerable. On a quantitative level, the inclusion of medical consultations would add several volumes to the five published by Adelmann. Even without a precise estimate of how many volumes could have been added to the correspondence, however, it is not difficult to assess the sheer size of this corpus of writings and its import: it covers a huge portion of Malpighi's activity and daily practice. Moreover, the geographical provenance of the letters of referral, the social conditions of the patients, and the nature of the diseases give us a vivid picture of Malpighi's sphere of action and status, as well as of the conceptualization and taxonomy of disease in the late seventeenth century. Several documents suggest the existence of different therapeutic practices and traditions at different places. It would be interesting to investigate whether such local traditions are reflected in the geographical provenance of letters of referrals and how physicians coped with this phenomenon of differing local traditions. The recent work by Marco Bresadola is beginning to shed light on these important questions, by showing that most requests for consultation to Malpighi came from the Papal States, for example, from physicians who had a link with Bologna University and Malpighi personally.[17]

12.3 Structure and Contents of Malpighi's Consultations

Overall, consultations do not enjoy a great reputation as a literary genre because they are considered to be rather stifled, standardized, and wholly conventional, a form of remunerated professional duty rather than a locus for creative reflection and innovation. However, they provide the medical historian with a wealth of written evidence that would not have otherwise survived. At times the secondary literature indulges in retrospective diagnoses or accounts of the therapeutic history and uses of this or that preparation. Here, by contrast, I am especially interested in the structure of Malpighi's consultations and their relationship to anatomical research for the insights they provide into Malpighi's pathological thinking. Medical consultations offer concrete evidence of how female patients were diagnosed and treated compared to male ones, thus adding a significant dimension often lacking from other texts. Based on a partial count, about one-third of Malpighi's consultations concerned women. In many cases dealing with a number of organs from the brain to the genitals, his consultations complement his anatomical tracts in a way that is mutually enriching. The texts of the 1713 editions were arranged by the editor starting from diseases of the head, chest, belly, and genitalia, ending with fevers.[18] I follow a seventeenth-century terminology and taxonomy of disease, even if they were quite different from ours; what they meant by dropsy, epilepsy, and hysteric affections generally bears no relation to the corresponding modern notions and can be quite difficult to reconstruct in their own terms, let alone modern ones.

Malpighi's consultations are highly structured and generally follow a common pattern, although I shall add some qualifications below. They are written in both Italian and Latin and are more frequently addressed to the attending physician than to the patient.[19] Stylistically they rely so heavily on anatomical knowledge that they seem to be conceived for a physician even when they are addressed to a patient. Malpighi explained anatomical matters without referring to the relevant anatomical or pathological literature. Although consultations rely on the letters of referral for the description of the case, unfortunately such letters are rarely included in printed collections. Letters of referral usually provided information on the age, sex, and constitution of the patient, followed by the case history, including symptoms, at times a causal explanation of those symptoms, and a list of cures attempted to that point with their results. Besides the form of salutation and the praise and summary of the letter of referral, Malpighi's texts included an etiology of the disease, *indicationes*, and the suggested therapy with a list of *remedia*. I now examine these aspects in turn.

It was customary for the consultation to start with praise to the author of the letter of referral for his thoroughness in the summary of the case and course of action taken. By seeking advice, the attending physician was implicitly acknowledging the authority of the physician whose help was being sought; therefore, the praise was also a form of thanks. Malpighi's consultations are no exception; only rarely does one find a critical note of the analysis in the letter of referral, as in the following case: "The deductions drawn from the compression of the lymphatic vessels are not entirely persuasive, because their bulk, the weight of the lymph and its force of compression are much smaller than [those of] the almost innumerable arteries, whose fluid move faster and with greater impetuosity."[20]

After the opening salutation, praise, and brief summary of the case, Malpighi moved to an etiological account based on anatomy. At times he started with the words[21] "The cause of the various symptoms that have been observed . . . ," highlighting that his chief concern was with causal explanation. This portion of the consultation could be quite long and elaborate, relying on the new understanding of the body provided by recent anatomical findings, especially his own. Although the overall emphasis was on pathology, at times Malpighi expressed his views and doubts on anatomical matters, as in the following example on heart palpitation:[22] "Since the mechanics with which the heart moves in its natural state is still unknown, the cause of heart palpitation is obscure." In another instance he attributed a patient's dropsy to heredity, admitting that although this phenomenon was already known to Hippocrates, its cause was obscure.[23] He frequently supported his causal analysis of the disease with the result of previous postmortems; those passages, dealing with epilepsy, headache, phthisis, bloody sputum, and retention of urine, for example, are of considerable interest for the link they establish between symptoms in living patients and corresponding anatomical lesions in cadavers, a topic explored by Théophile Bonet that was to become the focus of Morgagni's research.[24]

Malpighi's concept of disease hinges on the role of the glands as filtration devices.

Typically the improper operations of the intestine or other causes produce an imbalance in the blood, which often contains salty particles or is too acid. This imbalance leads to damage in various body parts, such as glands or the capillary network in the lungs. An exemplary illustration of Malpighi's views comes from the following case of abdominal dropsy or ascites (swelling of the abdomen):[25]

> From the symptoms exposed in the very learned account we understand that the œconomy of the body of the Noble Patient is upset, since the fluids are corrupted and the viscera damaged, especially in the lower belly; since from the very bad nutrition received in his infancy, the universal fluids, especially those of the glands, were contaminated. Hence tumors, abscesses and erosions occurred, first in the mouth, because of the corrupted saliva, then in the other parts, because of the fault in the lymph that has become coagulative, erosive and analogous to *aqua fortis*. Now, from the universal fluid, the fault has been communicated to the large glands, especially the liver and the spleen; they, having been made full of tartar and obstructed, cannot repair the bile and the splenetic humor, and therefore chyle is not sweetened; consequently, blood remains infected and full of those erosive salts that lacerate the lymphatic vessels and open violently the vessels' pores, whereby the serum sweats into the cavity and forms the ascites.

We see here represented some key ingredients of Malpighi's pathological thinking, with its strong chymical bend and the interplay between fluids and filtering glands. Malpighi was relying on his 1668 *De viscerum structura* in identifying the liver and spleen as glandular, despite the fact that he was unable to detect any humor secreted by the spleen.

Another interesting case is bloody sputum, displaying similar processes affecting other body parts:[26]

> For it is clear that active, sharp and sylvestrian[27] salts are present in gross excess in the noble patient's fluids, in which the mass of the humors is dissolved. Consequently, when the passages in the glands are relaxed, an abundance of abnormal saliva burst forth. This is made so salty and annoying by the admixed erosive particles that the gums are eroded as if by *aqua fortis* and the teeth become carious. Befouled by the same particles as the blood approaches the passageways in the lungs, it gradually produces congestion, for particles not thoroughly mixed in are intercepted, and, when they become lumped together, they impede the circulation until under the impulsion of the rest of the blood the delicate little membranes in the lungs are taken apart and the blood bursts forth into the pulmonary vesicles and the ends of the trachea, and shows itself as bloody sputum.

Malpighi here was relying on his own 1661 anatomical discoveries of the microstructure of the lungs, as well as Steno's findings on the salivary glands and duct, published in the 1662 *Observationes anatomicae*. In another case, this time of melancholia, Malpighi relied on his work on the cerebral cortex, which he believed consisted of glands that filter a fluid flowing through the nerves. Cartesian themes occur here in the account of the soul perceiving the external world and being excited by passions. From a quartan fever two years previously and a sedentary life, Malpighi inferred an excess of acid juices that were not eliminated and ended up in the brain; possibly the corrupt glands of the lower belly and especially the pancreas contributed to this excess:[28]

> The structure of the brain seems to consist in a network of fibers, or small cavities filled with a juice that is separated in the cerebral cortex, whence they originate. These fibers, moved by means of the nervous connections by the external organs and in succession by the external objects, receive a trembling motion that is transmitted in an ineffable fashion to the soul and excites the passion in it. Since the same movements can be communicated without the external objects by the motion or quality of the nervous juice, or by other bodies touching those fibers, it follows that without external motions the soul can feel and those very passions that external objects produce through the senses can be excited in us.

It is not always easy to detect what difference a patient's sex makes to pathology and therapy within Malpighi's mechanistic framework. In certain cases it seems to make no difference, as one would expect if the disease was unrelated to the reproductive organs: in some rare cases in which the patient's sex is not explicitly indicated, one may have a hard time figuring it out. However, the issue is more complex, since lack or defect of menstruations had broader ramifications. Malpighi conceived menstruation as a fermentation, a cleansing and purification process similar to that occurring in the glands; thus, a problematic menstrual cycle led to an excess of salty, corrosive, and acid particles in the blood traveling to the rest of the body. Indeed, at one point Malpighi claimed that hypochondriac melancholia, namely, the feeling of melancholia due to the malfunction of the viscera of the lower belly, is called in women hysteric affection. This means that a defective menstrual cycle could lead to affections in all other parts of the body, such as primarily blood, the brain, the heart, and also the lungs. In a case of periodic bloody sputum of a lady, Malpighi attributed the cause to the suppression or diminution of the menstrual flux. Information on the link between menstrual cycle, blood, and pulmonary disease can be found not only in the consultations but also in Malpighi's very first publication, *Epistolae de pulmonibus*; pathological and therapeutic observations in Malpighi's anatomical treatises were rather common.[29]

The next portion of Malpighi's consultations is a short section giving the *indi-*

cationes, which play a pivotal role in his text. In his *Risposta* to Sbaraglia Malpighi mentioned *indicationes* several times and provided a definition: the indication "is knowledge and invention of the remedy to help nature, in order to remove the cause of the disease, mitigate the accidents, and preserve the body's natural economy."[30] It is especially noteworthy that the first feature of the *indicationes* is that their aim is to remove the cause of the disease and only later to mitigate the effects. Malpighi's causal and rational approach to disease and therapy could not be more explicit. *Indicationes* is a *terminus technicus* occurring in Galen, *De methodo medendi*. According to Galen, *indicationes* do not rely on trial and error, but on the causal knowledge of the disease and on logical deduction. Much like Galen, Malpighi emphasized that *indicationes* do not descend from experience, but are derived a priori, by which he meant "from physiological and pathological notions."[31] The central role of *indicationes* in the structure of Malpighi's consultations can be seen even by their typographic layout: they stand out at the beginning of a new paragraph, as if the page reflected Malpighi's stages of thinking about disease. *Indicationes* are the point of the consultation where pathology leaves the field to therapy. Typical examples of *indicationes* are the following: "Therefore the *indicationes* are to remove the fixations and irritations, as much as possible, sweetening the fluids, and consolidate the parts that are shredded or too open, to nourish the body afresh, and corroborate the ferments of the viscera"; or "Therefore, the *indicationes* will be to perfect as much as possible the first coction [digestion], cleanse the hypochondria from the acids, and corroborate the lower viscera and their ferments."[32] Occasionally the case was so serious that Malpighi omitted the *indicationes* altogether and did not dare offering remedies; such an example occurs in a consultation in which he talked of "the most serious symptoms observed in the Most Illustrious Patient," suggesting a negative prognosis. Generally Malpighi's texts do not give a prognosis, unless it is a negative one; in his *Risposta* to Sbaraglia he quoted a passage by the Altdorf physician Caspar Hofmann arguing that since prognosis is often fallacious, it is wiser to avoid it than to be ashamed of it.[33]

The *indicationes* were followed by a list of *remedia*, suggestions on how to prepare them or where to purchase them, when to take them, and for how long. Malpighi's favorite apothecary in Bologna was the Spezieria del Sole in the Galliera quarter, run by his close friend Angelo Michele Cantoni. Malpighi's familiarity with Cantoni's pills and potions, shown in several consultations, suggests a degree of intellectual cooperation besides friendship, as one may expect from two people in the medical world. In a consultation of 1692, when Malpighi was in Rome, he recommended a vegetable powder against hemorrhoids made by Cantoni. Malpighi's great closeness to an apothecary suggests a degree of interest in therapeutics, chymical analysis, and *materia medica*.[34]

Typically, Malpighi's cures involved the purification of blood and the cleansing of glands and required long periods, of the order of several weeks or even months. I have already mentioned some among the ingredients he recommended that look more colorful to us. In addition, he frequently suggested cow and jenny-ass milk, as well as several iron preparations and chalybeate wine and broth; at times he recommended quenching hot stones or metals in wine or broths;[35] he also relied on venesection; occasionally he had recourse to new remedies such as tea, coffee, and *china-china*;[36] enemas were less common than purgation. Unlike his pathological thinking, his pharmacopœia seems to have been quite traditional and included some chymical remedies, such as the hysterical water of Quercetanus and Pierre de la Poterie's antimonial preparations, in line with what one would expect in mainstream medicine of the second half of the seventeenth century.[37]

Despite Malpighi's emphasis on a priori reasoning, there are passages suggesting some degree of trial and error on his part, as in a consultation on bloody sputum where he states,[38] "I know that some authors praise decoctions, hoping to extirpate the erosive parts that flow out of the glands and the mass of blood. With all the attention and observation possible, I did try everything and till now I have not encountered the happy outcome described, but rather have seen an increase in the accidents and the outflow of humors." He went on to argue that his judgment and that of other physicians with whom he had discussed the case were based "on experiment and reason." This is probably the closest Malpighi got to Thomas Sydenham's radical empiricism and skepticism toward causal explanations and the therapeutic significance of anatomy. Interestingly, however, he mentioned Sydenham in a consultation about colic pains and also Gaspari quoted "the English Hippocrates" in the preface to the reader, possibly suggesting that his views were not considered to be opposed to those of the rational school.[39]

Malpighi usually concluded his consultations in an entirely traditional fashion by submitting his analysis to the better judgment of the attending physician. Together with the final form of salutation one often finds a brief prayer or wish, "Faxit Deus" or "May God be favorable." I wish to emphasize that I have provided only a representative structure of a consultation; some of them, such as those Vallisneri called sketches, were much shorter and lacked the structure we have been investigating.

A rather peculiar case occurred in September 1682, when Malpighi was asked to offer his opinion on whether the healing of a nun, Sister Serafica Francesca, who suffered from asthma, was miraculous. Another physician had argued that the cure was not miraculous, because the nun's affection was hysterical, and one can recover from such affections in a short time. By contrast, Malpighi claimed that the disease had lasted for so many months that the solid parts must have been affected. Since they require a long time to heal, the instantaneous cure of Sister Serafica Francesca must indeed have been miraculous and was probably due to the intercession of St. Ignatius of Loyola.[40]

Unfortunately, as a result of the process of copying through which many consultations have come down to us, dates have usually been removed. Although some letters on therapeutics can be found in the correspondence with Borelli,[41] Malpighi's earliest surviving consultations date from the 1670s; therefore, we do not know to the extent we would like whether and to what degree his thinking about disease and therapies changed with time and with novel anatomical discoveries. His anatomical publications, however, provide us with a rich source of information on his thinking about pathology and therapy from the very beginning of his career. At the end of his very first publication on the lungs, thirty-three-year-old Malpighi explained that after the ingestion of vegetables and pulses, he regularly felt tension in the lungs as a result of the fermentation of the blood and announced to himself and those with him that a palpitation was forthcoming—not exactly the liveliest conversation at a dinner party. However, it is possible that Malpighi was referring to his students here, because he lectured at the university in the early afternoon, in the second hour *post prandium*. Similarly, *De externo tactus organo* explains that southerly winds induce a general laxity, a swelling of the flesh, and diarrhea. Hopefully this was not another topic of conversation at the dinner table, although it would have been appropriate in class. In hectics, he continued in the epistle, blood fermentation is more powerful and induces a fever after eating, exactly the symptoms described in some of his consultations. On the basis of this analysis, he argued that in order to inhibit this excessive fermentation one has to administer remedies inhibiting motion, such as milk, tisane, snails, baths, and oils.[42]

12.4 Curing with the Pen: Francesco Redi

Redi's consultations, like Malpighi's, appeared posthumously, first in 1726 thanks to the publisher Giuseppe Manni, with a dedication to Michelangelo Tilli, prefect of the botanic garden at Pisa, and then at Venice as part of Redi's *Opere*, which were often reprinted. In the preface to his edition Manni thanked all those who had sent him Redi's texts, especially Vallisneri, who thus appears as a key figure behind both cases discussed in this chapter. Unlike the Padua editions of Malpighi's consultations, Redi's were not ordered according to disease from head to toe, but seem to be arranged rather arbitrarily; indices of diseases and of the main subjects allow the reader to search the book. In recent times Carla Doni has produced an excellent critical edition of Redi's *Consulti medici*, based on the original manuscripts, but excluding those published by Manni, since the manuscripts in his possession could not be retrieved.[43]

Often the texts edited by Doni, however, can be more properly called medical letters rather than consultations, because they were written to patients he knew very well, such as his family members, or were reports written from the bedside. In comparing different documents it is crucial to deal with homogeneous texts: formal consultations written for a patient whom the author had never seen may differ quite dramatically in tone and content from letters written to a friend or a family member. In 1666 Redi replied to his friend and member of the Accademia del Cimento Lorenzo Magalotti, who was seeking advice for a friend with a spinning head:[44]

> Your friend's spinning head originated from the causes given and hinted at in his letter; but even if it had come from the highest peaks of Babel's tower, what the heck would it be? Would it be anything but a spinning head? And who is that man that at times does not have a spinning head? Nuns, who have a head capable of spinning, in order to prevent this, bandage it; tell your friend to do the same and to laugh about all this nonsense.

He then added the exhortation "Si faccia un[a] serviziale," or "Let him have an enema." To a marquis who had been shocked by an earthquake, he recommended locking himself up in a cellar and drinking ten or twelve large cups of the most generous wine. Writing to his sister-in-law, who was suffering from an unknown complaint, Redi urged her to be happy, and in order to combine divine and human remedies, he included the "measure of the head of the Most Miraculous San Renieri," which he had received from the Grand Duchess, urging his sister-in-law to keep it with her faithfully. He also included some cotton wool in which San Reineri's bones were kept, urging her again to show faith and devotion, expressing his certainty in the intercession "of this great Saint."[45]

The structure of Redi's consultations, much like the structure of the book, differs from the one we have seen above. Overall, Redi was considerably less generous with anatomical details and only rarely did he mention *indicationes*; he did so most frequently in the early texts in Latin.[46] Even if the opening of the consultation at times contained a pathological section about the causes and location of the disease, Redi then provided a series of remedies without offering a rigorous causal account as in Malpighi's case. The remedies he offered were frequently justified by previous successful experience and were often disjoined from anatomical theory. He loved listing an endless series of increasingly powerful therapies taken by the patient over exceedingly long periods, at times several years, using with biting irony the summary of the case from the letter of referral to show the absurdity of the situation. Frequently Redi stated that his therapeutic principle was to have the patient live a long life rather than fighting stubbornly every disease; only very serious diseases deserved the full onslaught of his therapeutic arsenal, whereas minor ailments were best left alone. He liked to compare disease to a viper; vipers had been associated with medicine since antiquity, to strengthen the body's natural heat and to prepare theriac, but Redi's association was different. He argued that although vipers are poisonous, if they are left alone they usually do not attack man. Similarly with the disease, Redi thought that it was more dangerous to attempt a cure than to leave it alone. Writing to the physician Domenico David, who was suffering from hypochondria, Redi stated:[47]

> I rejoice with you that you are a good hypochondriac. Oh oh how much I rejoice with you! If I rejoice with you it is because I use to say and I see it every day verified by experience that if a professor of medicine becomes hypochondriac he lives a long, super long life; and the reason for the length of his life is that a hypochondriac physician knows how to live following continuously a good rule and knows to abstain from all those concoctions of medications that doctors as veritable swindlers use to prescribe to others, but never gulp down themselves.

Redi developed a highly personal approach to therapies, attacking traditional remedies and advocating, not only as a rhetorical gambit, simplicity and attention to diet rather than fancy concoctions to be bought at great expense from the apothecary's shop. In a memorable consultation for a case of hypochondria, with excessive flatulence, he argued that the patient had taken so many medications that they would have been able to heal, or kill, all the sick people at the Santo Spirito Hospital, and the Lateran one as well. If all those medications had not killed him to that point, they would in the future, Redi argued, echoing Molière's *Malade imaginaire*. After a rather cursory survey of the causes of the disease that was localized in the blood, Redi listed the remedies, starting with a recommendation to live happily and serenely.[48] In another consultation, also memorable, on the "most obstinate obstruction of the vein in the uterus of a lady," Redi stated:[49]

> This Most Illustrious Lady started taking medications at the age of approximately thirteen or fourteen, and thence to the present time, when she is in her thirty-sixth year, she has always been taking medications and has been affected by illnesses from which (as one finds written in the report) "the cause has not been entirely expunged and overcome." This cause is believed by the most prudent and vigilant assisting Physician to be a contumacious obstruction in the veins of the uterus, due to mixed humors, and for the most part bilious and hot ones. Now I say, if in the period of 22 or rather 24 years the cause of the diseases of this Lady has not been expunged and overcome with so many medications, how ever can it be expunged and overcome from now onwards by means of new medications? I for my part would believe that a healthy and very profitable advice for this Lady would be from now on to ban completely all, all medications from the apothecary's shop, and to

> leave the matter of her health to nature, helped by a long and good mode of living: *Naturae morborum medicatrices.*

The quotation from Hippocrates reminds the reader that the author was a physician, not a medical iconoclast; possibly his authority as archiater to the Medici enabled him to adopt an especially unconventional style. In this case Redi prescribed enemas and an appropriate diet. Similar recommendations can often be found in his consultations; the frequency with which he prescribed enemas—even every other day—sets apart his therapies from Malpighi's. In this case, and more generally, Redi was opposed to wine, a peculiar feature for the author of the dithyramb *Bacco in Toscana.*[50] He also recommended drinking water, especially mineral waters from Tuscan wells such as the Tettuccio, ignoring common concerns about its effects on the stomach, and chicken broth, skimmed of fat and without salt, to be taken early in the morning; the patient was then required to lay in bed for an hour and sleep if possible.[51] Redi was a firm believer in the therapeutic properties of tea, but he was opposed to coffee and even wrote a poem about it; he was also opposed to the fashionable spirit of hartshorn, then popular in Rome.[52] There was clearly a social stratification in the recommended *materia medica,* since not every patient would have been able to add ground pearls to the morning and evening milk, as Redi recommended to a cardinal.[53]

Of course, one should not overemphasize differences. Redi and Malpighi were exact contemporaries with common backgrounds. In Redi's case, unlike Malpighi's, we have rather early texts, some dating from the 1660s and other important documents dating from the second half of the 1640s, when Redi was still a student or had recently finished his studies. These texts are valuable to reconstruct his intellectual itinerary, showing him taking notes on the art of writing consultations. Generally early works and works in Latin are closer to Malpighi's. They are structured in four main parts: the essence and cause of the affection, the *indicationes*, the prognosis, and the therapy. It is clear from these texts how much Redi departed from tradition in his later texts, forging his own unique style.[54]

Both Redi and Malpighi shared an interest in postmortems: Malpighi dissected several corpses at the Hospital of Santa Maria della Vita in Bologna and elsewhere, Redi at the Hospital of Santa Maria Novella in Florence.[55] However, passages mentioning postmortem dissections are more frequent in Malpighi than in Redi, as one would expect. In a consultation with a Malpighi flavor, Redi wrote to the physician Gilberto Gualtiero establishing a link between symptoms and anatomical reports:[56] "In many cadavers I found in the heart's ventricles and in the vessels that come out from it, some of those gross matters that I have mentioned, which are called by modern authors heart polyps, and rest assured that almost all those dead bodies, when they were alive, were affected by the same accidents by which is affected our Signor

Cavaliere." This is a reference to recent anatomical literature, notably Malpighi's *De polypo cordis*. In a consultation on a case of hypochondria, for example, Redi provided a causal account based on the glands of the cerebral cortex; in another, on a case of an irregular heartbeat, he provided an explanation based on the glands of the liver. These examples testify to how deeply Malpighi's anatomical works had shaped pathology.[57]

In many consultations Redi complained that the main result of extremely elaborate medications was to enrich apothecaries. In the preface to the 1726 edition, Manni reports that Redi had devoted considerable efforts to reforming the Florence pharmacopœia, abolishing many old and damaging preparations, and simplifying many others. Redi probably had a major supervisory role both in his capacity of archiater to the Medici and as head of the *Spezieria e Fonderia* of the Grand Duke. In several medical letters Redi referred to preparations from the *Spezieria e Fonderia*, appearing here in a role analogous to Cantoni's *Spezieria del Sole* in Malpighi's consultations. He often prescribed the "pillole del Redi."[58] Several pages of Redi's *Osservazioni intorno alle vipere* (Florence, 1664) and *Esperienze intorno a' sali fattizi*, first published in the *Giornale de' letterati di Roma* for 1674, testify to some of his activities in this area and give us a feeling for the conversations taking place between the walls of the *Spezieria*. In a medical consultation discussing the merits of diaphoretic versus diuretic remedies, Redi came down squarely in support of the latter, arguing that sweat is less effective than urine in removing salts and other impurities from blood. He then proceeded to support his claim by reporting on some experiments presumably performed at the *Spezieria*:[59] "Together with urine, a great quantity of salt, both fixed and volatile, exits our body, as I could excellently learn by means of often repeated anatomies of urine, which I have performed at different times and in different people." Here by "anatomies" Redi meant chymical distillation. In some cases Redi recommended artificial salts or *sali fattizi*, made from the ashes of vegetables of all sorts. Initially the Grand Duke had surmised that such salts with their radically different shapes would retain the properties of the original vegetables, but it turned out that this was not the case. In the *Osservazioni intorno alle vipere* and *Esperienze intorno a' sali fattizi*, as well as in medical consultations, Redi stated that all vegetable salts have the same properties and he used them as purgatives. In one instance the effects of such salts were tried on several subjects of dubious standing, in an early documented example of medical experimentation on humans.[60]

Redi argued that according to Democritus and Hippocrates the uterus was the principal source of all the diseases of women, which were attributed to the lack of, or insufficient, menstruations. The stagnation of blood in the uterus would result in an improper fermentation and acidity, leading to all possible evils.[61] Later Redi seems to have somewhat changed his views on the matter; in a brief theoretical essay on the causes of menstruation introducing a consultation on the periodic pains afflicting a lady, he held it unlikely that blood would stagnate in the uterus, arguing instead that its fermentation would occur throughout the body, as testified by the fact that during menstruation women are affected and feel pain in all parts of the body.[62]

An unusual case challenging the standard sexual taxonomy is mentioned in a letter to the son of Grand Duke Ferdinand II, Francesco Maria. Redi was replying to the letter asking for his opinion on a nun, who seemed to have a very large clitoris grooved and pierced at the top, whence a fluid similar to semen came out. Redi admitted that these were not features of the clitoris, but surmised that the person who had inspected it may have imagined it so and that the fluid may have been flowing outside the clitoris and originated elsewhere. He was not surprised that the nun felt pleasure in that part since he argued that most women feel more pleasure in titillating the clitoris than any other body part. Redi stated that if he could inspect the case directly, he would be able to offer a more certain report, as he had done in the past about another woman who after a long time in a convent was taken out because they said she had become a man. He expressed his disbelief, however, that the nun in the case at hand had become a man and yet still had menstruations. In all probability Redi was regularly called to assess unusual cases of all sorts in his capacity of court physician.[63]

The contrast between Malpighi's and Redi's consultations has revealed a wide range of styles and therapeutic strategies. Their texts were, in some regard, both extreme: Malpighi's for his staunch rationalism, Redi's for his iconoclasm. A broader look at some of their contemporaries and followers will enrich our understanding of this genre and of the interplay between theory and practice, patients and physicians. Moreover, the previous considerations have shown that theoretical justifications for therapies were far from univocal even within the neoteric camp. Tommaso Cornelio at Naples, for example, did not wish to practice phlebotomy, to which Malpighi was not opposed. Lucantonio Porzio, member of the Accademia degli Investiganti at Naples, published a whole treatise on the subject. Redi relied extensively on purgation, a practice used with much greater caution by Malpighi, especially in acute illnesses.[64]

12.5 A Broader Look at Medical Consultations: Vallisneri and Morgagni

Medical consultations were a popular genre in the eighteenth century in terms of actual letters written and their manuscript circulation in collections of various nature and in print. A systematic study of the fortunes of consultations in the first half of the eighteenth century is a fertile topic for further research but lies beyond the scope of my project; some preliminary reflections, however, can be helpful for gaining insights into the transformations of anatomy and medicine at the time. Malpighi's student

Vallisneri was engaged in both 1713 editions of his teacher's consultations and was himself an active collector and prolific author in this genre; a collection of his own consultations was published posthumously in 1733. The recent outstanding edition of a large portion of his *Consulti medici* by Benedino Gemelli shows him as a follower of the rational school who took Malpighi as a model. Vallisneri shared Malpighi's theoretical framework, terminology linked to a corpuscular and chymical understanding of disease, emphasis on the rational role of *indicationes*, and therapeutic strategy. At the same time, Vallisneri was also an admirer and collector of Redi's consultations and asked his students to recite them by heart. He admired Redi both for specific therapies, such as a predilection for enemas, and for the jokes and use of laughter typical of the archiater to the Grand Duke, although he rarely displayed the same verve.[65]

Morgagni, Vallisneri's colleague at Padua, also collected and composed several consultations, which highlight his rational understanding of the causes of disease and therapy based on anatomy and *indicationes*, which were often subdivided into different subsections, such as preventive, surgical, and therapeutic; occasionally Morgagni, much like Malpighi, referred to his own anatomical findings. Unlike Malpighi, however, Morgagni was leaner on the theoretical part and at one point expressed his preference for experience over "ingenious ratiocination," but he loved to refer extensively to an impressive list of cases and remedies reported in the literature across the centuries and regardless of the authors' philosophical affiliation, from Hippocrates and Sydenham to Bellini and the antimechanist Georg Ernst Stahl: as he put it to a correspondent, most patients are like the Empirics in believing that it is not the cause but the cure that matters. The title he added to his name went hand in hand with his display of erudition: whereas at times he used the simple acronym "P.P.P.," for "Primario Professore Patavino," on other occasions he signed as "Professore Pubblico Primario all'Università di Padova, e Membro delle Accademie Reali di Parigi e di Londra, delle Accademie Imperiali degli Scienziati e di Pietroburgo," not exactly a "most humble" servant, but one whose authority and presumably fee were thus enhanced.[66]

Several consultations have a surgical nature and were signed together with surgeons. Noteworthy is the case of a child who bit his tongue so violently as to partially detach it: the letter of referral contained a piece of red cloth representing the tongue with a slit showing the location and size of the wound.[67] Several texts contain valuable references to contemporary practices of physical examination: in one case Morgagni suggested a vaginal examination with the fingers by a surgeon, if the patient allowed it; on another occasion, the letter of referral reports on the state of the internal hemorrhoidal veins of a patient. An intriguing case concerned a probable case of phthisis for which Morgagni deduced from uroscopy that the patient had no fever.[68] An especially interesting episode can be found in the exchange with the Pesaro physician Giorgio

Giorgi, who reported to his "praeceptor" an unusual case of a hernia in a seventy-year-old man due to the descent of the bladder into the scrotum. Morgagni urged Giorgi to investigate the matter further and to make sure to dissect the cadaver, when the time came. Alas, the relatives did not allow the dissection, but the episode highlights the "scavenger" attitude of Morgagni and his associates.[69] In the case of a tumor of the thyroid Morgagni referred to the shape of similar tumors he had observed in cadavers.[70]

A further example concerned a child with a tumor in the left clavicle: expecting the worst, Morgagni doubted his chances of seeing the structure of the tumor,[71]

> because the love of the mother toward the patient is so tender that I consider it almost impossible—such being the nature of women—that she should grant permission to touch the cadaver. At the same time Your Excellency can be certain that if I ever can find a way of attaining our end, I myself shall do nothing more willingly than to seek or cause to be sought with all diligence the true innermost structure of things, in order to send you the most reliable information and to send the clavicle itself, in order to combine it with other bones that I am keeping for similarly laudable reasons.

Although Morgagni's collection reminds us of Ruysch's museum and Malpighi's specimens donated to the Aldrovandi Museum, there were also significant differences: Morgagni's collection was private, whereas Ruysch's was open to a fee-paying public and the Aldrovandi Museum was open to visitors; many of Ruysch's items were the result of his injection techniques and highlighted normal structures as opposed to peculiar pathological cases; further, whereas Malpighi was eager to interpret every bony excrescence or petrified part in terms of his glandular anatomy or his mechanical explanation of growth and Ruysch was eager to display the vascular structures he had uncovered, Morgagni appeared more concerned with an accurate survey of the specimens without immediately enlisting them into a philosophical system.[72]

The publication date and authorship of some key collections of consultations are of considerable interest. At the end of the second volume of Armillei's *Consulti medici* in 1745, the publisher Giuseppe Corona announced the forthcoming very accurate edition of authentic consultations by Malpighi and Lancisi, which he had obtained from "un Amico Professore." It is plausible to identify behind this edition the hand of Morgagni, whose links with Corona date at least from the 1712 *Nova institutionum medicarum idea* and who had copied many of Malpighi's consultations. These observations point to a growing interest in the link between medical practice and pathology in circles stretching from Bologna to Padua. A thread unites Malpighi's systematic anatomical analyses of the causes of diseases and Morgagni's *De sedibus, et causis morborum per anatomen indagatis* (1761), in which he sought to establish a systematic link between symptoms in the patients and lesions found in their postmortems. In the preface to *De sedibus*, Morgagni claimed that he was stimulated to compose his work in the 1740s; however, the first project of his magnum opus dates from his reading of Théophile Bonet's *Sepulchretum* in the early eighteenth century, an attempt at a systematic collection of medico-anatomical cases well known to Malpighi, who mentioned it in his *Risposta* to Sbaraglia. In an important letter dated from Padua 15 May 1707—less than two years after his rejoinder to Sbaraglia—Morgagni asked his Bolognese correspondent to keep all reports of postmortems for a work he had planned on diagnoses based on the dissection of cadavers performed by himself, performed by his friends, or found in the literature. In this respect the editions of medical consultations represent an intellectual bridge among three medical traditions: the dissection of healthy bodies providing knowledge of anatomical structures, the dissection of diseased bodies in search of the seats of lesions, and the study of therapies based on recent anatomical knowledge and documented by case histories and consultations. Unlike Sydenham, therefore, Morgagni anchored securely the course of the disease and therapy to causal explanations and the detection of anatomical lesions in postmortems. Thus, medical consultations in the tradition of Malpighi, far from being a routine exercise devoid of interest, were seen by major anatomists and clinicians as valuable repositories of medical knowledge and represent a key stepping stone toward the investigation of the seats and causes of diseases investigated through anatomy.[73]

Medical consultations, far from being standardized exercises devoid of interest, present a wide stylistic range reflecting profound differences about the theory and practice of medicine. Consultations form a continuum with other forms of medical publications and provide readers with precious information on the interplay among anatomical research, pathology, therapeutics, and the relationships among physicians and between physicians and patients.

Malpighi's rigidly structured texts relying on etiology and the key term of *indicationes* contrast with Redi's less structured and often iconoclastic letters bristling with biting irony. In his consultations Malpighi relied on anatomical knowledge and frequently referred to postmortems before moving to medicaments, thus tying pathology to anatomy on the one hand and to therapy on the other. It seems plausible that attending physicians and patients would have found his display of knowledge and emphasis on rational explanations based on recent anatomical findings both convincing and reassuring.

The shift in the publication of Malpighi's works in the eighteenth century corresponds to a broader shift in the medical world of the time toward clinical medicine, including the study and classification of disease. Within the wide range of anatomical research, the first decades of the eighteenth century witnessed a growing suspicion toward problematic interventionist methods, such as microscopy and injections, and an increased interest in postmortems as a way to tie anatomy to the study of disease and therapy. With their systematic effort to locate diseases and their causes by means of anatomy, Malpighi's texts attracted the attention of Morgagni, whose consultations are replete with a careful survey of the literature displaying the knowledge and authority of the Padua professor. It is revealing that Morgagni had conceived the plan for a work on diagnoses based on the systematic investigation of disease through postmortems by 1707. Thus, from this perspective medical consultations, especially Malpighi's, are a crucial link to *De sedibus*.

Epilogue

The notions of mechanism, experiment, and disease were central to the history of anatomy, science, philosophy, and medicine; I will now briefly reexamine them in turn. A word of caution is in order here: my work should not be seen as a history of anatomy. On the one hand, I have omitted areas—such as myology—that were central to anatomy, although not to Malpighi's work; on the other hand, I have included topics—such as the investigation of plants and medical consultations—that would hardly figure in a history of anatomy. I hope readers found the structure I have adopted as instructive as I found it congenial.

The notion of mechanism was central to anatomy in the second half of the seventeenth century: the very term "mechanism," meaning a material mechanical device, gained common currency from Henry More's 1659 *The Immortality of the Soul*. Mechanistic anatomists argued that in order to grasp how an organ works, it was necessary to understand its structure and how its components operate separately and together. Mechanisms progressively replaced the immaterial faculties of the soul, which were either dismissed altogether or, more subtly, constrained within the bounds of the body's organs, whose structure alone offered a way to understand their operations. The challenge to the notion of faculty led anatomists to search with renewed determination how organs work in their structures and microstructures. Mechanists sought to replace notions like selective attraction and sanguification with others more amenable to their philosophy, such as filtration and fermentation.

Malpighi's mechanistic understanding of nature was tied to his belief in her regularity and uniformity regarding not just physical laws but also the actual devices she employs, which display increasing complexity from plants to lower and higher animals. Analogous mechanisms can be found in different organisms, with variations in size, structure, and mode of operation, although their purpose is unchanged: the vascular network in the lungs of frogs is not especially fine, enabling the anatomist to observe the anastomoses of arteries and veins; the optic nerve of the swordfish is so large that it is folded like a cloth; the heart of the silkworm consists of several segments along the length of the animal, and in the late larval and subsequent stages the direction of its beat is periodically reversed. According to Malpighi, however, changes across organisms proceed continuously, step by step, without violating nature's fundamental uniformity, because her ways of operating are the same everywhere. This conception—partly a metaphysical assumption, partly a working hypothesis that Malpighi saw confirmed in his daily investigations—played a key role in anatomical research across the dichotomy between the individual specimen under examination and the general case: the identity of nature's laws and the analogies among her mechanisms justified Malpighi's having recourse to the microscope of nature, mining for stones, and the microscope of disease, moving across different species, healthy and diseased states, and even jokes of nature and monsters, as he stated in *De polypo cordis*. It is not surprising that Malpighi saw the views of Giovanni Battista Trionfetti and Filippo Buonanni, who cautioned against hasty generalizations, as a challenge to a fundamental tenet of his worldview, destroying the true method of philosophizing by making—as he put it—everything uncertain and at the same time every bizarre thing plausible.

Ultimately, however, the fortunes of mechanistic anatomy were decided on the dissecting table or under the microscope, not simply at the philosophical level. In understanding why nettles sting, how the "beards" of oats curl, or how coded arsmart discharges seeds, magnification revealed the mechanisms at play: tiny spines erupting from a sac of irritating liquid acting like a syringe, twisted components with different textures, and a coiled membrane working like a spring ready to unleash the seeds. Other processes proved more challenging.

In the case of respiration, Malpighi's 1661 findings about the microstructure of the lungs led him and Borelli to an interpretation that everyone soon found untenable. Their account was all the more surprising in view of Malpighi's awareness that air turns blood bright red. By the end of the decade a wide consensus had been reached about respiration, accommodating Malpighi's structural findings with striking experiments and knowledge of the chymistry of the air: Richard Lower and Robert Hooke enacted respiration in a dead animal, showing that one of the key operations associated with life involved purely chymical and mechanical processes.[1] In other cases, the discovery of new ducts and passageways led to major transformations in anatomy and medicine: Jean Pecquet's thoracic duct revealed chyle's path and challenged the traditional role of the liver; Nicolaus Steno's glands and ducts in the mouth led to a different understanding of the origin of phlegm and of the production of saliva, removing the brain from the process; Malpighi's discovery of air vessels or tracheas in plants led him to argue that plants breath.

Matters, however, were far from straightforward: in no other area than the study of living organisms was the notion of mechanism as crucial and at the same time as problematic. In some instances, as with Descartes' views on the optic nerve involving threads free to shake and vibrate and his notion that the brain consists of tense or lax fibers, or with Lorenzo Bellini's account of microscopic vessels in the kidneys, mechanisms explaining the operations of an organ seemed at hand; subsequent research by Malpighi and Steno, however, undermined them without providing clear alternatives on how vision, the brain, and glands actually work. Ligatures to the vessels of the salivary glands enabled Anton Nuck to claim that arterial blood was a more likely source of saliva than nervous juice: although the microscope and syringe allowed deeper insights into the microstructure of many organs, the mechanism of gland secretion was the biggest puzzle and no clear light was shed on it through either instrument. Malpighi believed in a mechanistic explanation, although he did not know exactly which; Paolo Mini and Giovanni Girolamo Sbaraglia questioned his belief; and Frederik Ruysch argued in despair that secretion was due to a force introduced by God at the moment of creation. Around the turn of the century mechanistic anatomy took different forms, for example, with a growing emphasis on quantification in Britain, while antimechanistic schools grew with François Boissier de Sauvages at Montpellier and Georg Ernst Stahl at Halle.[2]

In addition to structures, operations such as nutrition, growth, generation, sense perception, and motion required mechanistic explanations at a more complex level: the issue was no longer to explain the purpose of an organ or its parts but to account for complex processes traditionally associated with the soul and its faculties. Corpuscularism served as a privileged framework for understanding nutrition; the notion of unfolding and weaving analogies provided a linguistic and conceptual analogue for growth, especially in the case of bones and teeth; generation remained a puzzle that most mechanists—although not Descartes—preferred to push aside and subsume under growth; sense perception proved a challenge that Malpighi met at the level of structures in the tongue and skin and Carlo Fracassati bravely sought to meet with Raffaello Magiotti's hydrostatic device; motion and the contraction of muscles, including the heart, were tackled in a variety of ways involving chymistry, mechanics, and mathematics. Pecquet and Borelli were among the first to rely on elasticity and periodic oscillations of a pendulum and of springs to account for anatomical phenomena such as the expansion and contraction of arteries, the motion of chyle, and the beating of the heart. Pecquet sought to account for the motion of chyle without attraction.[3] Even when borrowing the expression "plastic virtue" from Robert Boyle, Malpighi defended a thoroughly mechanistic program in all areas of anatomy, struggling to unravel the operations of the body by anchoring his philosophical stance to the anatomical detail.

Several scholars, from Sam Westfall to Alan Gabbey, have questioned the status of mechanistic anatomy and of the mechanical philosophy more generally, arguing that mechanical explanations were unsuitable to "biological facts" and that the explanations provided by the mechanical philosophy often bore a resemblance to those of the peripatetics, with the size and shape of particles replacing relevant virtues, such as the dormitive power of opium: in both cases empty words unrelated to empirical investigations provided the explanations. Time is ripe for a critical reassessment of these views. The issue is quite complex, since our critical analysis should be based on the standards of the time rather than on present-day knowledge.[4]

Unquestionably, mechanistic explanations frequently did not live up to the expectations of many, in that key issues remained unresolved. Mechanistic anatomists made assumptions that were unsupported by empirical evidence, first among many that glands secrete fluids depending on the size and shape of pores and particles that had never been observed. Theirs, however, were not simply abstract pronouncements: Malpighi and other anatomists embarked on extensive and painstaking investigations involving a panoply of techniques and an astounding range of animals in the hope of unlocking the structure of many organs. Antonio Oliva, Carlo Fracassati, and Francesco Redi embarked on the study of the correlation between taste and the shape of salts. Even if the correlation failed to materialize in the form in which it had been originally envisaged, and proved elusive, the investigation involved an experimental program and microscopy. *Pace* Gabbey, the mechanistic program was emphatically not armchair philosophy but involved cutting-edge research and striking techniques that were problematic precisely for their novelty. The new anatomy led to the identification of microstructures in the lungs, kidneys, liver, and brain and the study of the composition of blood, just to mention some notable—if at times highly problematic—examples. The localization of microstructures responsible for specific operations was especially significant from a mechanistic standpoint, in that it challenged the notion of an undifferentiated parenchyma and questioned the existence of immaterial faculties associated with it. For example, Malpighi revealed the microstructure of the lungs, identified in the tongue and skin microscopic nervous terminations in which taste and touch occurred, and located in the liver and kidneys structures—glands and glomerules—in which secretion occurred. Other relevant examples involve insects: the works by Hooke, Redi, Malpighi, and Swammerdam radically transformed traditional views and led to a new understanding of their microstructure, including the respiratory, circulatory, digestive, nervous, and reproductive systems. Westfall's claim that "iatromechanics made no significant discovery whatsoever" appears problematic from the standards of any time. Reflections on the notion of experiment also call his claim into question.[5]

Experiment was central to the research of anatomists and physicians: their investigative practices and settings were among the most remarkable of the time in any discipline. Some experiments enabled anatomists to visualize structures, others to investigate actions and purposes, but often they were used in such creative ways as to defy expectations and classifications. Following a tradition going back to antiquity, Aselli used vivisection to study the actions of the laryngeal nerves and the motion of the diaphragm, unexpectedly finding some structures—the milky veins—so ephemeral as to have escaped earlier detailed study. Generally the microscope was used to detect structures, although it also made visible the motion of blood in opposite directions in the arteries and veins in the lungs of frogs. Conversely, ligatures were usually applied to study directionality in the motion of blood, lymph, and bile, for example, although they were also employed to enlarge body parts and make their minute structures visible, as Malpighi did in his study of the kidneys, or to investigate the respective roles of nerves and arterial blood in secretion, as Nuck did for saliva.

The revival of vivisection played a key role in anatomical research. Although it was not a novel technique, vivisection was associated with some remarkable and innovative experimental designs and results, from Gasparo Aselli, William Harvey, and Johannes Walaeus to Reinier de Graaf and Anton Nuck. Studying the live animal enabled the anatomist to grasp the actions of the parts and avoid the pitfalls of the rapid decay of delicate parts. Richard Lower's and Robert Hooke's vivisection experiments to uncover the purpose of respiration were both creative and cruel, although other cases were remarkable as well: de Graaf's collection of pancreatic juice, Nuck's investigation of bladder stones and the site of fecundation, Malpighi's ligatures of tree trunks to study the motion of lymph, and his removal of cotyledons stand out as among the most striking and original examples. At times even entirely traditional methods led to novel findings and the broad reconfiguration of entire fields: Swammerdam's study of snails led to the realization of their hermaphroditism and to a new perception of sexuality in nature, setting the horizon for a new understanding of an entirely different field with Rudolph Jakob Camerarius's remarkable experiments on the sexual reproduction of plants.

Often, however, novel findings were due to technical innovations involving primarily microscopy and injections. Especially after midcentury anatomy witnessed a mushrooming of novel but also problematic techniques, including the microscope of nature, the microscope of disease, reliance on mechanical models, the study of peculiar clinical cases and monsters, mining for stones, chymical analysis, and the excision of organs. Their rise and creative use in the second half of the seventeenth century were due to several motives: prominent among them—especially in Malpighi's case—was the search for visible mechanisms accounting for hidden causes and traditional faculties. In some cases, such as microscopy, it seems more appropriate to talk of a cluster of techniques, involving not only different instruments with variations in lighting and magnification but also elaborate preparations such as boiling, delamination, the fixation of body parts, staining, and injections. Thanks especially to works by Catherine Wilson, Marian Fournier, and Edward Ruestow, we can now appreciate in much greater depth the richness of early microscopic investigations. Similarly, works such as Francis Cole's classic study and Harold Cook's recent analyses have shed light on the practice of anatomical injections.[6] Here, however, I have emphasized the interrelatedness and complexity of those techniques and the problems resulting from combining them: Malpighi was especially concerned with providing evidence for the glandular structure of many organs evinced from a wide range of techniques. At times, however, those techniques led to a destructive rather than constructive interference, notably—but not only—with Malpighi and Ruysch.[7]

This broader picture of investigative practices in seventeenth-century anatomy sheds light on the fortunes of individual techniques and also on their changing patterns over time, which can now be seen as part of a larger canvas. While some were broadly accepted, others—such as advanced microscopy, some types of injections, mining for stones, and the microscope of disease—became embroiled in controversies over their reliability, especially when they were applied in highly interventionist fashion. Contradictory results were not the only source of problems; at times data proved recalcitrant to generalization and anatomists questioned whether what happened in a specific case contained a lesson about other cases as well.

Debates on seventeenth-century experimentation have focused primarily on the physical sciences and on issues such as the tension between individual contrived experiments on the one hand and generalized experiences on the normal course of nature on the other.[8] Focusing on anatomy and natural history enriches and problematizes this picture. First, we have encountered a tension between minimalist and interventionist techniques affecting the object of study: in *Micrographia* Hooke drew a distinction between microscopy of the water gnat, whose transparent outer shell allows easy access to its interior, and animal vivisection, which in his view altered the state of the animal.[9] Malpighi's microscopy, however, was highly interventionist, and his preparations, so effectively satirized by Gideon Harvey, led to results that Nuck, Ruysch, and others were soon to find problematic. Similarly, Ruysch's vascular injections under high pressure led him to a vascular view of the body that Manget and Boerhaave found questionable.

Secondly, we have encountered a dicothomy between individual cases and the need to generalize, even within the same species, which is not surprising, given the differences among living organisms; even the same organism could exhibit different features depending on age, geographical location, and time of the year. Unusual or even monstruous occurrences, from the seemingly impossible conservation of the body of Caterina Vigri to the healing of Sister Serafica Francesca, were deemed miraculous. Francesco Torti's egg, encapsulating in its yolk two smaller eggs, like Russian dolls, raised problems concerning the process of generation in that it seemingly supported Swammerdam's views that the first egg of every species encapsulated all future eggs. Malpighi, however, warned against premature generalizations from what he called a monstrous case, which in this case did not reveal nature's normal operations.

Thirdly, we have encountered the problem of transferring knowledge among species, including humans. Relying on a species in order to draw conclusions about another was an exceedingly useful yet problematic and contested research practice: Malpighi relied on frogs and snails to draw general conclusions on the lungs and liver of all animals, and de Graaf transferred anatomical features of the reproduction in rabbits to other species; Ruysch, however, argued that transverse fibers in the calf spleen were not found in humans, and Camerarius noticed with irritation the anomalous behavior of horsetail and hemp in his study of the sexual reproduction of plants.

Fourthly, we have encountered the dichotomy between dead and living organisms. Hooke occupies a peculiar position in this regard, since he questioned the reliability of vivisection because it alters the state of the organism yet was himself involved in some of the most gruesome experiments of the century. Further debates included the glands of the cerebral cortex, about which Morgagni reminded Sbaraglia that their rapid decay required vivisection.

Fifthly, we have faced a tension between healthy and diseased states, as when Ruysch challenged Malpighi's identification of glands based on mining for stones and the microscope of disease, arguing that greater caution was needed to infer the structure of body parts such as the liver, cerebral cortex, and spleen from diseased states, whether directly witnessed by Malpighi or reported in the literature.

Sixthly and lastly, we have examined several themes on the female and male body, from silkworms and sharks to rabbits and snails. Up to a point, however, female and male bodies work in the same ways in both healthy and diseased states. Malpighi, for example, found exactly the same malformation in the kidneys of Antonio Davia and his sister Pentesilea. However, we have also encountered differences in humans to do with menstruation, which was thought to remove salty, acidic, and corrosive particles from blood. A defect in menstruation was thought to lead to a range of diseases in all parts of the body.

Anatomical experimentation involved a range of novel techniques, but these were not always presented with a view toward enabling replication, but rather were often introduced at best somewhat vaguely and at times actively concealed. Whereas a century later anatomists addressed and discussed the role of problematic instruments and techniques, around 1700 many anatomists retreated from highly interventionist methods to more traditional and secure ones. Despite his prowess with the microscope, Malpighi was no Johannes Müller, the head of the Berlin anatomical institute who taught and formed a generation of microscopists.[10] Readers expecting Malpighi's immediate followers to take on Ruysch's challenge and investigate by means of injections and microscopy the structure of the cerebral cortex, liver, kidneys, and spleen would be disappointed. In fact, advanced injection techniques seem to have been absent from Italy and even refined microscopy dwindled after Malpighi, with the possible exceptions of Giorgio Baglivi and Antonio Vallisneri. The issue of the transmission of techniques and training in the generation after Malpighi and Ruysch still awaits careful study.

The centrality of experimentation in the anatomy and natural history of the seventeenth century, however, can no longer be in dispute. Nor was this research peripheral to the intellectual world of the time: the extensive collaborations among anatomists, physicians, experimental philosophers, and practitioners of the physico-mathematical discipline; the pervasive presence of physicians in the learned societies of the time and in the intellectual world more widely; the profound philosophical implications of anatomical research; and the wide circulation of the vast anatomical literature, enshrined in the *Bibliotheca anatomica*, all testify to the wide impact of the experiments discussed in this work. They were far better known among seventeenth-century intellectuals than they are among historians of the Scientific Revolution. The recognition of the centrality of this body of work calls for a drastic reassessment of traditional views about seventeenth-century experimentation.

It is a central claim of this work that disease was at the same time a subject and a tool of anatomical investigation, albeit a problematic one; therefore, the historical study of anatomy cannot ignore disease even when the main focus is the healthy organism. In this respect my work differs from most histories of anatomy and physiology, which exclude disease as irrelevant to their enterprise.[11] Although I have focused on pathology and therapy mainly in chapters 2, 11, and especially 12, the study of disease has been interwoven with my narrative all along, and at times has taken a central role, as with Malpighi and heart polyps. His publications routinely had a pathological and therapeutic dimension alongside an anatomical one; even the machines representing different organs in his *Risposta* to Sbaraglia, such as the camera obscura, the filled bladders, the artificial thorax, and the chymical in vitro experiments, had a pathological role.

It is not surprising that patients and disease were never too distant from, and at times central to, anatomical research, since the overwhelming number of the protagonists of all these investigations were physicians, whose profession and daily activities involved patients and healing alongside anatomy and natural history. Malpighi tasted the morbidly abundant fluid from the pericardium of a deceased patient to ascertain its origin, using disease in order to collect an amount sufficient for assaying. By contrast, Lower cautioned against relying on the fluid from the pericardium of chronic patients, whose blood is much depleted, arguing that such investigations are best carried out on those who have died of sudden death and whose blood is unaffected. This example testifies to the problematic nature of disease as a tool of chymical and anatomical investigation.

New anatomical findings joined forces with a novel philosophical perspective to offer a rational account of disease based largely on hydraulic and chymical causes and relying on plumbing, the excess of corrosive fluids, and the obstruction of glands. As anatomists relied on the localization of ducts and glands as a key tool for providing a mechanistic understanding of bodily processes, so too did mechanistic pathology and therapy depend on localization as a key tool for conceptualizing and treating disease. The changing knowledge of the body's plumbing, from Pecquet's thoracic duct bypassing the liver to Steno's glands and ducts in the mouth cutting out the brain, and the identification of glands in a large number of organs transformed the anatomical understanding of those organs and the pathology associated with them. This was a key feature of Malpighi's *Risposte* to Lipari and Sbaraglia. The structure and style of Malpighi's medical consultations suggest that the reliance on novel anatomical findings and the strong emphasis on rational explanations were likely components of the appeal of therapies inspired by mechanistic views.[12]

A few recent readers have felt a sense of discomfort with the medical practice of the seventeenth century, especially the pharmacopœia. Some found it especially problematic to reconcile the new experimental philosophy with the use of remedies that appear at best bizarre to modern eyes; clearly, tradition played a large role in therapeutics. Against such anachronistic views, however, we should remember that several remedies were based on experimental practices in vitro, as those mentioned by Malpighi in *De polypo cordis*, and at times on direct experience with patients—despite the fact that few modern readers would wish to experience them. Rational and mechanistic medicine included experimentation. Moreover, we have seen that Redi relied on parallel trials in his study of the generation of insects and then used similar techniques on animals to study the properties of a French styptic water, coming to question its efficacy. Further, the practice of medical consultations at a distance has seemed at odds with the empiricism of the time; in this case too, however, matters are more complex in that the authors of the consultations routinely relied on the better judgment of attending physicians, who did see the patients, and had presumably encountered similar cases in their own practice. Lastly, we have to recognize that the notion of what constitutes a "successful" therapy has a historical dimension that should not be overlooked: it would be anachronistic to assess past therapies by the standards of a later age.[13]

Approaching the turn of the century, we witnessed a growing concern with the reconceptualization and classification of disease across Europe by physicians of different persuasions, from Sydenham to Sbaraglia: different forms of empiricism and a disenchantment or downright opposition to the panoply of dubious highly interventionist techniques associated with the anatomy of minute parts, and at times anatomy itself, were common traits among many of them. The interests of Bellini, Malpighi's close friend and fellow collaborator of Borelli, also shifted toward disease and therapy: in his *De urinis et pulsibus, de missione sanguinis, de febribus, de morbis capitis, et pectoris* of 1683, he expressed his project to reform medical practice, and *Opuscula aliquot*, of 1695, includes a section on bloodletting.[14] Anatomy was an extraordinarily rich and vibrant field in the eighteenth century, yet it also differed profoundly from earlier times with regard to key methods of inquiry and philosophical outlook. Our reflections call for a fresh study of the interactions between the transformations of anatomy and the understanding of disease in the Enlightenment.

The interplay among the notions of mechanism, experiment, and disease in the works by Malpighi and his contemporaries played a crucial role in the history of anatomy, science, philosophy, and medicine. I hope that my work, while building bridges among those areas, will contribute to forging a novel picture of the intellectual world of the seventeenth century. If any reader will feel stimulated to carry out further research as a result of reading this book, one of its main aims will have been fulfilled.

ABBREVIATIONS

AIHS	*Archives internationales d'histoire des sciences*
AIMF	*Annali dell'Istituto e Museo di Storia della Scienza di Firenze*
AS	*Annals of Science*
BHM	*Bulletin of the History of Medicine*
BJHS	*The British Journal for the History of Science*
BW	Boyle, Robert. *Works*. Edited by Michael Hunter and Edward B. Davis. 14 vols. London and Brookfield, VT: Pickering & Chatto, 1999–2000.
DBI	*Dizionario biografico degli italiani*. Rome: Istituto della Enciclopedia Italiana, 1960–.
DNB	*Dictionary of National Biography*. Edited by Henry C. G. Matthew and Brian H. Harrison. 60 vols. Oxford: Oxford University Press, 2004.
DOAT	Descartes, René. *Oeuvres*. Edited by Charles Adam and Paul Tannery. 12 vols. Paris: Léopold Cerf, 1897–1913, rev. ed. Paris: Vrin, 1964–76.
DSB	*Dictionary of Scientific Biography*. Edited by Charles C. Gillispie. 16 vols. New York: Scribner, 1970–90. With 8 additional volumes (*New DSB*), edited by Noretta Koertge. Detroit: Thomson-Gale, 2008.
ESM	*Early Science and Medicine*
GLI	*Giornale de' letterati d'Italia*. Rome: Tinassi, 1668–81.
GOF	Galilei, Galileo. *Opere*. Edited by Antonio Favaro. 20 vols. Florence: G. Barbèra, 1890–1909, reprinted 1968.
HS	*History of Science*
JAMA	*Journal of the American Medical Association*
JHI	*Journal of the History of Ideas*
JHM	*Journal of the History of Medicine*
MAP	Bertoloni Meli, Domenico, ed. *Marcello Malpighi, Anatomist and Physician*. Florence: Olschki, 1997, Biblioteca di *Nuncius*, XXVII.
MC	Miscellanea curiosa
MCA	Malpighi, Marcello. *Correspondence*. Edited by Howard B. Adelmann. 5 vols. Ithaca: Cornell University Press, 1975.
MH	*Medical History*
MOB	Malpighi, Marcello. *Opere scelte*. Edited by Luigi Belloni. Torino: UTET, 1967.

MOO	Malpighi, Marcello. *Opera omnia.* 2 vols. London: Robert Scott & George Wells, 1686. Facsimile edition, Hildesheim: Olms, 1975.
MS	*Medicina nei secoli*
NRRS	*Notes and Records of the Royal Society of London*
OCH	Oldenburg, Henry. *Correspondence.* Edited by A. Rupert Hall and Marie Boas Hall. 1965–86. 13 vols.; vols. 1–9, Madison: University of Wisconsin Press; vols. 10–11, London: Mansell; vols. 12–13, London: Taylor and Francis.
PS	*Perspectives on Science*
PT	*Philosophical Transactions of the Royal Society of London*
QS	*Quaderni storici*
RQ	*Renaissance Quarterly*
RS	*Renaissance Studies*
SHPB	*Studies in History and Philosophy of Biological and Biomedical Sciences*

Introduction

1. Bylebyl, in "Disputation," 224–27, argued that for Jean Fernel, unlike Vesalius, anatomy was limited to structures. Cunningham, "Fabricius"; "Pen and Sword." Schiefsky, "Galen's Teleology." Pomata, "*Praxis*."

2. Du Laurens, *Historia anatomica humani corporis* (Paris, 1600); Bauhin, *Theatrum anatomicum* (Frankfurt, 1605). On the term "chymistry," capturing the features of the discipline as it was practiced and conceived in the seventeenth century, see Newman and Principe, "Origins."

3. Portions of the *Bibliotheca anatomica* appeared also in English translation together with other works as *Bibliotheca anatomica, medica, chirurgica, etc.* Harvey's *De motu* (Leiden, 1639) and Aselli's *De lactibus* (Leiden, 1640) were published by Johannes Maire. Thomas Bartholin's *Anatomia* (1641) was a notable textbook that included also the celebrated letters by Johannes Walaeus in support of Harvey. Highmore's *Disquisitio* (The Hague, 1651) was a notable mid-century treatise. Aselli's, Harvey's, and Walaeus's works were included in the 1645 *Opera omnia* of the Padua anatomist Adriaan van der Spiegel, owing to the Dutch physician and professor of medicine Johannes van der Linden; on them and Walaeus see Lindeboom, *Dutch Medical Biography*, 1858–59, 1200–1203, 2117–19.

4. Van Beverwijck, *De calculo renum*, 20–24. Pagel, *Harvey's Biological Ideas*, 99–102. Keynes, *Harvey*, 271–73. Bylebyl, "Medical Side," esp. 32–33. *Bibliotheca anatomica*, vol. 1, preface (2r. not numbered), 957. I find French's claim of a diminution of human dissection in the late seventeenth century unwarranted; *Dissection*, 269.

5. *Bibliotheca anatomica*, 1:359; all references are to the second edition (1699). Baglivi, *Correspondence*, 133–36, at 136, Manget to Baglivi, 30 December 1693.

6. Frati, *Bibliografia*. Bartholin's treatise appeared also at Leiden in 1672. Malpighi's work was reissued twice in 1666, in Padua together with Lorenzo Bellini's *De structura et usu renum* and in Amsterdam as an appendix to Johann Vesling's *Syntagma anatomicum* by Gerardus Blasius; his reprint in Vesling's 1666 *Syntagma*, 444–55, is not mentioned in Frati. Starting from the 1677 Jena edition, Malpighi's *Epistolae* were incorporated in *De viscerum structura*. Later editions of *De viscerum structura*, including the *Epistolae* and at times other works, appeared in Frankfurt in 1678, in Bologna in 1680, in Toulouse and Frankfurt in 1682 (not mentioned in Frati; see Garisenda Libri e Stampe, Artelibro, Bologna, 21–24 Settembre 2007, no. 49), and again in Frankfurt in 1683. The French translation was reissued in 1687. The *Epistolae* appeared

in Malpighi's *Opera omnia* (London, 1686; Leiden, 1687) and in the 1685 and 1699 editions of the *Bibliotheca anatomica*. In addition, they were excerpted in successive editions of anatomical compendia and textbooks. Malpighi, *Discours anatomiques sur la structure des visceres* (Paris, 1683, 1687); for the information on surgeons I used the 1687 edition, *avertissement*.

7. *Bibliotheca anatomica*, vol. 1, *De thorace*, 807–1072, at 983–84; the section includes also *De polypo cordis*. Bertoloni Meli, "Additions," 305. An aspect of Swammerdam's work is briefly discussed in sec. 11.3; review in *PT* 2 (1667), 534–35. On Mayow see Frank, *Harvey*, 224–32; *PT* 3 (1668), 833–35; 4 (1669), 1142. Chartier, *Cultural Uses; Forms and Meanings; Order.*

8. In his address to the reader, at xxii, Adelmann curiously traced a lineage going from Malpighi, Valsalva, Morgagni, Scarpa, and others to himself through his teacher Benjamin Freeman Kingsbury. *Embryology* omits anatomical illustrations unrelated to generation; written in an often celebratory fashion, it is also unwieldy, opening with a one-hundred-folio-page history of Bologna University. It was awarded the Pfizer medal in 1967. Belloni's contemporary edition of Malpighi's *Opere scelte* is a subtler work.

9. Adelmann, *Embryology*, 136–37, 165, 174, 204, 294. Pomata, *Promessa*, provides a rich picture of the Bologna medical scene.

10. Malpighi, *Vita*, 4. *MOB*, 20. *MCA*, 3:930 and 966–67. Adelmann, *Embryology*, 471–75. Martinotti, "L'insegnamento," 30–47. Pamela Smith, "Science and Taste"; *Body*, chap. 6. Palmer, "Pharmacy," 115–17.

11. *MCA*, *ad indicem*, under "Malpighi, summer residences of"; 3:1289–93, Malpighi to Bellini, Ronchi di Corticella, August 1687, at 1290. Adelmann, *Embryology*, 1, 457.

12. *MCA*, 1:190–91, Borelli to Malpighi, 21 December 1663. Adelmann, *Embryology*, 157–59. Frank, *Harvey*, 182, provides similar remarks for Thomas Willis. Gascoigne, "Reappraisal."

13. Ferrari, "Anatomy," 76. Richter, *Theater*, 55–62. Martinotti, "Anatomia," 122ff. Contemporary reports on public anatomies at Bologna can be found in Bertoloni Meli, "Additions," letter 986A (Antonio Maria Valsalva reports on Paolo Mini's anatomy), and its *Appendix*, letters I and II (Francesco d'Andrea reports on Sbaraglia's anatomy). Gallassi, "Malpighi." Cavazza, "Uselessness." Rupp, "Matters."

14. Malpighi, *Vita*, in *Opera posthuma*, 4. Findlen, "Controlling the Experiment." Baldwin, "Snakestone." Keynes, *Harvey*, 97 and 323. Malpighi's reports have been edited by Münster, "Anatomica." A partial translation is in *MOB*, 411–48; four dissections took place in hospitals (418, 419, 446, 447). No names are given of the people whose bodies were dissected in hospitals; on the important role of hospitals see Conforti and De Renzi, "Sapere." Harley, "Political Post-Mortems." On different attitudes toward the dead body in Italy and Northern Europe, as well as toward social status, see Park, "Life"; "Criminal"; *Secrets*.

15. Malpighi, *Vita*, 55; Adelmann, *Embryology*, 210–11, 330–31. On his artistic trips see sec. 7.3. On Pignataro see Gentilcore, *MAP*, 85. Fisch, "Academy"; Torrini, "Accademia." Some of the issues in the Lake Agnano controversy involve traditional doctrines and resonate with themes touched on by Borelli's *Delle cagioni*.

16. *MOB*, 334, 338, 345, 349, 351, 352. Cavazza, tables between 132 and 133, in *MAP*. Busacchi, "Iconografia," 49–50. On the Bologna Museum see Findlen, *Nature*, 28 and passim; Tribby, "Body/Building." Siraisi, *History*. Malpighi, *Anatome plantarum*, dedication. Legati, *Museo cospiano*. A painting dated 1683 in Malpighi's honor in the Archiginnasio shows a resemblance to his medal: Mercury, representing eloquence and virtue, reclining over Malpighi's

publications stacked on an altar, passes on to Eternity a sheet of paper with Malpighi's name, while Medicine looks on.

17. Bertoloni Meli, "Archive." Frank, *Harvey*, 53–54. Palmer, "Pharmacy," 105–6.

18. Garber and Ayers, *Cambridge History*. But see also Duchesneau, *Modèles*.

19. See, e.g., Wilson, *Invisible*; Aucante, *Philosophie*; Hatfield, "History"; Duchesneau, *Modèles*; and the essays in issue 11.4 of *Perspectives on Science*, 2003, edited by Fisher, on "Early Modern Philosophy and Biological Thought"; and in Justin Smith, *Problem*. Manning, "Machines."

20. Power, *Experimental Philosophy*, 5; Hooke, *Micrographia*, 152; More, *Immortality*, 220, preface, b6r, b8r. Descartes, *Passions*, par. 13. Newman, *Atoms*, 178n48. Manning, "Machines."

21. Steno, *Discours*, 32–33; I have modified the English translation in Scherz, *Lecture*, 139.

22. Galen, *Usefulness*, 222. Hero, *Spiritalia*. Des Chene, *Spirits*, chap. 4. Canguilhem, "Machine." Veneziani, *Machina*. Roux, "Quelle machines." Bertoloni Meli, *Thinking*. Hall, *Physiology*, 218–29. Frank, *Harvey*, 90–93.

23. Grew, *Anatomy*, 188. Dear, *Intelligibility*. Descartes, *Principes*, 4:203. Gaukroger, *Biography*, 269–90.

24. Park, "Organic Soul." Bloch, *Gassendi*, 362–76, 397–400. Garber, "Soul and Mind." Von Staden, "Teleology." Berryman, "Automata." Galen, *Faculties*. Des Chene, "Mechanisms."

25. Steno, *De glandulis oculorum*, in *Opera*, 1:81. Machamer, Darden, and Craver, "Mechanisms." Lennox, "Teleology." Shiefsky, "Galen's Teleology." Siraisi, "Teleology." *MOO*, 2:128. *MOB*, 207. See the section "Ursache und Notwendigkeit" in Kullman and Föllinger, *Biologie*, especially Gill, "Material Necessity," and von Staden, "Teleology and Mechanism." Marcialis, "Immagine," 304.

26. Bertoloni Meli, "Collaboration." Ogilvie, *Science*.

27. *GOF*, 10:348–53, Galileo to Vinta, 7 May 1610. Bertoloni Meli, *Thinking*, 70–72. Baldini, "Animal Motion before Borelli," *MAP*, 203–8 and 212–14. Büttner, "Pendulum," 227. *MCA*, 1:305–7, Capucci to Malpighi, Crotone, 4 April 1666, at 306.

28. Brown, *Mechanical Philosophy*. French, *Harvey*.

29. Bertoloni Meli, "Collaboration," sec. 8.

30. Scriba, "Autobiography," 39–40. Wallis was also present at Lower's experiments on blood transfusion at Oxford in February 1665; Lower, *De corde*, 174. Bertoloni Meli, "Authorship"; "Collaboration." Despite the distinguished tradition from Webster, *Great Instauration*, to several works by Hal Cook and Anita Guerrini advocating an integrated approach to the histories of medicine and science, the histories of medicine and anatomy are not integrated into the main picture of the Scientific Revolution, and we lack studies of medical and anatomical experimentation; see in this regard Shapin, *Revolution*. Hunter, "Boyle."

31. See secs. 7.2 and 8.4. Siraisi, "Vesalius." Maehle, *Kritik*. Dear, *Discipline*. Daston, "Baconian Facts." Newman, *Ambitions*, chap. 5.

32. Hooke, *Micrographia*, 186. Guerrini, "Ethics"; *Experimenting*.

33. See sec. 3.2; *MOB*, 24–25. Maehle, *Kritik*, 15–70.

34. French, *Harvey*, chap. 11. Cole, "Injections." Ruestow, *Microscope*. Wilson, *Invisible World*. Fournier, *Fabric*. Bennett, "Microscope." See sec. 10.5.

35. Cook, *Decline*. Pagel, *Van Helmont*, 141–98. Giglioni, *Immaginazione e malattia*, 68–75. Cunningham and Williams, *Laboratory*.

36. Canguilhelm, *Normal and Pathological*. Cunningham, introduction to *Adenographia*.

Wear, "Practice." Webster, "Harvey." Cook, "New Philosophy"; *Decline*. Bertoloni Meli, "Pathology and Therapy." *MOB*, 543. Shapiro, "Health." Steno, *De glandulis oris*, in *Opera*, 1:27–28. See sec. 2.4.

37. Boyle, *Correspondence*, 2:1–5, Lower to Boyle, 18 January 1662, at 3; 78–81, Lower to Boyle, 4 June 1663. Dewhurst, "Willis and Steno," 43–44, quotation at 43. Despite my respect and admiration for Adelmann, *Embryology*, and Frank, *Harvey*, these works do tend to push medical matters and concerns under the carpet.

38. These were fundamental processes explored by Descartes in the posthumous *De la formation du fœtus*, *DOAT*, 11:217–90, esp. 252–86. Hooke, *Micrographia*, 123–24, 133, 193–94. Fournier, *Fabric*, 54–55.

Chapter 1 • *The New Anatomy, the Lungs, and Respiration*

1. On bedside practice see Bylebyl, "School," 339. Adelmann, *Embryology*, 125–29, 131–35. Malpighi, *Vita*, in *Opera posthuma*, 2, 21, and the entry in *DBI* by B. Giovannozzi. In his *Vigiliae*, 564–65, Malpighi's rival Sbaraglia attacked Giacchini, thus confirming his role in Bolognese controversies over several decades. Malpighi, *Opera posthuma*, 1–2. Schmitt, "Aristotle."

2. *Collegium privatum, Observationes anatomicae*, 5, 22, 27–30, 32; *Pars altera*, 7, 13, 15–16, 17–18, 27. Cole, *Comparative Anatomy*, 330–40.

3. Adelmann, *Embryology*, 120–36, esp. 126–27; Cavazza, *Settecento*, 31–78, esp. 42–44. Malpighi, *Vita*, in *Opera posthuma*, 1.

4. Bonaccorsi, *Polsi*, 45. Pagel and Poynter, "Harvey's Doctrine," 427–28. Eales, "Lymphatic System," 280–82.

5. On Capponi see *MCA*, 1:111–12n2, Borelli to Malpighi, 13 January 1662.

6. Cohen, *Studies on William Harvey*, is a useful collection of earlier material. Frank, *Harvey*, chaps. 1–2. Bylebyl, ed., *William Harvey and His Age*; "Nutrition"; "Boyle and Harvey." Wear, "William Harvey." French, *Harvey's Natural Philosophy*, esp. chaps. 1–5. Bates, "Game of Truth"; "*Machina ex deo*"; "Harvey's Account." Galen, *On the Natural Faculties*, I.13, III.4.

7. Belloni, "Circolazione," was unknown to French, whose *Harvey's Natural Philosophy*, chaps. 6–11, provides an extensive although often unreliable account of Harvey's reception.

8. Walaeus, in Caspar Bartholin, *Institutiones anatomica* (Leiden, 1645), 443–88, at 452, 472. Bartholin, *Epistolae*, II:583–92, by Bartholin's student Heinrich van Moinichem to Bartholin, Padua, 18 February 1655, at 583, and 599–601, 12 April 1655, at 601; van Moinichem later became Royal physician in Denmark. Cole, "Injections," 291. French, *Harvey*, 162n39; at 161, and in *Dissection and Vivisection*, 241, French erroneously states that Walaeus punctured the crural artery rather than the vein. Schouten, "Walaeus." Pagel and Poynter, "Harvey's Doctrine," 419–29. Gamba and Ongaro, "*Anatomes peritissimus*." Bonaccorsi, *Della natura*, 44–45, 51–52. Bonaccorsi confusingly joined the discovery of the chyliferous ducts with that of the circulation, attributing them to "Arveo pisano," by which he may have meant Andrea Cesalpino. Harvey, *Circulation*, 60.

9. Bertoloni Meli, "Collaboration"; *Thinking*, 130 and chap. 8. Mani, "Jean Riolan," 132–42.

10. Pecquet, *Experimenta*. Rudbeck's work was reprinted together with Pecquet's and Bartholin's in the second edition of Hemsterhuis, *Messis aurea*.

11. *MOB*, 464 and 488. Secret, "Cassiano dal Pozzo."

12. Auberius, *Testis examinatus*; see also Thomas Bartholin, *Institutiones* (1677), 224. Highmore, *Disquisitio*, 90–91. *MCA*, vol. 1, Borelli to Malpighi. Malpighi, *Opera posthuma*, 2 and 4. On Borelli see the entry by Ugo Baldini in the *New DSB*. Lüthy, "The Fourfold Democritus." Gómez, "Atomism," 175. In *De musculis et glandulis* Steno stated that the heart is a muscle. Steno, *Opera*, 1:167–69, 181–82. See also the letter to Bartholin of 30 April 1663, in Bartholin, *Epistolae*, 4:414–21, at 417–18.

13. Borel, *Observationum microscopicarum centuria*, observations 85–86, reproduced in Belloni, "Introduction" to Malpighi, *De pulmonibus*, 12–13. See also Borel, *Historiarum*. Ruestow, *Microscope*, 6–60, at 38. Wilson, *Invisible World*, chap. 3. Lüthy, "Atomism." Pighetti, "Odierna," especially 329ff. *MOB*, 13–14. At a later date Malpighi owned works by Odierna: *MCA*, 1:374, Malpighi to Oldenburg, 1 April 1668. Dollo, "Astronomia," 241. Severino, *Zootomia*, 139. Freedberg, *Lynx*, esp. 151–94. Fournier, *Fabric*, 24–29. Harvey, "Circulation," 28, 77.

14. *MCA*, 1:87–88, Borelli to Malpighi, 13 May 1661; 159–60, Borelli to Malpighi, Pisa, 12 April 1663. Redi, *Esperienze*, 25, 117, 142, 170, 180, 191, 192, 196.

15. *MCA*, 1:21–23 (7 November 1659) and 47 (6 November 1660). Malpighi, *Vita*, 2–4. Adelmann, *Embryology*, 155–57. Park, "Organic Soul." On the social standing of surgeons in Italy see Gentilcore, "Organisation," 95–101.

16. Malpighi, *De omento, pinguedine, et adiposis ductibus*, in *MOO*, 2:48. See also Belloni's introduction to *MOB* and "Anatomical Microscopy."

17. Borelli, *De motu animalium*, II, prop. 96. Brown, *Mechanical Philosophy*, 83–86 and 396. A curious reference to Cartesius in Malpighi's *Vita* has now been identified as a misreading for Cortesius, the Bologna anatomist Giovanni Battista Cortesi; Bertoloni Meli, "Archive," 112.

18. Quoted in Adelmann, *Embryology*, 806. Frank, *Harvey*, 90–93. *MCA*, 1:72–76, Borelli to Malpighi, Pisa, 4 February 1661, at 73 and n5.

19. *MCA*, 1:58–59, at 58, Malpighi to Borelli, 18 January 1661; Borelli to Malpighi, 87–88, 13 May 1661; see also 177, 17 August 1663. Adelmann, *Embryology*, 369–70, 2236. *MOB*, 76, 79, 95n9, 97nn11–12. Malpighi, *Vita*, 31–32.

20. *MOB*, 75, 82–88, 87n15, 89–90. Harvey, *De motu cordis*, in *Circulation*, 13. Frank, *Harvey*, 15.

21. *MCA*, 1:72–76, Borelli to Malpighi, Pisa, 4 March 1661, at 73. *MOB*, 92, 94–98. On the notion of "action" see Schiefsky, "Galen's Teleology," 381–88. In his lecture notes, Adelmann, *Embryology*, 159, Malpighi stated that his finding was contrary to what Harvey had believed: "Harveus et alij censent nulla adesse anastomosim sed et tracheam et vasa sanguinea hiare in carnem pulmonum in qua recipiatur sanguis et ab eadem propagetur in venas." Adelmann, *Embryology*, 194–95.

22. *MOB*, 98–99. *MCA*, vol. 1, letters by Borelli to Malpighi: 65–68, 18 February 1661, at 67; 80–83, 24 March 1661, fermentation is discussed at 82; 83–85, 8 April 1661. Galluzzi, "Cimento." Gómez López, "Atomism," 179–80, argues that Borelli's suggestions were related to his research on capillarity at the Cimento exactly in those years; *Passioni*, 58–59. Borelli, *De motu*, vol. 2, prop. 116. Bertoloni Meli, "Posthumous Dispute," 271–72; "Authorship."

23. *MCA*, 1:54–56, 4 January 1661, on 55: "Ne perche le cose sono assai piccole si dovranno stimare difficili a disegnarsi, et intagliarsi, perche Vostra Signoria puó fare le cose in grande protestandosi, che per maggior' chiarezza è necessario alterar' le dimensioni di detti lobuli, o membrane, e loro siti: una cosa simile fece il Cartesio nella sua filosofia, e meteora, il quale con

quel suo bello, et artificioso modo di spiegarsi, e dichiararsi ha affascinato non pochi huomini da bene." Compare this passage with Malpighi's account of his art of representation in *Anatome plantarum* discussed in sec. 9.2. On the status of icons of Malpighi's figures see J. C. Sournia, *The Illustrated History of Medicine* (London, 1992, translation of the original 1991 French edition), 271. Baigrie, "Scientific Illustration." Wilkin, "Figuring." Des Chene, *Spirits*, 72–78. Kemp, "Mark." See also secs. 3.2, 7.5, and 9.2.

24. Camerota, *Galileo*, 387–89. *MAP*, 281–82, at 182, Borelli to Malpighi, Pisa, 5 March 1659 [*more pisano*=1660]. See also sec. 2.3. Steno, *De glandulis oris*, in *Opera*, 1:35–36. Wilson, *Epicureanism*, 42–43.

25. Frank, *Harvey*, 255–56. Gassendi, *Syntagma*, in *Opera omnia*, 2:310b–318a. On 314a Gassendi begins his list of objections to Harvey: "Imprimis, si arteriae monstrari alicubi possent oscula iungere cum venis, haberi res posset non modo verisimilis, sed etiam pene dicam, confecta: at ipsemet Harvaeus cogitur fateri, ne suspicari quidem licere, ubi-nam fiat talis coniunctio." French, *Harvey*, 152–53, 328–33. Gassendi's *Discours* was published under the initials S. S. (for Samuel Sorbiére). Bloch, *Gassendi*, 440–41. *MCA*, 1:81, Borelli to Malpighi, 24 March 1661. *MOO*, 2:143. *MOB*, 98.

26. *MCA*, 1:65–69, at 68, Borelli to Malpighi, 18 February 1661. See the entry on Thomas Bartholin in *DSB* by Charles D. O'Malley. Bartholin, *Epistolae*, I:21–27ff.

27. Adelmann, *Embryology*, 220–21, briefly summarizes Bartholin's work. Borelli found it objectionable; *MCA*, 1:172–75, Borelli to Malpighi, Florence, 20 July 1663. Bartholin, *De pulmonum substantia*, 20–21, 30, 44–45. Besides Munierus, also Sibold Hemsterhuis published Bartholin's work (together with Pedquet's and Rudbeck's) at Leiden in 1654 and Heidelberg in 1659.

28. *MCA*, 1:407–8n11. Bartholin, *Epistolae*, IV, *ad indicem*; 113–17, Bartholin to Steno, Copenhagen, 7 September 1662, at 117; 348–59, Steno to Bartholin, 5 March 1663, at 348–51 (the reference to Swammerdam is at 351); 359–63, Bartholin to Steno, Copenhagen, 7 April 1663, at 363. *MOB*, 85. Frank, *Harvey*, 155. Harwood, "Rhetoric," 124–25.

29. Thruston, *De respirationis usu primario, diatriba* (London, 1670), in *Bibliotheca anatomica*, 1:1000–1057, at 1000–1001, 1008a, 1032a; Thruston's original dissertation is at 1000–1019. Frank, *Harvey*, 241–43. Adelmann, *Embryology*, 362–64. Malpighi, "An Extract of a Latin Letter," *PT* 6 (1671), 2149–50. *OCH*, 7:448–51, Malpighi to Oldenburg, 10 February 1671. On the disputations as vehicles for novelties see French, "Harvey and the Circulation," especially 54–58.

30. Frank, *Harvey*, 157–59, 242. Walmsley, "Locke on Respiration."

31. Lower, *De corde*, 167–68, Franklin's translation. Colombo, *De re*, 224. Frank, chap. 7. Lower, *Vindicatio*, 117–18. Frank, *Harvey*, 200–201, 315. *PT* 2 (1667), 509–16. Chapman, *Leonardo*, chaps. 2 and 6.

32. Lower, *De corde*, 165–67. Frank, *Harvey*, 214–15.

33. Harvey, *Circulation*, Harvey to Schlegel, 26 March 1651, 140–45, at 140–41; the letter was first published in the 1766 edition of Harvey's work with a reference to Ent's role, 611–19. French, *Natural Philosophy*, 279–85. Cole, "Injections," 290–91. Harvey's experiment was probably known to Lower and Hooke, but it involved only the plumbing of the heart and lungs.

34. Frank, *Harvey*, chaps. 9–10. Lower, *De corde*, 168–71. Boyle had published recently *Experiments and Considerations Touching Colours* (London, 1664).

Chapter 2 · Epidemic Fevers and the Challenge to Galenism

1. Morgagni, *Opera postuma* (Rome, 1964), vols. 2–3, contains his lessons on theoretical medicine and commentary on Galen. Guinther's edition of *On Anatomical Procedures* was a landmark in the history of anatomy; Sylvius's edition was less significant but provided an amended translation and commentary to Niccolò da Reggio's translation. Vesalius, *Fabrica*, preface to Charles V. Bylebyl, "Disputation," 224–26.

2. Mani, "Jean Riolan II." Brown, "College"; "Physiology." Cook, *Decline*.

3. Cook and Lux, "Closed Circles." Beretta, Clericuzio, and Principe, eds., *Cimento*.

4. *Bibliotheca anatomica*, 1:62. Adelmann, *Embryology*, 290.

5. Bertoloni Meli, "Neoterics," 60, 62, 66; *MCA*, 1:122–23, Borelli to Malpighi, 1 April 1662; 124–25, the Messina Senate to Malpighi, 25 April 1662. On Tuscany and the Cimento see Beretta, Clericuzio, and Principe, eds., *Cimento*. Cavazza, *Settecento*.

6. *MOB*, 15; the reference to the "iatromechanical manifesto" is one of the very few points of disagreement I have with Belloni. On Santorio see the entry by Mirko D. Grmek in the *DSB*. Borelli, *Delle cagioni*, 31, 141.

7. Borelli, *Delle cagioni*, 8, 129; here he relied on color. Stolberg, "Uroscopy."

8. Borelli, *Delle cagioni*, 53–54; *De vi percussionis*, prop. 8, quoted in Bertoloni Meli, *Thinking with Objects*, 229; *De motu animalium*, 2:96.

9. Borelli, *Delle cagioni*, 89–91. On Borelli see the entry by Ugo Baldini in *New DSB*. Siraisi, *Medieval and Early Renaissance Medicine*, 16, 67–68, 123, 128–29, 134–36.

10. Borelli, *Delle cagioni*, 113. Nutton, "Reception," especially 203, 233. Harvey, *Circulation*, 147, 155–56, 162. Beretta, "Revival"; "Scienziati"; "Lucretius," 11. Trabucco, "Corpuscolarismo."

11. Borelli, *Delle cagioni*, 115. Galileo's views were put forward in *The Assayer*. Camerota.

12. Borelli, *Delle cagioni*, 127. Castelli, *Praservatio*, 4–5, 22–25. Dollo, *Modelli*, 148. Camerota, *Galileo*, 365–86. Here, too, Borelli relied on color.

13. Borelli, *Delle cagioni*, 128, 158–59. Belloni, "Circolazione." French, *William Harvey*, is unaware of Borelli's treatise and Belloni's essay.

14. Castelli, *Praeservatio*, 13–16, 26. Borelli, *Delle cagioni*, 138–39. Jean Béguin, *Tyrocinium chymicum* (Venice, 1643); *Epistolae medicinales* in the *Opuscula medica* by Baudouin van Ronss (Leiden, 1618); *Praxis chymiatrica* by Johann Hartmann (including a treatise by Hamer Poppius, this being the author referred to by Borelli; Geneva, 1635); Oswald Croll, *Basilica chymica* (Frankfurt, 1608); Johann Daniel Mylius, *Basilica chymica* (Frankfurt, 1618–30); Andreas Libavius, *Syntagma selectorum alchymiae arcanorum* (Frankfurt, 1611); Petrus Severinus, *Idea medicinae philosophicae* (Erfurt, 1616); Johann Rhenanus, *Antidotarium pestilentiale* (Frankfort, 1613). I identified the *Idea* by "Pietro Senario" as a work by Petrus Severinus; my identification of the different editions is conjectural. Dollo, *Modelli*, 149–55. Trabucco, "Delle cagioni," 254–55. A medicament based on sulphur had been recommended by the Roman physician Cintio Clementi to Galileo's correspondent Virginio Cesarini; *GOF*, 13:88–89, Cesarini to Galilei, 7 May 1622; I thank Antonio Clericuzio for this reference.

15. Borelli, *Delle cagioni*, 140–41, 143, 149–53. *MCA*, 1:27–28, Borelli to Malpighi, 18 March 1660. For his use of mercury-based medications see also *MCA*, 1:95–96 and 96–97, Borelli to Malpighi, 21 October 1661, and Malpighi to Borelli, 8 November 1661. Camerota, "Lucrezio." Castelli, *Responsio*, 10–17, at 11 and 13.

16. Borelli, *Delle cagioni*, 154, 157, 172, 177, 214. Lonie, "Fever Pathology," at 39–40. Pagel, *Van Helmont*, 159–61.

17. *MCA*, 1:97–98, Borelli to Malpighi, 18 November 1661; 104–5, Malpighi to Borelli, December 1661. Conforti, "Experimenters," 34. Adelmann's treatment of the correspondence between Borelli and Malpighi on the Pisa epidemics is uncharacteristically brief; *Embryology*, 199. Borelli discussed fevers in *De motu*, vol. 2, chap. 22, including phlebotomy.

18. *MCA*, 1:101–3, Borelli to Malpighi, 9 December 1661; 109–10, Borelli to Malpighi, 29 December 1661.

19. *MCA*, 1:98–100, Borelli to Malpighi, 25 November 1661, at 99; 101–3, Borelli to Malpighi, 9 December 1661, at 103; 100–101, Malpighi to Borelli, early December 1661; 104–5, Malpighi to Borelli, mid-December 1661. For a reference to the 1654 epidemic at Pisa see below.

20. *MCA*, 1:98–100, Borelli to Malpighi, 25 November 1661, at 99; 105–9, Borelli to Malpighi, 20 December 1661, at 106. A useful collection on fevers is Bynum and Nutton, eds., *Theories of Fevers*, especially the entries on "fever" in the index.

21. *MCA*, 1:104–5, Malpighi to Borelli, mid-December 1661; 105–9, Borelli to Malpighi, 20 December 1661, at 106. In order to make sense of the entire passage, I interpret the term "speculative" as used by Borelli as meaning "on the contrary." On Giacchini's and Malpighi's views on purging see the following section.

22. Galileo, *Assayer*, par. 48, 309. Middleton, *Experimenters*, 234–37, 362, 264. *MCA*, 1:105–9, Borelli to Malpighi, Pisa, 20 December 1661, at 106–8. Bertoloni Meli, "Color." Clericuzio, "Cimento," 24. The Scottish physician James Prim(e)rose had referred to the change of color of tincture of roses when spirit of vitriol is added in *Exercitationes, et animadversiones in librum, De motu cordis, et circulatione sanguinis*, reprinted in the 1639 Leiden edition of Harvey, *De motu cordis*, 80. French, *Harvey*, 114–21.

23. Adelmann, *Embryology*, 211; Malpighi, *Memorie*, 15. *MCA*, vol. 1.

24. Moscheo, "Real Edition." Dollo, *Modelli*, 290. On the medical scene in Italy at the time see Gentilcore, "Medical Practice"; "Protomedici."

25. I have slightly altered Adelmann's translation in *Embryology*, 269n3. Adelmann, *Embryology*, 200. Dollo, *Modelli*, 286–87, conclusion 29. Malpighi, *Consultationes medicae*, 22: "parum enim differt secundum causam materialem Apoplexia ab Epilepsia."

26. Dollo, *Modelli*, 283–89, provides a full transcription of the Latin text: "Hinc naturales partium constitutiones, et actiones, earumdemque morbosas dispositiones facillime, et commode explicamus, tamquam eas, quae machinis quibusdam et organis constant, et mecanice, anima dirigente, perficiuntur." Adelmann, *Embryology*, 271–75, provides a free English rendering with commentary. I have slightly altered his translation at 271.

27. Borelli, *De motu animalium*, 2:37. Lonie, "Hippocrates." See secs. 10.3 and 11.3.

28. Dollo, *Modelli*, 284. Adelmann, *Embryology*, 272. Galen, *Usefulness*, IV, 20; V, 2; VI, 4; VIII, 14; XI, 5; XVI, 2. Galen also claimed that some glands produce a juice to moisten their surroundings; *Usefulness*, 683n5.

29. Dollo, *Modelli*, 285–86; Adelmann, *Embryology*, 273.

30. For Malpighi's definition of *indicationes* see *MOB*, 564; Bertoloni Meli, "Mechanistic Pathology," 176. Adelmann, *Embryology*, 273–75, and 274n5. Dollo, *Modelli*, 286–89. Steno, *De glandulis oris*, in *Opera*, 1:27–28. Pagel, *Helmont*, 162–70, 183–85. *MCA*, 1:147–48, Borelli to Malpighi, 15 February 1663. Descartes, *L'homme*, 203–4.

31. Lindeboom, *Dutch Medical Biography*, 148–50. Cook, *Exchange*, 271–77. *MCA*, 1:255n6. Moscheo, "Real Edition."

32. Dollo, *Modelli*, 290–304, quotation at 299. Avicenna, *Canon*, 7r. Newman, "Roger Bacon," 320–21. Adelmann, *Embryology*, 276–80. Gentilcore, "*Organisation*," 90. Moscheo, "Real Edition," 314.

33. Bertoloni Meli, "Neoterics," 76–79.

34. Fisch, "Investigators," 525–34, at 534n78; Bartoli's *Artis medicae dogmatum examen* appeared in 1666 as part of the controversy over the fevers around Lake Agnano. Dollo, *Modelli*, 177. Torrini, "Investiganti." *MCA*, 1:286–88, Borelli to Malpighi, 4 November 1665, at 287. See also *MCA*, 1:302, Borelli to Malpighi, 13 February 1666. Malpighi, *Risposta* to Lipari, 83.

35. Malpighi, *Risposta* to Lipari, *Opera posthuma*, 25–26. Zinato, *Vero*, 59. Piccolino, *Zufolo*, 141. *MCA*, 1:305–7, Capucci to Malpighi, Crotone, 4 April 1666, at 306.

36. Malpighi, *Risposta* to Lipari, 20–21, 26, 62–63. *MCA*, 1:269–70, Borelli to Malpighi, 1 August 1665. Adelmann, *Embryology*, 277, 287. For Malpighi's shifting attitudes on the certainty of medicine and the a priori compare sec. 11.3.

37. Malpighi, *Risposta* to Lipari, 51–63, at 53, 54, 61.

38. Malpighi, *Risposta* to Lipari, 32, 36, 40–43. On the great significance of this simple observation, later published by Fracassati, see Frank, *Oxford Physiologists*, 205–7. *MOB*, 192–93. See also *Vita*, 3, where Malpighi reported chymical experiments on the change of color in blood.

39. Malpighi, *Opera posthuma, Vita*, 2; *Risposta* to Lipari, 40–41 and 76–79 (separate pagination). See also *MCA*, 1:14–16, Borelli to Malpighi, 20 September 1659. Bertoloni Meli, "New Anatomy," 33 and 35. Adelmann, *Embryology*, 135. On purgation see also *MCA*, 1:14–16, Borelli to Malpighi, 20 September 1659; 152–53, Borelli to Malpighi Pisa, 22 March 1663. Adelmann, *Embryology*, 275; Dollo, *Modelli*, 288–89, 43rd conclusion. On Giacchini see the entry in the *DBI* by B. Giovannozzi.

40. Malpighi, *Risposta* to Lipari, 75–81.

41. Malpighi, *Risposta* to Lipari, 45, 50–51. Dollo, *Modelli*, 298. Brown, *Mechanical Philosophy*, 152ff., esp. 158–59. On Willis see Frank, *Harvey*. Isler, *Willis*, 45–83. Hughes, *Thomas Willis*, 51–53. Bates, "Thomas Willis." Malpighi, *Risposta* to Lipari, 64–70, at 66.

Chapter 3 • The Anatomy of the Brain and of the Sensory Organs

1. Gabbey, "Mechanical Philosophies." For a more nuanced view see Meinel, "Atomism."

2. *De omento* is discussed in chap. 4. *Bibliotheca anatomica*, 2:1–3 and 318–19; the portion of Willis's treatise on the nerves is included in pt. 4 of the *Bibliotheca. Bibliotheca anatomica* includes also Malpighi's later *De cerebri cortice*, which is discussed in chap. 4, and Wepfer's work on a girl born without a brain: "De puella sine cerebro nata," *Miscellanea curiosa*, 3:175–203; *Bibliotheca anatomica*, 2:99–112; *MOB*, 346.

3. *Tetras anatomicarum epistolarum* included *De omento*, ostensibly anonymous but also due to Malpighi, discussed in chap. 5. Adelmann, *Embryology*, 283–84. Belloni, "Neuroanatomia."

4. Rossetti, *Antignome*, 19, states that Malpighi's works were often bound together. Frati, *Bibliografia malpighiana*, #14. Sbaraglia's copies at the Biblioteca of the Archiginnasio in Bologna are bound together. Bernard Quaritch Catalogue 1332, item 69.

5. Willis, *Anatomy of the Brain*, 72–73 and 82–86. Cole, "Anatomical Injections," at 292.

Hooke, *Micrographia*, 144. Gibson, "Wren." Bennett, "Respiration." Glisson, *Anatomia hepatis*, in *Bibliotheca anatomica*, 1:295b. Boyle, *Correspondence*, 2:1–5, Lower to Boyle, 18 January 1662, quotation at 3; 78–81, Lower to Boyle, 4 June 1663. Frank, *Harvey*, 17–18, 170–75. Bertoloni Meli, "Collaboration," sec. 4.

6. Willis, *Anatomy of the Brain*, chaps. 1, 5, 8, 10, 20; pp. 86, 91, 96; quotations at 85, 96. For the nervous juice see Boyle, *Correspondence*, 2:1–5, Lower to Boyle, 18 January 1662, at 3. *MCA*, 1:236, Borelli to Malpighi, 6 October 1664. Isler, *Willis*, 86–112. Hughes, *Willis*, chap. 8.

7. Willis, *Anatomy of the Brain*, 51, 106. Bynum, "Anatomical Method," 447–59. *MCA*, 1:242–43, Borelli to Malpighi, 19 December 1664. Descartes, *Treatise on Man*, 86–92.

8. Bartholin, *Epistolae*, IV:103–13, at 113, Steno to Bartholin, Leiden, 26 August 1662; also 348–59, at 358, Steno to Bartholin, Leiden, 5 March 1663; 359–63, at 360, Bartholin to Steno, 7 April 1663. Steno, *Epistolae*, vol. 1, Steno to Leibniz, November 1677, 366–69, at 367, states that he came to doubt Descartes' views through the study of the heart. C. Bartholin, *Institutiones* (Leiden, 1641), 261, 279–84. Clarke, "Brain Anatomy," 27–29. Dewhurst, "Willis and Steno." Steno's work on glands is discussed in the following chapter; for his work on muscles see Bertoloni Meli, "Collaboration," sec. 8. Brown, *Mechanical Philosophy*, 91–99. Faller, "Die Hirnschnitt-Zeichnungen," 118. My interpretation relies on and underscores Wilkin, "Figuring." Des Chene, *Spirits*, 74. Baigrie, "Scientific Illustrations," 89, is very confused, attributing Schuyl's 1662 engraving to Descartes, while his fig. 3.4 shows a later rendering of Schuyl's drawing, which he attributes to the 1664 French edition by Clerselier. Ragland, "Experimenting," contains useful material on Schuyl.

9. Steno, *Opera*, 2:4–7, 16. Scherz, *Nicholas Steno's Lecture on the Anatomy of the Brain*, provides a useful introduction, 61–103; captions to Steno's drawings are at 202–7. The 1671 Latin edition of Steno's lecture omitted the figures altogether. Faller, "Hirnschnitt-Zeichnungen," 126–35. Lindeboom, *Dutch Medical Biography*, 743–44. Clerselier attacked Schuyl in the preface to *L'homme*. Wilkin, "Figuring."

10. Steno, *Opera*, 2:8–12, 24–26. At the end of his treatise Steno quoted passages mostly on the pineal gland from Descartes, *L'homme*, notably from pp. 11, 12, 63, 65, 72, 77, 78. Aucante, *Philosophie*, 229–47. Wilkin, "Figuring."

11. *MCA*, 1:154–56, Borelli to Malpighi, 30 March 1663, at 156; 237–38, Malpighi to Bonfiglioli, 23 October 1664. Malpighi, *De cerebro*, in *Opera*, 2:4–6, separate pagination. Belloni, "Neuroanatomia." Several of Malpighi's consultations deal with apoplexy and epilepsy: *Centuria prima*, 25–34, 38–39, 41–42, 49–51. *Consultationes medicae*, 18–24.

12. Gassendi, *Syntagma*, in *Opera omnia*, 2:369–82, at 369b–370b. On Gassendi and vision see Bloch, *La philosophie de Gassendi*, chap. 1. Dennis, "Graphic Understanding." Harwood, "Rhetoric."

13. *MCA*, 1:207 (28 March 1664) and 212–13 (16 May 1664). *Opera posthuma*, 1–8 (separate pagination). Adelmann, *Embryology*, 227–32. *DOAT*, 6:109ff. Wolf-Devine, *Descartes on Seeing*. Cunningham, "Fabricius," 216–17. Malpighi, *De cerebro*, in *Opera*, 2:8–12, separate pagination. *PT* 2 (1666), 491–92. Aucante, *Philosophie*, 269–77.

14. *MCA*, 1:298, 8 January 1666, Borelli to Malpighi. Adelmann, *Embryology*, 240–41.

15. *MCA*, 1:221–23, 19 July 1664, Borelli to Malpighi, at 222.

16. *MCA*, 1:229–31, at 230, Borelli to Malpighi, 15 August 1664. *MOB*, 113. Adelmann, *Embryology*, 258–59.

17. *MOB*, 108–9, 110. Luigi Belloni's annotated translation of *De lingua* and *De externo*

tactus organo is especially valuable for the additional plates based on his reproduction of Malpighi's observations and experiments following his original techniques. According to Sbaraglia, *Vigiliae*, 71, Niccolò Massa had already studied the tongue by boiling it.

18. *MOO*, 2:14–15 and 19; *MOB*, 110–11 and 120–21. Fracassati, *De lingua*, in *Bibliotheca anatomica*, 2:323b–324a.

19. *MOO*, 2:17–18; *MOB*, 116–17; Wharton, *Adenographia*, chap. 22, esp. 147; du Laurens, *Opera*, 758–59.

20. *MOO*, 2:18; *MOB*, 118; Galileo, *Assayer*, in *GOF*, 6:349; Plato, *Timaeus*, 65–66. Gassendi, *Syntagma*, 356b–358b. Aucante, *Philosophie*, 261.

21. *MOO*, 2:19; *MOB*, 121. Aristotle, *De anima*, II.10 and 421a 16–16; *De sensu* (part of *Parva naturalia*), IV and 441a 1–4. Johansen, *Aristotle*, 178–225.

22. *MOO*, 2:22; *MOB*, 130–31. The footnotes can be found also in the 1669 Amsterdam edition and *Opera omnia*.

23. *MOO*, 2:23–27, 28, 30; *MOB*, 132–42, 144, 149; on *MOO*, 2:26 and *MOB*, 141, as in *De lingua*, Malpighi referred to Plato's *Timaeus*.

24. *MOO*, 2:27–29, 29–30; *MOB*, 142–46, 147–48. Johansen, *Aristotle*, 199–212. Gassendi, *Syntagma*, in *Opera*, 2:353–56, on 354a; 377b. Sabra, *Light*, 54–55; Descartes, *L'homme*, 50. Aucante, *Philosophie*, 256–61.

25. Gómez López, "Atomism," 182–83.

26. *MCA*, 1:205–6, at 205, Fracassati to Malpighi, 24 March 1664; 245–46, Borelli to Malpighi, 24 January 1665, at 246; 248–49, Borelli to Malpighi, 7 March 1665; 250–51, Fracassati to Malpighi, 20 March 1665, at 250; 277–78, Fracassati to Malpighi, 15 August 1665. Fracassati, *De cerebro*, in *Bibliotheca anatomica*, 2:70b. On Fracassati see the entry in *DBI* by Gabriella Belloni Speciale.

27. Fracassati, *De lingua*, in *Bibliotheca anatomica*, 2:323b–324b, 325b, 326a, 332a. Aristotle, *De sensu et sensibilis*, IV, 441a–b.

28. Fracassati, *De lingua*, in *Bibliotheca anatomica*, 2:327a–328b, 329b.

29. Fracassati, *De lingua*, in *Bibliotheca anatomica*, 2:326a, 328b, 331a–332a, 333a–334a. Tribby, "Club Medici," at 232–33. Bernardi, "Teoria e pratica," at 19n. Baldini, *Libertino*. Images of the shapes of different salts were published in the plate of M. Lister, *De fontibus medicatae Angliae* (York, 1682), and subsequent editions.

30. Fracassati, *De cerebro*, in *Bibliotheca anatomica*, 2:71b. Magiotti, *Renitenza certissima dell'acqua alla compressione* (Rome, 1648), is reprinted in Belloni, "Schemi e modelli," 271–82; it was dedicated to Lorenzo de' Medici, uncle of Grand Duke Ferdinando II. The reference to Borelli, as well as to Harvey, is at 286. Borelli, *De motu*, 2:23. Cavazza, "Vis irritabilis," 65.

31. Fracassati, *De cerebro*, in *Bibliotheca anatomica*, 2:76b, 79a. Bertoloni Meli, "The Color of Blood." *PT* 2 (1666), 492.

32. Fracassati, *De cerebro*, in *Bibliotheca anatomica*, 2:67b–68b, 76a–b, 79b, 80b–82b. *MCA*, 1:248–49, Borelli to Malpighi, 7 March 1665, at 249; 279–80, Bonfiglioli to Malpighi, 5 September 1665, at 280. Borelli, *De motu animalium*, vol. 2, prop. 224. Frank, *Harvey*, 173–76. *PT* 1 (1665), 128–30; 2 (1666), 490–91. Cook, *Matters of Exchange*, 271–78. Peumery, *Origines*, 5, 8–9, 11–12. Marinozzi and Conforti, "Blood."

33. Fracassati, *De lingua*, in *Bibliotheca anatomica*, 2:334b. Bellini, *Gustus organum*, in *Bibliotheca anatomica*, 2:357a. Adelmann, *Embryology*, 242–43. *MCA*, 1:224–26, Borelli to Malpighi, 2 August 1664, at 225.

34. Bellini, *Gustus organum*, in *Bibliotheca anatomica*, 2:337a–338b, 341a–b, 344a, 345a. Aristotle, *De sensu*, IV, 442b, 20–24. Johansen, Aristotle, 182–88, at 186. Sailor, "Moses," at 8–13. McGuire and Rattansi, "Newton," 130, 134.

35. Bellini, *Gustus organum*, in *Bibliotheca anatomica*, 2:358b–359b, 361a–b, 362a–363b.

36. Rossetti, *Antignome*, 14, 16–20, 30–31. Aristotle, *De anima*, II.6, II.11, 422b, 18–34. Two years later, in *Osservatione*, Borelli tried to "decompose" sight too, arguing on the basis of some experiments that the left eye sees objects larger and more distinctly than the right eye. Conforti, "Medicina."

37. Atti, *Notizie*, 115. Adelmann, *Embryology*, 333–37.

Chapter 4 · The Glandular Structure of the Viscera

1. Du Laurens, *Historia anatomica*, 171–73; his views were entirely traditional.

2. Christie's, London, sale Einstein-7269, lot 8: Wharton, *Adenographia* (Nijmegen, 1664); Steno, *De glandulis oris* (Leiden, 1662); de Graaf, *De succo pancreatico* (Leiden, 1664); Highmore, *Exercitationes duae* (Amsterdam, 1660). The first three treatises form a remarkable cohesive collection on glands, while the last—focusing on diseases—reminds us that there were no boundaries between physicians and anatomists, given that de Graaf's treatise had a medical dimension as well. Jeremy Norman, catalogue number 7815: de Graaf, *Du suc panreatique* (Paris, 1666); Malpighi, *De viscerum structura* (Amsterdam, 1669); Steno, *De musculis et glandulis* (Amsterdam, 1664); and *Collegium privatum* of Amsterdam, *Observationes anatomicae selectiores* (Amsterdam, 1667). Christie's, New York, Haskell Norman sale, 1998, lot 474 joins de Graaf, *De succo pancreatico* (Leiden, 1664), and Malpighi, *De viscerum structura* (Amsterdam, 1669). The other items in the *Sammelband* are Wolferdus Senguerdius, *Tractatus physicus de tarantula* (Leiden, 1668); György Kovács Tatai, *Epilepsiae vera dignotio* (Leiden, 1670); and Charles Drelincourt, *De partu octimestri* (Leiden, 1668).

3. Cunningham, "Historical Context," xxxiv–xxxviii. Giglioni, "Atheist." See the *DNB* entry on Glisson by Guido Giglioni.

4. Glisson, *Anatomia*, in *Bibliotheca anatomica*, 1:306b; 295b–296b, 329a, 339a. Eriksson, *Vesalius*, 165. Harvey, *Circulation*, 108.

5. Glisson, *Anatomia*, in *Bibliotheca anatomica*, 1:251, 283a–284a, 293a–294a, 337a.

6. Glisson, *Anatomia*, in *Bibliotheca anatomica*, 1:343b–358b. Cunningham, *English Manuscripts*, 149, 163–75, 181–91; "Historical Context," xlv–xlviii.

7. Wharton, *Adenographia*, 44.

8. Glisson, *Anatomia*, in *Bibliotheca anatomica*, 1:346b, 353b. Cunningham, *English Manuscripts*, 129, 181. Wharton, *Adenographia*, 1–3, 23.

9. Wharton, *Adenographia*, chap. 21, at 130–32.

10. Wharton, *Adenographia*, 125–27, 140–46, esp. 140, 143; 76, 79, 81.

11. Sylvius, *De lienis et glandularum usus* (Leiden, 1660); in *Opera medica* (Amsterdam, 1680), 21–24, at 23. Steno, *Opera*, 1:20–22; 2:182–92, at 182–83. His main work is *De musculis et glandulis* (1664).

12. Steno, *Opera*, 1:25–26, 35–36, 38–39. Ruestow, *Microscope*, 89.

13. Adelmann, *Embryology*, 291–93.

14. *MCA*, 1:324–25, Borelli to Malpighi, 11 September 1666. *De hepate*, *MOO*, 2:75, refers to a postmortem dissection dated 6 August 1666, *MOB*, 415. Bertoloni Meli, "Blood."

15. *MOO*, 2:53–55, 111. Adelmann, *Embryology*, 296–98, 310. On the respective merits of verbal and visual representations see Ekholm, "Representations."

16. Bartholin, *Exequiae*. *MOO*, 2:59–60; French, 12–13. Wepfer had had a similar intuition as well: see sec. 10.4. *Bibliotheca anatomica*, 1:359.

17. Harvey, *Generation*, 116, chap. 19. *MOO*, 2:58–60, 62–63, 75. Adelmann, *Embryology*, 298–300. *Risposta* to Sbaraglia, *MOB*, 514.

18. *MOO*, 2:65, *De hepate*, sec. 4.

19. *MOO*, 2:67. Steno, *De glandulis oris*, in *Opera*, 1:42–43; *Responsio ad vindicias*, in the 1662 *Observationes anatomicae*, in *Opera*, 1:59–73.

20. De Bils, *Epistolica dissertatio*, 4. *MOO*, 2:68–69.

21. *MOO*, 2:69–72. Adelmann, *Embryology*, 299. De Back, *Dissertatio*, 76–89. Lindeboom, *Dutch Medical Biography*, 50–51. Pechlin, *De purgantium medicamentorum facultatibus*, 502–3. The reference to de Back's work is of broader interest for Malpighi's formation because his treatise was published together with a reprint of Harvey's *De motu cordis*.

22. Dodoens, *Medicinalia* (1621), 105–6, on a case of "Lapides in tunica hepatis reperti"; the text says: "iocinera vidisse, adeo dura & lapillis undequaque plena." Malpighi referred to Dodoens as a practitioner of postmortems in his *Risposta* to Lipari, 58. *MOO*, 2:61, 63. Glisson, *Anatomia*, in *Bibliotheca anatomica*, 1:263b–264a. Malpighi may have referred to Möbius, *Fundamenta medicinae physiologica* (Jena, 1657, 1661^2), a text he quoted in the *Risposta* to Sbaraglia, *MOB*, 537, and also referred to in his lectures: Adelmann, *Embryology*, 157n. Unlike Adelmann, *Embryology*, 209, I see no evidence that he met Möbius in Rome rather than seeing his book. Siraisi, *Clock*, 153–58. *MCA*, 1:376–80, Capucci to Malpighi, Crotone, 24 July 1688, at 378. Daston and Park, *Wonders*, chaps. 6–7; for the bezoar stone and its remarkable properties see 75, 155, 161, 167.

23. *MOO*, 2:62–63, 75.

24. *MOO*, 2:63. Münster, "Anatomica," 185–86. *MOB*, 415–16, 466–67 (discussed in chap. 12). *MCA*, 1:105–9, Borelli to Malpighi, 20 December 1661, at 108. Malpighi, *Centuria*, 67; see also 66. Nutton, "Galen," 8, 11. Lower, *De corde*, 6, spells out the link between origin of the fluid and behavior when heated.

25. Bertoloni Meli, "New Anatomy," 46. Clarke and Bearn, "Brain 'Glands.'"

26. Steno, *Opera*, 1:190; 2:232–33.

27. The *rete mirabile* exists in sheep and ungulates, however, not in humans, as Berengario and later Vesalius pointed out. Willis, chap. 9, 87; see also 83. Gibson, "Wren." Bennett, "Respiration." *MOO*, 2:82. Adelmann, *Embryology*, 300–302.

28. Fracassati, *De cerebro*, in *Bibliotheca anatomica*, 2:70a. Velthuysen, *Tractatus duo*, 62–64 (first pagination) and 166–69 (second pagination). Lindeboom, *Dutch Medical Biography*, 2038–39. *MOO*, 2:77, 83. Adelmann, *Embryology*, 301.

29. Wharton, *Adenographia*, 9.

30. *MOO*, 2:77–79, 81–82, 85. *MOB*, 39–40.

31. *MOO*, 2:85. *MOB*, 183–87.

32. Willis, *Brain*, 91, 96. *MCA*, 1:236, Borelli to Malpighi, 6 October 1664. *MOO*, vol. 2, chap. 4, esp. 85. Descartes, *Homme*, 69–70.

33. Bertoloni Meli, "Authorship," 78–81. An excellent characterization of Bellini's treatise is in *Bibliotheca anatomica*, 1:389.

34. *MOO*, 2:88; *MOB*, 164–65.

35. *MCA*, 1:131–34, Borelli to Malpighi, 5 August 1662. Grondona, "Esercitazione," 447–51, 454, 460.

36. Grondona, "Esercitazione," 455–62.

37. *MOO*, 2:89–90. *MOB*, 159, 166–67, 169.

38. *MOO*, 2:90–91, 93. *MOB*, 169, 171, 174. Fournier, *Fabric*, 58.

39. *MOO*, 2:98–100. *MOB*, 183–87.

40. Bertoloni Meli, "Blood." *MCA*, 1:348–49, Malpighi to Sampson, Bologna (May 1667). Malpighi is quite explicit that the preliminary version of *De viscerum structura* had been published in 1666 without *De liene*.

41. *MOO*, 2:102: "Membranarum generatio, proventus, & usus adeo in natura familiaris est, ut non tantum singula corporis membra quoquomodo alicui dicata muneri circumducantur, sed quae naturam ipsam oppugnant, eamque a recta operandi ratione deflectere cogunt, tam providè membranis ipsis donentur, sicuti in tumoribus praecipue accidit, ut suspicari possimus aberrantem etiam naturam membranarum texturam non dedisceret."

42. *MOO*, 2:104, 107. On the appreciation of Malpighi's work on the spleen in the Netherlands see Ruestow, *Microscope*, 82.

43. *MOO*, 2:109–12. Bosco, *De facultate*, 13–14. Malpighi reported that Bartholin had found round bodies in a dolphin's spleen. *MOO*, 2:102–3, 111–12, 116–17.

44. *MOO*, 2:121–22. *MOB*, 419–20, 466. Malpighi, *Centuria*, 116–19; *Consilia*, 61–62.

45. The French translation has expanded on this important passage: Malpighi, *Discours*, 210–11. *MOO*, 2:108: "Praecipuae in animalium operibus officinae sunt viscera, quae naturae micrologiam depraedicant, & fovent, unde lippientibus imponere possunt, quin, & affuso sanguine, illuc copiosè irrumpente, veluti tenebricoso quodam velo naturae mysteria obscurant; quare tot de Lienis substantia disceptationes prodiere, quae caliginem hucusque non detersere." *MOO*, 2:114. Webster, "George Thompson."

46. *MOO*, 2:117, 119, 121. Ragland, "Experimenting." Adelmann, *Embryology*, 311. Palmer, "Odour."

Chapter 5 • Fat, Blood, and the Body's Organization

1. Thomas Aquinas, *De principiis naturae*, chap. 4. Buonamici, *De motu*, 605.

2. *Bibliotheca anatomica*, 1:957. See sec. 3.4.

3. Descartes, *L'homme*. *DOAT*, 11:217–86. Descartes, *The World*, 170–205; Gaukroger's statement at xxxvi that *The Description of the Human Body* first appeared in Schuyl's edition (Leiden, 1662) is inaccurate. Gaukroger, *Biography*, 269–90. Ekholm, "Generation," provides a clear taxonomy of different approaches. Des Chene, *Spirits*, chap. 2. On generation see chap. 8 of this book.

4. *DOAT*, 11:246–50. Descartes, *L'homme*, 1–2, 7–8; *Description*, at 170–72, 182–85. Lennox, "Material and Formal," 168. Aristotle, *Parts of Animals*, 651a20–651b19; 672a1–26.

5. Descartes, *Description*, 184–85; *L'homme*, 7–8, with La Forge's comments on 195–96. Gaukroger, *Descartes' System*, 192–96.

6. Descartes, *The Description*, in *The World*, 185. *L'homme*, 3–4, 7–8, 32–33. Galen, *Faculties*, esp. book 1, secs. 8–12. Galen, *Usefulness*, book 4, secs. 3–4. See Hall's note 19 in Descartes, *Treatise of Man*, 9. Descartes addressed problems with teleology and disease in a celebrated passage from the *Meditations*, *DOAT* 7:143–45.

7. Malpighi, *Vita*, 25; Adelmann, *Embryology*, 212, 265. *MCA*, 1:147–48, Borelli to Malpighi, Pisa, 15 February 1663; 169–70, Borelli to Malpighi, Florence, 24 June 1663. *MOO*, 2:34–37.

8. *MOO*, 2:33 (my emphasis): "Non enim acquiescebam congeri pinguedinem transudante oleosa materia a sanguineis vasis, *ex natura necessitate, et vasorum conditione*, ad fovenda scilicet víscera alimentis destinata, cum non assuescat Natura, incongrua, et inutilia in corporis penitioribus cumulare; quin & in ipsis praecordiis, de quibus praecipue solicita fuit: Nec adeo caecutivit natura in separandis, miscendisque corporibus, quibus conflatur sanguinea moles, vel adeo defecit in densandis vasorum tunicis, ut delata continuo per vasa alimoniae pars in itinere elaberetur."

9. *MOO*, 2:41 (my emphasis): "Circa pinguedinis generationem assentiri nequeo statuentibus sanguinem quasi fortuito e vasis exsudantem in adipis materiam facessere; nam si hoc contingeret *ex natura materiae, et vasorum continentium*, hoc esset perpetuum in singulis partibus, ad quas appellunt sanguinea vasa; in pulmonibus tamen, vesica, meningibus, aliisve partibus membranosis, venis arteriisque minimis, & gracilibus refertis, pinguedinis collectionem non observamus."

10. *MOO*, 2:38–39. See sec. 4 of the introduction.

11. Adelmann, *Embryology*, 262–63. *MOO*, 2:37 (my emphasis): "Postremo haesitari potest, an haec corpora sint communiones striarum factae a pinguedine, quae calore fusa cuniculosas sibi vias inter membranes efformet, *sola materiae necessitate, nullo intercedente naturae fine*. Hoc autem destrui potest, si consideremus haec corpora, ubi solum rete vasorum absque membranis observatur, ut in histrice, sinuoso tractu, quin & lateraliter propagates ramis, hoc rete per longum spatium excurrere, & in insignem altitudinem, si in calenti adhic animali observentur, elevari." Heinemann, "Geschichte," 9.

12. *Anatome plantarum*, *MOO*, 1:56, and sec. 9.2 in this book. The interpretation I provide here differs from the one I offered in "Blood," 520.

13. Adelmann, *Embryology*, 264–65. *MOO*, 2:44 (my emphasis): "Quoniam tamen non tantum calori tribuendum censeo, quantum vulgo jactatur, et in animantibus fortasse *sola materiae necessitate* excitatur; unde calorem morborum facilius auctorem observamus, quam tranquilla vitae opificem."

14. *MOO*, 2:43. Galen, *Usefulness*, 214–15.

15. *MOO*, 2:34–35, 46–47, 48.

16. [Manolessi], *Relazione*, 2. Peumery, *Origines*, 1–12, 38–41. Webster, "Blood Transfusion." Frank, *Harvey*, 176 and 335n108. Marinozzi and Conforti, "Blood." Guerrini, "Ethics," 402–6. Elsholtz, *Clysamatica editio secunda*, 24–26, 59–60.

17. [Manolessi], *Relazione*, 2, 52, 69–70. Peumery, *Origines*, 16–17, 19–22, 34. Frank, *Harvey*, 178, 202; on King see the entry in the *DNB* by Robert L. Martensen.

18. [Manolessi], *Relazione*, 26, 54–55. Frank, *Harvey*, 176. Boyle, *Correspondence*, 3:182–86. Peumery, *Origines*, 24–26.

19. [Manolessi], *Relazione*, 25–26, 36, 41, 61, 63–64.

20. Ibid., 28, 32, 66–67, 69–74. Peumery, *Origines*, 27–36, 42–44, 47–75. Frank, *Harvey*, 202. Conforti and De Renzi, "Ospedali," 455–56. A journalistic account is in Moore, *Blood and Justice*.

21. Anonymous, "Relazione del successo di alcune trasfusioni di sangue, fatte negli animali." [Manolessi], *Relazione*, 69–74. Cavazza, *Settecento*, 44–51.

22. *MOB*, 415–22. The third report is not fully transcribed by Belloni. The missing part

contains references to the spleen and can be found, together with the original Latin of all reports, in Münster, "Anatomica," 204. Malpighi's manuscripts, Bologna, Biblioteca Universitaria, 2085 vol. 12, fol. 61v. Münster, "Anatomica," 194. *MOB*, 419–20. Malpighi, *De polypo*, Forrester translation, 488; *MOO*, 2:129, where the crucial passage is identified by the words "ut in Virgine novissimè deprehendi."

23. Brunner, *Exercitatio*, 12, referred to in *MOB*, 24n8.

24. *MOO*, 2:125–26, 130–31. *MOB*, 198–201, 210, 213. Malpighi, *Risposta* to Lipari, 66. Bertoloni Meli, "Blood," 521–22.

25. *MOO*, 2:123 (my emphasis): "Morbosas constitutiones, quas Naturae ludentis, vel vi morbi aberrantis frequenter in animalium corporibus excitatas miramur, plurimùm lucis pro rimanda ejusdem genuina operandi norma, & methodo conferre perpetuò credidi, quandam enim *materiae necessitatem, & determinatam inclinationem* demonstrant, quae in compingenda animalium mole elucescit, ità ut monstra, caeterísque errores faciliùs, & tutiùs nostram erudiant insipientiam, quàm mirabiles, & perpolitae Naturae machinae: hinc plura didicit praesens hoc saeculum insecta, pisces, primáque & rudia nascentium animalium *stamina* lustrans, quàm anteactae priscorum aetates circa sola perfectorum corpora solicitae."

26. See sec. 9.2. Galen, *Faculties*, 1:7 and 2:3. Newman, *Ambitions*. Von Staden, "Telelology." Cheung, "Fibra," 66–77. Grmek, "Fibre."

27. Findlen, "Jokes of Nature"; *Nature*, 26, 28–29. Daston and Park, *Wonders*. Harvey also used a weaving analogy when he wrote of "the first threads of Nature's weaving" in *De generatione*, 42, quoted in Adelmann, *Embryology*, 2:937.

28. *MOO*, 2:128 (my emphasis): "Nec levia haec mirabitur, qui in assidua animalium sectione naturae industriam in morbosis tumoribus, aequè ac in partium ducendis *staminibus* deprehendit, cùm ferè eadem incedat methodo. Ità memini ferream acum extra gallinae carnosum ventriculum erumpentem bino membranoso, hócque valido involucro, inducta etiam pinguedine, munitam vidisse; & dubitare insuper possumus *sola materiae necessitate, & motu haec omnia contingere, nullo in animalis usum dirigente motore*: ità in quibusdam tumoribus in pulmonibus, hepate, & alibi excitatis circumvolutiones seu multiplices vesicae, quarum amplior minorem continet, et ità successivè coagmentantur, consimilium tumorum, conglobatio ad Polyposam naturam reduci potest, cum eadem probabiliter materia, & productionis ratio in utrisque consimilis sit, nam ex *filamentorum* implicatione ex consueta *naturae lege* multiplices tunicae efformari possunt, inter quas si mediet non concrescibilis, sed aquosus ichor, qualis in hujusmodi tumoribus abundat, solutae undequaque tunicae permanere valent."

29. Harvey, *Exercitatio anatomica*, in Harvey, *Circulation*, 96–97. Bylebyl, "Medical Side," 66–67n24. See also *A Second Essay to Jean Riolan*, 125–28. Harvey, *Circulation*, chaps. 9, 11, 16. De Renzi, "Medical Competence."

30. Fontenelle, *Histoire*, 11: "Mais telle partie dont la structure est dans le Corps humain si délicate ou si confuse qu'elle en est invisible, est sensible & manifeste dans le corps d'un certain Animal. Delà vient que les Monstres même ne sont pas à negliger. La Mechanique cachée dans une certaine espece ou dans une structure commune se développe dans une autre espece, ou dans une structure extraordinaire, & l'on diroit presque que la Nature à force de multiplier & de varier ses ouvrages, ne peut s'empêcher de trahir quelquefois son secret." Daston and Park, *Wonders*, 204, refer to this passage in dating the establishment of a new role of monsters, outlined by Malpighi in *De polypo cordis*, to the beginning of the eighteenth century.

31. *MOO*, 2:130–32. *MOB*, 214–16. Malpighi later expressed reservations about Cornelio;

MCA, 2:886, Malpighi to Ferrarini, 12 December 1682. Torrini, *Cornelio*. The role of niter is discussed in Frank, *Harvey*, esp. 106–13 and chap. 5. Borrelli, "Medicina." Newman, *Ambitions*. Bensaude-Vincent and Newman, *Artificial*.

32. Malpighi, *Vita*, 44–45; *MOB*, 417–18 (erroneously dated 1677 in *Vita*, 45), 435–36, 436–37, 512. Kerckring, *Spicilegium*, obs. 73. *Bibliotheca anatomica*, 1:962a–3b. Malpighi, "*De polypo*," 485, 492.

33. Pomata, "Malpighi." Findlen, *Nature*, 28. On Serafica Francesca see chap. 12.

34. Malpighi, *Vita*, 45; Albertini is discussed in sec. 11.6.

Chapter 6 • The Structure of Glands and the Problem of Secretion

1. Kerckring, *Spicilegium*, 177–79. Adelmann, *Embryology*, 369–70; *MCA*, 2:590–91, Capucci to Malpighi, 21 April 1671.

2. Pechlin, *De medicamentorum purgantium facultatibus*, 491–515, esp. 509–14. Pages 505–15 were excerpted in *Bibliotheca anatomica*, 1:184–86. The work attributed to Pechlin is Janus Leonicenus Veronensis, *Metamorphosis Aesculapii & Apollinis Pancreatici* [Leiden], 1682. Cobb, *Generation*, 181–82. Swammerdam, *De respiratione*, dedicatory poem. Lindeboom, *Dutch Medical Biography*, 1505–7. *Collegium privatum, Observationes selectiores, pars altera*, 18, 27; see the introduction by Lindeboom to the facsimile edition, 20–22. Ragland, "Experimenting."

3. Peyer, *De glandulis intestinorum*, 61, 75, 85. Heinemann, "Frügeschichte," 231–32. A useful summary of Peyer's main results is in *Bibliotheca anatomica*, 1:150, where the editors point out that Peyer's intestinal glands had already been noticed by Pechlin; see also ibid., 163–66, for the extensive annotations on medical issues and case histories. Peyer, *Parerga anatomica et medica septem*, whose *Exercitatio II* is also in *Bibliotheca anatomica*, 1:173–80, at 174b. Joos-Renfer et al., *Beobachtungen*, 10–17. Stiftung von Schnyder von Wartensee, *Aus den Briefen*, provides an overview of the Swiss school.

4. Peyer, *De glandulis intestinorum*, 41. The fact that Peyer's patches, as they were later called, are now considered to be aggregates of lymphatic nodules does nothing to challenge the experimental creativity of his and Brunner's work. Herrlinger, "Rolle."

5. Peyer, *De glandulis intestinorum*, 29–35; preface to the reader and 99–136; *MOB*, 24–25; Belloni, "Scoperta," *Simposi Clinici*, 1964, ix–xiv.

6. Harder, *Prodromus physiologicus*. My account follows the ten chapters of this work.

7. Wepfer, *Cicutae historia*, 113–34, at 121–24 and 129. The *Bibliotheca anatomica*, 1:188a–189b, reports the relevant abstract and the anatomy of the beaver, *Ventriculi castoris descriptio*, at 186–88a, originally published in the *Miscellanea curiosa*. Maehle, *Wepfer* and *Drugs on Trial*. In addition, the *Bibliotheca anatomica* includes the correspondence with the Zurich physician Wilhelm Muralt.

8. Wepfer, *Cicutae historia*, 202–7. Maehle, *Wepfer*.

9. Brunner, *Experimenta*, 39, 45, 55, 58, 59–60 (mercury injection into the pancreatic duct), 118 (for the use of technē), 166; von Staden, "Physis," esp. 38–42. Later Brunner published a brief *Exercitatio anatomica de glandulis in intestino duodeno hominis detectis* (Heidelberg, 1687) in which he challenged Malpighi's belief that secretions occur only in glands; Heinemann, "Frügeschichte," 232–36, at 235. Major, "Brunner." Foster, *Physiology*, 162–64.

10. Brunner, *Experimenta*, dedication, 5r.

11. Brunner, *Experimenta*, 15.

12. See, e.g., Brunner, *Experimenta*, 6, 8, 11, 26. Webster, "George Thomson." Nuck, *Operationes et exprimenta chirurgica*, Leiden, 1692; on Ruysch see Lindeboom, *Dutch Medical Biography*, 1700–1701.

13. Brunner, *Experimenta*, 120.

14. Brunner, *Experimenta*, 32, 89. On thirst and urination see 9, 12, 15, 52; Brunner, *Glandulae duodeni*.

15. Brunner, *Experimenta*, 89, 110, 116, 119.

16. Brunner, *Experimenta*, 95, 97, 118, 121.

17. Cole, *De secretione*, in *Bibliotheca anatomica*, 2:853–91, at 866a, 869b.

18. Cole, *De secretione*, in *Bibliotheca anatomica*, 2:872a, 873b, 874b.

19. On Cole see Frank, *Oxford Physiologists*. Heinemann, "Frügeschichte," 155–56.

20. Lindeboom, *Dutch Medical Biography*, 807. De Heide, *Anatome mytuli*, 175–80. The observation on congealed blood had already been made by Malpighi and reported in print by Fracassati, as we have seen in sec. 1.3; in 1686 de Heide published at Amsterdam *Experimenta circa sanguinis missione*. Malpighi mentioned de Heide's study of blood also in the *Risposta* to Sbaraglia, *MOB*, 504.

21. Lindeboom, *Dutch Medical Biography*, 1403–4. I disagree with Adelmann's identification of Muys's work, since he refers to the 1685 edition and 1685 is clearly too late, and of the second observation of the fourth decade, which was added to that edition; Malpighi was referring to the first decade of observations already published in 1683. The work by Muys was translated into English as *A Rational Practice of Chyrurgery* (London, 1686), 6–11.

22. *MCA*, 3:1043–44, Malpighi to Bellini, 17 April 1685, in response to Bellini's request in ibid., 1029–31, Bellini to Malpighi, 26 March 1685, at 1030. Borelli, *De motu animalium*, pt. 2, chaps. 9–10, esp. prop. CXLIV.

23. *MCA*, 2:892–93, Malpighi to [Ferrarini], 27 February 1683.

24. *MCA*, 3:1229–30, 10 December 1686, Malpighi to Bellini. Ferrari, "Public anatomy lessons." Bayle, *Dissertationes physicae*, 181–83. *MOB*, 558. Sbaraglia, *Vigiliae*, 384.

25. Malpighi, *De structura glandularum*, in *Opera posthuma*, 4–5; *MOB*, 338, 340.

26. *MCA*, 3:1296–98, Bellini to Malpighi, 7 September 1687.

27. *MCA*, 3:1289–91, Malpighi to Bellini, August 1687. *MOB*, 330–31. Adelmann, *Embryology*, 449–50.

28. Malpighi, *De structura*, in *Opera posthuma*, 7, 8–9; *MOB*, 348, 354–55. Adelmann, *Embryology*, 515 has shown that Malpighi started carrying out research on the pericardium following his receipt of Bellini's letter of 7 September 1687. Lower, *De corde*, 3–6.

29. Malpighi, *De structura*, in *Opera posthuma*, 3–4; *MOB*, 336, 338.

30. Malpighi, *De structura*, in *Opera posthuma*, 6; *MOB*, 346.

31. Malpighi, *De structura*, in *Opera posthuma*, 6, 9; *MOB*, 347, 356–57 (the insects were presumably of the genus *gasterophilus*).

32. Malpighi, *De structura*, in *Opera posthuma*, 2–3, 9; 1–2; 3–4; 6; *MOB*, for ink injections see 334–35, 356; for maceration in water see 332, 333, 336–37 (to highlight the lymphatics), 339; 344–45. *MOO*, 1:25; *MOB*, 298–99.

33. Malpighi, *De structura*, in *Opera posthuma*, 7–8; *MOB*, 351, 353. Lower, *De corde*, 6–7.

34. Malpighi, *De structura*, in *Opera posthuma*, 4–5; *MOB*, 339–40; curiously, Malpighi seemed to rely on the fluid from the scrotal sac in order to test the properties of pulmonary dropsy.

35. Biblioteca Universitaria, Bologna, Ms 936, II, A, cc. 20–21. The figure is reproduced in *MOB*, plate XIX, although the accompanying text has not so far been studied.

36. Ruestow, "Vascular Secretion," 268–69, 275; quotations at 283, 279. *MCA*, 3:979–80, Chirac to Malpighi, 18 August 1684. *MOB*, 327–29. Cook, *Matters of Exchange*, 293–98.

37. Nuck, *De ductu salivali novo*, Leiden, 1685. Lindeboom, *Dutch Medical Biography*, 1442–45. *MCA*, 4:1569, Bohn to Malpighi, 16 January 1690, on Nuck's alleged error in mistaking an artery for an eye duct; 1833, Malpighi to unidentified correspondent, 22 November 1692. *Bibliotheca anatomica*, 2:797, 848–50, at 849b–850a. Lindeboom, "Dog and Frog." Luyendijk-Elshout, "*Oeconomia animalis*," 302–3.

38. Nuck, *Sialographia*, in *Bibliotheca anatomica*, 2:808b, 809a.

39. Nuck, *Sialographia*, in *Bibliotheca anatomica*, 2:810a–811b, 813a. The case histories occupy almost half his work.

40. Nuck, *Adenographia*, in *Bibliotheca anatomica*, 2:829b–830a, 830b, 838a–b. Cole, "Injections," 312–13. Ruestow, "Vascular Secretion," 274 and n61.

41. Nuck, *Adenographia*, in *Bibliotheca anatomica*, 2:835a–836a. In his posthumous *Vita*, 101, Malpighi attempted a response dealing mainly with structural features. Hienemann, "Frügeschichte," 222–27, provides a German translation of and commentary on Malpighi's reply to Nuck.

42. Nuck, *Adenographia*, in *Bibliotheca anatomica*, 2:839a.

43. Nuck, *Adenographia*, in *Bibliotheca anatomica*, 2:838b, 839b–840a. Needham, *Embryology*, 144.

Chapter 7 • The Challenge of Insects

1. *MAP*, 283, Malpighi to Redi, 9 October 1668. Malpighi sent Redi a copy of *De bombyce*; Adelmann, *Embryology*, 350. Redi, *Insetti*, 77–78, 145–47; Swammerdam, *Histoire*, 58–59, 198–99. Cobb, *Generation*, 90–92. Findlen, *Kircher*. Stolzenberg, *Great Art*.

2. Aristotle, *Parts of Animals*, IV, 5, 682a, 10–22. Ruestow, *Microscope*, 106.

3. Ruestow, *Microscope*, 219; see also 216–17, on spontaneous generation.

4. Mazzolini, *Non-Verbal Communication*. Jones and Galison, *Picturing Science*. Baigrie, *Picturing Knowledge*. Kusukawa and Maclean, *Transmitting Knowledge*. See also the Focus section on *Science and Visual Culture* in *Isis* 97 (2006), 75–220. Anderson, "Eye" and the special issue of *AS*, 67.3 (2010).

5. I refer to the first edition of the *Vocabolario*, 1612. Ruestow, *Microscope*, 105–6.

6. Hooke, *Micrographia*, 163–64, 180–82. Bertoloni Meli, "Representation." Harwood, "Rhetoric." Turner, "Hooke." Neri, "Observation."

7. Hooke, *Micrographia*, 173, 185–87. Harvey, *De motu cordis*. Cobb, *Generation*, 63–68, on generation at the Royal Society.

8. Hooke, *Micrographia*, 123–24.

9. Ibid., 133, 193–94. Fournier, *Fabric*, 54–55. Ruestow, *Microscope*, 202–3, 209–11. Birch, *History*, 3:420. Regius, *Fundamenta*, 216–17.

10. Redi, *Insetti*, 12n13, 18–25, 115–25, 146–47. Ruestow, *Microscope*, 201–3.Cobb, *Generation*, 82–93.

11. Redi, *Insetti*, 83, 117, 155, 170, 180, 186, 191, 192, 196.

12. Ibid., 196, 153–55; Steno, *Opere scientifiche*, 1:50 and 53. Franceschini, "Galileo."

13. Tongiorgi Tomasi, "Immagini," 309–10, 313–14, and colored plates; Tongiorgi Tomasi and Tongiorgi, "Naturalista," 39. *MOB*, 24–25. Freedberg, *Eye*, 193–94. Bertoloni Meli, "Representation."

14. Redi, *Insetti*, 128–30. Newman, *Atoms*, 105.

15. Redi, *Insetti*, 86–89n36, 91.

16. The original pamphlet, [Pascal], *Récit*, was reprinted in Pascal, *Traitez*, and is also available at http://gallica.bnf.fr/. Middleton, *Barometer*, 24, 46–52; *Experimenters*, 129–33. Dear, *Discipline*, 196–201. Licoppe, *Formation*, chap. 1, esp. 31–33. Shea, *Games and Chance*, 105–16 and 120–22. Bertoloni Meli, "Mountain." Boschiero, "Natural."

17. Redi, *Opere*, 4:269–78. Fournier, *Microscope*, 173–77. Ruestow, *Microscope*, 219. Bernardi's claim in Redi, *Esperienze*, 15, that Redi's procedure was entirely novel is not justified in the light of Périer's Puy-de-Dôme's trial. Middleton, *Experimenters*, on Thévenot. *Giornale de' Letterati*, 1673, 115–19, "Esperienze fatte in Francia, e in Inghilterra"; 119–24, "Esperienze fatte dal Signor Francesco Redi." Subsequent editions of Redi's works omit the initial essay with the reports from France and England, thus curtailing the context of Redi's trials. Stroup, *Scientists*, 161–62: see also Malpighi's experiments on plants reported in the posthumous *Vita* and discussed in sec. 7.2. Bertoloni Meli, "Mountain," sec. 3. Wear, "Medicine," 227–29. Maehle, *Drugs*.

18. Redi, *Insetti*, 25–32, 41–48, 135, 158–86; 164n320. Bernardi et al., *Natura*. Bertoloni Meli, "Neoterics," 64–65. *BW*, 4:78.

19. Adelmann, *Embryology*, 327, 338–39. *MCA*, 1:354–57, Oldenburg to Malpighi, 28 December 1667/7 January 1668. Malpighi, *Vita*, 56.

20. *MCA*, 2:577–79, Bonfiglioli to Malpighi, Rome, 21 March 1671, at 578–79. At 577 Bonfiglioli reports that he had ordered microscopes from Eustachio Divini, which Malpighi had requested in January; *MCA*, 2:498.

21. *MCA*, 1:398–99 and 402–4. However, see Bertoloni Meli, "Representations," secs. 6–8.

22. *MOO*, 2:15, 38; Malpighi, *Bacofilo*, 40–41, 80. An extensive account can be found in Cole, *Comparative Anatomy*, 184–97, extensively quoted by Adelmann, *Embryology*, 339–44.

23. *MOO*, 2:9, 18, 23; *Bacofilo*, 31–32, 46, 54.

24. *MOO*, 2:10; *Bacofilo*, 32–33.

25. *MOO*, 2:16–17, 20; *Bacofilo*, 42–43, 49.

26. *MOO*, 2:27; Münster, "Anatomica," 185; *MOB*, 415.

27. Malpighi, *De bombyce*, 35 and plate X; *MOO*, 2:19–20; Adelmann, *Embryology*, 342–43. The eggs appear to us the same shape as red blood cells. Cole, *Anatomy*, 194–96. Hooke, *Micrographia*, preface, 181–82. Hackman, "Natural Philosophy," 182–83. Kawaguchi et al., "Polygonal Patterns," 440, fig. A.

28. Cobb, "Malpighi," 115. Cole, *Comparative Anatomy*, 190. Bologna, Biblioteca Universitaria, Ms 2085 II, 63r.

29. Redi, *Insetti*, 128–30. *MOO*, 2:13; *Bacofilo*, 37.

30. *MOO*, 2:13–14; *Bacofilo*, 38.

31. *MOO*, 2:14–15; *Bacofilo*, 38–39.

32. *MOO*, 2:23; *Bacofilo*, 54.

33. *MOO*, 2:37, 42–43; *Bacofilo*, 77–78, 87–88; Adelmann, *Embryology*, 856–58.

34. Boyle, *Experiments and Considerations Touching Colours* (London, 1664), was translated

into Latin in the following year, *Experimenta et considerationes de coloribus* (London, 1665). By the end of November 1668 the physician Giovanni Battista Capucci at Crotone had a copy of the latter; *MCA*, 1:391–93, Capucci to Malpighi, 29 November 1668, at 393. *BW*, 4:32–33, 40–41, 52–53.

35. *De bombyce*, *MOO*, 2:2 (wrongly numbered 66), 7, 20; Cobb, "Colorful Silkworm," 119–21. *MCA*, 1:388–89, Malpighi to Antonio Ruffo, 24 November 1668. Smith, "Science and Taste." Adelmann, *Embryology*, 346, 467–68.

36. *MCA*, 3:1247–48, Piccoli to Malpighi, 14 January 1687; 1255–56, Malpighi to Piccoli, 27 January 1687. Adelmann, *Embryology*, 513. *MOB*, 46–48. Faucci, "Polemica."

37. Swammerdam, *De respiratione*, 114. Swammerdam, *Histoire*, 61–62. Ruestow, *Microscope*, 109–10, 123. Fournier, *Fabric*, 63. Bertoloni Meli, "Additions," 305. See the review in *PT* 5 (1670), 2078–80, at 2080.

38. Cobb, "Silkworm"; *Generation*, 140–52. Swammerdam, *Histoire*, 124–25.

39. The full title in French, *Histoire generale des insectes ou l'on expose clairement la maniere lente & presqu' insensible de l'accroissement de leurs membres, & ou l'on decouvre evidemment l'erreur ou l'on tombe d'ordinaire au sujet de leur prétendué transformation*, differs somewhat from the original Dutch: *Historia insectorum generalis, ofte, Algemeene verhandeling van de bloedeloose dierkens waar in, de waaragtige gronden van haare langsaame aangroeingen in leedemaaten, klaarelijk werden voorgestelt: kragtiglijk, van de gemeene dwaaling der vervorming, anders metamorphosis genoemt, gesuyvert: ende beknoptelijk, in vier onderscheide orderen van veranderingen, ofte natuurelijke uytbottingen in leeden, begreepen*. *PT* 5 (1670), 2078–80; the pagination of this issue is somewhat in disarray.

40. Harvey, *Generation*, 203–4, and the entire sec. 45; 175–76; at xxxiv–xxxv Whitteridge discusses Harvey's classification of animals. Swammerdam, *Histoire*, 9–10, 17–18, 32–38, 40.

41. Swammerdam, *Histoire*, 159–62.

42. Swammerdam, *Histoire*, 74–75, 212. Ruestow, *Microscope*, 30–31, 61, 112, 127n87, 143; Fournier, *Fabric*, 147.

43. Swammerdam, *Histoire*, 171, 101–5. Ruestow, *Microscope*, 131n101, 134. Cobb, "Malpighi," 131, 135, 145n129; "Reading." Hooke, *Micrographia*, 185–87.

44. Swammerdam, *Histoire*, 4, 54, 132, 143, 164–65, 167, 179, 202. See also Swammerdam, *De respiratione*, 119, where Descartes is referred to as most noble and subtle. The *Sammelband* with Descartes and Swammerdam was Jeremy Norman & co. #38087. Ruestow, *Microscope*, 244–45 and n90, presents a different picture of Swammerdam's relations to the mechanical philosophy and Descartes. Neumann, "Machina."

45. *MCA*, 2:483–86, Malpighi to Oldenburg, Bologna, 20 November 1670; at 484 Malpighi stated that he had seen Swammerdam's figures, but he may have seen the entire book; 487–92, Malpighi to Bonfiglioli, 1 December 1670, at 488n10. Harvey, *Generation*, 136. Adelmann, *Embryology*, 858–59. Malpighi, "Alcune osservazioni," 170.

46. Swammerdam, *Miraculum*, 16–17. Cobb, "Malpighi," 127. Adelmann, *Embryology*, 342, quoting Cole, *Comparative Anatomy*.

47. Adelmann, *Embryology*, 399. Malpighi, *Vita*, 59–62, at 62. Swammerdam, *Biblia*, plate XIV. Cobb, "Silkworm," 137, 141–46.

Chapter 8 • Generation and the Formation of the Chick in the Egg

1. Adelmann, *Embryology*, 747, 757, 1087. Roger, *Life*. Bernardi, *Metafisiche*.

2. Adelmann, *Embryology*. Roger, *Life*. Bernardi, *Metafisiche*, 60. Pinto-Correira, *Ovary*, esp. 1–47.

3. Adelmann, *Embryology*, 871–85, provides a useful discussion of epigenesis and preformation.

4. *Bibliotheca anatomica*, 1:412–544. Lennox, "Comparative Study," and Ekholm, "Harvey and Highmore" and "Representations" frame Harvey's work within the Aristotelian and mid-seventeenth-century contexts.

5. Keynes, *Harvey*, 334. Harvey, *Generation*, 1, 20, 189, 293, 344, 395–96.

6. Harvey, *Generation*, 20. The copy of Fabricius's treatise owned and annotated by Harvey survived the 1666 fire of London and is now at the Lilly Library, Indiana University, Bloomington. Ekholm, "Harvey and Highmore"; "Representations."

7. Harvey, *Generation*, 11–13, 19–20. Ekholm, "Harvey and Highmore." Wear, "Harvey," 237–39. Keynes, *Harvey*, 333, missed the philosophical import of the lack of pictures in *De generatione*. Biagioli, *Instruments*, 136–43.

8. Harvey, *Generation*, 18, 282–92, at 285. Lennox, "Comparative Study," 36–38.

9. Harvey, *Generation*, 154, 294, 296; references to Aristotle are pervasive: see, e.g., 17, 19, 59, 145, 209. Lennox, "Comparative Study," 38–39. Smith, "Introduction" to *Animal Generation*, 6–7.

10. Descartes, *World*, 186–205, esp. 196. Ekholm, "Harvey and Highmore." Wilkin, "Essaying." Smith, "Imagination." Aucante, "Generation of Animals."

11. *PT* 3 (1668), 843–44; *MCA*, 2:483–84, Malpighi to Oldenburg, 20 November 1670. Adelmann, *Embryology*, 155. De Graaf, *Human*, 21, 129. Borelli, *De motu*, props. 166, 168. Cole, "Injections," 297–301.

12. De Graaf, *Human*, 81, 45–46. On de Graaf and his subsequent polemics see Cobb, *Generation*, 97–124.

13. De Graaf, *Human*, 31, 35–36, 55, 59; on 145 reference to the microscope. Adelmann, *Embryology*, 723–26. Cobb, *Generation*, 161–87.

14. De Graaf, *Human*, 148–49; *Mulierum*, 237–44. Van Leeuwenhoek, "Observationes"; "Concerning the Animalcula"; "Generation." *MCA*, 3:1219–21, Bellini to Malpighi, 8 October 1686, at 1219; Adelmann, *Embryology*, 420, 506, 868. *Bibliotheca anatomica*, 1:800–803; both Martin Lister and the editors of the *Bibliotheca* found van Leeuwenhoek's findings and theory objectionable. Franceschini, "Generazione," 151–61. *PT* 12 (1677), 1040–45. Lancisi, *Consulti medici*, 102. Wilson, *Invisible*, 131–37. Ruestow, *Microscope*, 171–72. Cobb, *Generation*, 193–216.

15. Steno, *Opere*, vol. 1, Introduzione, 1–105, at 47–57; *Opera*, 1:202–3, 2:152–54, 159–79. On Steno's work on the chick in the egg see also his 1664 *De vitelli in intestina pulli transitu*, in Tallmadge May, "On the Passage." Franceschini, "Generazione," 169, 178–83. De Graaf, *Human* (Latin paginantion is in brackets), 81, 139 [204–5], 133–34 [179, 182], 146–48 [231–41], 168 [319], 181–83. Ruestow, *Microscope*, 47n71. Schrader, *Observationes*, preface [xiv]. Descartes, *De homine*, 11. Wilkin, "Figuring," 47. Cobb, *Generation*, 169–70, points out that rabbits are unusual in that mating induces ovulation. Laqueur, *Sex*, 157–59.

16. Malpighi, *Vita*, 86. *MOB*, 311–13. De Graaf, *Mulierum*, 177–78. *Human*, 132–33n173.

The *corpus luteum* is not actually quite as yellow as the name suggests; it is a temporary structure developing in the ovary from the ovarian follicle at specific stages depending on the animal species. Adelmann, *Embryology*, 2:780–81, 865.

17. De Graaf, *Human*, 132–33 and n170, 166 and n276, 170.

18. Swammerdam, *Miraculum naturae*, 37–38. Cole, "Injections," 301–3. Cobb, "Silkworm," 131–35; *Generation*, 165–67. Cook, "Time's Bodies." Bertoloni Meli, "Collaboration."

19. De Graaf, *Human*, 149; *Mulierum*, 242–44. Harvey, *Generation*, 227–28, 443, 448, 463. Malpighi, *Anatome plantarum*, 62. Borelli, *De motu*, vol. 2, prop. 186. Adelmann, *Embryology*, 318, 2:841, 852, 869, 918.

20. Borelli, *De motu*, vol. 2, props. 169, 170, 173, 184. Adelmann, *Embryology*, 2:858–59; Adelmann relied on Malpighi's letter to Spon to claim that Malpighi would attribute fecundation to an immaterial agent, adding that he would be in the company of Fabricius, Harvey, and de Graaf; *MOB*, 316–17. Adelmann's claim, however, is inaccurate in that Malpighi attributed fecundation to "volatile particles"—as Adelmann himself reports—that cannot be immaterial. Bernardi, *Metafisiche*, 103–4.

21. Borelli, *De motu*, props. 179, 180, 185–86; *Movement*, 387, 390, 397. Bernardi, *Metafisiche*, 99–112; at 103, 106–9 he discusses the occurrence of the expression "plastic virtue" in Borelli. Boyle, *A Defence of the Doctrine Touching the Spring and Weight of the Air*, in *BW*, 3:84. According to Gal, *Foundations*, 127, the relevant passage was due to Hooke.

22. Ekholm, "Chick Generation."

23. Schrader, *Observationes*; the preface consists of twenty-three unnumbered pages. *Bibliotheca anatomica*, 1:544–51. *MCA*, 4:1510–11, 16 September 1689. *PT* 18 (1694), 150–52. Lindeboom, *Dutch Medical Biography*, 1141–42, 1173–74. Roger, *Life*, 268. Bernardi, *Metafisiche*, 70–71. On van Leeuwenhoek and plants see Ruestow, *Microscope*, 251–52.

24. Swammerdam, *Historia*, 51–52; *Miraculum naturae*, 21–22; *Bibliotheca anatomica*, 651b. Adelmann, *Embryology*, 2:907–9. Ruestow, *Microscope*, 181, 226–28. Schierbeck, *Swammerdam*, 112–22. Ekholm, "Harvey's and Highmore's." Pyle, "Malebranche." Cobb, *Generation*, 236–37. Smith, "Imagination." Wilkin, "Essaying." Roger, *Life Sciences*, 267–75.

25. Adelmann, *Embryology*, 882n2. *MCA*, 4:1659–64, Torti to Malpighi, Modena, 25 January 1691; Malpighi to Torti, 28 January 1691; Torti to Malpighi, 5 February 1691; the relevant passage is at 1662 and n2. Bernardi, *Metafisiche*, 92, reports the relevant passage, but I disagree with his translation of Malpighi's memorandum.

26. *MOB*, 86–88. Adelmann, *Embryology*, 2:833, where he refers to other instances in which Malpighi used the same technique; 853–55. *MOB*, 237, 254, 260, 263. On techniques such as the usage of alcohol as a fixative see also Adelmann, *Embryology*, 5:2238. Cook, *Exchange*, 280. *Bibliotheca anatomica*, 1:724–25, quoting from *Anatome plantarum*, 1:81–82. Bernardi, *Metafisiche*, 79–99.

27. Adelmann, *Embryology*, 984–85; I have slightly modified the translation.

28. Harvey, *Generation*, 11–13, 19–20. Bertoloni Meli, "Additions," 289–90, Malpighi to Redi, 30 December 1673.

29. I have modified the translation in Adelmann, *Embryology*, 2:935; *MOB*, 223.

30. See sec. 4 of the introduction.

31. Adelmann, *Embryology*, 840–70. *MOB*, 224. Similar findings were reported at the same time to the Royal Society by William Croone; Birch, *History*, 3:30–40. Bernardi, *Metafisiche*, 72–76. The letter to Spon is translated and commented on in *MOB*, 301–19.

32. Adelmann, *Embryology*, 2:956–57; I have slightly modified the translation at 957. See also ibid., 848, for a comparison with the study of the silkworm.

33. I have slightly altered the translation from Adelmann, *Embryology*, 2:944–45; for the notion of compendium see 844, 867. *MOB*, 236, 263–64, 317–18, 577, 580. Malpighi, *De formatione*, in *MOO*, 2:6, separate pagination. Harvey, *Generation*, 73. Bernardi, *Metafisiche*, 112–19.

34. *MOO*, 2:8–9; 1:4, 9. *MOB*, 242–44, 245, 263, 276–77.

35. *MOO*, 2:5, 6–8; *MOB*, 235, 237, 238, 240–41; Adelmann, *Embryology*, 2:958.

36. *MOO*, 2:5; 1:3, 7, 8, 9, 10 (separate pagination). *MOB*, 235, 258, 272, 276, 277, 279.

37. *MOO*, vol. 1, letter to Spon, 26, 34. *MOB*, 301 (omitting the term "plastic"), 317–18. Harvey, *Generation*, 209. Adelmann, *Embryology*, 866, 1162, 1164, 1170.

38. Adelmann, *Embryology*, 874; 571. Wilson, *Invisible*, 128. Giglioni, "Machines," 166.

39. *MOB*, 587–88.

Chapter 9 · The Anatomy of Plants

1. Ogilvie, *Science*. Sachs, *Geschichte*. Mägdefrau, *Geschichte*. Stroup, *Company*, 67, 131–35. Histories of botany mention works such as Johann-Daniel Major's 1665 *Dissertatio botanica de planta monstrosa* and Edme Mariotte's 1679 *De la végétation des plantes*, on the nature and motion of sap, nutrition, generation, and chymical composition of plants. Heller, "Mariotte."

2. Hooke, *Micrographia*, 116–21. Camerarius and Mauchart, *De herba mimosa seu sentiente*. Stroup, *Company*, 67–68. Gaukroger, *Descartes' System*, 187.

3. Hooke, *Micrographia*, 141, 142–45, 147–52. Fournier, *Fabric*, 102. Gremk, "Fibre."

4. Adelmann, *Embryology*, 901. *MCA*, 1:152–53, Borelli to Malpighi, 22 March 1663, at 153; *MCA*, 2:847–48, Malpighi to Corraro, 28 May 1680, and 849–50, Malpighi to Hooke, 20 November 1680. Malpighi, *Anatome plantarum*, vol. 2, preface. *Anatomes plantarum idea* was completed on 1 November 1671 and submitted for approval to the Royal Society.

5. *MCA*, 2:626–27, Malpighi to Oldenburg, 2 August 1672. *MAP*, 302–6, at 303, d'Andrea to Redi, 28 January 1673.

6. *MCA*, 2:778–79, Malpighi to Grew, 21 June 1678. Malpighi, *Anatome plantarum*, vol. 2, preface. See also Bertoloni Meli, "New Anatomy," 55–57, and "Additions," 302–3. Kemp, "Mark." Saunders, *Plants*. Möbius's translation provides a useful identification and description of all the plates in *Anatome plantarum*.

7. Malpighi, *Anatome plantarum*, preface and 1:40. Adelmann, *Embryology*, 1092–1103.

8. Malpighi, *Anatomes plantarum idea*, 3–4. Zanoni, *Istoria*, 144–53. Adelmann, *Embryology*, 681.

9. Malpighi, *Vita*, 39. The relevant passage is translated in *MOB*, 369–71. Malpighi, *Anatome plantarum*, 2:1; see also 1:79. *MOB*, 226, 263, 264.

10. Malpighi, *Anatome*, explanation of figures by Möbius, *Anatomie*, 127. For the role of weaving analogies see below.

11. Malpighi, *Anatome plantarum*, 1:7–8. Bertoloni Meli, "Additions," 300 and 303.

12. Malpighi, *Anatome plantarum*, 1:15–16.

13. Ibid., 1:19–20; *Vita*, 47–53. *MOB*, 369–85. Randelli, "Gagliardi."

14. Malpighi, *Anatome plantarum*, 1:22 and 28–30.

15. Ibid., 1:32 and 38–39; see also 1:20 and 2:71. Möbius, 45, is at a loss as to what Malpighi may have meant by the follicles or receptacles (*loculi*) in this case.

16. Malpighi, *Vita*, 15–16; *MOO*, 1:16; 2:69. See also the preface to *De viscerum structura*, *MOO*, 2:55. Middleton, *Experimenters*, 154–57. Gómez López, "Malpighi," 179–81. *Passioni*, esp. chap. 1.

17. Malpighi, *Anatome plantarum*, 1:45, 53–54, 56, 74. Möbius, *Anatomie*, 161n15. Camerarius, *Opuscula*, 58; *Geschlecht*, 11, discusses different etymologies for "amentaceus," including one provided by Ovidio Montalbani in *Dendrologia aldrovandiana* stemming from "mentum" or chin.

18. Malpighi, *Anatome plantarum*, 1:62, 81, and 82 (erroneously numbered 78). Adelmann, *Embryology*, 841, 849–55, 901 (tracing a note by Malpighi attributing to Cesalpino the view that the plant is in the seed as the animal is in the egg). Borelli, *De motu*, vol. 2, prop. CLXXXVI. Roger, *Life*, 285. The term "colliquament" was used by Harvey: Adelmann, *Embryology*, 1031–34.

19. Malpighi, *Anatome plantarum*, 2:34–36. Massalongo, "Galle." Basile, *Invenzione*, 125–44. Bertoloni Meli, "Additions," 304–12.

20. Malpighi, *Anatome plantarum*, 2:51; *Vita*, 93–96. *MOB*, 41–2 and 394–404.

21. Malpighi, *Anatomes plantarum idea*, 13–14; *Anatome plantarum*, 2:69–70. *MOB*, 320–21. Hooke, *Micrographia*, 116. Similar experiments were performed by Mariotte and Claude Perrault; Stroup, *Company*, 135.

22. Malpighi, *Anatome plantarum*, 2:71–72.

23. Malpighi, *Anatome plantarum*, 2:11–15.

24. Malpighi, *Anatome plantarum*, 2:15, 65–66. See also 1:16.

25. Adelmann, *Embryology*, 437–38; on the whole affair see 493–532. Malpighi, *Vita*, 102. Angeletti and Marinozzi, "Triumfetti." Ottaviani, "Dibattito," 298–303; "Scuola galileiana."

26. *MCA*, 3:1135–38, Malpighi to Ronchi, 6 February 1686, at 1137; 1241–42, Malpighi to unidentified correspondent. [Marsigli], *Relazione*, 55–56, 73–74; see also the anonymous review in the *Philosophical Transactions*. Swammerdam, *De respiratione*, 114, in *Bibliotheca anatomica*, 1:999b. Harder, *De ovis*. Fazzari, "Redi, Buonanni," esp. 114. Basile, *Invenzione*, 144–67.

27. *MCA*, 2:894–95, Bellini to Malpighi, 30 April 1683; 3:1061–62, Bellini to Malpighi, 16 July 1685; 4:1818–20, Malpighi to Luigi Ferdinando Marsili, 30 August 1692, at 1819 and n11. Malpighi, *Vita*, 63. Adelmann, *Embryology*, 439. [Buonanni], *Riflessioni*, 45–141; see esp. 74, 83–84, 92–93. Fazzari, "Redi," esp. 110–12, 123–24.

28. Trionfetti, *Observationes*, 11–12, 15, 24, 60. Malpighi, *Anatome plantarum*, 1:62; 2:66. Fazzari, "Redi, Buonanni," 114–15.

29. Trionfetti, *Observationes*, 25–26, 28–29, 31–33, 47–48, and 56; the reference to M. Dedu, *De l'ame des plantes*, is at 44. Mariotte, *Plantes*, 138–39. Stroup, *Company*, 147–49. Roger, *Life*, 275–76.

30. Trionfetti, Observationes, 49–52. His experiments were performed in winter.

31. Ibid., 56–57. Adelmann, *Embryology*, 439.

32. Trionfetti, *Observationes*, 56–61, 72–76. Adelmann, *Embryology*, 440.

33. Cestoni, "Lettera." See also Baglivi, *Correspondence*, 60–61n31. *MCA*, 3:1120–21, Cestoni to Malpighi, Livorno, 17 January 1686; 1267, Malpighi to Henrici, 13 May 1687. Adelmann, *Embryology*, 494–97, 520–21, 649.

34. Malpighi, *Vita*, 63–64. Adelmann, *Embryology*, 437–38, did not analyze the change in Malpighi's experimental procedures; 496–97, 653.

35. Malpighi, *Vita*, 64–69.

36. Bolam, "Botanical Works." Hunter, "Early Problems," and his entry in the *DNB*. LeFanu, *Grew*. Fournier, *Fabric*, 72–79, 121–28, and passim.

37. Grew, *Anatomy of Plants*, preface, 73, 117–18, 121; *An Idea*, 4. Arber, "Grew and Malpighi." Redi, "Sali fattizi." Stroup, *Company*, 98–100. Shapin, *Truth*. Daston, "Objectivity." Biagioli, "Etiquette."

38. Grew was closely associated with Hooke (they were both secretaries of the Royal Society) and was instructed by the Gresham professor in microscopy. Harwood, "Rhetoric," 258, 263.

39. Grew, *Anatomy of Plants*, 21, 126. See also Hooke, *Micrographia*, 21, 28–29. Stroup, *Company*, 136, 139–42. Mariotte, *Plantes*, 133–34.

40. Grew, *Anatomy of Plants*, 188–89. Bertoloni Meli, *Thinking*, 244.

41. Grew, *Anatomy of Plants*, 169–73, esp. at 171–72. Sachs, *Geschichte*, 412–14.

42. Camerarius, *Opuscula*, 61, 80–107; *Geschlecht*, 13, 28–46. Sbaraglia, *Vigiliae*, 171.

43. Malpighi, *Anatomes plantarum idea*, 2. Camerarius, *Epistola*, preface by Möbius, *Geschlecht*, iv.

44. These are identified in the Möbius translation; Camerarius, *Geschlecht*, 5–6, 16, 22, nn17, 19, 53, 64. Möbius also identified relevant passages from Grew and John Ray, who in *Historia* listed a number of plants producing "semen absque florem" or "florem absque semine," including many on which Camerarius later experimented; *Geschlecht*, 55n3.

45. Camerarius, *Opuscula*, 45, 51, 59; *Geschlecht*, 2, 6, 12. In modern terminology, they are hermaphrodite (with bisexual reproductive units or flowers), monoecious (the same plant has separate male and female reproductive units), and dioecious (the same species has separate male and female plants).

46. A brief account of sexuality in plants before Camerarius is in Sachs, *Geschichte*, 406–15. Marsigli's work was included in the 1688 Leyden edition of Malpighi's *Opera omnia*. Camerarius, *Opuscula*, 80, 85–89; *Geschlecht*, 27, 31–33.

47. Camerarius, *De sexu plantarum Epistola* [secunda], at 35.

48. Camerarius, *Opuscula*, 76, 107–12; *Geschlecht*, 24–25, 47–50. Sachs, *Geschichte*, 416–21.

49. See, e.g., Camerarius, *Opuscula*, 92–95; *Geschlecht*, 36, 38.

50. Camerarius, *De lolio temulento*, 1690; *De lolio temulento*, 1695, with a reference to Trionfetti's work on 133 of Camerarius, *Opuscula*; *De lolio temulento*, 1710. Trionfetti, *Vindiciarum veritatis*, 126. Camerarius's work appears to have been unknown to Sbaraglia, whose *Vigiliae*, 163, states that plants lack sexual reproduction.

Chapter 10 • The Fortunes of Malpighi's Mechanistic Anatomy

1. Robinet, *Iter*, 309–18, at 312.

2. *MCA*, 4:1533–36, Malpighi to Bellini, 17 October 1689, at 1535. See also 1498–1500 and 1561–63 for more information on Malpighi's state of mind at the time. *MCA*, 4:1671, Malpighi to Waller, 27 March 1691. For Waller's reply see *MCA*, 5:1926.

3. Adelmann, *Embryology*, 556–58. Malpighi, *Memorie di me*, 7–8, quoted in Bertoloni Meli, "Posthumous Dispute," 266.

4. Shapin, *Truth*, 107–25, esp. 124–25. Biagioli, "Etiquette."

5. Momigliano, *Development*, 89ff. and 103. Misch, *History*, vol. 1, esp. pt. 2. Volume 5 of *Epitome Operum Galeni* (Venice, 1548) contains Galen's *Vita*. Nutton, "Medical Autobiography." Cardano, *De Vita*. Buck, "Cardano." Grafton, *Cardano*, chap. 10, esp. 178–84. May, *L'autobiographie*. Sturrock, *Autobiography*. Dear, *Structure*.

6. The letter, dated from Bologna, 1 November 1681, appeared in *Philosophical Transactions* 14 (1684–85), 601–8 and 630–46. See also *MOO*, 1:21–35.

7. As a general guide, pp. 4–21 of the *Vita* deal with the lungs and respiration; 30–42 with the liver and kidneys; 47–55 with bones and teeth; 57–63 with the silkworm; 63–81 with plants; and 93–100 with hairs, feathers, and nails.

8. Malpighi, *Vita*, 27.

9. Adelmann, *Embryology*, 663n6 and 664. *MCA*, 5:1591, Malpighi to his nephew Antonio Fabri, 28 November 1694.

10. Malpighi, *Vita*, 2. Borelli, *De motu*, vol. 2, prop. 37. Malpighi, *Memorie*, 13. Gómez López, "Malpighi." Dollo, "Inediti." Malpighi, *Risposta*, in *Opera posthuma*, 15, 36.

11. *MCA*, 1:374, Malpighi to Oldenburg, 1 April 1668. Malpighi, *Vita*, 25, "L'ho anch'io fatta, ma peró la vista non m'aiuta."

12. The modern English translation not only rendered many passages inadequately but also presented the work in a rather insensitive fashion. See the review by Baldini. I have appropriately modified Maquet's translation, including the Latin text in the footnote. Malpighi, *Vita*, 11. Adelmann, *Embryology*, 333.

13. Borelli, *De motu*, vol. 2, prop. 107, 206–7; *Movement*, 309. Malpighi, *Vita*, 11.

14. *MCA*, 1:80–83, Borelli to Malpighi, 24 March 1661. Borelli, *De motu*, vol. 2, props. 98–109.

15. Borelli, *De motu*, 2:225, 226: "Videtur automa umbratilem quandam similitudinem cum animalibus habere, quatenus ambo sunt corpora organica se muventia, quae legibus mechanicis utuntur, & ambo a facultatibus naturalibus moventur." On automata in *De motu* see 153, 157, 159, 160, 227, 389. Roux, "Machines." Des Chenes, "Mechanisms."

16. Borelli, *De motu*, 2:227. See props. 113–16. Willis, *Opera*, 2:133. Borelli, *De motu*, vol. 2, props. 97, 129, 179. For the third function of respiration see props. 128–29. Trabucco, "Willis," esp. 324. *MCA*, 2:740–41, Bellini to Malpighi, 20 November 1676, n3.

17. Malpighi, *Vita*, 6 and 11–17. Borelli, *De motu*, vol. 2, props. 13, 22ff., 26, 105–7.

18. Bertoloni Meli, "Additions," 300. Malpighi, *Risposta*, in *Opera posthuma*, 124, 142.

19. *MOB*, 193, 212–14. Adelmann, *Embryology*, 196–97. Malpighi, *Vita*, 16. Boyle wrote several short tracts on phosphorus; *BW*, *ad indicem*. Adelmann, *Embryology*, 552n3, reports Malpighi's notes following his meeting with Leibniz: "[Leibniz] Docuit materiam phosphori esse caput mortuum urinae humanae, unde coruscat, rubedinem communicat, et ignem concipit. An huiusmodi fiat a lumine solis communicato aeri, et pulmonibus separato an a magneticis efluviis adhuc ignoro et dubium valde est."

20. Grondona, *Bellini*, 455–62.

21. Borelli, *Movement*, 350; *De motu*, 289 (see also 290): "[Clarissimus anatomicus] confidenter pronunciat, 'obscurissimum esse, artificium quo urina in renibus separatur; licet (inquit) glandularum (renum nempe) ministerio totum hoc subsequi rationi sit consonum, quoniam tamen minima (nempe individualis) illa, simplexque meatum in glandulis structura nos latet, ideo quaedam tantum meditari possumus, ut huic quaesito probabiliter satisfaciamus: Necesse

est, hanc machinam interna configuratione separationis opus peragere; an vero iis, quae ad humanos usus usurpamus, quibus fere consimilia effingimus, consonet, dubium; licet enim occurrant analogae spongiae, incerniculi, fistularum, & cribrorum structurae, cui tamen ex his consimilis undequaque sit renum fabrica, difficillimum est assignare, & cum naturae operandi industria faecundissima sit, eiusdem ignotae nobis reperientur machinae, quas nec mente quidem assequi licet.' "

22. Borelli, *De motu*, vol. 2, props. 139–42; *Movement*, 342; *De motu*, 272: "Nomine fermentationis intelligitur motus intestinus partium corporis misti, quae propria vi motiva, vel alterius corporis advenientis agitantur." This definition was not challenged by Malpighi. Wear, "Medicine," 356.

23. Malpighi, *Vita*, 38. Borelli, *De motu*, vol. 2, props. 107ff., 129, 132, 135.

24. Malpighi, *Vita*, 3 (letter by Borelli of 20 September 1659) and 21–23. The 1659 letter contains an attempt to understand purging in mechanical terms, thus rejecting the *tractio electiva*. Borelli, *De motu*, vol. 2, chap. 22, esp. props. 224–27 and 233. For Malpighi's opinions on purging see *Risposta*, in *Opera posthuma*, 76, where he refers to the teachings of Leonardo Giachini. Guerrini, "Varieties."

25. Borelli, *Movement*, 386. Malpighi, *Anatome plantarum* (1675), 81–82; *De motu*, 2: 363–64.

26. *Anatome plantarum*, 1:79ff.; 2:17.

27. Borelli, *Movement*, 387, 390, 397; *De motu*, 2:366; *Anatome plantarum*, 1:38–39 and esp. 74: "Temporis tamen tractu, fermentatione media exaltatis & digestis succis, novus substantiae modus & sapor emergit, qui dum nostro occurrens palato gratam excitat titillationem, & ingestus facile evincitur, coctionem seu maturationem adeptus dicitur." Borelli, *De motu*, vol. 2, props. 179–80 and 185; on 175–77 and 181, Borelli refers to the Santorio thermometer and acknowledges having been guided in this investigation by his teacher Benedetto Castelli.

28. Malpighi, *Vita*, 70–77. Adelmann, *Embryology*, 437–38. Borelli, *De motu*, vol. 2, prop. 127.

29. Borelli, *De motu*, 2:286.

30. Borelli, *Delle cagioni*, 137ff. *MCA*, 1:28, Borelli to Malpighi, 18 March 1660. See also the second part of Borelli's *De motu*, prop. 189. King, *Philosophy*, 102–9. *Bibliotheca anatomica*, 2:892–95, at 893–94.

31. Brown, "College of Physicians," 24–25; *Mechanical Philosophy*, 179, 189, 215, 223, 242, 249. Guerrini, "Varieties," 120–24.

32. Mini, *Medicus*, 150–51. Adelmann has identified and transcribed several manuscripts by Malpighi pertaining to Mini's claims during public dissections performed by himself or others: *Embryology*, 88n (lists the names of the professors who performed public anatomies, including Fracassati, Mini, and Sbaraglia), 381–82, 387–88, 389–90, 412–13, 418–19, 465–67, 538–39. *MCA*, 2:800–802, Malpighi to Bellini, 20 March 1679, at 801. Bertoloni Meli, "Additions," 295–96, Valsalva to Malpighi, 24 January 1693.

33. Malpighi, *Vita*, 101–2. Adelmann, *Malpighi*, 533–34.

34. Malpighi, *Vita*, 17, 32, 36–38. Adelmann, *Embryology*, 539, 825. Paulli, *Anatome*, 93–101, contains Wepfer's letter of 20 June 1664, esp. 98. In his reply of 11 July 1664, 102–28, Paulli questioned Wepfer's tentative views by relying on traditional arguments derived from Wharton and Glisson.

35. Mini, *Medicus*, 94–95, 97, 102–6. Adelmann, *Embryology*, 412–13, 415.

36. Adelmann, *Embryology*, 414. *MOB*, 84, 86–87.

37. Mini, *Medicus*, 97. Malpighi, *Vita*, 18–20. The experiments described in Swammerdam, *De respiratione*, 20–21, 58–59, differ from Mini's; however, Swammerdam's teacher Franciscus Sylvius reports an experiment performed by Swammerdam in *Disputationes*, in *Opera*, 34. Malpighi, *Anatome plantarum*, 1:15. The same experiment was performed in May 1668 at the Royal Society; Frank, *Harvey*, 204.

38. Malpighi, *Vita*, 32–33. Malpighi had received a copy of Duverney's *Traité* from its author; *MCA*, 3:1000–1001. Bohn, *Circulus*, 195–96.

39. Mini, *Medicus*, 2–3, 7. Giglioni, *Machines*, esp. 167–70, provides a valuable analysis of Mini's philosophical underpinnings.

40. Malpighi, *Vita*, 38–39, 41.

41. *MOB*, 516. Malpighi's statement relies on the Aristotelian notion of the body as the instrument of the soul and the soul as the act of the body; *actu secundo*, namely, in action, the faculty has to conform to the material constraints of the body.

42. *MOP*, 39–42 at 42 (my emphasis): "Summi enim opificis industria animal fabrefecit, & particulis minimis plasmavit, quarum aliae mutua adaptatione consistunt, reliquae perpetuo fluunt pro functionibus edendis, hinc ad statutum tempus a Natura totum consistit, decrescit et solvitur; ita ut sensim moriamur. *Haec autem omnia necessitate materiae et motus contingunt.* Manutenetur vero talis compages, et ratio motus in viventibus a principio vitae; cujus ratio, licet nobis ignota sit, probabiliter tamen, ex perenni in animalibus sanguinis motu et nervei succi propagatione deducitur, juvantibus aere, & externis, unde ex placido & nativo hoc motu coercentur particulae evolare tentantes, & novae substituuntur; *& haec est radix facultatum, vel saltem instrumentum, quo anima in nobis utitur.* Caeterum si facultas, fluens ab anima, habeat (ut contenditur) potestatem superandi activitatem spirituum, & particularum evolare tentantium, non evincetur ejus vis & energia temporis tractu; sed talis manutenebitur, & ita in aevum durabunt animalia. At si admittatur repassio in facultate, necesse erit admittere in ipsa remissam activitatem in edendo motu, & coercendo eodem, quod in incorporeo nequeo concipere, cum in nobis anima esset radix mortalitatis. Quare sicut ratio mortalitatis est a compage, motu, & figura partium, quae talis perpetuo manuteneri non potest a sanguinis motu, & spirituum, ita conservatio vitae habebitur ab iisdem in brutis; in nobis autem a rationali anima, quae proprias habet operationes independenter a corpore, reliquas vero alligatas eidem."

43. *Bibliotheca anatomica*, 1:73–74. Marcialis, "Immagine," 308. Plutarch, *Lives*, I, Theseus, XXIII.

44. Malpighi, *Vita*, 91: "Succedunt autem in prima gagatis productione tot ova, non quia lapides ab ovo viventium more ortum necessario trahant, sed materiae necessitate."

45. Malpighi, *Vita*, 43–44. Bohn, *Circulus*, 294.

46. *Bibliotheca anatomica*, 1:63. Ruysch, *Observationes*, 60, 111, observations 45, 86.

47. *Bibliotheca anatomica*, 1:61. Ruysch, in response to J. Gaubius, *Epistola problematica prima* (Amsterdam, 1696), 6, 8–9.

48. *Bibliotheca anatomica*, 1:75–76.; Ruysch, in response to J. J. Campdomercus, *Epistola problematica quarta*. Bidloo, *Vindiciae*, 25. Ruysch, *Observationes*, 67–68, observation 51; *Responsio*, 10–13. It is curious to see a *Sammelband* including the early *Observationes* with the entire collection of *Epistolae problematicae*, documenting Ruysch's changing views between two covers: EOS Antiquariat, Zürich, #40204DB.

49. Cook, *Exchange*, 276–88. Wepfer, "De puella sine cerebro nata," *Miscellanea curiosa*,

3:175–203; *Bibliotheca anatomica*, 2:99–112. Boerhaave and Ruysch, *Opusculum*, 56–58, 66. Hansen, "Resurrecting." Luyendijk-Elshout, "Death." Truitt, "Balm."

50. Ruysch, Ettmüller, *Epistola problematica duodecima*, Amsterdam, 1699; Ruysch, *Thesaurus anatomicus sextus* (1705), 6–9. Ruysch, *Dilucidatio*.

51. On Bidloo see Lindeboom, *Dutch Medical Biography*, 135–39. Boerhaave and Ruysch, *Opusculum*, 66. Bertoloni Meli, "New Anatomy," 51.

52. Ruysch, *Thesaurus anatomicus primus*, 42–43. The address to the reader in the *Thesaurus anatomicus secundus* is not paginated; examples of untranslated passages are in the *Thesaurus secundus*, 62, and *Thesaurus anatomicus quartus*, 7, 27. On the pancreas see *Thesaurus anatomicus quartus*, 41.

53. Lindeboom, *Biography*, 1702. Cole, "Injections," 322 and 324 on Ruysch's pupil Abraham Vater. Biagioli, *Instruments*, raises issues for comparison to our case.

54. Boerhaave and Ruysch, *Opusculum*, 21, 27–28, 62.

55. Ibid., 48, 61, 63. Cole, "Injections," 320.

56. Boerhaave and Ruysch, *Opusculum*, 34–35. Manget, *Theatrum anatomicum*, 1:231b–232a; at 362b–364a Manget reports extensively on Ruysch's *Epistola problematica quarta* on the spleen, defends Malpighi's views, reviews the Ruysch-Bidloo controversy, and argues that the pressure of the injections may have destroyed the soft parts.

57. Boerhaave and Ruysch, *Opusculum*, 13, 30–31, 56–58. Heinemann, "Frügeschichte," 227–30. Ruysch, *Epistola problematica quarta*, 1714^2 (first published in 1696), 10.

58. *Thesaurus anatomicus primus*, 31–32, challenges the glands in the liver; *Thesaurus anatomicus secundus*, 65–66, expresses doubts on the existence of glands. *Thesaurus anatomicus tertius*, 39–40 and fig. 3 in plate IV; *Sextus*, 4 and 6; and *Decimus*, 36–38, all challenge the glands in the kidneys. Grondona, "Ruysch," 282–89.

59. Boerhaave and Ruysch, *Opusculum*, 51, 52, 63, 69–72, 77. Williams, *Montpellier*, 155, 170.

Chapter 11 • From the New Anatomy to Pathology and Therapy

1. Gideon Harvey, *Conclave*, 30–31; many of the techniques satirized by Harvey had been invented and employed by Malpighi. Hooke, *Micrographia*, 186, discussed above in sec. 5 of the introduction. Adelmann, *Embryology*, 558–59, highlights the analogy with Sbaraglia's *De recentiorum medicorum studio*, whereas his concerns in *Oculorum et mentis vigiliae* echo Harvey's criticism even more closely.

2. Malpighi, *Opera posthuma*, 93–98; *MOB*, 536, 623. Adelmann, *Embryology*, 586. Bologna, Biblioteca Universitaria, Ms 936, I, I. In addition, both Sbaraglia and Malpighi were art collectors, and—not surprisingly—in a letter to Bellini Malpighi questioned the taste of his rival in intellectual and artistic matters as well: *MCA*, 3:1229–30, Malpighi to Bellini, 10 December 1686, at 1230. Pamela Smith, *Body*, chap. 6.

3. An account of the dispute is in Zeno, "Relazione." Rossi, "Medicina." Cavazza, "Impegno." Pighetti, "Dialogo." Dini, "Difesa."

4. Adelmann, *Embryology*, 543–47, 556–58. *MCA*, 4:1477–79, Malpighi to Borghese, 13 July 1689; 1561–63, Malpighi to Agostino Marsili, and 1563–64, Malpighi to unidentified correspondent. *MOB*, 631; Malpighi reproduced Mini's theses in his *Vita*, 101–2. Cavazza, *Settecento*,

chap. 2, esp. 91, 100–104. Pighetti, "Dialogo." Guglielmini, *Pro theoria medica adversus empiricam sectam praelectio*, dated 2 May 1702, in *Opera*, 2:57–72.

5. Adelmann, *Embryology*, 539.

6. *MOB*, 500. In commenting on Sbaraglia's and Malpighi's work I refer to Belloni's edition in *MOB*. Marcialis, "Macchinismo," 3–14.

7. *MOB*, 540–41, 577, 595, 607.

8. Cunningham, "Sydenham." Wolfe, "Sydenham."

9. Malpighi, *Opera posthuma*, 87; *MOB*, 563.

10. Adelmann, *Embryology*, 620.

11. Bertoloni Meli, "Medical *Assayer*." Galen, *Opera*, 5:667–71. *MOB*, 615. Altieri Biagi and Basile, eds., *Scienziati*, 1169, identifies Galileo's distinction between primary and secondary qualities as Malpighi's source. Adelmann, *Embryology*, 1:585, does not report or comment on this passage. Surprisingly, Belloni also did not identify Galileo's *Assayer* as Malpighi's source for the structure and style of the *Risposta* to Sbaraglia.

12. Galen referred to the atomistic doctrine of primary and secondary qualities in *De elementis ex Hippocrate*, 1:2, in Galen, *Opera*, 1:417–18. Malpighi may have been inspired by Boyle, *A Defence of the Doctrine Touching the Spring and Weight of the Air*, in *BW*, 3:84. Bertoloni Meli, "Mechanistic Pathology," 179–80, with specific reference to the role of Antonio Baldigiani SJ. Dollo, "Inediti"; Malpighi's relevant letter is dated from Rome, 7 February 1693.

13. *MOB*, 533, 591–92. Fazzari, "Visioni," 21–22. Bertoloni Meli, *Thinking*, 84–85; Castelli's theorem states that the speeds of water at different points in a river are inversely proportional to the cross sections of the river at those points.

14. Malpighi, *De renibus*. Wear, "Medical practice," 300. *MOB*, 542–46, 611–12.

15. *MOB*, 512.

16. Reisel, "De statua humana circulatoria," 232–33; "Statua humana circulatoria," at 9 and 12. Belloni, "Schemi e modelli," 292–96.

17. *MCA*, 3:1268–69, Malpighi to Tarantino, 29 March 1687. Canguilhem, *Normal*.

18. *MOB*, 513. On Cassini and Malpighi see Bertoloni Meli, "Images as experiments"; Verle, *Anatomia*. A copy of the book with the accompanying model of the eye from the Wellcome Library is on deposit at the Science Museum, London.

19. Harvey, *Second Essay to Riolan*, in *Circulation*, 124–25.

20. *MOB*, 512–16. Swammerdam, *De respiratione*, 30, 36–37. Schierbeek, *Swammerdam*, 67–71. See secs. 7.4 and 12.3.

21. Eriksson, *Vesalius*, 252–55.

22. *MOB*, 516, 596–97, 606–7. Duchesneau, "Malpighi," 113–14.

23. *MOB*, 516, 514.

24. Ibid., 593–95. Adelmann, *Embryology*, 582. Ragland, "Chemical Assaying," provides a detailed analysis of Malpighi's sources.

25. *MOB*, 575–76.

26. Ibid., 536–37, 570.

27. Sbaraglia's *Dissertatio epistolaris*, included in the 1693 edition of *Dissertationes*, ended as in the original edition. [Sbaraglia], *Dissertationes*, 52; Sbaraglia, *Exercitationes physico-anatomicae*, 29. Sbaraglia, *Vigiliae*, 204, states that the place of publication of the 1693 edition was not Naples but Vienna, with permission of Emperor Leopold.

28. *MCA*, 4:1702–3, Bohn to Malpighi, 26 September 1691; 1749–50, Malpighi to Albertini,

9 January 1692; 1756–57, Malpighi to Bohn, 15 January 1692. Sbaraglia, *Exercitationes*, 31–54, at 45, 50–51.

29. Sbaraglia, *Dissertatio secunda*, 55–171, at 123, 150–53, 169. See *MCA*, 5:1942, for Malpighi's rather crude comment on Sbaraglia's 1693 book.

30. Sbaraglia, *Appendix*, 173–200, at 177.

31. Malpighi, *Opera posthuma*, Amsterdam, 1698, preface, dated 1 December 1697, IV.–2r. Sbaraglia, *Appendix*, 188, 190–95. Cavazza, "Uselessness," 129. Williams, *Vitalism*.

32. Sbaraglia, *De vivipara generatione scepsis*, 329–30; *Vigiliae*, 456. Bernardi, *Metafisiche*, 112–19. On Sbaraglia and the soul see Marcialis, "Macchinismo," 7–14.

33. Sbaraglia, *Oculorum et mentis vigiliae*, 1–2, 552; Malpighi's name appears, however, as reported from a critical quotation by George Ent, 495.

34. Sbaraglia, *Vigiliae*, x–xiii, xxxi–xxxvii, 32–38, 53–54, 71, 365; 137, 449.

35. Ibid., 22, 103, 252–54; *MOB*, 516.

36. Sbaraglia, *Vigiliae*, 273, 328–29, 265–66; *MOB*, 518 (but see also 593–95).

37. Sbaraglia, *Vigiliae*, 523–27, at 527, and 341. Bertoloni Meli, *Thinking*, 80–82.

38. Sbaraglia, *Vigiliae*, 206, 557–96 at 558–63, and 574. On Sydenham see also ibid., 113, 254–57, 419. On Sydenham see the *DNB* entry by Harold J. Cook. Cunningham, "Sydenham." Cavazza, "Uselessness," 143–45; *MOB*, 590. Sbaraglia may have been referring to the Amsterdam 1683 edition of Sydenham's *Opuscula*, or the 1695 Leipzig edition of *Opuscula universa*.

39. Sbaraglia, *Vigiliae*, 597–643, at 602. Sbaraglia's statement that his colleague then taught the same subject at Padua supports the identification of Montanari, who also made microscopes for Malpighi.

40. Sbaraglia, *Vigiliae*, 599 and passim, 606, 612–19, 622, 626. Griendel, *Micrographia nova*, was published in Nuremberg in 1687 in German and Latin.

41. Morgagni, *Opera posthuma*, 1:14–16. *MCA*, 5:1892–93, 6 July 1693. Cavazza, *Settecento inquieto*, chap. 5, 186–87. [Morgagni]-Terranova, *Epistola*, 54–53; [Morgagni]-Floriani, *Epistola*, 175; see also a reference to Valsalva's *De aure humana*, on 162, a treatise to which Morgagni claimed to have contributed. The publisher of Morgagni's and Pacchioni's works was Francesco Buagni. Dini, "Difesa." Cavazza, "Impegno." Conforti and De Renzi, "Ospedali," 459–68. *MCA*, 5:1924–25n3; Conforti, "Il moto fermentativo." In 1696 Sandri published *De naturali & praeternaturali sanguinis statu*, dealing with the composition of blood.

42. [Morgagni]-Florianis, *Epistola*, 35; 85–87; 83–85; 153–57; 166–69; 82, 184–86. Cavazza, *Settecento*, 193–94.

43. Morgagni, *Adversaria prima*, 1; the three parts are at 1–12, 12–19, and 19–39, respectively. Casini, "Morgagni," 200–202.

44. Mosca, *Morgagni*, 34–35, quoted in *MOB*, 42. Morgagni, *Adversaria*, 1706, fig. H in plate III; *Adversaria tertia*, 39. Malpighi's opinion is reported by Lancisi in Eustachio, *Tabulae*, ix–x.

45. *MCA*, 4:1687–88n1. *MCA*, 4:1626–27n1. Generali, *Vallisneri*, 271–307, esp. 273–74 and 295. *MCA*, 4:1473–74n1. Albertini, *Animadversiones*, in Jarcho, *Heart*, chap. 18. A collection of medical letters by Albertini is in *Clinical Consultations and Letters*. Morgagni, *Consultations*, 392n597. Nicholson, "Morgagni." *MOB*, 254n1.

Chapter 12 • Medical Consultations

1. Agrimi and Crisciani, *Consilia*, 19–21. Lockwood, *Benzi*, esp. 47–138 and 323–24. Siraisi, *Alderotti*, chap. 9; *Clock*, 201–7. *MOB*, 483. Redi and Doni, *Consulti*. Cook, "New Philosophy." Bresadola, "Pazienti," 25–26, 36. De Renzi, "Natura." Malpighi and Lancisi, *Consultationes*, 9–10, 11, 49, 54; the text by Lancisi has separate pagination, 4, 80 (this case refers to a cure effected by a woman). See also *De consiliis medicis conscribendis ad tyrones praeloquium*, by Eusebio Sguari, in ibid., five unnumbered pages prefixed to the volume. Morgagni, *Consultations*, l, 85, 241. See Baglivi, *Correspondence*, 118, for a reference to Malpighi's refusal to accept compensation for a consultation he wrote for Manget.

2. Basile, "Consulti"; review of Doni; *L'invenzione del vero*, chap. 3, 91; 186–88. Piccolino, *Zufolo*, 159. Bertoloni Meli, "Redi," 77. Bresadola, "Pazienti," 18–19.

3. Malpighi, *Consulti*, 3:44–45. On the use of human excrement in antiquity see von Staden, "Women," at 11–12. Morgagni, *Consultations*, 391n590. Malpighi, *Centuria*, 32, 34, 36, 49; *Consulti*, 2:202; 3:12. Malpighi and Lancisi, *Consultationes*, 20, 56. Morgagni, *Consultations*, 31, 174. *BW*, 3:386–88; *Correspondence*, vol. 2, Lower to Boyle, 24 June 1664. Albertini, *Consultations, sub indice*. On the use of vipers as late as the second half of the eighteenth century see the essay by Enrico Benassi, "Morgagni's use of vipers," in Morgagni, *Consultations*, 316–18. *MOB*, 459, 455, 475. Red coral is also referred to among the astringent remedies taken by a lady afflicted by sterility in a consultation by Redi, *Opere*, 9:261. See the excellent glossary in Vallisneri, *Consulti*. Pressman, *Resort*.

4. On the lack of physical inspection see Bynum, "Health," at 212–13. Morgagni, *Consultations*, lii. *MOB*, 458, 460, 473, 484, 489.

5. G. A. di Porcia, *Notizie*, 47–48, quoted in Vallisneri, *Consulti*, xxxii.

6. Bakhtin, *Rabelais*, chap. 1, esp. 67–68; see also 161–62, 179–80, and 360–61; *Problems*, chap. 4, esp. 100–13.

7. Vallisneri, *Epistolario*, 1:252. See also Gaspari's preface to the reader in Malpighi's *Centuria*. A consultation by Malpighi had been published by Manget in the second volume of the *Bibliotheca medico-practica*, 1100–1101. Baglivi, *Correspondence*, 108–9.

8. See, e.g., the letter by Giorgio Baglivi to Manget of 1 April 1694, Baglivi, *Correspondence*, 147. Basile, "Consulti," 171. Atti, *Malpighi*, 421–47.

9. Vallisneri, *Epistolario*, 1:301–9. The coded reference to Sbaraglia is at 306: "qualche parziale del pur morto malpighiano antagonista."

10. Ibid., 1:304. Frati, *Bibliografia*, 31–32, no. 40. Atti, *Notizie*, 421–47, provides an extensive reconstruction of the affair.

11. Bertoloni Meli, "Blood."

12. Armillei, *Consulti medici*; *Consultationum medicarum centuria*, with occasional overlaps. Both are omitted by Frati, *Bibliografia*, or Adelmann, *Embryology*. Malpighi and Lancisi, *Consultationes*, includes sixty-three consultations by Malpighi and fifty by Lancisi, whose style was close to Malpighi's. Lancisi's collection frequently includes letters of referral: Lancisi, *Consulti*, separate pagination, 2, 4, 13, 15, 17, 24, etc.; 81–87; 90–95; 101–2; 115–17. Lancisi's edition contains several dissertations, such as a short treatise on generation, on the properties of salts of hartshorn, on phlebotomy, and on a case of presumed infanticide. All these editions were due to Giuseppe Corona. There is a modern edition of Malpighi's consultations, *Consulti*. Vallisneri, *Consulti*, xxix, n44.

13. *MCA*, 1:xiv. In the index, under "Malpighi" and "medical practice," Adelmann added a subsection on "his distaste for," 5:2199.

14. Malpighi, *Consulti*, 1:70–71 ("facendole umilissimo inchino, mi confermo per sempre etc."); 2:132 ("Excellentissime vir"); 3:64 ("excellentisimo et praeclarissimo viro domino Joseph Maria Zamperio").

15. Vallisneri, *Epistolario*, 2:307. Del Papa, *Consulti*, 2:206–7.

16. Agrimi and Crisciani, *Consilia medicaux*, 19–21.

17. Redi, *Opere*, 9:318. See sec. 2.3 on different geographic traditions. Bresadola, "Pazienti," 24–25.

18. Bresadola, "Pazienti," 26–27. On how Malpighi arranged his own manuscripts see Bertoloni Meli, "Archive," 110–11.

19. Examples of letters written by patients are in Malpighi, *Consulti*, 1:171–73; 2:182, 313 (Malpighi's reply is at 314), 376–78 (Malpighi's reply is at 378–79); 3:118–19. In the last letter, by Erminia Santa Croce Lancelotta, the patient challenges the therapies prescribed and reports contrasts among the attending physicians.

20. *MOB*, 479. Malpighi, *Consultationes*, 28. For a critical comment of the letter of referral see also Lancisi, *Consulti*, 20.

21. Malpighi, *Centuria*, 84. *MOB*, 462.

22. Malpighi, *Consultationes*, 51. *MOB*, 478. However, see also Jarcho, *Concept*, 193. In *De motu animalium*, Borelli discussed at length the cause of heart movement, providing a series of explanations without offering a clear solution: Giglioni, "Machines," 173–74.

23. Armillei, *Centuria*, 121.

24. See, e.g., Malpighi, *Centuria*, 19, 21, 28, 35, 40, 41, 47, 70 (*MOB*, 459), 81 (*MOB*, 461), 128, 143 (*MOB*, 468); *Consultationes*, 23, 40, 51, 69. Cunningham, "Pathology."

25. Malpighi, *Centuria*, 127. *MOB*, 466.

26. Malpighi, *Centuria*, 75. Jarcho, *Concept*, 202. A similar account is in *MOB*, 460–61.

27. Vallisneri, *Consulti*, xl–xliii.

28. Bertoloni Meli, "Archive," 119–20. See also *MOB*, 556.

29. *MOO*, 2:139; *MOB*, 89. Malpighi, *Centuria*, 82–83, 152–59; *Consultationes*, 66, 75–76, 78–80. *MOB*, 470, 484, 485, 489–90. On sex differences in Descartes' mechanistic anatomy see Stuurman, *Invention*, chap. 3. Laqueur, *Sex*, 134–63. On a debate on shifting understandings of male and female anatomy see Stolberg, "A Woman"; Laqueur, "Sex in the Flesh"; Schiebinger, "Skelettestreit."

30. *MOB*, 564.

31. *MOB*, 609; see also 516, 527, 536, 543, 544, 555, 558, 563, 572, 586. Galen, *Therapeutic Method*, II.7.1–2, 63–64; see the commentary by Harkinson, 202–6. On the fortunes of Galen's text see Kudlien and Durling, *Galen's Method*.

32. Malpighi, *Centuria*, 71, 84. *MOB*, 459, 462.

33. Malpighi, *Centuria*, 43. *MOB*, 455, 573.

34. Bertoloni Meli, "Archive," 116, 119, 120. *MCA*, 4:1830. See also *MCA*, 2:513–15, at 412, Malpighi to Bonfiglioli, 4 February 1671, referring to chymical experiments on medicinals. Gentilcore, "Apothecaries." Issue 45 of *Pharmacy in History* is devoted to "The World of the Italian Apothecary."

35. Malpighi, *Centuria*, 127, 143. *MOB*, 467, 469.

36. Malpighi, *Consultationes*, 37, 76. *MOB*, 486, 522. Malpighi, *Centuria*, 179. Armillei, *Centuria*, 2, 11, 13, 96.

37. Malpighi, *Centuria*, 92, 163; *Consultationes*, 77. *MOB*, 464, 473, 488.

38. Malpighi, Centuria, 82. *MOB*, 461.

39. Malpighi, *Centuria*, 93. Malpighi, *Centuria*, signature B. Cunningham, "Sydenham."

40. *MCA*, 2:882–84, Malpighi to Bentivoglio, 21 September 1682. Adelmann, *Embryology*, 459–62.

41. *MAP*, 281–82. *MCA*, 1:9, 11–12, 14, etc.

42. *MOO*, 2:138–39. *MOB*, 88, 150, 458. Adelmann, *Embryology*, 174.

43. Redi, *Consulti medici. All'illustrissimo Michelangelo Tilli* (Florence: G. Manni, 1726, 1729), 2 vols.; *Opere. In questa nuova edizione accresciute e migliorate* (Venice: Gio. Gabriello Hertz, 1712–1730), 7 vols.; the *Consulti medici* occupy vols. 6–7; Redi, *Opere*, 9:xix. Redi, *Consulti*, ed. Carla Doni.

44. Redi, *Consulti*, 163–64.

45. Redi, *Opere*, 9:229–31; *Consulti*, 269–70, Florence, 22 May 1688.

46. Redi, *Consulti*, 438, 442, 446, 448, 455, 459, 464, 469.

47. Redi, *Consulti*, 134–35; see 122, 131, and 338 for the use of viper's meat.

48. Redi, *Opere*, 9:98–105.

49. Ibid., 9:82–85, at 82–83.

50. See, e.g., ibid., 9:110, 112, 132.

51. Ibid., 9:103. Water from Tettuccio was later known as Montecatini spa.

52. Ibid., 9:35, 143, 314–15, 318. Basile, *Invenzione*, 114–18.

53. Redi, *Opere*, 9:166.

54. Redi, *Consulti*, 33 and 435–73.

55. *MOB*, 418–19. Luigi Guerrini, "Contributo," 49n3.

56. Redi, *Consulti*, 163; see also *Opere*, 9:324.

57. Redi, *Consulti*, 343; *Opere*, 9:221.

58. See, e.g., Redi, *Consulti*, 138, 272, 372, 385, 451; *Opere*, 9:174–75, 366.

59. Redi, *Opere*, 9:xv and 76–78, at 78.

60. Bernardi, "Teoria," 18–19. Redi, *Opere*, 9:33–34 for a case of gout, and 39 for a case of constipation; 4:269–78. Tribby, "Cooking." Perifano, "L'alchemie." P. Galluzzi, "Motivi paracelsiani."

61. Redi, *Opere*, 9:53–58: periodic pains in the lower belly, at 55; 167–78: uterine epilepsy, lack of menstruation, sterility, at 167–68.

62. Redi, *Opere*, 9:245–47.

63. Redi, *Consulti*, 231, Redi to Francesco Maria Medici, 16 May 1684. De Graaf, *Human*, 89–90, 163–64. Daston and Park, *Wonders*, 203. Fend, *Grenzen*, chap. 2, esp. 19–25. Krämer, "Individualisierung."

64. Basile, *Invenzione*, chap. 3. *MCA*, 1:336–38, Capucci to Malpighi, 21 January 1667, at 337. Redi, *Consulti*. Porzio, *Erasistratus*.

65. Vallisneri, *Opere*, 3:483–558; Vallisneri, *Consulti medici*, introduction by Gemelli, xliii, liv, lix, lxv, lxxi.

66. Morgagni, *Consultations*, 64, 77–78, 128, 234, 238, 251, 270 (refers to the little channels Morgagni found in the urethra); quotation at 161. See also 189 for a defense of rational medicine. On Stahl see Chang, "Fermentation"; "Motus tonicus."

67. Morgagni, *Consultations*, consultation 6 and 345n54.
68. Ibid., 24, 68, 134, 226. Nicholson, "Morgagni."
69. Armillei, *Consulti*, 2:174–79.
70. Morgagni, *Consultations*, 197; see also 212.
71. Ibid., 47–58, at 57–58.
72. Findlen, *Possessing Nature*, 28. Guerrini, "Duverney's Skeletons."
73. Morgagni, *De sedibus*, preface. *MOB*, 45, 508, 589. Cavazza, *Settecento*, 201. Armillei, *Consulti*, 2:196. Cunningham, "Pathology."

Epilogue

1. Frank, *Harvey*, chaps. 8–10.
2. Brown, *Mechanical Philosophy*, chaps. 4–6. Guerrini, "Newton." Williams, *Vitalism*. Wolfe, *Vitalism*. Chang, "Fermentation"; "*Motus tonicus*."
3. Bertoloni Meli, *Thinking*, 70–72, 116, 130, 224–25, and passim; "Collaboration." Bastholm, *Muscle*. Fournier, *Fabric*, 198.
4. Westfall, *Construction*, 104. Gabbey, "Mechanical Philosophies." Meinel, "Atomism."
5. Westfall, *Construction*, 104.
6. Wilson, *Invisible*. Fournier, *Fabric*. Ruestow, *Microscope*. Cole, "Injections." Cook, *Exchange*, 276–88; "Time's Bodies."
7. Ruestow, "Doctrine."
8. The *locus classicus* is Dear, *Discipline*.
9. Hooke, *Micrographia*, 186.
10. Otis, *Müller*. On training see Ruestow, *Microscope*, 293–99. Lawrence, "Monro," 211.
11. Foster, *Lectures*. Hall, *History*. Frank, *Harvey*.
12. Brown, *Philosophy*; "College." Cook, *Decline*. Wear, "Practice."
13. Canguilhem, *Normal*. Pressman, *Last Resort*.
14. Guerrini, "Varieties," 120–24. Brown, *Mechanical Philosophy*, 203–11. Müller, *Iatromechanische*. Cunningham and French, *Enlightenment*.

REFERENCES

Primary Sources

Albertini, Ippolito Francesco, and Francesco Torti. *Clinical Consultations and Letters.* Translated and annotated by Saul Jarcho. Boston: Francis A. Countway Library of Medicine, 1989.

Altieri Biagi, Maria Luisa, and Bruno Basile, eds. *Scienziati del Seicento.* Milano-Napoli: Ricciardi, 1980.

Anonymous. "An Account of a Book Entituled *Relatione de ritrovamento dell'uova di chiocciole.*" *PT* 13 (1683), 356–58.

———. "An Account of the Rise and Attempts, of a Way to Conveigh Liquors Immediately into the Mass of Blood." *PT* 1 (1665), 128–30.

———. "Esperienze fatte dal Signor Francesco Redi alla presenza del Serenissimo Gran Duca di Toscana intorno ad un acqua, che si dice dotata della suddetta virtù." *GLI* 7 (1673), 119–24.

———. "Esperienze fatte in Francia, e in Inghilterra dal Signor Denis, Medico ordinario del Ré Christianissimo, e da altri, intorno a quell'acqua, che si dice, che stagna subito tutti quanti i flussi di sangue." *GLI* 7 (1673), 115–19.

———. "Relazione del successo di alcune trasfusioni di sangue, fatte negli animali." *GLI* 1 (1668), 91–93.

Aristotle. *Generation of Animals.* Translated by Arthur Leslie Peck. Cambridge, Mass.: Harvard University Press, 1990.

———. *On the Soul. Parva naturalia. On Breath.* Translated by Walter Stanley Hett. Cambridge, Mass.: Harvard University Press, 1957.

Armillei, Gaetano. *Consultationum medicarum centuria.* Venice: Giuseppe Corona, 1744.

———. *Consulti medici di vari professori spiegati con le migliori dottrine moderne, e co' le regole più esatte della scienza meccanica.* 2 vols. Venice: Giuseppe Corona, 1743–45.

Aselli, Gasparo. *De lactibus sive lacteis venis, novo invento.* Milan, 1627. Facsimile edition with introduction by Pietro Franceschini, Milan, 1972. Leiden: ex officina Johannis Maire, 1640.

Avicenna, *Canon.* Venice: in edibus Luce Antonii Iunta Florentini, 1527. Facsimile edition, Brussels: medicinae historia, 1971.

Back, Jacobus de. *Dissertatio de corde,* in William Harvey, *Exercitatio anatomica de motu cordis & sanguinis.* Rotterdam: ex officina Arnoldi Leers, 1648.

Baglivi, Giorgio. *The Correspondence from the Library of Sir William Osler.* Ithaca: Cornell University Press, 1974.

Bartholin, Caspar, Thomas Bartholin, and Johannes Walaeus. *Anatome quartum renovata: non tantum ex institutionibus b. m. parentis, Caspari Bartholini, sed etiam ex omnium cùm veterum, tum recentiorum observationibus: ad circulationem Harveianam, & vasa lymphatica directis.* Leiden: Joan Ant. Huguetan, & Soc., 1677.

———. *Institutiones anatomicae.* Leiden: Apud Franciscum Hackium, 1645 (Walaeus's letters first included in 1641 edition).

Bartholin, Thomas, ed. *Acta philosophica et medica hafniensia.* 5 vols. Copenhagen: Sumptibus P. Haubold, 1673–80.

———. *De pulmonum substantia & motu diatribe. Accedunt Marcelli Malpighij "De pulmonibus observationes anatomicae."* Copenhagen: Typis Henrici Gödiani, 1663.

———. *Epistolarum medicinalium centuria I–IV.* 3 vols. Copenhagen: Typis Matthiæ Godicchenii, impensis Petri Haubold, 1663–67.

Bauhin, Caspar. *Pinax theatri botanici.* Basel: Ludovicus Regius, 1623.

———. *Theatrum anatomicum.* Frankfurt/M: Thomas Becker, 1605.

Bayle, François. *Dissertationes physicae.* Toulouse: Ex Officina C. L. Colomeri & H. Poysvel, 1677.

Bidloo, Govart. *Anatomia corporis humani.* Amsterdam: Sumptibus viduae Joannis à Someren, haeredum Joannis à Dyk, Henrici & viduae Theodori Boom, 1685.

———. *Vindiciæ quarundam delineationum anatomicarum, contra ineptas animadversiones Fred. Ruyschii.* Leiden: Jordanus Luchtmans, 1697.

Birch, Thomas. *The History of the Royal Society of London.* 4 vols. London: Millar, 1756–57.

Boerhaave, Herman, and Frederik Ruysch. *Opusculum anatomicum de fabrica glandularum in corpore humano, continens binas episotlas.* Leiden: Petrus vander Aa, 1722.

Bonaccorsi, Bartolommeo. *Della natura de polsi.* Bologna: Giacomo Monti, 1647.

Borel, Pierre. *De vero telescopii inventore.* The Hague: Adrian Vlacq, 1655.

———. *Historiarum, et observationum medicophysicarum centuriae IV.* Paris: Apud Ioannem Billaine et Viduam Mathurini Dupuis 1656–57[2].

Borelli, Giovanni Alfonso. *Delle cagioni delle febbri maligne della Sicilia. Negli anni 1647 e 1648.* Cosenza: Gio. Battista Rosso, 1649.

———. *De motu animalium.* Rome: Angeli Bernabò, 1680–81. 2 vols. English translation by Paul Maquet, *On the Movement of Animals.* Berlin: Springer, 1989.

———. "Osservatione intorno alla virtù ineguale degli occhi." *GLI* 2 (1669), 11–12.

———. *Theoricae Mediceorum planetarum ex causis physics deductae.* Florence: Ex typographia S[erenissimi] M[agni] D[ucis], 1666.

Bosco, Ippolito. *De facultate anathomica per breves lectiones cum quibusdam observationibus.* Ferrara: Victorius Baldinus, 1600.

Boyle, Robert. *Correspondence.* Edited by Michael Hunter, Antonio Clericuzio, and Lawrence M. Principe. 6 vols. London and Brookfield, Vt.: Pickering & Chatto, 2001.

Brunner, Johann Conrad. *Exercitatio anatomico-medica de glandulis in intestine duodeno hominis detectis.* Heidelberg: Buchta, 1688.

———. *Experimenta nova circa pancreas.* Amsterdam: Wetstenius, 1683.

———. *Glandulae duodeni seu pancreas secundarium.* Frankfurt and Heidelberg: Maximilian von Sand, 1715.

[Buonanni, Filippo]. *Riflessioni sopra la "Relazione del ritrovamento dell'uova delle chiocciole."* Bologna: Eredi di Antonio Pisari e Domenico Antonio Ercole, 1695. Originally printed in Rome: per il Varese, 1683.

Camerarius, Rudolph Jakob. *De sexu plantarum epistola.* Tübingen: typis viduae Rommeii, 1694. German translation by Martin Möbius, *Ueber das Geschlecht der Planzen.* Leipzig: Engelmann, 1899.

———. "De sexu plantarum Epistola [secunda]." *MC* 33 (1695–96), Appendix, 31–36.

———. *Opuscula botanici argumenti.* Prague: Carulum Barth, 1797.

Camerarius, Rudolph Jakob, and Johann David Mauchart. *Disquisitio botanica de herba mimosa seu, sentiente.* Tübingen: Typis G. Henrici & J. C. Reisl, 1688.

Camerarius, Rudolph Jakob, and Georg Burckhard Seeger. *De lolio temulento.* Tübingen: Typis vidua Georg-Henrici Reisl, 1710.

Cardano, Gerolamo. *De vita propria liber. The Book of My Life.* Translated by Jean Stoner. London: Dent, 1931.

Castelli, Pietro. *Praeservatio corporum sanorum ab imminenti lue ex aeris intemperie hoc anno 1648.* Messina: Apud haeredes Petri Breae, 1648.

———. *Responsio chimica. De effervescentia, et mutatione colorum in mixtione liquorum chimicorum.* Messina: Typis Haeredum Petri Breae, 1654.

Cestoni, Diacinto. "*Lettera . . . se l'alga marina faccia il fiore, ed il seme, o se nasca dalla putredine, o spontaneamente ne fondi del mare.*" *Galleria di Minerva* 2 (1697), 121–24.

Collegium privatum Amstelodamense. Observationes anatomica selectiores. Amsterdam: Commelin, 1667, 1673. Facsimile edition with introduction by Gerrit Arie Lindeboom, Nieuwkoop: de Graaf, 1675.

Colombo, Realdo. *De re anatomica libri XV.* Venice: Nicolaus Bevilacqua, 1559.

De Graaf, Regnier. *De succi pancreatici natura et usu, exercitatio anatomico-medica.* Leiden: Officina Hackiana, 1664; 3rd ed. as *Tractatus anatomico-medicus de succi pancreatici natura et usu.* Leiden: Officina Hackiana, 1671.

———. *On the Human Reproductive Organs. An Annotated Translation of "Tractatus de virorum organis generationi inservientibus" (1668) and "De mulierum organis generationi inservientibus tractatus novo" (1672),* by H. D. Jocelyn and Brian P. Setchell. Oxford: Blackwell, 1972.

De Heide, Antonius. *Anatome mytuli, Belgicè mossel.* Amsterdam: Janssonio Waesbergios, 1683 (also 1684).

Del Papa, Giuseppe. *Consulti medici.* 2 vols. Rome: G.M. Salvioni, 1733.

Descartes, René. *De homine.* Translated with an introduction and figures by Florentius Schuyl. Leiden: P. Leffen & F. Moyardum, 1662.

———. *L'homme. De la formation du foetus,* with remarks by Louis de la Forge. Paris: Girard, 1664.

———. *The World and Other Writings.* Edited by Stephen Gaukroger. Cambridge: Cambridge University Press, 1998.

Dodoens, Rembert. *Medicinalium observationum exempla rara.* Cologne: Apud Maternum Cholinum, 1581; Harderwijk: Henricus Laurentius, 1621.

Du Laurens, André. *Historia anatomica humani corporis.* Paris: Marcum Orry, 1600.

Duverney, Guichard Joseph. *Traité de l'organe de l'ouie.* Paris: Estienne Michallet, 1683.

Elsholtz, Johann Sigismund. *Clysmatica nova.* Berlin: typographia Rungiana, 1665. Facsimile edition with a preface by Akitomo Matsuki and commentary by John W. R. McIntyre, Tokio: Iwanami, 1995.

———. *Clysmatica nova. Editio secunda.* Kölln: ex officina Goergii Schultzii, 1667. Facsimile edition with a preface by Heinz Goerke, Hildesheim: Olms, 1665.

Eustachio, Bartolomeo. *Tabulae anatomicae.* Rome: F. Gonzaga, 1714.

Fontenelle, Bernard le Bovier de. *Histoire du renouvellement de l'Académie Royale des Sciences.* Amsterdam: Pierre de Coup, 1709.

Galen. *On the Natural Faculties.* Translated by Arthur John Brock. Cambridge, Mass.: Harvard University Press, 1916.

———. *On the Therapeutic Method: Books and I and II.* Translated with an introduction and commentary by R. J. Hankinson. Oxford: Clarendon Press; New York: Oxford University Press, 1991.

———. *On the Usefulness of the Parts of the Body.* Translated with an introduction by Margaret Tallmadge May. 2 vols. Ithaca: Cornell University Press, 1968.

———. *Opera omnia.* Edited by Karl Gottlob Kühn. Leipzig: Carl Cnobloch, 1821–33. Reprinted in Hildesheim: Olms, 1964.

Gassendi, Pierre. *Opera omnia.* Lyon: Anisson and Devenet, 1658, reprinted Stuttgart Bad Cannstatt: F. Frommann, 1964, 6 vols., introduction by Tullio Gregory.

Giraldi, Giovanni Battista. *Historia medica.* Bologna: typis HH. Antonij Pisarij, n.d.

Grew, Nehemiah. *The Anatomy of Plants, with an Idea of a Philosophical History of Plants, and Several Other Lectures, Read before the Royal Society.* London: W. Rawlings, 1682.

Griendel, Johann Franz. *Micrographia nova.* Nuremberg: Johann Ziegers, 1687.

Guglielmini, Domenico. *Opera omnia: mathematica, hydraulica, medica, et physica.* 2 vols. Geneva: Cramer, Perachon & socii, 1719.

Harder, Johann Jacob. *De ovis et genitalibus cochlearum.* Augsburg: sumptibus Theophili Goebelii bibliopolae, literis Leonhardi Zachariae, 1684.

———. *Prodromus physiologicus naturam explicans humorum nutritioni et generationi.* Basel: Jacob Bertsch, 1679.

Harvey, Gideon. *The Conclave of Physicians.* London: James Partridge, 1683.

Harvey, William. *De motu cordis et sanguinis in animalibus, anatomica exercitatio.* Leiden: ex officina Johannis Maire, 1639.

———. *Exercitatio anatomica de circulatione sanguinis.* Cambridge: Roger Daniels, 1649. Translated by Kenneth J. Franklin in Harvey, *Circulation,* 96–139.

———. *Exercitatio anatomica de motu cordis et sanguinis in animalibus.* Frankfurt: Wilhelm Fitzer, 1628. Translated by Kenneth J. Franklin with introduction by Andrew Wear as *The Circulation of the Blood and Other Writings.* London: Everyman, 1993.

———. *Exercitationes de generatione animalium.* London: Typis Du Gardianis, impensis Octaviani Pulleyn, 1651. Translated with introduction and notes by Gweneth Whitteridge. *Disputations Touching the Generation of Animals.* Oxford: Blackwell, 1981.

Hemsterhuis, Siboldus. *Messis aurea, exhibens anatomica, novissima et utilissima experimenta. Huic editioni accesserunt De vasis lymphaticis tabulae Rudbeckianae.* Heidelberg: typis Adriani Wyngaerden, 1659.

Hero of Alexandria. *Spiritalium liber. A Federico Commandino Urbinate, ex Graeco, nuper in Latinum conversus.* Urbino: [Domenico Frisolino], 1575.

Highmore, Nathaniel. *Corporis humani disquisitio anatomica.* The Hague: Samuel Broun, 1651.

Hooke, Robert. *Micrographia; or Some Physiological Descriptions of Minute Bodies Made by the Magnifying Glasses with Observations and Inquiries Thereupon.* London: John Martyn and James Allestry, 1665. New York: Dover, 1961.

Le Clerc, Daniel, and Jean-Jacques Manget, eds., *Bibliotheca anatomica.* 2 vols. 2nd ed., Geneva: Johan. Anthon. Chouët & David Ritter, 1699.

———. *Bibliotheca anatomica, medica, chirurgica, etc.* 3 vols. London: John Nutt, 1711–14.

Lipari, Michele. *Galenistarum triumphus.* Cosenza: Apud Io. Baptistam Russo, 1665.

Lower, Richard. *Tractatus de corde, item de motu et colore sanguinis et chyle in eum transitu.* London: Typis Jo. Redmayne impensis Jacobi Allestry, 1669. Facsimile edition with introduction and translation by Kenneth J. Franklin, in *Early Science in Oxford,* ed. Robert T. Gunther, vol. 9. Oxford: Oxford University Press, 1932.

Malpighi, Marcello. "Alcune osservazioni intorno à vermi da seta." *GLI* 3 (1670), 166, 71.

———. *Anatome plantarum.* 2 vols. London: John Martyn, 1675–79. Partial translation by Martin Möbius, *Die Anatomie der Pflanzen.* Leipzig: W. Engelmann, 1901.

———. *Consultationum medicinalium centuria prima.* Padua: Jo. Manfré, 1713. A corrected version appeared in the same year and with the same title.

———. *Consulti. 1675–1694.* Edited by Giuseppe Plessi and Raffaele Bernabeo. 3 vols. Bologna: Istituto per la storia dell'università, 1992.

———. "*De polypo cordis.* An Annotated Translation," by John M. Forrester. *MH* 39 (1995), 477–92.

———. *De pulmonibus observationes anatomicae* and *Epistola altera.* Bologna: Gio. Battista Ferroni, 1661. Reprinted with translation and introduction by Luigi Belloni. Messina-Palermo: Società italiana di istochimica, 1658.

———. *Discours anatomiques sur la structure des visceres.* Paris: Laurent d'Houry, 1687^2.

———. *Memorie di me a i miei posteri fatte in villa l'anno 1689.* Bologna: Zanichelli, 1902.

———. *Opera posthuma.* London: A. & J. Churchill, 1697.

———. "Sul baco da seta. Dissertazione epistolare," trans. E. Regonati. *Il bacofilo italiano,* 2 (1860), 17–90.

Malpighi, Marcello, and Giovanni Maria Lancisi. *Consultationum medicarum, nonnullarumque dissertationum collectio.* Venice: Giuseppe Corona, 1747.

Manget, Jean-Jacques. *Bibliotheca medico-pratica.* 4 vols. Geneva: J. A. Chouët and D. Ritter, 1695–97.

———. *Theatrum anatomicum.* 3 vols. Geneva: Cramer & Perachon, 1717.

[Manolessi, Emilio Maria]. *Relazione dell'esperienze fatte in Inghilterra, Francia, ed Italia intorno alla celebre, e famosa trasfusione del sangue.* Bologna: Manolessi, 1668.

Mayow, John. *Tractatus duo, quorum prior agit de respiratione, alter de rachitide.* Oxford: excudebat Hen: Hall, 1668; Leiden: Apud Felicem Lopez de Haro, Cornelium Driehuysen, 1671.

Mini, Paolo. *Medicus igne, non cultro necessario anatomicus.* Venice: Apud Joannem Franciscum Valvasensem, 1678.

More, Henry. *The Immortality of the Soul.* London: J. Flesher for William Morden, 1659.

Morgagni, Giovanni Battista. *Adversaria anatomica omnia.* 6 parts. Padua: Josephus Cominus, 1717–19.

———. *Adversaria anatomica prima.* Bologna: Typis Fernandi Pisarri, 1706.

———. *The Clinical Consultations*. The edition of Enrico Benassi (1935). Translated and revised by Saul Jarcho. Boston: Francis A. Countway Library of Medicine, 1984.

———. *De sedibus, et causis morborum per anatomen indagatis*. Venice: Ex typographia Remondiniana, 1761.

———. *Opera posthuma*. 6 vols. Rome: Istituto per la storia della medicina, 1964–82.

[Morgagni, Giovanni Battista]. Horatius de Florianis, *Epistola, qua plus cento & qinquaginta errors ostenduntur*. Luca Terranova, *Altera epistola in illud idem argumentum*. Rome: Typis Ioannis Francisci Buagni, 1705.

Mosca, Giuseppe. *Vita di Giovambattista Morgagni*. Napoli: Vincenzo Manfredi, 1764.

[Pascal, Blaise]. *Récit de la grande experience*. Paris: Charles Saureux, 1648, reprinted in *Traitez de l'eqvilibre des liqvevrs, et de la pesantevr de la masse de l'air*. Paris: Guillame Desprez, 1663.

Paulli, Jacok Henrik. *Anatomiae bilsianae anatome*. Strasburg: Apud Simonem Paulli, 1665.

Pechlin, Johann. *De medicamentorum purgantium facultatibus*. Leiden and Amsterdam: Apud Danielem, Abrahamum & Adrianum à Gaasbeek, 1672.

[Pechlin, Johann, attributed to]. Jani Leoniceni Veronensis [pseud.] *Metamorphosis Aesculapii & Apollinis Pancreatici*. Gratianopoli [Leiden]: Apud Orlandum Bon Tempi, 1672.

Pecquet, Jean. *Experimenta nova anatomica*. Paris: Sebastian and Gabriel Cramoisy, 1651. Second expanded edition, 1654.

Peyer, Johann Conrad. *Exercitatio anatomico-medica de glandulis intestinorum*. Schaffausen: Impensis Onophrii a Waldkirch, typis Alexandri Riedingii, 1677.

Plato, *Timaeus*. Translated by Donald J. Zeyl. Indianapolis: Hackett, 2000.

Plutarch. *Lives*. I. Translated by Bernadotte Perrin. Cambridge, Mass.: Harvard University Press, 1914.

Porzio, Lucantonio. *Erasistratus, sive de sanguinis missione*. Rome: Angelo Bernabò, 1682.

Redi, Francesco. *Consulti medici*. Critical edition by Carla Doni. Florence: Centro Editoriale Toscano, 1985.

———. *Esperienze intorno alla generazione degli insetti*. Firenze: All'insegna della Stella, 1668. Edited with an introduction by Walter Bernardi, Firenze: Giunti, 1996.

———. *Opere*. 9 vols. Milan: Società tipografica de' classici italiani, 1809–11.

Regius, Henricus. *Fundamenta physices*. Amsterdam: Apud Ludovicum Elzevirium, 1646.

Reisel, Salomon. "De statua humana circulatoria." In *Miscellanea curiosa*, decuria I, anni IV–V, 1673–74, Frankfurt and Leipzig, 1688, 232–33.

———. "Statua humana circulatoria." In *Miscellanea curiosa*, decuria I, anni IX–X, 1678–79, Nuremberg, 1693, 1–22.

Rossetti, Donato. *Antignome fisico-matematiche*. Livorno: Giovanni Vincenzo Bonfigli, 1667.

Ruysch, Frederik. *Dilucidatio valvularum in vasis lymphaticis et lacteis*. The Hague: Harman Gael, 1665. Facsimile edition with an introduction by Anoine M. Luydendijk-Elshout. Nieuwkoop: De Graaf, 1964.

———. *Epistola anatomica, problematica prima*—[*sexta et decima*]. Amsterdam: Joann Wolters, 1696–1714.

———. *Observationum anatomico-chirurgicarum centuria*. Amsterdam: Henricus & vidua Theodori Boom, 1691.

———. *Responsio ad Godefridi Bidloi, libellum, cui nomen vindiciarum inscripsit*. Amsterdam: Joann Wolters, 1697.

———. *Thesaurus anatomicus primus—[decimus]*. Amsterdam: Johann Wolters, 1701–16.

Sbaraglia, Giovanni Girolamo. *Exercitationes physico-anatomicae*. Bologna: Pietro Maria Monti, 1701.

———. *Oculorum et mentis vigiliae*. Bologna: Pietro-Maria Monti, 1704.

[Sbaraglia, Giovanni Girolamo]. *Dissertationes quibus moderna medicorum studia . . . ventilantur*. Naples: [s.n.], 1693.

Schrader, Justus. *Observationes de generatione animalium*. Amsterdam: Typis Abrahami Wolfgang, 1674.

Severino, Marco Aurelio. *Zootomia democritea*. Nurenberg: Literis Endterianis, 1645.

Steno, Nicolaus. *De musculis et glandulis observationum specimen*. Copenhagen: Literis Matthiæ Godicchenii and Amsterdam: Apud Petrum le Grand, 1664.

———. *Epistolae*. 2 vols. Copenhagen: Busck; Freiburg: Herder, 1952.

———. *Lecture on the Anatomy of the Brain*. Introduction by Gustav Scherz. Copenhagen: Nyt Nordisk Forlag. Arnold Busck, 1965.

———. *Observationes anatomicae, quibus varia oris, oculorum, & narium vasa describuntur, novique salivæ, lacrymarum & muci fontes deteguntur*. Leiden: Apud Jacobum Chouët, 1662.

———. *Opera philosophica*. Edited by Vilhelm Maar. 2 vols. Copenhagen: Vilhelm Tryde, 1910.

———. *Opere scientifiche*. Edited by Enrico Coturri. 2 vols. Prato: Cassa di Risparmi e Depositi, 1986.

Swammerdam, Jan. *Biblia naturae*. 2 vols. Leiden: Apud Isaacum Severinum, Balduinum vander Aa, Petrum vander Aa, 1737–38.

———. *Ephemeri vita*. Amsterdam: Abraham Wolfgang, 1675.

———. *Historia insectorum generalis*. Utrecht: Meinardus van Dreunen, 1669. French translation as *Histoire generale des insectes*. Utrecht: Guillaume de Walcheren, 1682.

———. *Miraculum naturae, sive uteri muliebris fabrica*. Leiden: Apud S. Matthaei, 1672.

———. *Tractatus physico-anatomico-medicus de respiratione usuque pulmonum*. Leiden: Apud Danielem, Abrahamum & Adrianum à Gaasbeeck, 1667.

Sylvius, Franciscus. *Opera medica*. Amsterdam: Daniel Elzevier and Abraham Wolfgang, 1679.

Thruston, Malachi. *De respirationis usu primario*. London: Apud Johannem Martyn, 1670; Leiden: Apud Felicem Lopez de Haro, Cornelium Driehuysen, 1671.

Trionfetti, Giovanni Battista. *Observationes de ortu ac vegetatione plantarum*. Rome: A Herculis, 1685.

———. *Vindiciarum veritatis a castigationibus quarundam propositionum quae habentur in opusculo De ortu et vegetatione plantarum*. Romae: Ex Typographia Antonii de Rubeis, 1703.

Vallisneri, Antonio. *Consulti medici*. Edited by Benedino Gemelli. Florence: Olschki, 2006.

———. *Epistolario*. Edited by Dario Generali. 2 vols. Milan: Franco Angeli, 1991.

Valsalva, Antonio Maria. *De aure humana tractatus*. Bologna: Typis Constantini Pisarii, 1704.

van Beverwijck, Jan. *De calculo renum et vesicae*. Leiden: ex officina Elseviriurum, 1638.

van Leeuwenhoek, Antoni. "An Abstract of a Letter about Generation by an Animalcule in the Male Seed." *PT* 15 (1683), 347–55.

———. "Concerning the Animalcula in *Semine humano*." *PT* 21 (1699), 301–8.

———. "Observationes de natis a semine genitali animalcules." *PT* 12 (1678), 1040–43.

Verle, Giovanni Battista. *Anatomia artifiziale dell'occhio umano*. Florence: per il Vangelisti stampatore arcivescovale, 1679.

Vesling, Johannes. *Syntagma anatomicum*. Edited by Gerardus Blasius. Amsterdam: Joannem Janssonium a Waesberge, & Elizeum Weyerstraet, 1666, first published at Padua: Frambotti, 1641.

Wepfer, Johann Jakob. *Cicutae aquaticae historia et noxae*. Basel: Joh. Rodolphum Konig, 1679.

Willis, Thomas. *Cerebri anatome: cui accessit nervorum descriptio et usu*. London: Tho. Roycroft, Impensis Jo. Martyn & Ja. Allestry, 1664.

———. *Of the Anatomy of the Brain and the Description and Use of the Nerves*. In *The Remaining Works of Dr. Thomas Willis*, trans. Samuel Pordage, 51–192. London: T. Dring, C. Harper, J. Leigh, and S. Martyn, 1681. Facsimile edition edited by William Feindel. Montreal: McGill University Press, 1965.

———. *Opera omnia*. 2 vols. Amsterdam: Apud Henricum Wetstenium, 1682.

Zanoni, Giacomo. *Istoria botanica*. Bologna: Gioseffo Longhi, 1675.

Zeno, Apostolo. "Relazione della Controversia." *Giornale de' Letterati d'Italia* 4 (1710), 263–92.

Secondary Sources

Adelmann, Howard B. *Marcello Malpighi and the Evolution of Embryology*. 5 vols. Ithaca: Cornell University Press, 1966.

Agrimi, Jole, and Chiara Crisciani. *Les consilia medicaux*. Turnhout: Brepols, 1994.

Anderson, Nancy. "Eye and Image: Looking at Visual Studies of Science." *Historical Studies in the Natural Sciences* 39 (2009), 115–25.

Angeletti, Luciana Rita, and Silvia Marinozzi, "Giovanni Battista Triumfetti e la rinascita dell'orto medico di Roma." *MS* 12 (2000), 439–75.

Arber, Agnes. "Nehemiah Grew (1641–1712) and Marcello Malpighi (1628–1694): An Essay in Comparison." *Isis* 34 (1942), 7–16.

Atti, Gaetano. *Notizie edite ed inedite della vita e delle opere di Marcello Malpighi e di Lorenzo Bellini*. Bologna: Tipografia Governativa alla Volpe, 1847.

Aucante, Vincent. "Descartes's Experimental Method and the Generation of Animals." In *The Problem of Animal Generation*, ed. Justin Smith, 65–79.

———. *La philosophie médicale de Descartes*. Paris: Presses Universitaires de France, 2006.

Baigrie, Brian S. "Descartes's Scientific Illustrations and 'La grand mécanique de la Nature.'" In *Picturing Knowledge*, ed. Brian S. Baigrie. Toronto: University of Toronto Press, 1996.

Bakhtin, Mikhail. *Problems of Dostoevski's Poetics*. [Ann Arbor]: Ardis, 1973.

———. *Rabelais and His World*. Cambridge, Mass.: MIT Press, 1965.

Baldini, Ugo. *Un libertino accademico del Cimento: Antonio Oliva*. *AIMF*, Monografia 1, 1977.

———. Review of Giovanni Alfonso Borelli, *On the Movement of Animals*, trans. Paul Maquet. *AIHS* 42 (1992), 192–96.

Baldwin, Martha. "The Snakestone Experiments. An Early Modern Medical Debate." *Isis* 86 (1995), 394–418.

Basile, Bruno. "I consulti medici di Francesco Redi," Atti dell'accademia delle scienze dell'istituto di Bologna. Classe di scienze morali. Rendiconti, vol. 70 (1981–82), 171–99.

———. *L'invenzione del vero*. Rome: Salerno editrice, 1987.

———. Review of Redi and Doni, *Consulti*. In *Studi e problemi di critica testuale* 32 (1986), 197–200.

Bastholm, Eyvind. *The History of Muscle Physiology*. Copenhagen: Ejnar Munksgaard, 1950.

Bates, Donald G. "Harvey and the Game of Truth." *HS* 36 (1998), 213–32 and 245–67.

———. "Harvey's Account of His Discovery." *MH* 36 (1992), 361–78.

———. "*Machina ex deo:* William Harvey and the Meaning of Instrument." *JHI* 61 (2000), 577–93.

———. "Thomas Willis and the Fevers Literature of the Seventeenth Century." In *Theories of Fevers*, ed. Bynum and Nutton, 45–70.

Belloni, Luigi. "La dottrina circolazione del sangue e la scuola Galileiana 1636–61." *Gesnerus* 28 (1971), 7–34.

———. "Marcello Malpighi and the Founding of Anatomical Microscopy." In *Reason, Experiment, and Mysticism*, ed. Righini Bonelli and Shea, 95–110.

———. "La neuroanatomia di Marcello Malpighi." *Physis* 8 (1966), 253–66.

———. "Schemi e modelli della macchina vivente nel Seicento." *Physis* 5 (1963) 259–98.

Bennett, James A. "Malpighi and the Microscope." *MAP*, 63–72.

———. "A Note on Theories of Respiration and Muscular Action in England c. 1660 (Christopher Wren)." *MH* 20 (1976), 59–69.

Bensaude-Vincent, Bernadette, and William R. Newman, eds. *The Artificial and the Natural. An Evolving Polarity*. Cambridge, Mass.: MIT Press, 2007.

Beretta, Marco. "Gli scienziati e l'edizione del *De rerum natura*." In *Lucrezio. La natura e la scienza*, ed. Beretta and Citti, 177–224.

———. "Lucretius as Hidden *Auctoritas* of the Cimento." In *The Accademia del Cimento*, ed. Beretta, Clericuzio, and Principe, 1–16.

———. "The Revival of Lucretian Atomism and Contagious Diseases during the Renaissance." *MS* 15 (2003), 129–54.

Beretta, Marco, and Francesco Citti, eds. *Lucrezio. La natura e la scienza*. Florence: Olschki, 2008.

Beretta, Marco, Antonio Clericuzio, and Lawrence M. Principe, eds. *The Accademia del Cimento and Its European Context*. Sagamore Beach: Science History Publications, 2009.

Bernardi, Walter. *Le metafisiche dell'embrione*. Florence: Olschki, 1986.

———. "Teoria e pratica della sperimentazione biologica nei protocolli sperimentali rediani." In *Francesco Redi*, ed. Bernardi and Guerrini, 13–30.

Bernardi, Walter, and Luigi Guerrini, eds., *Francesco Redi. Un protagonista della scienza moderna*. Florence: Olschki, 1999.

Bernardi, Walter, et al. *Natura e immagine. Il manoscritto di Francesco Redi sugli insetti delle galle*. Pisa: ETS, 1997.

Berryman, Sylvia. "Ancient Automata and Mechanical Explanation." *Phronesis* 48 (2003), 344–69.

Bertoloni Meli, Domenico. "Additions to the Correspondence of Marcello Malpighi." *MAP*, 275–308.

———. "The Archive and Consulti of Marcello Malpighi." In *Archives of the Scientific Revolution*, ed. Michael Hunter, 109–20. Woodbridge: Boydell Press, 1998.

———. "Authorship and Teamwork around the Cimento Academy." *ESM* 6 (2001), 65–95.

———. "Blood, Monsters, and Necessity in Malpighi's *De polypo cordis.*" *MH* 45 (2001), 511–22.

———. "The Collaboration between Anatomists and Mathematicians in the Mid-Seventeenth Century." *ESM* 13 (2008), 665–709.

———. "The Color of Blood: Between Sensory Experience and Epistemic Significance." Forthcoming in Lorraine Daston and Elizabeth Lunbeck, eds., *Histories of Scientific Observation*. Chicago: University of Chicago Press, 2010.

———. "Francesco Redi e Marcello Malpighi: ricerca anatomica e pratica medica." In *Francesco Redi*, ed. Bernardi and Guerrini, 73–86.

———. "Images as Experiments: Steno's Myology, Viviani, and Galileo." Forthcoming.

———. "A Lofty Mountain, Putrefying Flesh, Styptic Water, and Germinating Seeds." In *The Accademia del Cimento*, ed. Beretta, Clericuzio, and Principe, 121–34.

———. "Mechanistic Pathology and Therapy in the Medical *Assayer* of Marcello Malpighi." *MH* 51 (2007), 165–80.

———. "The Neoterics and Political Power in Spanish Italy: Giovanni Alfonso Borelli and His Circle." *HS* 34 (1996), 57–89.

———. "The New Anatomy of Marcello Malpighi." *MAP*, 17–60.

———. "The Posthumous Dispute between Borelli and Malpighi." *MAP*, 245–73.

———. "The Representation of Insects in the 17th Century: A Comparative Approach." *AS*.

———. "Shadows and Deceptions: From Borelli's *Theoricae* to the *Saggi* of the Cimento." *BJHS* 31 (1998), 383–402.

———. *Thinking with Objects: The Transformation of Mechanics in the Seventeenth Century*. Baltimore: Johns Hopkins University Press, 2006.

Biagioli, Mario. "Etiquette, Interdependence, and Sociability in Seventeenth-Century Science." *Critical Inquiry* 22 (1996), 193–238.

———. *Galileo's Instruments of Credit. Telescopes, Images, Secrecy*. Chicago: University of Chicago Press, 2006.

Bloch, Olivier R. *La philosophie de Gassendi*. La Haye: Nijhoff, 1971.

Bolam, Jeanne. "The Botanical Works of Nehemiah Grew, F.R.S. (1641–1712)." *NRRS* 27 (1973), 219–31.

Borrelli, Antonio. "Medicina e atomismo a Napoli nel secondo Seicento." In *Atomismo e continuo nel XVII secolo*, ed. Festa and Gatto, 341–60.

Boschiero, Luciano. "Natural Philosophizing inside the Late Seventeenth-Century Tuscan Court." *BJHS* 35 (2004), 383–410.

Bresadola, Marco. "Pazienti e curatori nella pratica medica di Marcello Malpighi." In *Storia, Scienza e Società*, ed. Paola Govoni, 17–45. Bologna: Dipartimento di Filosofia, 2006.

Bröer, Ralf. *Salomon Reisel (1625–1701). Barocke Naturforschung eines Leibarztes im Banne der mechanischen Philosophie*. Halle: Acta Historica Leopoldina, 1996.

Brown, Theodore M. "The College of Physicians and the Acceptance of Iatromechanism in England, 1665–1695." *BHM* 44 (1970), 12–30.

———. *The Mechanical Philosophy and the "Animal Œconomy."* New York: Arno, 1981.

———. "Physiology and the Mechanical Philosophy in Mid-Seventeenth Century England." *BHM* 51 (1977), 25–54.

———. "Reflections on a Changing Field." *MAP*, 3–17.

Buchdahl, Gerd. *Metaphysics and the Philosophy of Science. The Classical Origins: Descartes to Kant*. Oxford: Blackwell, 1969.

Buck, August. "Girolamo Cardano's *De Propria Vita*." *Annali d'Italianistica. Autobiography* 4 (1986), 80–90.

Busacchi, Vincenzo. "Iconografia malpighiana." Università di Bologna, *Celebrazioni malpighiane*, 45–55.

Büttner, Jochen. "The Pendulum as a Challenging Object in Early-Modern Mechanics." In *Mechanics and Natural Philosophy before the Scientific Revolution*, ed. Walter Roy Laird and Sophie Roux, 223–37. Dordrecht: Springer, 2008.

Bylebyl, Jerome J. "Boyle and Harvey on the Valves in the Veins." *BHM* 56 (1982), 351–67.

———. "Disputation and Description in the Renaissance Pulse Controversy." In *The Medical Renaissance*, ed. Wear, French, and Lonie, 223–45.

———. "The Medical Side of Harvey's Discovery: The Normal and the Abnormal." In *William Harvey and His Age*, ed. J. J. Bylebyl, 28–102.

———. "Nutrition, Quantification and Circulation." *BHM* 51 (1977), 369–85.

———. "The School of Padua: Humanistic Medicine in the Sixteenth Century." In *Health, Medicine and Mortality in the Sixteenth Century*, ed. Charles Webster, 335–70. Cambridge: Cambridge University Press, 1979.

———, ed. *William Harvey and His Age: The Professional and Social Context of the Discovery of the Circulation*. Baltimore: Johns Hopkins University Press, 1979.

Bynum, William F. "The Anatomical Method, Natural Theology, and the Functions of the Brain." *Isis* 64 (1973), 445–68.

———. "Health, Disease and Medical Care." In *The Ferment of Knowledge: Studies in the Historiography of Eighteenth-Century Science*, ed. George S. Rousseau and Roy Porter, 211–53. Cambridge: Cambridge University Press, 1980.

Bynum, William F., and Vivian Nutton, eds. *Theories of Fevers from Antiquity to the Enlightenment*. London: Wellcome Institute for the History of Medicine, 1981, *Medical History*, suppl. 1.

Bynum, William F., and Roy Porter, eds. *Medicine and the Five Senses*. Cambridge: Cambridge University Press, 1993.

Camerota, Michele. *Galileo Galilei e la cultura scientifica nell'età della Controriforma*. Rome: Salerno Editrice, 2004.

———. "Galileo, Lucrezio e l'atomismo." In *Lucrezio. La natura e la scienza*, ed. Beretta and Citti, 141–75.

Canguilhem, Georges. "Machine et organisme." In *La connaissance de la vie*, 101–27. Paris: Vrin, 1992.

———. *On the Normal and the Pathological*. Dordrecht: Reidel, 1978.

Cappelletti, Vincenzo, and Federico Di Trocchio, eds. *"De sedibus, et causis." Morgagni nel centenario*. Rome: Istituto della enciclopedia italiana, 1986.

Casini, Paolo. "Morgagni nella stampa periodica." In "*De sedibus*," ed. Cappelletti and Di Trocchio, 199–207.

Cavazza, Marta. "L'impegno del giovane Morgagni per la riforma dell'Accademia degli Inquieti e in difesa della tradizione malpighiana." In "*De sedibus*," ed. Cappelletti and Di Trocchio, 91–103.

———. *Settecento inquieto*. Bologna: il Mulino, 1990.

———. "The Uselessness of Anatomy: Mini and Sbaraglia versus Malpighi." *MAP*, 129–45.

———. "*Vis irritabilis* e spiriti animali: Una disputa settecentesca sulle cause del moto muscolare." In *Neuroscienze controverse*, ed. Marco Piccolino, 49–74. Torino: Bollati Boringhieri, 2008.

Chang, Ku-Ming (Kevin). "Fermentation, Phlogiston, and Matter Theory: Chemistry and Natural Philosophy in Georg Ernst Stahl's *Zymotechnia Fundamentalis.*" *ESM* 7 (2002), 31–64.

———. "*Motus Tonicus*: Georg Ernst Stahl's Formulation of Tonic Motion and Early Modern Medical Thought." *BHM* 78 (2004), 767–803.

Chapman, Allan. *England's Leonardo: Robert Hooke and the Seventeenth-Century Scientific Revolution.* Bristol and Philadelphia: Institute of Physics Publishing, 2005.

Cheung, Tobias. "Omnis fibra ex fibra: Fibre Œconomies in Bonnet's and Diderot's Models of Organic Order." *ESM* 15 (2010), 66–104.

Clarke, Edwin S. "Brain Anatomy before Steno." In *Steno and Brain Research*, ed. Gustav Scherz, 27–34.

Clarke, Edwin S., and J. G. Bearn. "The Brain 'Glands' of Malpighi Elucidated by Practical History." *JHM* 23 (1968), 309–30.

Clerizuzio, Antonio. "The Other Side of the Accademia del Cimento: Borelli's Chemical Investigations." In *The Accademia del Cimento*, ed. Beretta, Clericuzio, and Principe, 17–30.

Cobb, Matthew. *Generation. The Seventeenth-Century Scientists Who Unraveled the Secrets of Sex, Life, and Growth.* New York and London: Bloomsbury, 2006.

———. "Malpighi, Swammerdam, and the Colorful Silkworm: Replication and Visual Representation in Early Modern Science." *AS* 59 (2002), 111–47.

———. "Reading and Writing 'The Book of Nature': Jan Swammerdam (1637–1680)." *Endeavour* 24 (2000), 122–28.

Cohen, I. Bernard, ed. *Studies on William Harvey.* New York: Arno Press, 1981.

Cole, Francis Joseph. "The History of Anatomical Injections." In *Studies in the History and Method of Science*, ed. Charles Joseph Singer, 2 vols., vol. 2, 285–343. Oxford: Clarendon Press, 1917–21.

———. *A History of Comparative Anatomy.* London: Macmillan, 1944.

Conforti, Maria. "The Experimenters' Anatomy." In *The Accademia del Cimento*, ed. Beretta, Clericuzio, and Principe, 31–44.

———. "La medicina nel *Giornale de' letterati* di Roma (1668–1681)." *MS* 13 (2001), 59–91.

———. "Il *Moto fermentativo prima origine della vita*: dibattiti sulla natura del sangue in Italia tra Sei e Settecento." *MS* 15 (2003), 269–90.

Conforti, Maria, and Silvia De Renzi, "Sapere anatomico negli ospedali romani. Formazione dei chirurghi e pratiche sperimentali (1620–1720)." In *Rome et la science moderne entre Renaissance et Lumière*, ed. Antonella Romano, 433–72. Rome: École française de Rome, 2008.

Conrad, Lawrence I., Michael Neve, Vivian Nutton, Roy Porter, and Andrew Wear. *The Western Medical Tradition, 800 BC to AD 1800.* Cambridge: Cambridge University Press, 1995.

Cook, Harold J. *The Decline of the Old Medical Regime in Stuart London.* Ithaca: Cornell University Press, 1986.

———. *Matters of Exchange. Commerce, Medicine, and Science in the Dutch Golden Age.* New Haven: Yale University Press, 2007.

———. "The New Philosophy and Medicine in Seventeenth-Century England." In *Reappraisals of the Scientific Revolution*, ed. Lindberg and Westman, 397–436.

———. "Physicians and Natural History." In *Cultures of Natural History*, ed. Nicholas Jardine, James Secord, and Emma Spary, 91–105. Cambridge: Cambridge University Press, 1996.

———. "Time's Bodies. Crafting the Preparation and Preservation of Naturalia." In *Merchants and Marvels. Commerce, Science, and Art in Early Modern Europe*, ed. Pamela H. Smith and Paula Findlen, 223–47. New York: Routledge, 2002.

Cook, Harold J., and Davis S. Lux, "Closed Circles or Open Networks? Communicating at a Distance during the Scientific Revolution." *HS* 36 (1988), 179–211.

Cunningham, Andrew. *English Manuscripts of Francis Glisson*. Cambridge: Wellcome Unit for the History of Medicine, 1993.

———. "Fabricius and the 'Aristotle Project' in Anatomical Teaching and Research at Padua." In *The Medical Renaissance*, ed. Wear, French, and Lonie, 195–222.

———. "The Historical Work of Wharton's Work on Glands." In Thomas Wharton, *Adenographia*, xxvii–lii. Oxford: Clarendon Press, 1996.

———. "Pathology and the Case-History in Giambattista Morgagni's *On the Seats and Causes of Diseases Investigated Through Anatomy*." *Medizin, Gesellschaft und Geschichte* 14 (1996), 37–61.

———. "The Pen and the Sword: Recovering the Disciplinary Identity of Physiology and Anatomy before 1800." *SHPB* 33 (2002), 631–65; 34 (2003), 51–76.

———. "Thomas Sydenham: Epidemics, Experiment and the 'Good Old Cause.'" In *The Medical Revolution*, ed. French and Wear, 161–90.

Cunningham, Andrew, and Roger French, eds. *The Medical Enlightenment of the Eighteenth Century*. Cambridge: Cambridge University Press, 1990.

Cunningham, Andrew, and Perry Williams, eds. *The Laboratory Revolution in Medicine*. Cambridge: Cambridge University Press, 1992.

Daston, Lorraine, "Baconian Facts, Academic Civility, and the Prehistory of Objectivity." *Annals of Scholarship* 8, 1991, 337–63.

Daston, Lorraine, and Katharine Park. *Wonders and the Order of Nature. 1150–1750*. New York: Zone Books, 1998.

Dear, Peter. *Discipline and Experience. The Mathematical Way in the Scientific Revolution*. Chicago: University of Chicago Press, 1995.

———, ed. *The Literary Structure of Scientific Argument: Historical Studies*. Philadelphia: University of Pennsylvania Press, 1991.

Dennis, Michael A. "Graphic Understanding: Instruments and Interpretation in Robert Hooke's *Micrographia*." *Science in Context* 3 (1989), 309–64.

De Renzi, Silvia. "Medical Competence, Anatomy and the Polity in Seventeenth-Century Rome." *RS* 21 (2007), 551–67.

———. "La natura in tribunale. Conoscenze e pratiche medico-legali a Roma nel XVII secolo." *QS* 108 (2001), 799–822.

Des Chene, Dennis. "Mechanisms of Life in the Seventeenth Century: Borelli, Perrault, Régis." *SHPB* 36 (2005), 245–60.

———. *Spirits and Clocks. Machine and Organism in Descartes*. Ithaca: Cornell University Press, 2001.

Dewhurst, Kenneth. "Willis and Steno." In *Steno and Brain Research*, ed. Gustav Scherz, 43–48.

Dini, Alessandro. "La difesa della 'medicina razionale' e il giovane Morgagni." In "*De sedibus*," ed. Cappelletti and Di Trocchio, 147–54.

———. *Filosofia della natura, medicina, religione. Lucantonio Porzio (1639–1724)*. Milano: Franco Angeli, 1985.

Dollo, Corrado. "Astronomia e profetismo nel *Nunzio del secolo cristallino* di Giovanni Battista Odierna." In *La scuola galileiana. Prospettive di ricerca*, ed. Gino Arrighi et al., 241–53. Firenze: La nuova Italia, 1979.

———. "Inediti per l'epistolario malpighiano." *Rivista di storia della filosofia* 39 (1984), 537–50.

———. *Modelli scientific e filosofici della Sicilia spagnola*. Napoli: Guida, 1984.

Dooley, Brendan. "La scienza in aula nella Rivoluzione Scientifica: dallo Sbaraglia al Vallisneri." *Quaderni per la storia dell'Università di Padova* 21 (1988), 23–41.

Duchesneau, François. "Malpighi, Descartes, and the Epistemological Problems of Iatromechanism." In *Reason, Experiment, and Mysticism*, ed. Righini Bonelli and Shea, 111–30.

———. *Les modèles du vivant de Descartes à Leibniz*. Paris: Vrin, 1998.

Eales, Nellie B. "The History of the Lymphatic System, with Special Reference to the Hunter-Monro Controversy." *HJM* 29 (1974), 280–94.

Ekholm, Karin J. "Fabricius's and Harvey's Representations of Animal Generation." *AS* 67 (2010), 329–52.

———. "Harvey's and Highmore's Accounts of Chick Generation." *ESM* 13 (2008), 568–614.

Eriksson, Ruben. *Andreas Vesalius' First Public Anatomy at Bologna. 1540. An Eyewitness Report by Baldasar Heseler*. Uppsala and Stockholm: Almqvist & Wiksell, 1959.

Faller, Adolf. "Die Hirnschnitt-Zeichnungen in Stensens *Discours sur l'anatomie du cerveau*." In *Steno and Brain Research*, ed. Gustav Scherz, 115–45.

Faucci, Ugo. *La Polemica Bonomo-Lancisi sull "origine acarica della scabbia."* Livorno: S. Belforte, 1937.

Fazzari, Michela. "Incredibili visioni: Roma e i microscopi alla fine del '600." In *From Makers to Users. Microscopes, Markets, and Scientific Practices in the Seventeenth and Eighteenth Centuries*, ed. Dario Generali and Marc J. Ratcliff, 3–42. Florence: Olschki, 2007.

———. "Redi, Buonanni, e la controversia sulla generazione spontanea: una rilettura." In *Francesco Redi*, ed. Bernardi and Guerrini, 97–127.

Fend, Mechthild. *Grenzen der Männlichkeit*. Berlin: Reimer, 2003.

Ferrari, Giovanna. "Public Anatomy Lessons and the Carnival: The Anatomy Theatre of Bologna." *Past and Present* 117 (1987), 50–106.

Festa, Egidio, and Romano Gatto, eds. *Atomismo e continuo nel XVII secolo*. Napoli: Vivarium, 2000.

Findlen, Paula, ed. *Atanasius Kircher. The Last Man Who Knew Everything*. New York: Routledge, 2004.

———. "Controlling the Experiment: Rhetoric, Court Patronage and the Experimental Method of Francesco Redi." *HS* 31 (1993), 35–64.

———. "Jokes of Nature and Jokes of Knowledge: The Playfulness of Scientific Discourse in Early Modern Europe." *Renaissance Quarterly* 43 (1990), 292–331.

———. *Possessing Nature. Museums, Collecting, and Scientific Culture in Early Modern Italy.* Berkeley: University of California Press, 1994.

Fisch, Max H. "The Academy of the Investigators." In *Science, Medicine, and History,* ed. E. Ashworth Underwood, 2 vols., vol. 1, 521–63. Cambridge: Cambridge University Press, 1953.

Fisher, Saul. "Early Modern Philosophy and Biological Thought." *PS* 11 (2003), 373–77.

Foster, Michael. *Lectures on the History of Physiology.* Cambridge: Cambridge University Press, 1901. New York: Dover, 1970.

Fournier, Marian. *The Fabric of Life. Microscopy in the Seventeenth Century.* Baltimore: Johns Hopkins University Press, 1996.

Franceschini, Pietro. "Il secolo di Galileo e il problema della generazione." *Physis* 6 (1964), 141–204.

Frank, Robert G., Jr. *Harvey and the Oxford Physiologists.* Berkeley: University of California Press, 1980.

Frati, Carlo. *Bibliografia malpighiana.* London: Dawsons of Pall Mall, n.d. Reprint of "Bibliografia delle opere a stampa di Marcello Malpighi." In *Marcello Malpighi e l'opera sua,* ed. Ugo Pizzoli, 281–334. Milan: Vallardi, 1897.

Freedberg, David. *The Eye of the Lynx.* Chicago: University of Chicago Press, 2002.

French, Roger K. *Dissection and Vivisection in the European Renaissance.* Aldershot: Ashgate, 1999.

———. "Harvey in Holland: Circulation and the Calvinists." In *The Medical Revolution,* ed. French and Wear, 46–86.

———. *William Harvey's Natural Philosophy.* Cambridge: Cambridge University Press, 1994.

French, Roger K., and Andrew Wear, eds. *The Medical Revolution of the Seventeenth Century.* Cambridge: Cambridge University Press, 1989.

Gabbey, Alan. "Mechanical Philosophies and Their Explanations." In *Late Medieval and Early Modern Corpuscular Matter Theories,* ed. Christoph Lüthy, John E. Murdoch, and William R. Newman, 441–65. Leiden: Brill, 2001.

Gal, Ofer. *Meanest Foundations and Nobler Superstructures: Hooke, Newton and the "Compounding of the Celestiall Motions of the Planet."* Dordrecht: Kluwer, 2002.

Gallassi, Augusto. "Malpighi e la funzione pubblica dell'anatomia a Bologna." *Rivista di storia delle scienze mediche e naturali* 41, 1950, 7–28 (suppl. to no. 1).

Galluzzi, Paolo. "L'Accademia del Cimento: 'gusti' del principe, filosofia e ideologia dell'esperimento." *QS* 16 (1981), 788–844.

Gamba, Antonio, and Giuseppe Ongaro, "*Anatomes peritissimus*: Johann Georg Wirsung's Unknown Experiments on the Circulation of the Blood." *Physis,* n.s. 30 (1993), 231–42.

Garber, Daniel, "Soul and Mind: Life and Thought in the Seventeenth Century." In *The Cambridge History of Seventeenth-Century Philosophy,* ed. Garber and Ayers, 759–95.

Garber, Daniel, and Michael Ayers, eds. *The Cambridge History of Seventeenth-Century Philosophy.* Cambridge: Cambridge University Press, 1998.

Gascoigne, John. "A Reappraisal of the Role of the Universities in the Scientific Revolution." In *Reappraisals of the Scientific Revolution,* ed. Lindberg and Westman, 207–60.

Gaukroger, Stephen. *Descartes. An Intellectual Biography.* Oxford: Oxford University Press, 1995.

———. *Descartes' System of Natural Philosophy.* Cambridge: Cambridge University Press, 2002.

Generali, Dario. *Antonio Vallisneri. Gli anni della formazione e le prime ricerche.* Florence: Olschki, 2007.

Gentilcore, David. "'All that Pertains to Medicine': Protomedici and Protomedicati in Early Modern Italy." *MH* 38 (1994), 121–42.

———. "Apothecaries, 'Charlatans,' and the Medical Marketplace in Italy, 1400–1750." *Pharmacy in History* 45 (2003), 91–94.

———. "The Organisation of Medical Practice in Malpighi's Italy." *MAP*, 75–110.

Gibson, William C. "The Bio-Medical Pursuits of Christopher Wren." *MH* 14 (1970), 331–41.

Giglioni, Guido. "Anatomical Atheist? The 'Hylozoistic' Foundations of Francis Glisson's Anatomical Research." In *Religio medici. Medicine and Reigion in Seventeenth-Century England*, ed. Ole Peter Grell and Andrew Cunningham, 115–35. Aldershot: Scolar Press, 1996.

———. *Immaginazione e malattia. Saggio su Jan Baptiste van Helmont.* Milano: FrancoAngeli, 2000.

———. "The Machines of the Body and the Operations of the Soul in Marcello Malpighi's Anatomy." *MAP*, 149–74.

Gill, Mary Louise. "Material Necessity and *Meteorology* IV 12." In *Aristotelische Biologie*, ed. Kullmann and Föllinger, 145–61.

Gómez López, Susana. "Marcello Malpighi and Atomism." *MAP*, 175–89.

———. *Le passioni degli atomi.* Florence: Olschki, 1997.

Grafton, Anthony. *Cardano's Cosmos. The Worlds and Works of a Renaissance Astrologer.* Cambridge, Mass.: Harvard University Press, 1999.

Grmek, Mirko. "La notion de la fibre vivante chez les médicins de l'ecole iatrophysique." *Clio Medica* 5 (1970), 297–318.

Grondona, Felice. "L'esercitazione anatomica di Lorenzo Bellini sulla struttura e funzione dei reni." *Physis* 5 (1963), 423–63.

———. "La struttura dei reni da F. Ruysch a W. Bowman." *Physis* 7 (1965), 281–316.

Guerrini, Anita. "Duverney's Skeletons." *Isis* 94 (2003), 577–603.

———. "The Ethics of Animal Experimentation in Seventeenth-Century England." *JHI* 50 (1989), 391–407.

———. *Experimenting with Humans and Animals: From Galen to Animal Rights.* Baltimore: John Hopkins University Press, 2003.

———. "Isaac Newton, George Cheyne and the *Principia Medicinae*." In *The Medical Revolution*, ed. French and Wear, 222–45.

———. "The Varieties of Mechanical Medicine: Borelli, Malpighi, Bellini, and Pitcairne." *MAP*, 111–28.

Guerrini, Luigi. "Contributo critico alla biografia rediana. Con uno studio su Stefano Lorenzini e le sue 'Osservazioni intorno alle torpedini.'" In *Francesco Redi*, ed. Bernardi and Guerrini, 47–69.

Hackmann, Willem D. "Natural Philosophy Textbook Illustrations, 1600–1800." In *Non-Verbal Communication in Science prior to 1900*, ed. Mazzolini, 169–96.

Hall, Thomas S. *History of General Physiology. 600 B.C. to A.D. 1900.* 2 vols. Chicago: University of Chicago Press, 1975^2.

Hansen, Julie V. "Resurrecting Death: Anatomical Art in the Cabinet of Dr. Frederik Ruysch." *The Art Bulletin* 78 (1996), 663–79.

Harley, David. "Political Post-Mortems and Morbid Anatomy in Seventeenth-Century England." *Social History of Medicine* 7 (1994), 1–28.

Harwood, John T. "Rhetoric and Graphic in *Micrographia*." In *Robert Hooke. New Studies*, ed. Michael Hunter and Simon Schaffer, 119–47. Woodbridge: Boydell, 1989.

Hatfield, Gary. "Essay Review: The Importance of the History of Science for Philosophy in General." *Synthese* 106 (1996), 113–38.

———. "Metaphysics and the New Science." In *Reappraisal of the Scientific Revolution*, ed. Lindberg and Westman, 93–166.

Heinemann, Käthe. "Aus der Frühgeschichte der Lehre von den Drüsen im menschlichen Körper." *Janus* 45 (1941), 137–65 and 219–40.

———. "Zur Geschichte der Entdeckung der roten Blutkörperchen." *Janus* 43 (1939), 1–41.

Heller, René. "Mariotte et la physiologie végétale." In *Mariotte savant et philosophe*, 185–203, with a preface by Pierre Costabel. Paris: Vrin, 1986.

Herrlinger, Robert. "Die Rolle der Idee und Technik in der Geschichte der Anatomie." *Sudhoffs Archiv* 46 (1962), 1–16.

Hughes, J. Trevor. *Thomas Willis. 1621–1675. His Life and Work*. London: Royal Society of Medicine Services Limited, 1991.

Hunter, Michael. "Boyle versus the Galenists: A Suppressed Critique of Seventeenth Century Medical Practice and Its Significance." *MH* 41 (1997), 322–61.

———. "Early Problems in Professionalizing Scientific Research: Nehemiah Grew (1641–1712) and the Royal Society." *NRRS* 36 (1982), 189–209.

Isler, Hansruedi. *Thomas Willis, 1621–1675: Doctor and Scientist*. New York-London: Hafner, 1968.

Jarcho, Saul, ed. *The Concept of Heart Failure from Avicenna to Albertini*. Cambridge, Mass.: Harvard University Press, 1980.

Johansen, T. K. *Aristotle on the Sense-Organs*. Cambridge: Cambridge University Press, 1997.

Joos-Renfer, Susi, Marie-Louise Portmann, and Heinrich Buess. *Pathologisch-anatomische Beobachtungen bedeutender Schweizer Ärtze 1670–1720*. Basel/Stuttgart: Benno Schwabe, 1961.

Kawaguchi, Yutaka, Yutaka Banno, Katsumi Koga, Hiroshi Doira, and Hiroshi Fujii. "Polygonal Patterns on Eggshells of Giant Egg Mutant and Large Eggs Induced by 20-Hydroxyecdysome in *Bombyx mori*." *Journal of Insect Physiology* 39 (1993), 437–43.

Kemp, Martin. " 'The Mark of Truth': Looking and Learning in Some Anatomical Illustrations from the Renaissance and Eighteenth Century." In *Medicine and the Five Senses*, ed. Bynum and Porter, 85–121.

Keynes, Geoffrey. *The Life of William Harvey*. Oxford: Clarendon Press, 1966.

King, Lester S. *The Philosophy of Medicine*. Cambridge, Mass.: Harvard University Press, 1978.

Krämer, Fabian. "Die Individualisierung des Hermaphroditen in Medizin und Naturgeschichte des 17. Jahrhunderts." *Berichte zur Wissenschaftsgeschichte* 30 (2007), 49–65.

Kudlien, Fridolf, and Richard J. Durling, eds. *Galen's Method of Healing*. Leiden: Brill, 1991.

Kullman, Wolfgang, and Sabine Föllinger, eds. *Aristotelische Biologie*. Stuttgart: Franz Steiner, 1997.

Kusukawa, Sachiko, and Ian Maclean, eds. *Transmitting Knowledge: Words, Images, and Instruments in Early Modern Europe*. Oxford: Oxford University Press, 2006.

Laqueur, Thomas. *Making Sex. Body and Gender from the Greeks to Freud.* Cambridge, Mass.: Harvard University Press, 1990.

———. "Sex in the Flesh." *Isis* 94 (2003), 300–306.

Lawrence, Christopher. "Alexander Monro 'Primus' and the Edinburgh Manner of Anatomy." *BHM* 62 (1988), 193–214.

LeFanu, William. *Nehemiah Grew, M.D., F.R.S.; a Study and Bibliography of His Writings.* Winchester: St. Paul's Bibliographies, 1990.

Lennox, James G. "The Comparative Study of Animal Development." In *The Problem of Animal Generation*, ed. Justin Smith, 21–46.

———. "Material and Formal Natures in Aristotle's *De partibus animalium.*" In *Aristotelische Biologie*, ed. Kullman and Föllinger, 163–81.

———. "Teleology." In *Keywords in Evolutionary Biology*, ed. Evelyn Fox Keller and Elisabeth A. Lloyd, 324–33. Cambridge, Mass.: Harvard University Press, 1992.

Licoppe, Christian. *La formation de la practique scientifique; le discours de l'expérience en France et en Angleterre (1630–1820).* Paris: Éditions la Découverte, 1996.

Lindberg, David C., and Robert S. Westman, eds. *Reappraisals of the Scientific Revolution.* Cambridge: Cambridge University Press, 1990.

Lindeboom, Gerrit A. "Dog and Frog: Physiological Experiments at Leiden during the Seventeenth Century." In *Leiden University in the Seventeenth Century*, ed. Lunsingh Scheurleer and Posthumus Meyjes, 279–94.

———. *Dutch Medical Biography.* Amsterdam: Rodopi, 1984.

Lockwood, Dean P. *Ugo Benzi, Medieval Philosopher and Physician, 1376–1439.* Chicago: University of Chicago Press, 1951.

Lonie, Iain, M. "Fever Pathology in the Sixteenth Century: Tradition and Innovation." In *Theories of Fevers from Antiquity to the Enlightenment*, ed. William F. Bynum and Vivian Nutton, 19–44.

———. "Hippocrates the Iatromechanist." *MH* 25 (1981), 113–50.

Lunsingh Scheurleer, Theodoor H., and Guillame H. M. Posthumus Meyjes, eds., *Leiden University in the Seventeenth Century: An Exchange of Learning.* Leiden: E. J. Brill, 1975.

Lüthy, Christoph. "Atomism, Lynceus, and the Fate of Seventeenth-Century Microscopy." *ESM* 1 (1996), 1–27.

———. "The Fourfold Democritus on the Stage of Early Modern Science." *Isis* 91 (2000), 443–79.

Luyendijk-Elshout, Antonie M. "Death Enlightened: A Study of Frederik Ruysch." *JAMA* 212 (1970), 121–26.

———. "*Oeconomia Animalis*, Pores and Particles." In *Leiden University in the Seventeenth Century*, ed. Lunsingh Scheurleer and. Posthumus Meyjes, 294–307.

Machamer, Peter, Lindley Darden, and Carl F. Craver. "Thinking about Mechanisms." *Philosophy of Science* 67 (2000), 1–25.

Maehle, Andreas-Holger. *Drugs on Trial: Experimental Pharmacology and Therapeutic Innovation in the Eighteenth Century.* Amsterdam: Rodopi, 1999. *Clio medica*, 53.

———. *Johann Jakob Wepfer (1620–1695) als Toxikologe.* Aarau: Sauerländer, 1987.

———. *Kritik und Verteidigung des Tierversuchs.* Stuttgart: Franz Steiner, 1992.

Mägdefrau, Karl. *Geschichte der Botanik.* Stuttgart: Fischer, 1992^2.

Major, Ralph H. "Johann Conrad Brunner and His Experiments on the Pancreas." *MH*, 3rd series, III (1941), 91–100.

Mani, Nikolaus. "Jean Riolan II (1580–1657) and Medical Research." *BHM* 42 (1968), 121–44.

Manning, Gideon. "Descartes' Healthy Machines and the Human Exception." Forthcoming in *The Mechanization of Natural Philosophy*, ed. Sophie Roux and Dan Garber.

Marcialis, Maria Teresa. "L'immagine della natura nel *De motu animalium* di Giovanni Alfonso Borelli." In *Descartes e l'eredità cartesiana*, ed. Marcialis and Crasta, 295–309.

———. "Macchinismo e unità dell'essere nella cultura italiana settecentesca." *Rivista critica di storia della filosofia* 37 (1981), 3–38.

Marcialis, Maria Teresa, and Francesca Maria Crasta, eds. *Descartes e l'eredità cartesiana nell'Europa sei-settecentesca*. Lecce: Conte, 2002.

Marinozzi, Silvia, and Maria Conforti. "Blood as Therapy, Therapy through the Blood." *MS* 17 (2005), 695–719.

Martinotti, Giovanni. "L'insegnamento dell'anatomia in Bologna prima del secolo XIX." *Studi e memorie per la storia dell'Università di Bologna* 2 (1911), 1–146.

Massalongo, Caro. "Le galle nell'*Anatome plantarum* di M. Malpighi." *Malpighia* 11 (1898), Commentario, 43 pages.

May, Georges. *L'autobiographie*. Paris: Presses Universitaires de France, 1979.

Mazzolini, Renato G., ed. *Non-Verbal Communication in Science prior to 1900*. Florence: Olschki, 1993.

McGuire, James E., and Piyo M. Rattansi. "Newton and the Pipes of Pan." *Notes and Records of the Royal Society of London* 21 (1966), 108–43.

Meinel, Christoph. "Early Seventeenth-Century Atomism. Theory, Epistemology, and the Insufficiency of Experiment." *Isis* 79 (1988), 86–103.

Middleton, W. E. Knowles. *The Experimenters. A Study of the Accademia del Cimento*. Baltimore: Johns Hopkins University Press, 1971.

———. *The History of the Barometer*. Baltimore: Johns Hopkins University Press, 1966.

Misch, Georg. *A History of Autobiography in Antiquity*. Translated by E. W. Dickes and Georg Misch. 2 vols. Cambridge, Mass.: Harvard University Press, 1951.

Momigliano, Arnaldo. *The Development of Greek Biography*. Cambridge, Mass.: Harvard University Press, expanded edition, 1993.

Moran, Bruce T. *Distilling Knowledge. Alchemy, Chemistry, and the Scientific Revolution*. Cambridge, Mass.: Harvard University Press, 2005.

Moscheo, Rosario. "The *Galenistarum triumphus* by Michele Lipari: A Real Edition, Not Merely a Bibliographic Illusion." *MAP*, 313–15.

Müller, Ingo Wilhelm. *Iatromechanische Theorie und ärztliche Praxis*. Stuttgart: Franz Steiner, 1991.

Münster, Ladislao. "Anatomica sive in cadaveribus sectis observationes." In Università di Bologna, *Celebrazioni malpighiane*, 170–228.

Needham, Joseph. *A History of Embryology*. New York: Abelard-Schuman 1959^2.

Neri, Janice. "Between Observation and Image: Representations of Insects in Robert Hooke's *Micrographia*." In *The Art of Natural History: Illustrated Treatises and Botanical Paintings, 1400–1850*, ed. Therese O'Malley and Amy R. W. Meyers, 83–107. New Haven: Yale University Press, 2008.

Neumann, Hanns-Peter. "Machina machinarum. Die Uhr als Begriff und Metapher zwischen 1450 und 1750." *ESM* 15 (2010), 122–91.

Newman, William R. *Atoms and Alchemy. Chymistry and the Experimental Origins of the Scientific Revolution*. Chicago: University of Chicago Press, 2006.

———. "An Overview of Roger Bacon's Alchemy." In *Roger Bacon and the Sciences*, ed. Jeremiah Hackett, 317–36. Leiden: Brill, 1997.

———. *Promethean Ambitions. Alchemy and the Quest to Perfect Nature*. Chicago: University of Chicago Press, 2004.

Newman, William R., and Lawrence M. Principe. *Alchemy Tried in the Fire: Starkey, Boyle, and the Fate of Helmontian Chymistry*. Chicago: University of Chicago Press, 2002.

———. "Alchemy vs. Chemistry: The Etymological Origins of a Historiographic Mistake." *ESM* 3 (1998), 32–65.

Nicholson, Malcolm. "Giovanni Battista Morgagni and Eighteenth-Century Physical Examination." In *Medical Theory, Surgical Practice*, ed. Christopher Lawrence, 101–34. London: Routledge, 1992.

Nutton, Vivian. "Galen and Medical Autobiography." *Proceedings of the Cambridge Philological Society* 198, n.s. 18 (1972), 50–62.

———. "Galen at the Bedside: The Methods of a Medical Detective." In *Medicine and the Five Senses*, ed. William Bynum and Roy Porter, 7–16.

———. "The Reception of Fracastoro's Theory of Contagion." *Osiris* 9 (1990), 196–234.

Ogilvie, Brian W. *The Science of Describing*. Chicago: University of Chicago Press, 2006.

Ottaviani, Alessandro. "Il dibattito sulla generazione delle piante imperfette tra sei e settecento." *MS* 15 (2003), 291–317.

———. "Scuola galileiana e cartesianesimo nella polemica tra Marcello Malpighi e Giovan Battista Trionfetti." In *Descartes e l'eredità cartesiana*, ed. Marcialis and Crasta, 261–76.

Pagel, Walter. *Joan Baptista van Helmont, Reformer of Science and Medicine*. Cambridge: Cambridge University Press, 1982.

———. *William Harvey's Biological Ideas*. Basel: Karger, 1967.

Pagel, Walter, and Frederick N. L. Poynter. "Harvey's Doctrine in Italy: Argoli (1644) and Bonaccorsi (1647) on the Circulation of the Blood." *BHM* 34 (1960), 426, 429.

Palmer, Richard. "In Bad Odour: Smell and Its Significance in Medicine from Antiquity to the Seventeenth Century." In *Medicine and the Five Senses*, ed. Bynum and Porter, 61–68.

———. "Pharmacy in the Republic of Venice in the Sixteenth Century." In *The Medical Renaissance*, ed. Wear, French, and Lonie, 100–117.

Park, Katharine. "The Criminal and Saintly Body: Autopsy and Dissection in Renaissance Italy." *RQ* 47 (1994), 1–33.

———. "The Life of the Corpse: Division and Dissection in Late Medieval Europe." *JHM* 50 (1995), 111–32.

———. "The Organic Soul." In *The Cambridge History of Renaissance Philosophy*, ed. Charles B. Schmitt and Quentin Skinner, 464–84. Cambridge: Cambridge University Press, 1988.

———. *Secrets of Women: Gender, Generation, and the Origins of Human Dissection*. New York: Zone Books, 2006.

Peumery, Jean-Jacques. *Les origines de la transfusion sanguine*. Amsterdam: Israël, 1975, first published in *Clio Medica*, 9, 1974.

Piccolino, Marco. *Lo zufolo e la cicala*. Turin: Bollati-Boringhieri, 2005.

Pighetti, Clelia. "Un dialogo di Domenico Guglielmini resituito alla critica da Giambattista Morgagni." In "*De Sedibus*," ed. Cappelletti and Di Trocchio, 125, 33.

———. "Giovan Battista Odierna e il suo discorso su *L'occhio della mosca*." *Physis* 3 (1961), 309–35.

Pinto-Correia, Clara. *The Ovary of Eve. Egg and Sperm and Preformation*. Chicago: University of Chicago Press, 1997.

Pomata, Gianna. "Malpighi and the Holy Body: Medical Experts and Miraculous Evidence in Seventeenth-Century Italy." *RS* 21 (2007), 568–86.

———. "*Praxis historialis*: The Uses of *Historia* in Early Modern Medicine." In *Historia: Empiricism and Erudition in Early Modern Europe*, ed. Gianna Pomata and Nancy Siraisi, 105–46. Cambridge, Mass.: MIT Press, 2005.

———. *La promessa di guarigione. Malati e curatori in antico regime*. Bari: Laterza, 1994. Translated as *Contracting a Cure: Patients, Healers, and the Law in Early Modern Bologna*. Baltimore: Johns Hopkins University Press, 1998.

Pressman, Jack D. *Last Resort: Psychosurgery and the Limits of Medicine*. Cambridge: Cambridge University Press, 1998.

Pyle, Andrew. "Malebranche on Animal Generation. Preexistence and the Microscope." In *The Problem of Animal Generation*, ed. Justin Smith, 194–214.

Ragland, Evan. "Chemical Assaying and the Products of the Body." Forthcoming.

———. "Experimenting with Chemical Bodies: Reinier de Graaf's Investigations of the Pancreas." *ESM* 13 (2008), 615–64.

Randelli, Mario. "La *Anatome ossium* di Domenico Gagliardi." *Physis* 2 (1960), 223–31.

Ratcliff, Marc J. *The Quest for the Invisible. Microscopy in the Enlightenment*. Farnham: Ashgate, 2009.

Richter, Gottfried. *Das anatomische Theater*. Berlin: E. Ebering, 1936.

Righini Bonelli, Maria Luisa, and William R. Shea, eds. *Reason, Experiment, and Mysticism in the Scientific Revolution*. London: Science History Publications, 1975.

Roger, Jacques. *The Life-Sciences in Eighteenth-Century French Thought*. Edited by Keith R. Benson and translated by Robert Ellrich. Stanford: Stanford University Press, 1997. Originally published as *Les sciences de la vie dans la pensée française du XVIIIe siècle; la génération des animaux de Descartes à l'encyclopédie*. Paris: Colin, 1963.

Rossi, Paolo A. "Medicina empirica e medicina razionale nel pensiero di Marcello Malpighi." In *Miscellanea filosofica 1978*, 118–50. Firenze: Le Monnier, 1979.

Rothschuh, Karl E. *Physiologie. Der Wandel ihrer Konzepte, Probleme, und Methoden von 16. bis 19. Jahrhundert*. Freiburg/München: Karl Alber, 1968.

Roux, Sophie. "Quelles machines pour quels animaux? Jacques Rohault, Claude Perrault, Giovanni Alfonso Borelli." Forthcoming.

Ruestow, Edward G. *The Microscope in the Dutch Republic*. Cambridge: Cambridge University Press, 1996.

———. "The Rise of the Doctrine of Vascular Secretion in the Netherlands." *JHM* 35 (1980), 265–87.

Rupp, Jan C. C. "Matters of Life and Death: The Social and Cultural Conditions of the Rise of Anatomical Theatres, with Special Reference to Seventeenth Century Holland." *HS* 28 (1990), 263–87.

Sabra, Abdelhamid I. *Theories of Light from Descartes to Newton*. Cambridge: Cambridge University Press, 1981[2].

Sachs, Julius. *Geschichte der Botanik vom 16. Jahrhundert bis 1860*. München: R. Oldenbourg, 1875; New York, Johnson Reprint Corp. [1965].

Sailor, Danton B. "Moses and Atomism." *JHI* 25 (1964), 3–16.

Saunders, Gill. *Picturing Plants. An Analytical History of Botanical Illustration*. Berkeley: University of California Press, in association with the Victoria and Albert Museum, London, 1995.

Scherz, Gustav, ed. *Steno and Brain Research in the Seventeenth Century*. Oxford: Pergamon Press, 1968. Vol. 3 of *Analecta Medico-Historica*.

Schiebinger, Londa. "Skelettenstreit." *Isis* 94 (2003), 307–13.

Schiefsky, Mark. "Galen's Teleology and Functional Explanation." *Oxford Studies in Ancient Philosophy* 33 (2007), 369–400.

Schierbeek, Abraham. *Jan Swammerdam. His Life and Works*. Amsterdam: Sweets & Zeitlinger, 1974.

Schmitt, Charles B. "Aristotle among the Physicians." In *The Medical Renaissance*, ed. Wear, French, and Lonie, 1–15.

Schouten, Jan. "Johannes Walaeus (1604–1649) and His Experiments on the Circulation of the Blood." *JHM* 29 (1974), 259–79.

Scriba, Christoph J. "The Autobiography of John Wallis." *NRRS* 25 (1970), 17–46.

Secret, François, "Le Commendator Cassiano dal Pozzo et le medecin chymique Pierre Potier d'Angers, fixe à Bologne." *Chrysopoeia* 5 (1992–96), 697–701.

Shapin, Steven. *The Scientific Revolution*. Chicago: University of Chicago Press, 1996.

———. *The Social History of Truth: Civility and Science in Seventeenth-Century England*. Chicago: University of Chicago Press, 1994.

Shapiro, Lisa. "The Health of the Body-Machine? Or Seventeenth-Century Mechanism and the Concept of Health." *PS* 11 (2003), 421–42.

Shea, William R. *Designing Experiments of Games and Chance. The Unconventional Science of Blaise Pascal*. Philadelphia: Science History Publications, 2003.

Siraisi, Nancy G. *The Clock and the Mirror: Girolamo Cardano and Renaissance Medicine*. Princeton: Princeton University Press, 1997.

———. *History, Medicine, and the Traditions of Renaissance Learning*. Ann Arbor: University of Michigan Press, 2007.

———. *Medieval & Early Renaissance Medicine. An Introduction to Knowledge and Practice*. Chicago: University of Chicago Press, 1990.

———. *Taddeo Alderotti and His Pupils: Two Generations of Italian Medical Learning*. Princeton: Princeton University Press, 1981.

———. "Vesalius and Human Diversity in *De humani corporis fabrica*." *Journal of the Warburg and Courtauld Institutes* 57 (1994), 60–88.

———. "Vesalius and the Reading of Galen's Teleology." *RQ* 50 (1997), 1–38.

Smith, Justin E. H. "Imagination and the Problem of Heredity in Mechanistic Embryology." In *Problem of Animal Generation*, ed. Justin E. H. Smith, 80–99.

———. ed. *The Problem of Animal Generation in Early Modern Philosophy*. Cambridge: Cambridge University Press, 2006.

Smith, Pamela H. *The Body of the Artisan*. Chicago: University of Chicago Press, 2004.

———. "Science and Taste. Painting, the Passions, and the New Philosophy in Seventeenth-Century Leiden." *Isis* 90 (1999), 420–61.

Stiftung Schnyder von Wartensee. *Aus den Briefen herrvorragender Schweizer Ärtze des 17. Jahrhunderts*. Basel: Benno Schwabe & Co., 1919.

Stolberg, Michael. "The Decline of Uroscopy in Early Modern Learned Medicine (1500–1650)." *ESM* 12 (2007), 313–36.

———. "A Woman Down to Her Bones. The Anatomy of Sexual Difference in the Sixteenth and Early Seventeenth Centuries." *Isis* 94 (2003), 274–99.

Stroup, Alicc. *A Company of Scientists. Botany, Patronage, and Community at the Seventeenth-Century Parisian Royal Academy of Sciences*. Berkeley: University of California Press, 1990.

Sturrock, John. *The Language of Autobiography*. Cambridge: Cambridge University Press, 1993.

Stuurman, Siep. *François Poulain de la Barre and the Invention of Modern Equality*. Cambridge, Mass.: Harvard University Press, 2004.

Tallmadge May, Margaret. "On the Passage of Yolk into the Intestines of the Chick." *JHM* 5 (1950), 119–43.

Tongiorgi Tomasi, Lucia. "L'infinitamente piccolo. Immagini al microscopio di Redi e al tempo di Redi." In *Francesco Redi*, ed. Berardi and Guerrini, 305–15.

Tongiorgi Tomasi, Lucia, and Paolo Tongiorgi. "Il naturalista e il cappellano. Osservazione della natura e immagini 'dal naturale' in Francesco Redi." In *Natura e immagine. Il manoscritto di Francesco Redi sugli insetti delle galle*, ed. Walter Bernardi et al., 29–47.

Torrini, Maurizio. "L'Accademia degli Investiganti. Napoli 1663–1670." *QS* 16 (1981), 845–81.

———. *Tommaso Cornelio e la ricostruzione della scienza*. Napoli: Guida, 1977.

Trabucco, Oreste. "Il corpuscolarismo nel pensiero medico del primo Seicento." In *Atomisto e continuo nel XVII secolo*, ed. Festa and Gatto, 321–39.

———. " 'Delle cagioni delle febbri maligne' di G.A. Borelli: una lettura contestuale." *Giornale critico della filosofia italiana* 20 (2000), 236–80.

———. "Thomas Willis e l'Italia: iatrochimica e biologia cartesiana." In *Descartes e l'eredità cartesiana*, ed. Marcialis and Crasta, 311–25.

Tribby, Jay. "Body/Building: Living the Museum Life in Early Modern Europe." *Rhetorica* 10 (1992), 139–63.

———. "Club Medici: Natural Experiment and the Imagineering of 'Tuscany.' " *Configurations* 2 (1994), 215–35.

———. "Cooking (with) Clio and Cleo: Eloquence and Experiment in Seventeenth-Century Florence." *JHI* 52 (1991), 417–39.

Truitt, Elly R. "The Virtues of Balm in Late Medieval Literature." *ESM* 14 (2009), 711–36.

Turner, Gerard L. E. "The Impact of Hooke's *Micrographia* and Its Influence on Microscopy." In *Robert Hooke and the English Renaissance*, ed. Paul Kent and Allan Chapman, 124–45. Gracewing: Anthony Rowe Ltd., 2005.

Università di Bologna, *Celebrazioni malpighiane. Discorsi e scritti*. Bologna: Azzoguidi, 1966, vol. 60 for the year 1965 of *L'archiginnasio*.

Veneziani, Marco. *Machina*. Florence: Olschki, 2005.

von Staden, Heinrich. "Physis and Technē in Greek Medicine." In *The Artificial and the Natural*, ed. Bernadette Bensaude-Vincent and William R. Newman, 21–49.

———. "Teleology and Mechanism: Aristotelian Biology and Hellenistic Medicine." In *Aristotelische Biologie*, ed. Kullman and Follinger, 183–208.

———. "Women and Dirt." *Helios* 19 (1992), 7–30.

Walmsley, Jonathan. "John Locke on Respiration." *MH* 51 (2007), 453–76.

Wear, Andrew. "Medical Practice in Late Seventeenth- and Early Eighteenth-Century England: Continuity and Union." In *The Medical Revolution*, ed. French and Wear, 293–320.

———. "Medicine in Early Modern Europe, 1500–1700." In *The Western Medical Tradition*, ed. Lawrence I. Conrad et al., 207–361.

———. "William Harvey and the 'Way of the Anatomists.'" *HS* 21 (1983), 223–49.

Wear, Andrew, Roger K. French, and Iain M. Lonie, eds. *The Medical Renaissance of the Sixteenth Century*. Cambridge: Cambridge University Press, 1985.

Webster, Charles. *The Great Instauration: Science, Medicine, and Reform, 1626–1660*. London: Duckworth, 1975.

———. "The Helmontian George Thomson and William Harvey: The Revival and Application of Splenectomy to Physiological Research." *MH* 15 (1971), 154–67.

———. "The Origins of Blood Transfusion: A Reassessment." *MH* 15 (1971), 387–92.

———. "William Harvey and the Crisis of Medicine in Jacobean England." In *William Harvey*, ed. Jerome J. Bylebyl, 1–27.

Westfall, Richard S. *The Construction of Modern Science*. Cambridge: Cambridge University Press, 1971.

Wilkin, Rebecca M. "Essaying the Mechanical Hypothesis: Descartes, La Forge, and Malebranche on the Formation of Birthmarks." *ESM* 13 (2008), 533–67.

———. "Figuring the Dead Descartes: Claude Clerselier's *Homme de René Descartes* (1664)." *Representations* 83 (2003), 38–66.

Williams, Elizabeth A. *A Cultural History of Medical Vitalism in Enlightenment Montpellier*. Aldershot: Ashgate, 2003.

Wilson, Catherine. *Epicureanism and the Origins of Modernity*. Oxford: Oxford University Press, 2008.

———. *The Invisible World. Early Modern Philosophy and the Invention of the Microscope*. Princeton: Princeton University Press, 1995.

Wolf-Devine, Celia. *Descartes on Seeing. Epistemology and Visual Perception. Journal of the History of Philosophy Monograph Series*. Carbondale: Southern Illinois University Press, 1993.

Wolfe, Charles T., ed. *Medical Vitalism in the Enlightenment*, special issue of *Science in Context*, 21 (2008), 461–664.

Wolfe, David E. "Sydenham and Locke on the Limits of Anatomy." *BHM* 35 (1961), 193–220.

Zinato, Emanuele. *Il vero in maschera: Dialogismi galileiani*. Napoli: Liguori, 2003.

INDEX